# Springer Undergraduate Texts in Mathematics and Technology

*Series Editors:*
J.M. Borwein
H. Holden
V.H. Moll

More information about this series at http://www.springer.com/series/7438

Arlie O. Petters • Xiaoying Dong

# An Introduction to Mathematical Finance with Applications

Understanding and Building Financial Intuition

 Springer

Arlie O. Petters
Department of Mathematics
Duke University
Durham, NC, USA

Xiaoying Dong
Department of Mathematics
Duke University
Durham, NC, USA

ISSN 1867-5506          ISSN 1867-5514   (electronic)
Springer Undergraduate Texts in Mathematics and Technology
ISBN 978-1-4939-8137-3          ISBN 978-1-4939-3783-7   (eBook)
DOI 10.1007/978-1-4939-3783-7

Mathematics Subject Classification (2010): 91Gxx

Printed on acid-free paper

This Springer imprint is published by Springer Nature
The registered company is Springer International Publishing AG Switzerland

# Preface

## Rationale and Aim

Given the increasing intricacies and interconnectedness of financial firms' activities and the potential opportunities and risks to which they expose themselves and the world's economy, the next generation of financial engineers needs to master an extensive array of mathematical financial models. Indeed, one of the current challenges in finance is that the complexity of modern securities and markets has forced modelers to employ increasingly sophisticated mathematical tools to address financial issues, creating a widening gap between the qualitative and quantitative approaches to finance.

Our book seeks to address this gap by introducing the quantitative aspects of finance to students with either a qualitative background or no background in the subject. At a firm the traders, risk managers, etc. employ proprietary analytical and numerical models custom made to the needs of their firm. However, since open access to such models is prohibited, the book instead strives to give students a fundamental understanding of key financial ideas and tools that form the basis for building more realistic models, including those of a proprietary nature.

## Distinctive Features and Benefits

This book is distinct in how it emphasizes and pedagogically conveys in an accessible manner the theoretical understanding and applications of the mathematical models forming key pillars of modern finance.

First, the book keeps a good balance between mathematical derivation and description for the sake of providing an adequate level of rigor and depth in mathematics and maintaining accessibility to the reader, which in turn adds flexibility of material selection for the instructor (e.g., Chapter 7 may be taught earlier). Specifically, this book addresses the gap between textbooks that offer a theoretical treatment without many applications and those that simply present and apply formulas without appropriately deriving them. Indeed, theoretical understanding is incomplete without enough practice in applications, and applications are risky without a rigorous theoretical understanding. To accomplish this, the book contains numerous carefully chosen examples and exercises that reinforce a student's conceptual understanding and develop a facility with applications. Indeed, the exercises are divided into conceptual, application, and theoretical problems that probe the material deeper.

Second, beyond a few required undergraduate mathematics courses (see Prerequisites below), this book is essentially self-contained. The large number of necessary financial terminologies and concepts can be overwhelming to a student new to finance. For this reason, after introducing some central, big-picture financial ideas in the first chapter, we present the financial minutia along the way as needed. We have tried to make the book self-contained in this regard through thoughtfully chosen illustrative applications starting at the ground level with simple interest. *We then gradually increase the difficulty as the book develops, ranging across compound interest, annuities, portfolio theory, capital market theory, portfolio risk measures, the role of linear factor models in portfolio risk attribution, binomial tree models, stochastic calculus, derivatives, the martingale approach to derivative pricing, the Black-Scholes-Merton model, and the Merton jump-diffusion model.*

Third, the book is also useful for students preparing either for higher level study in mathematical finance or for a career in actuarial science. For example, the syllabi for the actuarial Financial Mathematics Exam (Exam 2/FM) and Models of Financial Economics Exam (Exam 3F/MFE) include many topics covered in the book.

## Prerequisites

The required mathematics consists of introductory courses on multivariable calculus, probability, and linear algebra. Along the way, we introduce additional mathematical tools as needed—e.g., some measure theory is presented from scratch.

*No background in finance is assumed.* As noted above, the necessary financial concepts and tools are introduced in the text, with the first chapter giving an overview of several common finance terminologies associated with securities and securities markets.

Our book does *not* require computer programming. In our experience, finance courses based on computer programming are best taken after students have developed a fundamental understanding of the theoretical architecture of financial models.

## Audience

The text is aimed at *advanced undergraduates* and *master's degree students* who are either new to finance or want a more rigorous treatment of the mathematical models used in finance. The students typically are from economics, mathematics, engineering, physics, and computer science.

We also believe that a faculty member who is teaching finance for the first time will find this introduction readily manageable. Professionals working in finance who would like a refresher or even clarification on some of the theoretical and conceptual aspects of mathematical finance will benefit from the text.

## Scope and Guide

The chapters are organized naturally into four parts and range over the following topics:
- Part I (Chapters 1 and 2):
  *introduction to securities markets and the time value of money*
- Part II (Chapters 3 and 4):
  *Markowitz portfolio theory, capital market theory, and portfolio risk measures*
- Part III (Chapters 5 and 6):
  *modeling underlying securities using binomial trees and stochastic calculus*
- Part IV (Chapters 7 and 8):
  *derivative securities, BSM model, and Merton jump-diffusion model*

The material was tested in courses offered to upper-level undergraduates and master's degree students. Below are two examples of possible topics that may serve as a guide for **semester-long courses**:

- *Introduction to Mathematical Finance*: securities markets (Chapter 1), the time value of money (Chapter 2), Markowitz portfolio theory, capital market theory, and portfolio risk measures (Chapters 3–4), binomial security pricing (Chapter 5, omit most derivations), Itô's formula and geometric Brownian motion (Sections 6.8 and 6.9), forwards, futures, and options (Sections 7.2, 7.3, and 7.5), and call option pricing with applications (Sections 8.3, 8.2.2, 8.5, and 8.6.2).

- *Introduction to Financial Derivatives*: modeling underliers in discrete time (Sections 5.1–5.3), stochastic calculus and modeling underliers in continuous time (Section 5.4 and Chapter 6), general aspects of forwards, futures, swaps, and options, including trading strategies (Chapters 7), the Black-Scholes-Merton (BSM) model, BSM p.d.e. approach to pricing European-style options, risk-neutral approach to pricing European-style options, applications to warrants, delta hedging, managing portfolio risk, and extension of the BSM model to the Merton jump-diffusion model (Chapter 8).

A **year-long course** on introductory mathematical finance can be based on the entire book. The book can also be used as a reference for students enrolled in a mathematical finance **independent study course**.

## Acknowledgments

Specials thanks to the following individuals for their feedback and assistance:

| | | |
|---|---|---|
| Daniel Aarhus | Lu Liu | Chi Trinh |
| Amir Aazami | Ruisi Ma | Dan Turtel |
| Stanley Absher | Tanya Mallavarapu | Kari Vaughn |
| Vibhav Agarwal | Xavier Mela | Robert Vanderbei |
| Hengjie Ai | Vadim Mokhnatkin | Kevin Wan |
| Mitesh Amarthaluru | Julia Ni | Chenyu Wang |
| Vlad Bouchouev | James Nolen | David Williams |
| Michael Brandt | Vivek Oberoi | Chao Xu |
| Esteban Chavez | Feng Pan | Hangjun Xu |
| Rui Chen | Chloe Peng | Lu Xu |
| Kyuwon Choi | Hal Press | Junkai Xue |
| Qian Deng | Hui Qi | Chao Yang |
| Christian Drappi | Zhaozhen Qian | Jiahui Yang |
| Zachary Freeman | Hayagreev Ramesh | Ashley Yeager |
| Tingran Gao | Emma Rasiel | Jeong Yoo |
| William Grisaitis | Tianhua Ren | Yanchi Yu |
| Xiaosheng Guo | Chelsea Richwine | Yunliang Yu |
| Zhonglin Han | Irving Salvatierra | Javier Zapata |
| John Hyde | Andrew Schretter | Xiaodong Zhai |
| Huseyin Kortmaz | Yuhang Si | Biyuan Zhang |
| Baolei Li | John Sias | Yang Zhang |
| Junchi Li | Maxwell Stern | Bowen Zhao |
| Li Li | Lingran Sun | Ruiyang Zhao |
| Nan Li | Alberto Teguia | Xiaoyang Zhuang |
| Qiao Li | Nicholas Tenev | |
| Li Liang | Dominick Totino | Zilong Zou |

We are also thankful to *Elizabeth Loew* of Springer for her support and guidance along the entire way and to *Lisa Goldberg* for her valuable comments and constructive suggestions. AP is indebted to Duke University for providing the financial support needed to hire many students who assisted with writing computer codes, checking calculations, etc. He is also extremely grateful to his wife, *Elizabeth Petters*, for her patience, love, and steadfast encouragement throughout the project. XD would like to express her gratitude to her husband *Xin Zhou* who saw her through this book and offered great suggestions.

Durham, NC, USA                                                         A.O. Petters
Durham, NC, USA                                                            X. Dong
2016

# Contents

**1 Preliminaries on Financial Markets** . . . . . . . . . . . . . . . . . . . . . . . . . . . . . . . 1
  1.1 A Primer on Banks and Rates . . . . . . . . . . . . . . . . . . . . . . . . . . . . . . . . . 2
     1.1.1 Banks and the Federal Funds Rate . . . . . . . . . . . . . . . . . . . . . . 2
     1.1.2 Short-Term and Long-Term Rates and Yield Curves . . . . . . . 4
  1.2 A Primer on Securities Markets . . . . . . . . . . . . . . . . . . . . . . . . . . . . . . . 6
     1.2.1 Securities Markets Organization . . . . . . . . . . . . . . . . . . . . . . . 6
     1.2.2 Professional Participants in Securities Markets . . . . . . . . . . . 8
     1.2.3 Bid-Ask Spreads and Market Liquidity . . . . . . . . . . . . . . . . . 8
     1.2.4 Trading Costs . . . . . . . . . . . . . . . . . . . . . . . . . . . . . . . . . . . . . . 9
  1.3 Economic Indicators That May Affect Financial Markets . . . . . . . . 10

**2 The Time Value of Money** . . . . . . . . . . . . . . . . . . . . . . . . . . . . . . . . . . . . . 13
  2.1 Time . . . . . . . . . . . . . . . . . . . . . . . . . . . . . . . . . . . . . . . . . . . . . . . . . . . . . 14
  2.2 Interest Rate and Return Rate . . . . . . . . . . . . . . . . . . . . . . . . . . . . . . . 15
     2.2.1 Interest Rate . . . . . . . . . . . . . . . . . . . . . . . . . . . . . . . . . . . . . . . 15
     2.2.2 Required Return Rate and the Risk-Free Rate . . . . . . . . . . . 16
     2.2.3 Total Return Rate . . . . . . . . . . . . . . . . . . . . . . . . . . . . . . . . . . . 17
  2.3 Simple Interest . . . . . . . . . . . . . . . . . . . . . . . . . . . . . . . . . . . . . . . . . . . . 20
  2.4 Compound Interest . . . . . . . . . . . . . . . . . . . . . . . . . . . . . . . . . . . . . . . . 21
     2.4.1 Compounding: Nonnegative Integer Number of Periods . . 22
     2.4.2 Compounding: Nonnegative Real Number of Periods . . . . 24
     2.4.3 Fractional Compounding Versus Simple Interest . . . . . . . . . 30
     2.4.4 Continuous Compounding . . . . . . . . . . . . . . . . . . . . . . . . . . . 31
  2.5 Generalized Compound Interest . . . . . . . . . . . . . . . . . . . . . . . . . . . . . 31
     2.5.1 Varying Interest and Varying Compounding Periods . . . . . 31
     2.5.2 APR Versus APY . . . . . . . . . . . . . . . . . . . . . . . . . . . . . . . . . . . 33
     2.5.3 Geometric Mean Return Versus Arithmetic Mean Return . . 35
  2.6 The Net Present Value and Internal Rate of Return . . . . . . . . . . . . . 38
     2.6.1 Present Value and NPV of a Sequence of Net Cash Flows . 38
     2.6.2 The Internal Return Rate . . . . . . . . . . . . . . . . . . . . . . . . . . . . . 41

2.6.3   NPV and IRR for General Net Cash Flows . . . . . . . . . . . . . .   42
2.7   Annuity Theory . . . . . . . . . . . . . . . . . . . . . . . . . . . . . . . . . . . . . . .   46
      2.7.1   Future and Present Values of Simple Ordinary Annuities. .   47
      2.7.2   Amortization Theory . . . . . . . . . . . . . . . . . . . . . . . . . . . . . . .   53
      2.7.3   Annuities with Varying Payments and Interest Rates . . . . .   56
2.8   Applications of Annuities . . . . . . . . . . . . . . . . . . . . . . . . . . . . . . .   59
      2.8.1   Saving, Borrowing, and Spending . . . . . . . . . . . . . . . . . . . .   59
      2.8.2   Equity in a House . . . . . . . . . . . . . . . . . . . . . . . . . . . . . . . . .   61
      2.8.3   Sinking Funds . . . . . . . . . . . . . . . . . . . . . . . . . . . . . . . . . . . .   62
2.9   Applications to Stock Valuation . . . . . . . . . . . . . . . . . . . . . . . . . .   63
      2.9.1   The Dividend Discount Model . . . . . . . . . . . . . . . . . . . . . . .   64
      2.9.2   Present Value of Preferred and Common Stocks . . . . . . . . .   65
2.10  Applications to Bond Valuation . . . . . . . . . . . . . . . . . . . . . . . . . .   66
      2.10.1  Bond Terminologies . . . . . . . . . . . . . . . . . . . . . . . . . . . . . . .   66
      2.10.2  Bond Prices Versus Interest Rates and Yield to Maturity . . .   69
2.11  Exercises . . . . . . . . . . . . . . . . . . . . . . . . . . . . . . . . . . . . . . . . . . . . .   72
      2.11.1  Conceptual Exercises . . . . . . . . . . . . . . . . . . . . . . . . . . . . . .   72
      2.11.2  Application Exercises . . . . . . . . . . . . . . . . . . . . . . . . . . . . . .   73
      2.11.3  Theoretical Exercises . . . . . . . . . . . . . . . . . . . . . . . . . . . . . .   78
References . . . . . . . . . . . . . . . . . . . . . . . . . . . . . . . . . . . . . . . . . . . . . . . . .   81

3   Markowitz Portfolio Theory . . . . . . . . . . . . . . . . . . . . . . . . . . . . . . . . .   83
3.1   Markowitz Portfolio Model: The Setup . . . . . . . . . . . . . . . . . . . . .   83
      3.1.1   Security Return Rates . . . . . . . . . . . . . . . . . . . . . . . . . . . . . .   85
      3.1.2   What About Multivariate Normality of Security Return
              Rates? . . . . . . . . . . . . . . . . . . . . . . . . . . . . . . . . . . . . . . . . . . .   87
      3.1.3   Investors and the Efficient Frontier . . . . . . . . . . . . . . . . . . .   87
      3.1.4   The One-Period Assumption, Weights, and Short Selling . .   88
      3.1.5   Expected Portfolio Return Rate . . . . . . . . . . . . . . . . . . . . . .   94
      3.1.6   Portfolio Risk . . . . . . . . . . . . . . . . . . . . . . . . . . . . . . . . . . . .   96
      3.1.7   Risks and Covariances of the Portfolio's Securities . . . . . . .   96
      3.1.8   Expectation and Volatility of Portfolio Log Return . . . . . . .  100
3.2   Two-Security Portfolio Theory . . . . . . . . . . . . . . . . . . . . . . . . . . .  104
      3.2.1   Preliminaries . . . . . . . . . . . . . . . . . . . . . . . . . . . . . . . . . . . .  105
      3.2.2   Efficient Frontier of a Two-Security Portfolio . . . . . . . . . . . .  107
      3.2.3   Reducing Risk Through Diversification . . . . . . . . . . . . . . . .  114
3.3   Efficient Frontier for $N$ Securities with Short Selling . . . . . . . . . . .  117
      3.3.1   N-Security Portfolio Quantities in Matrix Notation . . . . . . .  118
      3.3.2   Derivation of the $N$-Security Efficient Frontier . . . . . . . . . .  120
3.4   $N$-Security Efficient Frontier Without Short Selling . . . . . . . . . . . .  126
3.5   The Mutual Fund Theorem . . . . . . . . . . . . . . . . . . . . . . . . . . . . . .  128
      3.5.1   The Global Minimum-Variance Portfolio . . . . . . . . . . . . . . .  128
      3.5.2   The Diversified Portfolio . . . . . . . . . . . . . . . . . . . . . . . . . . .  130

3.5.3 The Mutual Fund Theorem ............................ 130
3.6 Investor Utility Function ..................................... 131
3.6.1 Utility Functions and Expected Utility Maximization ..... 131
3.6.2 Risk-Averse, Risk-Neutral, and Risk-Seeking Investors.... 133
3.7 Diversification and Randomly Selected Securities .............. 138
3.7.1 Mean Portfolio Variance and the Uniform Dirichlet
Distribution ......................................... 138
3.7.2 Mean Portfolio Variance using the NASDAQ............. 142
3.8 Exercises.................................................... 143
3.8.1 Conceptual Exercises................................. 143
3.8.2 Application Exercises ................................ 144
3.8.3 Theoretical Exercises ................................ 146
References...................................................... 149

4 Capital Market Theory and Portfolio Risk Measures .............. 151
4.1 The Capital Market Theory ................................... 152
4.1.1 The Capital Market Line (CML) ........................ 153
4.1.2 Expected Return and Risk of the Market Portfolio ........ 157
4.1.3 The Capital Asset Pricing Model (CAPM)................ 158
4.1.4 The Security Market Line (SML) ....................... 163
4.1.5 CAPM Security Risk Decomposition ................... 164
4.2 Portfolio Risk Measures ...................................... 165
4.2.1 The Sharpe Ratio ................................... 166
4.2.2 The Sortino Ratio ................................... 170
4.2.3 The Maximum Drawdown ............................ 172
4.2.4 Quantile Functions.................................. 174
4.2.5 Value-at-Risk ...................................... 177
4.2.6 Conditional Value-at-Risk ............................ 182
4.2.7 Coherent Risk Measures.............................. 183
4.3 Introduction to Linear Factor Models.......................... 185
4.3.1 Definition and Intuition .............................. 186
4.3.2 Portfolio Variance Decomposition ...................... 189
4.3.3 Factor Categorization ............................... 191
4.3.4 Alpha and Beta ..................................... 192
4.3.5 CAPM Beta Versus Linear Factor Beta .................. 195
4.3.6 Fama-French Three-Factor Model ...................... 196
4.4 Exercises.................................................... 199
4.4.1 Conceptual Exercises................................. 199
4.4.2 Application Exercises ................................ 201
4.4.3 Theoretical Exercises ................................ 204
References...................................................... 206

**5   Binomial Trees and Security Pricing Modeling** ..................... 209
   5.1   The General Binomial Tree Model of Security Prices ............ 209
   5.2   The Cox-Ross-Rubinstein Tree ................................ 218
        5.2.1   The Real-World CRR Tree ............................. 219
        5.2.2   The Risk-Neutral CRR Tree ........................... 230
   5.3   Continuous-Time Limit of the CRR Pricing Formula ............ 237
        5.3.1   The Lindeberg Central Limit Theorem .................. 237
        5.3.2   The Continuous-Time Security Price Formula ........... 241
   5.4   Basic Properties of Continuous-Time Security Prices ............ 246
        5.4.1   Some Statistical Formulas for Continuous-Time Security
                Prices .............................................. 246
        5.4.2   Some Probability Formulas for Continuous-Time
                Security Prices ...................................... 247
   5.5   Exercises ................................................... 249
        5.5.1   Conceptual Exercises ................................ 249
        5.5.2   Application Exercises ................................ 249
        5.5.3   Theoretical Exercises ............................... 251
   References ....................................................... 252

**6   Stochastic Calculus and Geometric Brownian Motion Model** ...... 253
   6.1   Stochastic Processes: The Evolution of Randomness ............ 253
        6.1.1   Notation for Probability Spaces ...................... 253
        6.1.2   Basic Concepts of Random Variables .................. 257
        6.1.3   Basic Concepts of Stochastic Processes ............... 260
        6.1.4   Convergence of Random Variables ..................... 265
        6.1.5   Skewness and Kurtosis .............................. 266
   6.2   Filtrations and Adapted Processes ........................... 268
        6.2.1   Filtrations: The Evolution of Information .............. 268
        6.2.2   Conditional Expectations: Properties and Intuition ....... 270
        6.2.3   Adapted Processes: Definition and Intuition ........... 273
   6.3   Martingales: A Brief Introduction ........................... 275
        6.3.1   Basic Concepts .................................... 275
        6.3.2   Martingale as a Necessary Condition of an Efficient Market 277
   6.4   Modeling Security Price Behavior ............................ 278
        6.4.1   From Deterministic Model to Stochastic Model ......... 278
        6.4.2   Innovation Processes: An Intuition ................... 279
        6.4.3   Securities Paying a Continuous Cash Dividend .......... 281
   6.5   Brownian Motion ............................................ 282
        6.5.1   Definition of Brownian Motion ....................... 282
        6.5.2   Some Properties of Brownian Motion Paths ............. 284
        6.5.3   Visualization of Brownian Paths ..................... 285
        6.5.4   Markov Property for Brownian Motion ................. 288
   6.6   Quadratic Variation and Covariation ......................... 289

6.6.1   Motivation, Definition, and Notation .................... 289
6.6.2   Basic Properties......................................... 291
6.6.3   Quadratic Variation and Covariation Properties of BM .... 293
6.6.4   Significance of Quadratic Variation ..................... 297
6.7   Itô Integral: A Brief Introduction ............................ 299
6.7.1   Importance of Itô Integral with Respect to BM........... 299
6.7.2   Basic Concepts ......................................... 299
6.7.3   A Famous Example ..................................... 301
6.8   Itô's Formula for Brownian Motion .......................... 302
6.8.1   Itô Processes .......................................... 302
6.8.2   Itô's Lemma for Brownian Motion ...................... 304
6.8.3   Risk-Neutral Probability Measure ...................... 309
6.8.4   Girsanov Theorem for a Single Brownian Motion ........ 311
6.9   Geometric Brownian Motion ................................ 314
6.9.1   GBM: Definition ....................................... 314
6.9.2   GBM: Basic Properties.................................. 315
6.9.3   Relation Between Binomial Tree Model and GBM Model .. 317
6.10  BM as a Limit of Simple Symmetric RW ..................... 320
6.11  Exercises .................................................. 322
6.11.1 Conceptual Exercises................................... 322
6.11.2 Application Exercises .................................. 324
6.11.3 Theoretical Exercises .................................. 324
References........................................................ 326

7   Derivatives: Forwards, Futures, Swaps, and Options ............... 329
7.1   Derivative Securities: An Overview .......................... 329
7.1.1   Basic Concepts ........................................ 329
7.1.2   Basic Functions of Derivatives ......................... 331
7.1.3   Characteristics of Derivative Valuation ................. 332
7.1.4   No-Arbitrage Principle and Law of One Price ........... 334
7.2   Forwards ................................................... 337
7.2.1   Basic Concepts ........................................ 337
7.2.2   Forwards on Assets Paying a Continuous Cash Dividend . 340
7.2.3   Forward Price Formula and the Spot-Forward Parity ..... 341
7.2.4   Forward Value Formula ................................ 343
7.3   Futures .................................................... 345
7.3.1   Evolution from Forwards to Futures..................... 345
7.3.2   Basic Concepts ........................................ 346
7.3.3   Impact of Daily Settlement: A Brief Discussion .......... 347
7.4   Swaps ..................................................... 348
7.4.1   A Brief Introduction ................................... 349

7.5   Options . . . . . . . . . . . . . . . . . . . . . . . . . . . . . . . . . . . . . . . . . . . . . . . . . . . 353
    7.5.1   Basic Concepts  . . . . . . . . . . . . . . . . . . . . . . . . . . . . . . . . . . . . . . . 353
    7.5.2   How Options Work . . . . . . . . . . . . . . . . . . . . . . . . . . . . . . . . . . . 357
    7.5.3   Terminal Payoff and Profit Diagrams . . . . . . . . . . . . . . . . . . . . 359
    7.5.4   Market Sentiment Terminologies and Option Moneyness . 363
    7.5.5   Option Strategies: Straddle, Strangle, and Spread . . . . . . . . . 365
    7.5.6   Put-Call Parity for European Options Revisited  . . . . . . . . . . 368
    7.5.7   Relation Among Put, Call, and Forward . . . . . . . . . . . . . . . . . 369
    7.5.8   Intrinsic Value and Time Value . . . . . . . . . . . . . . . . . . . . . . . . . 370
    7.5.9   Some General Relations of Options . . . . . . . . . . . . . . . . . . . . . . 372
    7.5.10 Put-Call Parity for American Options  . . . . . . . . . . . . . . . . . . . 373
    7.5.11 Boundary Conditions for European Options  . . . . . . . . . . . . . 376
7.6   Exercises . . . . . . . . . . . . . . . . . . . . . . . . . . . . . . . . . . . . . . . . . . . . . . . . . 377
    7.6.1   Conceptual Exercises . . . . . . . . . . . . . . . . . . . . . . . . . . . . . . . . . . . 377
    7.6.2   Application Exercises  . . . . . . . . . . . . . . . . . . . . . . . . . . . . . . . . . . 379
    7.6.3   Theoretical Exercises . . . . . . . . . . . . . . . . . . . . . . . . . . . . . . . . . . 380
References . . . . . . . . . . . . . . . . . . . . . . . . . . . . . . . . . . . . . . . . . . . . . . . . . . . . . . 381

8  **The BSM Model and European Option Pricing** . . . . . . . . . . . . . . . . . . . . 383
8.1   The BSM Model  . . . . . . . . . . . . . . . . . . . . . . . . . . . . . . . . . . . . . . . . . . . 384
    8.1.1   Marketplace Assumptions . . . . . . . . . . . . . . . . . . . . . . . . . . . . . . 385
    8.1.2   Money Market Account and the Underlier Model  . . . . . . . . 386
    8.1.3   Self-Financing, Replicating Portfolio . . . . . . . . . . . . . . . . . . . . 388
    8.1.4   Derivation of the BSM p.d.e. . . . . . . . . . . . . . . . . . . . . . . . . . . . 390
8.2   Applications of BSM Pricing to European Calls and Puts . . . . . . . . 392
    8.2.1   Solving the BSM p.d.e. for European Calls . . . . . . . . . . . . . . . 392
    8.2.2   BSM Pricing Formula for European Puts . . . . . . . . . . . . . . . . 395
    8.2.3   Delta and the Partial Derivative Relative to Strike Price . . . 396
    8.2.4   European Call and Put Deltas at Expiration . . . . . . . . . . . . . . 398
8.3   Application to Pricing Warrants . . . . . . . . . . . . . . . . . . . . . . . . . . . . . . 398
8.4   Risk-Neutral Pricing . . . . . . . . . . . . . . . . . . . . . . . . . . . . . . . . . . . . . . . . 400
    8.4.1   Review of Conditional Expectation . . . . . . . . . . . . . . . . . . . . . 400
    8.4.2   From BSM Pricing to Risk-Neutral Pricing . . . . . . . . . . . . . . 404
    8.4.3   The Fundamental Theorems of Asset Pricing . . . . . . . . . . . . 408
    8.4.4   Risk-Neutral Pricing of European Calls and Puts . . . . . . . . 410
8.5   Binomial Approach to Pricing European Options . . . . . . . . . . . . . . . 411
    8.5.1   One-Period Binomial Pricing by Self-Financing Replication 412
    8.5.2   One-Period Binomial Pricing by Risk Neutrality . . . . . . . . . 415
    8.5.3   Two-Period Binomial Pricing  . . . . . . . . . . . . . . . . . . . . . . . . . . 417
    8.5.4   $n$-Period Binomial Pricing . . . . . . . . . . . . . . . . . . . . . . . . . . . . 421
8.6   Delta Hedging  . . . . . . . . . . . . . . . . . . . . . . . . . . . . . . . . . . . . . . . . . . . . . 422
    8.6.1   Theoretical Delta Hedging for European Calls . . . . . . . . . . . 422
    8.6.2   Application of Delta Hedging to Selling European Calls  . . 427

8.7 Option Greeks and Managing Portfolio Risk .................... 433
  8.7.1 Option Greeks for Portfolios and the BSM p.d.e. .......... 433
  8.7.2 Delta-Neutral Portfolios ............................... 435
  8.7.3 Delta-Gamma-Neutral Portfolios ....................... 438
8.8 The BSM Model Versus Market Data ......................... 440
  8.8.1 Jumps in Security Prices ............................... 440
  8.8.2 Skewness and Kurtosis in Security Log Returns .......... 441
  8.8.3 Volatility Skews ...................................... 445
8.9 A Step Beyond the BSM Model: Merton Jump Diffusion ........ 448
  8.9.1 Poisson Processes ...................................... 449
  8.9.2 The MJD Stochastic Process ............................ 450
  8.9.3 Illustration of MJD Jump, Skewness, and Kurtosis ........ 456
  8.9.4 No-Arbitrage Condition and Market Incompleteness ..... 458
  8.9.5 Pricing European Calls with an MJD Underlier ........... 461
  8.9.6 MJD Volatility Smile ................................. 465
8.10 A Glimpse Ahead .......................................... 465
8.11 Exercises .................................................. 467
  8.11.1 Conceptual Exercises ................................. 467
  8.11.2 Application Exercises ................................. 467
  8.11.3 Theoretical Exercises ................................. 470
References ...................................................... 473

Index ......................................................... 477

# Chapter 1
# Preliminaries on Financial Markets

*Financial markets* are a part of a financial system. In a modern economy, a financial system consists of *financial institutions* (e.g., banks) that essentially are intermediaries among policy makers, companies, and consumers who from time to time alternate between their roles as savers and investors and channel savings into investments in financial markets through buying and selling *financial products*.

Just as an industrial society needs physical infrastructure for transporting goods, an information society needs financial infrastructure for innovations. Just as a modern society needs mechanical and electrical engineers to apply the principles of science and mathematics to build and maintain our mechanical and electronic systems, a world increasingly built on finance needs financial engineers to support financial systems. Financial engineers apply mathematical tools like modeling to support the details of investment decision-making, such as risk management,[1] and the machinery of market making, such as derivatives pricing.[2]

The field of mathematics that is concerned with financial markets is called *mathematical finance*, also known as *quantitative finance*. Although mathematical models derived from mathematical finance need not be compatible with either economic or financial theory, the student of mathematical finance will find it important, and sometimes necessary, to have a basic understanding of interest rates and financial markets, as well as of the financial system as a backdrop.

---

[1] See Chapters 3 and 4.

[2] See Chapters 6, 7 and 8.

© Arlie O. Petters and Xiaoying Dong 2016

A.O. Petters, X. Dong, *An Introduction to Mathematical Finance with Applications*, Springer Undergraduate Texts in Mathematics and Technology, DOI 10.1007/978-1-4939-3783-7_1

## 1.1 A Primer on Banks and Rates

Interest rates are a key concept in economics. The level of interest rates plays an extremely important role in a market economy and therefore in financial markets as well.

### 1.1.1 Banks and the Federal Funds Rate

Banks can be classified into central banks, investment banks, and commercial banks.

1. A *central bank* is the monetary authority in a nation.
   The central bank of the United States is the *Federal Reserve System* (also known as the *Federal Reserve* or simply as the *Fed*) which is structurally composed of the Federal Reserve Board (a central board of governors), the Federal Open Market Committee (FOMC), and twelve Federal Reserve Banks. The FOMC meets eight times a year and is in charge of setting the reserve ratio and targets for the federal funds rate and the federal discount rate:

   ➤ Most banks operate under a form called *fractional reserve banking*, in which banks maintain reserves of cash for only a fraction of the customer's deposits and lend the rest out. This fraction is referred to as the *reserve ratio*.
   ➤ The *federal funds rate* is the interest rate at which US banks lend money to each other overnight.
   Since sometimes banks try to stay as close to the reserve limit as possible (for instance, when they try to maximize their lending profit), they may go under the reserve requirement. If at the end of a day, a bank has a reserve deficit, it needs to borrow money (demand) to boost its reserve from another bank that has excess reserve (supply). The fed funds rate is determined by the market of such demand and supply. The FOMC influences the fed funds rate by buying or selling financial instruments (such as bonds issued by the U.S. government) on the open market.
   ➤ The *federal discount rate* is the interest rate at which the Federal Reserve lends money to banks overnight. This lending window is designed as the last resort for banks to borrow money to meet the reserve requirement. For this reason, the federal discount rate is higher than the federal funds rate.[3]

---

[3] Due to the Fed's quantitative easings (QEs) in the last few years, where QEs were programs for large-scale purchases of assets from the banks that drove up the volume of excess reserves to unprecedented levels, presently many US banks can meet their reserve requirements without borrowing.

2. *Investment banking* provides services such as securities underwriting, mergers and acquisitions advisory, and asset management and securities trading, primarily for institutional clients.
3. *Commercial banking* provides services mainly in deposit and lending activities (such as mortgages (e.g., home loans), auto loans, credit card services, and student loans) to individuals and small business.

While discussion of the separation of commercial and investment banking is beyond the scope of this book, we can say that those banks whose primary business is investment banking are *investment banks* and that those banks whose primary business is commercial banking are *commercial banks*. Commercial banks often have many local retail branches.

Although they both have commercial banking divisions, Goldman Sachs and Morgan Stanley are examples of investment banks. JPMorgan Chase, the Bank of America, and Wells Fargo have significant business in investment banking but are primarily examples of commercial banks.

**Example 1.1.** The following discussion is designed to help the reader understand how the fractional reserve banking system works and establish an intuition of the concepts of money, credit/debt, and leverage.

When JD1 (John Doe number 1) deposits $10,000 into his bank account, if the bank's reserve ratio is 10%, then the bank needs to reserve only $1,000 and may lend the remaining $9,000 out. Let's say JD2 (John Doe number 2) gets the $9,000 loan to buy a car, and the car dealer puts the $9,000 back into another bank. The second bank reserves $900 and lends out $8,100 to JD3. Theoretically speaking, this series of financial maneuvers can go on to JD4, JD5, and so on, forever. As the result, the original $10,000 (cash) could generate $100,000 (cash and credit/debt), for

$$10{,}000 \sum_{n=0}^{\infty} (90\%)^n = \frac{10{,}000}{1 - 0.9} = 100{,}000.$$

This is why "10% reserve (ratio)" is equivalent to "10:1 *leverage (ratio)*".  □

**Remark 1.1.**

1. The last example shows that the money needs to travel to generate the credit. Although there are times when "cash is king," there are also times when money doesn't count unless it is in motion.

---

While the discussion on developed world central banks starting to impose negative interest rates is beyond the scope of this book, it is worth noting that on June 5, 2014, the European Central Bank cuts its deposit rate for banks from 0 to -0.1% to encourage banks to lend rather than hold on to money; on January 29, 2016, the Bank of Japan announced that it would cut the interest rate, set at -0.1%, further into negative territory if necessary, that compared with -0.65% in Denmark, -1.1% in Sweden, and -0.75% in Switzerland already (Source: WSJ, 01/29/2016).

2. High risk in financial investing is often caused by high leverage. In fact, Lehman Brothers was leveraged 30.7:1 according to its last annual financial statement before filing for bankruptcy in September 2008. And "in October 2011, another Wall Street bank was taken down by bad bets financed by excessive leverage: MF Global. Their leverage ratio? 40:1."[4]

$\square$

### 1.1.2 Short-Term and Long-Term Rates and Yield Curves

Interest rates (and bond yields[5]) affect all financial markets. Among them, there are *short-term* (fewer than 12 months) *rates*, there are *long-term* (usually 10 years or longer) *rates*, and there are some in between.

Since *short-term interest rates* and *long-term interest rates* are determined by different mechanisms, we consider them separately.

**Short-Term Rates**

Short-term rates are administered by the FOMC in the USA (and by central banks in other nations).

The fed funds rate not only is important to banks but also has trickle-down effects that affect investors and consumers (or "end users"). This rate is also used as the benchmark for other short-term interest rates.

When pricing a loan, a lender often uses the formula

$$\{ \text{ index rate } \} + \{ \text{ spread (or margin)} \}$$
$$= \{ \text{ the interest rate which the borrower is to pay} \},$$

where the size of a spread (or margin) depends on how risky the lender feels the loan is; the riskier the loan, the bigger the spread (or the higher the margin). The often used *index rates* (or *base rates*) are the prime rate, LIBOR, or COFI, depending on which of the following loans is under consideration:

        variable-rate credit card credits,
        variable-rate auto loans,
        variable-rate student loans,
        home equity lines of credit (HELOC),
        adjustable-rate mortgages (ARM),
        small business loans,
        personal loans.

Here are the descriptions of prime rate, LIBOR, and COFI:

---

[4] Source: `http://azizonomics.com/2011/11/15/zombie-economics/`
[5] See Section 2.10.

➤ The *prime rate*, as a generic term, is the interest rate that the banks charge their most credit-worthy customers; it primarily refers to the Wall Street Journal Prime Rate which is the consensus prime rate published by the WSJ after polling ten of America's largest banks. The prime rate moves up or down in lock step with changes by the FOMC. Normally, it runs approximately 300 *basis points* (or 3 percentage points) above the fed funds target rate, since the break-even lending rate of interest for banks making loans is essentially the same as the prime lending rate.
The prime rate is the most popular index rate used in the USA.

➤ The *London Interbank Offered Rates (LIBOR)* can be described as a daily reference of the wholesale cost of money in the London interbank money market of 18 banks.[6] Loosely speaking, LIBOR is like an international counterpart of the fed funds rate. However, unlike the fed funds rate, there are many different LIBOR rates with maturities ranging from overnight to 12 months.

The LIBOR is frequently the basis of investments including interest swap agreements and forward contracts,[7] and for many adjustable mortgage loans as well. American banks use LIBOR because LIBOR is updated much more frequently than the US prime rate as we mentioned earlier; this has advantages particularly when global credit market conditions deteriorate rapidly.

➤ The *11th District Cost of Funds Index (COFI)* is an index rate primarily used in the western USA to set the cost of variable-rate loans and is computed by using data from three western states Arizona, California, and Nevada (which are covered by the 11th district).

**Long-Term Rates**

Long-term rates are determined by market forces.

This is to say that long-term interest rates are not an influencing factor (which short-term rates often are). Instead, they are determined by the economic fundamentals. In other words, they serve as a measure of how our economy is doing.

The fixed-rate mortgage (such as 15- or 30-year) rates are determined by the trading of mortgage securities on Wall Street.

---

[6] These 18 banks include Bank of America, Barclays, Credit Suisse, Deutsche Bank, HSBC, and JP Morgan Chase. For a complete list, visit the website at https://en.wikipedia.org/wiki/Libor.
[7] See Sections 7.2 and 7.4.1.

**Yield Curve Basics**

Since there are different interest rates based on various terms as we explained earlier, it is desirable to represent all these interest rates simultaneously. A graphical representation of where interest rates are today is called a *yield curve*. More precisely, a *yield curve* is the graph of $y = r(t)$ where $t$ represents time to maturity for bonds of the same asset class and credit quality (e.g., US Treasuries, LIBOR, zeros,[8] or AA-rated corporate bonds, and so on), and $y$ is the corresponding yield to maturity. In short, a yield curve is a plot of bond yields to maturity against times to maturity.

**Example 1.2.** In daily financial news, the *Treasury yield curve* is often shortened to *the yield curve*. The yield curve is considered a *leading indicator* of economic activity (see Section 1.3) and often used as a reference point for forecasting interest rates by investors.

The *zero-coupon yield curve* or the *spot-rate curve* is created by plotting the yields of zero-coupon treasury bills against their corresponding maturities. The primary use of zero (or spot) interest rates is to discount cash flows.[9]    □

## 1.2 A Primer on Securities Markets

Market structures are defined by the trading rules and trading systems which include what information (e.g., orders and quotations) traders can see. A concise interpretation of the financial jargons includes (but is not limited to) the following:

*Securities* are financial products[10] that can be traded on securities markets.
*Securities markets* are the trade execution venue.
*Clearing houses* are responsible for settling trades.
*Securities depositories* are responsible for holding security certificates.
*Brokers* arrange trades for their clients (including clearing and settlement).
*Dealers* trade with their clients and are obligated liquidity providers.

### 1.2.1 Securities Markets Organization

Securities markets may be classified into two levels. One is the *primary market* or *new issue market*,[11] and another is the *secondary market* or *after market*.

---

[8] Conceptually speaking, the most basic debt instrument is a *zero-coupon bond* or simply a *zero*, which is a bond with a single cash flow equal to face value at maturity.

[9] See Chapter 2 for the concept of discount cash flows.

[10] See Section 7.1.1 for more detailed explanation. Examples of securities are stocks and bonds.

[11] The market where new securities are issued.

Primary markets deal with the trading of newly issued securities, whereas secondary markets deal with the trading of securities that have already been issued in the primary market (i.e., existing securities). Thus, "after market" may be interpreted as "after new issue market."

Secondary markets are organized in two basic ways. One is the *exchanges* and another is the *over-the-counter (OTC)* markets.

Exchanges are highly organized and centralized markets where securities are traded. Exchanges started as physical places, *trading floors*, where trading took place (although the advent of ECNs, or electronic trading, has eliminated the need for such traditional floors), whereas OTC markets are less formal, have never been a physical place, and connect broker-dealers only electronically.

Examples of exchanges are the New York Stock Exchange (NYSE) and the NASDAQ Stock Exchange (NASDAQ), where the majority of larger US public companies are traded.

Examples of OTC markets are *over-the-counter bulletin board (OTCBB)*, an electronic trading service, and *pink sheets* (a quotation service). In general, stocks that are traded on OTC are considered to be more speculative than stocks that are listed on exchanges.

NYSE is an order-driven market (it functions like an auction market), whereas OTC, by contrast, is a quote-driven dealer market.

For a private company to go public, issue shares and be traded on an exchange thereafter, it needs[12] to choose an exchange on which to be *listed*, which means that it must be able to meet that exchange's *listing requirements* (among other satisfactions[13]).

If a listed stock, at a later point in time, fails to comply with the exchange's listing requirements, it will be *delisted*, i.e., removed from the exchange on which the stock was issued. After a stock is officially delisted, normally it will be traded on the OTC markets, mainly either on OTCBB or on pink sheets.[14]

Hence by the very organization of the secondary securities markets, an internal quality control mechanism has been put in place. Exchanges with listing requirements are motivated to ensure that only high-quality securities are traded on them and to uphold the exchanges' reputation among investors.

---

[12] Otherwise, a public company is traded on OTC and the stock is referred to as *unlisted stock*, whereas those stocks traded on exchanges are referred to as *listed stocks*.

[13] These satisfactions include paying both the exchange's entry and yearly listing fees. Listing requirements usually include minimum stockholder's equity, a minimum share price, and a minimum number of shareholders. The standards vary by exchange. For example, listing on the NASDAQ is considerably less expensive than listing on the NYSE, which partially explains why newer companies often opt for the NASDAQ if they meet its requirements.

[14] A stock traded on pink sheets is considered to be riskier than that on OTCBB. In general, a stock traded on pink sheets is very speculative.

### 1.2.2 Professional Participants in Securities Markets

Both clearing houses and securities depositories are important professional participants in securities markets. These central organizations provide the infrastructure to support trades.

Buyers and sellers do not deal directly with each other but with a clearing house. *Clearing* is the procedure by which an organization acts as a centralized (single) counterparty to all counterparties in trades.

Central securities depositories facilitate ownership transfer through a *book entry*, which is a system of tracking ownership of securities where no certificate is given to investors.

Other professional participants in securities markets include, but are not limited to, brokers, dealers, mutual funds, and hedge funds. Both brokers and dealers are intermediaries between investors and financial markets.

*Brokerage houses* or simply *brokers* are the first place to contact for almost any one willing to participate in financial markets directly (rather than through mutual funds, for instance) since setting up a brokerage account is required for using services from the broker to implement trading. Electronic brokers are automated order-driven execution systems, whereas others may have capacities for providing clients with investment information and advice. All brokers are compensated by commissions for their services.

*Securities dealers* or simply *dealers* are just like merchants in that they make money from the difference between bid prices and ask prices (bid-ask spreads), where the *bid/ask price* is the price at which traders are willing to buy/sell. Depending on different exchanges, dealers may be known by different names such as *specialists*, *market makers*[15] or *floor traders*, and so on. Most of the time dealers are passive traders as they have affirmative obligations to provide liquidity (see Section 1.2.3) to stabilize the market. Generally speaking, dealers face inventory and adverse-information risks. Therefore they are motivated to hedge their portfolios.[16]

### 1.2.3 Bid-Ask Spreads and Market Liquidity

A *limit buy/sell order* is a trade instruction with a limit bid/ask price and a *size* (i.e., a quantity). A *liquidity pool* usually consists of a large number of limit orders which cannot be matched currently. These orders are referred to as current *nonmarketable orders*. A new order is called nonmarketable with respect to such

---

[15] The US Securities and Exchange Commission (SEC) defines a "market maker" as "a firm that stands ready to buy and sell stock on a regular and continuous basis at a publicly quoted price." (Source: http://www.sec.gov/answers/mktmaker.htm)

[16] See Chapter 8 for details.

a pool if this order cannot be matched in the pool. As a result, this new order will be added to the pool and this is called *adding liquidity* to the pool. On the other hand, if the new order can be matched in the pool immediately, the new order is filled and the size of the pool is reduced and therefore this is called *taking liquidity* from the pool. Usually, liquidity is said to be high if the bid-ask spread is small, and the ask size and bid size are large.

Given time t during market hours, let b(t) and a(t) denote the best bid price and the best ask price at time t, respectively; the *bid-ask spread* at time t is $a(t) - b(t)$. In other words, the current *bid-ask spread* or simply the *spread* is the amount by which the current lowest ask price exceeds the current highest bid price.

**Example 1.3.** Suppose that dealer quotations for MSFT, the ticker symbol for Microsoft, show that the best bid price is $50 and the best ask price is $50.02, then the bid-ask spread is $0.02.[17] If you are an impatient trader, then you either have to buy at price $50.02 or have to sell at price $50 at best. If the market dealer can buy and/or sell one million shares of Microsoft at the best bid price and/or sell at the best ask price, then he or she will make $20,000 in a short period of time.                                                              □

Liquidity plays a central role in the functioning of securities markets. In fact, market liquidity is the single most important characteristic of well-functioning markets.

Although there is no specific liquidity formula, the size of the *(bid-ask) spread* may be used as a rule of thumb for a trader to measure *market liquidity*, since a maximum spread rule is the most common affirmative obligation of designated market makers. The smaller the spread and the larger the size, the more liquid the market is.

If the definition of a security market risk is the standard deviation of the security return, then liquid markets are less risky than illiquid ones as liquid markets are less volatile than illiquid ones.

### 1.2.4 Trading Costs

A list of trading costs includes, but is not limited to, the following:

1. Brokerage cost: commissions and fees

2. Liquidity cost: bid-ask spreads and price impacts to the market that cause further slippage[18]

---

[17] Note that the bid-ask spread is the ask price minus the bid price.

[18] Note: the smaller the spreads, the lower the trading costs resulting in a higher return for investors. To a certain extent, tightening spreads is the single most important factor in improving security returns.

3. Difference between the short-term capital gain rate and the long-term capital gain rate (as the former is usually much higher than the latter)

In practice, trading costs cannot be neglected at all and an active investor must scrutinize the per trade cost.

## 1.3 Economic Indicators That May Affect Financial Markets

An *economic cycle (or business cycle)* is the term used to describe certain patterns of wide fluctuations in economic activities followed by economies: an expansion, until a peak, followed by a contraction, and until a trough. It is called a cycle because this pattern repeats—the trough phase is then followed by expansion phase, peak phase, contraction phase, and trough phase again, to compose another cycle.

An *economic variable* is a random variable whose sample space consists of economic-related events. Often used economic variables are population, poverty rate, available resources, dividend yield, inflation rate, imports and exports, etc. An economic variable that reveals the direction in which the economy is moving (i.e., signs of contraction or expansion) is an *economic indicator*. Just as the tense of a verb group can be classified into future, present, or past, economic indicators can be classified into leading, coincident, and lagging indicators:

- Those that change before the economy changes are called *leading indicators*. For example, new factory orders for consumer durable goods and the difference between interest rates at two different maturities (e.g., term spread).
- Those that occur at the same time as the related economic activity are called *coincident indicators*, e.g., GDP (gross domestic product[19]), nonfarm payrolls, and retail sales.
- Those that only become apparent after the related economic activity are called *lagging indicators*. For example, CPI (consumer price index[20]) and the unemployment rate.

The US Census Bureau always releases economic indicators on schedule. Online economic calendars provide convenient access to many types of information, including economic indicator announcements with forecasts; the definition of each indicator, with prior, prior revised and actual numbers,

---

[19] Gross domestic product is the monetary value of all the finished goods and services produced within a country in a specific time period.

[20] Consumer price index is a measure of average change over time in the prices paid by urban consumers for a market basket of consumer goods and services, such as transportation, food, and medical care.

**Table 1.1** Example of a simplified version of Bloomberg's domestic economic calendar for the business week beginning January 28, 2013. Source: based on the Bloomberg table at `http://www.bloomberg.com/markets/economic-calendar`.

| Mon Jan 28 | Tue Jan 29 | Wed Jan 30 | Thu Jan 31 | Fri Feb 1 |
|---|---|---|---|---|
| Durable goods Orders 8:30am | FOMC Meeting Begins | MBA Purchase Applications 7:00am | Job-Cut Report 7:30am | Employment Situation 8:30am |
| Pending Home Sales Index 10:am | Store Sales 7:45am | ADP Empl Rept 8:15am | Jobless Claims 8:30am | PMI Mfg Index 8:58am |
| Dallas Fed Mfg Survey 10:30am | Redbook 8:55am | GDP 8:30am | Personal Income and Outlays 8:30am | Consumer Sentiment 9:55am |
| 4 week Bill Anncmnt 11:00am | Home Price Index 9:00am | Petroleum Status Rept 10:30am | Employment Cost Index 8:30am | ISM Mfg Index 10:00am |
| 3-Month Bill Auction 11:30am | Consumer Confidence 10:00am | 7-Yr Note Auction 1:00pm | Chicago PMI 9:45am | Construction Spending 10:00am |
| 6-Month Bill Auction 11:30am | Investor Confidence Index 10:am | FOMC Meeting Anncmnt 2:15pm | Consumer Comfort Index 9:45am | |
| 2-Yr Note Auction 1:00pm | 4 week Bill Auction 11:30am | | Natural Gas Rept 10:30am | |
| | | | Farm Prices 3:00pm | |
| | | | Fed Balance Sheet 4:30pm | |
| | | | Money Supply 4:30pm | |

highlights and reflections on daily stock market focus; bond auction information; and other relevant events. A simplified economic calendar example is provided in Table 1.1.

The effects of the release of economic indicators on financial markets depend on many factors such as (including, but not limited to) the indicator itself, the business cycle phase that the economy is in, the market's expectations, market momentum, and many other factors.

We end Chapter 1 on an important note: *modern financial markets do need and will continue to need mathematical tools, and the mathematical assumptions must be verified carefully to ensure these tools are implemented properly.*

# Chapter 2
# The Time Value of Money

You may have heard the expression, *"A dollar today is worth more than a dollar tomorrow,"* which is because a dollar today has more time to accumulate interest. The time value of money deals with this basic idea more broadly, whereby an amount of money at the present time may be worth more than in the future because of its earning potential. In this chapter, we discuss the valuing of money over different time intervals, which includes a study of the present value of future money and the future value of present money. The theory is laid out in a rigorous, detailed, and general framework and accompanied by numerous applications with direct relevance to personal finance.

To be self-contained for readers new to finance, Sections 2.1 to 2.5 introduce our conventions and terminologies associated with time, interest rates, required return rates, total return rates, simple interest, compound interest for integral and nonintegral periods, and generalized compound interest, where the interest rate and compounding period vary. Readers already familiar with these topics should skim those sections for our notational usage. In Section 2.6, we introduce the net present value and internal return rate, including Descartes's Rule of Signs. The theory of annuities is presented in Section 2.7 and includes amortization theory and annuities with varying payments and varying interest rates. Applications of annuity theory to saving, borrowing, equity in a house, sinking funds, the present value of preferred and common stocks, and bond valuation are given in Sections 2.8 to 2.10.

© Arlie O. Petters and Xiaoying Dong 2016

A.O. Petters, X. Dong, *An Introduction to Mathematical Finance with Applications*, Springer Undergraduate Texts in Mathematics and Technology, DOI 10.1007/978-1-4939-3783-7_2

## 2.1 Time

Before delving into the value of money over time, it is important to be clear about our conventions and notation for time.

*Throughout the book, the default unit of time is a year.* Unless stated to the contrary, assume that a year consists of 365 calendar days and 252 trading days.[1] When designating time, assume that there is a fixed starting time relative to which the other moments of time are defined. The explicit choice of starting time will depend on the context of the application, but we shall always represent it by 0. Note that the starting time need not be the current time.

We employ the following notation:

$$t_0 = \text{fixed current time,} \qquad t = \text{general moment of time.}$$

Note that a general moment of time $t > 0$ simultaneously designates the *number* of years of elapsed time from 0 to the given moment. For example, writing

$$t_0 = \frac{1}{4}, \qquad t = \frac{1}{2}$$

means that the current time is 3 months after the starting time and $t$ is 3 months from now. If October 1, November 1, and December 1 in 2015 mark the times $0$, $t_1$, and $t_2$, respectively, then

$$t_1 = \frac{1}{12}, \qquad t_2 = \frac{1}{6}.$$

We shall distinguish between an *interval of time*, say, $[t_0, t_f]$, and its *time span* $\tau$, which is the length of the interval:

$$\tau = t_f - t_0 = \text{number of years from } t_0 \text{ to } t_f, \qquad\qquad (t_0 \geq 0).$$

If a time interval is partitioned into equal-length subintervals, then the length of a subinterval is called a *period*. For example, a year has 12 monthly periods and 4 quarterly periods.

We shall employ the following abbreviations:

$$\text{mth} = \text{month(s),} \qquad \text{yr} = \text{year(s),} \qquad \text{prd} = \text{period(s).}$$

In particular, one year is written as "1 yr" and two years as "2 yr."

---

[1] Apart from being mindful of leap years, note that banks may use a 360-day year when computing their charge on loans. Any deviation from a 365-day year will be stated explicitly.

## 2.2 Interest Rate and Return Rate

### 2.2.1 Interest Rate

You are perhaps most familiar with interest as the rate a bank pays into your savings account (where you lend the bank money) or the rate a bank charges you for a loan (where the bank lends you money). Overall, *interest* is the cost of money. It is the compensation received for lending or investing money. The initial amount of money you lend or borrow is called the *principal* and will be denoted by $F_0$. Henceforth, assume that money invested—whether in a savings account or in a start-up company—is money lent with the expectation of receiving back more than the amount invested (principal plus interest).

The compensation for lending or the charge for borrowing a principal $F_0$ is typically expressed as a percent $r$ of $F_0$ per year:

$$\{\text{compensation or charge per year}\} = rF_0.$$

The percent $r$ is called the *annual interest rate* or the *quoted rate*—e.g., a 5% per annum interest means $r = 0.05$. By default, all interest rates will be on or converted to a per annum basis. For this reason, we sometimes refer to $r$ simply as the *interest rate* rather than the annual interest rate. Interest rates appear in numerous settings—savings accounts, certificates of deposit, credit cards, auto loans, mortgages, treasuries, bonds, etc.

**Remark 2.1.** Bear in mind that the interest rate used for lending need not equal the interest rate employed for borrowing. However, in later modeling, we shall assume that the two rates are equal (e.g., see page 84). □

We shall also switch freely between expressing $r$ as a percent and decimal. It is possible to have $r > 1$ (interest rate of over 100% per year) or $r < 0$, which can be interpreted as a bank charging you for holding your principal. For simplicity, however, we abide by the following:

*Unless stated to the contrary, assume that $r$ is a positive constant.*

Though $r$ is constant by default, later in the chapter (e.g., Section 2.5), we shall study models where $r$ varies discretely and continuously with time. When the interest rate $r$ is a function of time, it is common practice to express this as $r(t)$—an abuse of notation that should not cause undue confusion.

Interest rates can, of course, be quoted for any time span (week, month, etc.). For example, an interest rate of 12% per year is mathematically the same as 1% per month. More generally, if we divide a year into $k$ equal-size interest periods, then

$$\text{interest rate per interest period} = \frac{r}{k}.$$

**Exact Interest, Ordinary Interest, and Banker's Rule**

The *exact time* of a time interval measures the length of the interval in days, but excludes the first day. *Exact interest* is interest computed using 365 days in a year or 366 days for leap years. Credit card companies tend to use exact time and exact interest. *Ordinary interest* is interest calculated using 360 days in a year with 30 days in each month. Banks usually lend using exact time and ordinary interest, which has come to be known as *Banker's Rule*.

## 2.2.2 Required Return Rate and the Risk-Free Rate

We always assume that when an investor commits her money for a specific period of time, whether to a security, portfolio, or start-up, she expects to be compensated. An investor's *required rate of return* over an investment period is then the interest rate the investor demands as compensation for the following:

➤ *Opportunity cost:* Since lending prevents an investor from using that money for other investment opportunities, the investor requires compensation for her money being tied up.

➤ *Inflation:* Since inflation erodes the value of money, the investor requires compensation that covers the impact of inflation.

➤ *Risk:* Since there is a nonzero probability that earnings promised to the investor will not materialize or that the investor can lose some or all of her money, the investor requires compensation for the risks of the investment.

*Unless stated to the contrary, we assume that no compensation to cover taxes and transaction costs is part of a required return rate.* It is messy to include these items in an introduction to mathematical finance, not to mention that tax laws and transaction costs change. Readers are referred to Reilly and Brown [16, Chap. 1] for a detailed discussion of the required return rate.

In the absence of inflation and risk, the required return rate is called the *real risk-free rate* and denoted $r_{real}$. It is a compensation purely for opportunity cost. If there is no risk, but you have inflation and an opportunity cost, then the required return rate is termed the *nominal risk-free rate* or, simply, the *risk-free rate*. When the real risk-free rate is intended as opposed to the risk-free rate, we shall indicate so explicitly.

**Notation.** Let $r$ denote the risk-free rate.

There is a simple relationship among $r_{real}$, $r$, and the inflation rate $i$. Assume that you invest $F_0$ in a riskless asset over 1 year. Your required return rate is $r$, which compensates you for opportunity cost and inflation. Specifically, your compensation for opportunity cost a year from now is $r_{real} F_0$. However, a year from now, the value of your compensation $r_{real} F_0$ for opportunity cost will reduce by $i\left(r_{real} F_0\right)$ due to inflation. Furthermore, your initial investment will also reduce in value by $i F_0$ due to inflation. Your required return rate amount $r F_0$ beyond your initial investment should then be

$$r F_0 = i F_0 + r_{real} F_0 + i\left(r_{real} F_0\right).$$

Consequently, we obtain a formula for the real risk-free rate:

$$r_{real} = \frac{r - i}{1 + i}.$$

A common proxy (i.e., substitute or model) for the risk-free rate $r$ is the coupon rate of a US Treasury. The specific type of US Treasury chosen in applications depends on the time horizon over which an analysis is conducted. In the modeling of derivatives, however, traders typically choose LIBOR as a proxy for $r$ (see Hull [9, p. 74] for more).

When inflation constitutes a major portion of the market risk-free rate $r$, sometimes $r$ is even called the inflation rate. It is also possible for the inflation rate to be above the market risk-free rate, which, for instance, can be due to the government lowering interest rates to increase liquidity. Hence, one cannot always assume $r \geq i$, but would expect it to hold under normal market conditions.

### 2.2.3 Total Return Rate

Receiving an interest rate of 4% per year on a $20,000 investment means that over 1 year, say, starting at time 0, you get

$$r F_0 = 0.04 \times \$20{,}000 = \$800.$$

In other words, your investment would grow from $20,000 to $20,800 over 1 year. The return rate $R(0,1)$ on your investment over 1 year is the fractional percentage change

$$R(0,1) = \frac{\$20{,}800 - \$20{,}000}{\$20{,}000} = 0.04 = r.$$

If you put $F(t_0)$ today in an investment that does not pay you any income and the value of the investment at a future time $t_f = t_0 + \tau$ is $F(t_f)$, then the *total return rate* on your investment from time $t_0$ to $t_f$ is defined to be

$$R(t_0, t_f) = \frac{F(t_f) - F(t_0)}{F(t_0)}$$

with the *total return amount* defined by $R(t_0,t_f)\,F(t_0)$.

The total return rate will not necessarily equal the interest rate. First, on mathematical grounds the interest rate is always positive, while the total return rate can be negative. Second, on financial grounds, the return rate is concerned only with the initial and final values of the investment and so is a performance measure of the investment. However, it is possible to apply an interest rate during each period into which a time span is divided, i.e., the interest rate is involved with the evolution of $F(t_0)$ to the final value $F(t_f)$. This will be made explicit when we look at simple and compound interest.

In general, we formalize the total return rate on a per-unit basis and with a cash dividend, i.e., an income.[2] Suppose that your investment has a current per-unit market value of $V(t_0)$, e.g., the price of a stock per share, and per-unit market value $V(t_f)$ at a final time $t_f > t_0$. Assume that the value of an investment at any point in time is nonnegative, i.e., there is no liability:

$$V(t) \geq 0 \qquad\qquad (t \geq 0). \qquad\qquad (2.1)$$

Assume that the investment pays a per-unit cash dividend of $D(t_0,t_f)$ during the interval $[t_0,t_f)$—e.g., a cash payout per share by a company to shareholders.

Several clarifying remarks are needed about *cash dividends*:

➤ For simplicity, we do not include any cash dividend at $t_f$, but tally it as part of the subsequent time interval starting at $t_f$.[3]

➤ It is also common practice to assume that $D(t_0,t_f)$ excludes any income such as interest from the cash dividend during $[t_0,t_f)$. This is not a serious concern for sufficiently short investment time intervals. We also exclude complications like share splits and noncash payouts.

➤ When an investment pays out a cash dividend, it has lost value by the amount of dividend. The market value $V(t_f)$ is then the *ex-dividend* (without dividend) value and the *cum-dividend* (with dividend) value is

$$V^c(t_f) = V(t_f) + D(t_0,t_f).$$

➤ In the case of a cash dividend-paying stock, there is actually an *ex-dividend date*, which is the cutoff date to be eligible for a declared cash dividend. It is actually the close of trading on the trading day before the ex-dividend date. The stock is said to be traded *cum-dividend* before the ex-dividend date and *ex-dividend* after that date. For this reason, when modeling, the value of the

---

[2] A dividend does not have to be in the form of cash. It can be a stock dividend—e.g., a company can pay you additional (typically, fractional) shares for each share of company stock you own.

[3] This bookkeeping for the cash dividend makes it convenient mathematically when considering reinvesting dividends to buy more units of the investment over consecutive time intervals.

stock is adjusted downward by the dividend amount on the ex-dividend date, not on the payment date. Stock price data sets like Yahoo! Finance have a column for adjusted prices, where the adjustments are for cash dividends and stock splits. The ex-dividend dates of stocks typically do not coincide with an exact quarter and are not the same for all companies.

➤ For securities like stocks and bonds, the cash dividends flow in discretely— e.g., quarterly, semiannually, and even annually in some cases. The dividend stream for a sufficiently broad stock index is often modeled as continuous.

Expressing the per-unit total return amount on your investment from $t_0$ to $t_f$ as a percent $R(t_0, t_f)$ of the initial value $V(t_0)$, we obtain

$$\underbrace{R(t_0, t_f)\, V(t_0)}_{\text{return amount}} = \underbrace{V(t_f) - V(t_0)}_{\text{capital gain}} + \underbrace{D(t_0, t_f)}_{\text{cash dividend}}.$$

The spread $V(t_f) - V(t_0)$ is called a *capital gain*. Note that a negative capital gain is a *capital loss*. Equivalently,

$$R(t_0, t_f) = \underbrace{\frac{V(t_f) - V(t_0)}{V(t_0)}}_{\text{capital-gain return}} + \underbrace{\frac{D(t_0, t_f)}{V(t_0)}}_{\text{dividend yield}} = \frac{V^c(t_f) - V(t_0)}{V(t_0)}. \quad (2.2)$$

This is called the *total rate of return* or *holding-period return* of the investment from $t_0$ to $t_f$. We shall often refer to $R(t_0, t_f)$ simply as the *return rate* and at times will even refer to $R(t_0, t_f)$ as the *return* when it is clear from the context that a rate is intended as opposed to the return amount $R(t_0, t_f)\, V(t_0)$. Note that if your ownership in the investment consisted of $n$ units (shares), then the return rate is still given by (2.2) since the numerator and denominator of each term would be multiplied by $n$ and so $n$ would drop out.

**Notation.** When the return rate depends on the length $\tau$ of $[t_0, t_f]$ rather than on the location of $[t_0, t_f]$ on the positive time axis $[0, \infty)$, we set

$$R(t_0, t_f) = R(\tau).$$

The ratio $\frac{D(t_0, t_f)}{V(t_0)}$ in (2.2) is called the *dividend yield* and represents the per-unit cash dividend from the investment as a percent of the initially invested capital $V(t_0)$. Additionally, we refer to the ratio $\frac{V(t_f)}{V(t_0)}$ as the *gross return* from $t_0$ to $t_f$.[4] It expresses the final value $V(t_f)$ as a percent of the initial value $V(t_0)$.

---

[4] Some authors call $\frac{V(t_f)}{V(t_0)}$ the *return rate*, but we shall not abide by that usage.

**Example 2.1.** Suppose that after 1 year, the return rate on your investment is 50%. Then the gain to you, beyond your initial investment, is 50% of your initial investment. If the return rate is $-100\%$, then you have a complete loss. If the return rate is 200%, then your gain is twice the initial investment, i.e., your initial investment tripled in value over the year.                                            □

Finally, observe that the return rate becomes random if the future value $V(t_f)$ and/or the cash dividend $D(t_0, t_f)$ is random. Almost all the return rates we encounter in this chapter are nonrandom, while all the return rates in Chapter 3 are random.

## 2.3 Simple Interest

A principal of \$1,000 held for a year at a 12% interest rate has a simple interest of \$120 at the end of 1 year. This amount is the same as adding 12 monthly interests of \$10, each of which is obtained from a monthly interest rate of 1%. For a time span of $\tau$ years, if we assume that the interest rate $r$ is applied *only* to the principal $F_0$, then

$$\text{(simple interest amount earned or owed over } \tau \text{ years)} = r\tau F_0. \tag{2.3}$$

If an annual simple interest rate is applied over multiple years (or periods) to a principal, then at the end of each year (or period), interest is applied only to the principal and the entire balance is reinvested back into the account. In other words, all interest accrued at the end of each period or year is carried forward without gaining interest. Under *simple interest growth* at rate $r$, a principal $F_0$ increases to the following amount at $\tau$ years from the present:

$$F(\tau) = F_0 + r\tau F_0 = (1 + r\tau) F_0, \tag{2.4}$$

where, by a slight abuse of notation, we write $F(\tau)$ instead of $F(t_0 + \tau)$ since the value depends on the length $\tau$ of the time interval.

**Example 2.2.** Suppose that an account has \$700 and pays 4% per annum. Applying a 4% annual simple interest growth to the \$700 for 1 year yields an interest of $0.04 \times \$700 = \$28$ and a total amount accrued of

$$\$700 + (0.04 \times \$700) = \$728.$$

To obtain simple interest growth of \$700 over 2 years, we add to the principal a simple interest of $0.04 \times \$700$ at the end of the first year and simple interest of $0.04 \times \$700$ at the end of the second year:

$$\$700 + (0.04 \times \$700) + (0.04 \times \$700) = \$756,$$

or, equivalently,

$$\$756 = (1 + 0.04 \times 2)\,\$700. \tag{2.5}$$

Note that a 4% annual simple interest growth applied to $700 for 2 years is the same as applying 8% per 2 years. □

Investing $700 under simple interest growth of 4% per annum yields a *future value* of $756 2 years from now. Conversely, the *present value* of $756 under 4% annual simple interest discounting is $700. In general, if at the current time, you invest (or borrow) a principal $F_0$ under simple interest growth at an interest rate $r$ applied over $\tau$ years, then the amount of money you receive (or owe) at the end of the time span is called the *future value* of $F_0$ and given by

$$\left\{ \begin{array}{c} \text{future value of } F_0 \\ \text{at } \tau \text{ years from now} \end{array} \right\} = F(\tau) = (1 + r\tau)F_0. \tag{2.6}$$

The principal $F_0$ is called the *present value* of the future amount $F(\tau)$:

$$\left\{ \begin{array}{c} \text{present value of the amount } F(\tau), \\ \text{which occurs } \tau \text{ years in the future} \end{array} \right\} = F_0 = \frac{F(\tau)}{1 + r\tau}. \tag{2.7}$$

The quantity $(1 + r\tau)^{-1}$ is called a *discount factor* since it reduces the amount $F(\tau)$ at the end of the time interval to the amount $F_0$ at the start of the interval.

In the context of (2.6), we sometimes call the interest rate $r$ the *simple interest growth rate* of the principal $F_0$, while in the setting of (2.7), we call $r$ the *simple interest discount rate* on the future value $F(\tau)$. The return rate when $F_0$ grows under simple interest $r$ over $\tau$ years is then

$$R(\tau) = \frac{F(\tau) - F_0}{F_0} = r\tau. \tag{2.8}$$

## 2.4 Compound Interest

We saw above that under simple interest for 2 years, an account with $700 at 4% per annum will grow to

$$\$700 + \left(0.04 \times \$700\right) = \$728 \tag{2.9}$$

at the end of the first year and to $756 at the end of second year, after the interest of $28 for the second year is added. However, there is a way to accumulate more money over the same 2 years using the same simple interest rate. Assume that at the end of the first year, you withdrew the $728, closed the account, and immediately used the $728 as principal to open another simple interest account paying the same interest rate. Then a year later, i.e., at the end of the second year, the total you would accrue is

$$\$728 + (0.04 \times \$728) = \$757.12, \tag{2.10}$$

which is greater than the original total of $756! This type of growth is called *compound interest*. In fact, an annually compounded account earning 4% per annum over 2 years would earn you the latter amount without you needing to engage in the previous inconvenient strategy.

Using (2.9), we can rewrite (2.10) as

$$\$700 + \left(0.04 \times \$700\right) + \left(0.04 \times (\$700 + 0.04 \times \$700)\right) = \$757.12. \tag{2.11}$$

Equation (2.11) summarizes exactly how the growth process works: annual compounding of a principal of $700 over 2 years at an interest rate of 4% means that one applies 4% simple interest to $700 at the end of the first year and then applies 4% simple interest again at the end of the second year to the *entire balance* (principal plus interest) carrying forward from the end of the first year. Rewriting (2.11) as

$$\$757.12 = (1 + 0.04)^2 \times \$700 \tag{2.12}$$

yields the standard form for two annual compounds.

Let us extend (2.12) to a finite number of compoundings. In general, *compound interest* occurs when the time span is divided into multiple periods, and simple interest is applied over each period to the *balance* at the end of the period. *We assume that the entire balance at the end of each period is reinvested back into the total being accrued, i.e., no money is withdrawn and no extra money is added.* For mathematical modeling purposes, we also treat the end of a period as equivalent to the start of the next period.

### 2.4.1 Compounding: Nonnegative Integer Number of Periods

Assume that an account with an initial amount $F_0$ (principal) pays an interest rate of $r$. Divide a year into $k$ interest periods, each of equal length:

$$1 \text{ prd} = \frac{1}{k} \text{ yr.}$$

In a compound interest setting, the end of each period marks when interest is applied to the balance from the start of the period. Consequently, we shall refer to each such period as an *interest period, compound interest period* (when being explicit), or a *compounding period*.

> *Unless stated to the contrary, assume that the date when the principal is deposited coincides with the start of an interest period.*

Following the structure of (2.12), we now compute the future value to which the principal $F_0$ will grow under $k$-periodic compounding at interest rate $r$ over $n$ interest periods, where $n$ is a nonnegative integer. Since $n$ periods correspond to $n/k$ years, the future value at the end of the $n$th period is $F(n/k)$. However, in compound interest theory, the emphasis is on the number $n$ of periods over which compounding occurs, rather than the number of years. For this reason, the future value is written as a function of the number of periods as follows:

$$F\left(\frac{n}{k}\right) = F_n.$$

➤ At the end of the first period, apply simple interest to $F_0$ to obtain the future value $F_1$ to which $F_0$ grows over the first period:

$$F_1 = F_0 + \frac{r}{k}F_0 = \left(1 + \frac{r}{k}\right)F_0.$$

Now, do not take out any of the money. Instead, reinvest the entire amount $F_1$ in the account at the end of the first period until the end of the second period.

➤ At the end of the second period, apply simple interest to $F_1$ to get the future value $F_2$ to which $F_1$ grows over the second period:

$$F_2 = F_1 + \frac{r}{k}F_1 = \left(1 + \frac{r}{k}\right)^2 F_0.$$

Note that compound interest occurs since interest was added to the whole $F_1$, yielding interest on the principal $F_0$ and interest on the interest $(r/k)F_0$. Next, reinvest the entire amount $F_2$ in the account at the end of the second period until the end of the third period.

➤ Continuing the above process, at the end of the $n$th period, apply simple interest growth to $F_{n-1}$ to obtain the future value $F_n$ to which $F_{n-1}$ grows over the $n$th period:

$$F_n = F_{n-1} + \frac{r}{k}F_{n-1} = \left(1 + \frac{r}{k}\right)^n F_0, \qquad n = 0, 1, 2, \ldots.$$

We have established the following: *Under $k$-periodic compounding over $n$ interest periods at an interest rate $r$, a principal $F_0$ will increase to the value $F_n$ at the end of the $n$th interest period:*

$$F_n = \left(1 + \frac{r}{k}\right)^n F_0, \qquad n = 0, 1, 2, \ldots, \qquad (2.13)$$

where $\frac{r}{k}$ *is the periodic interest rate.* Observe that $F_n$ depends on the size of the time interval over which the compounding occurs. This is because the interest rate is constant for the $n$ periods.

In (2.13), we call $F_n$ the *future value* of $F_0$ under $k$-periodic compounding over $n$ periods at interest rate $r$. The *present value* of $F_n$ under the above compounding is defined to be $F_0$.

**Example 2.3.** Borrow $1,000 for a year at 12% interest rate. Applying this interest with monthly compounding yields a balance due of

$$F_{12} = \left(1 + \frac{0.12}{12}\right)^{12} \$1,000 = \$1,126.83.$$

The interest owed is then $\$1,126.83 - \$1,000 = \$126.83$, which is more than the $120 due when simple interest is applied.                                              □

**Example 2.4. (Money's Growth Under Different Compounding Periods)** Invest $1,000 at an interest rate of 7% and consider monthly, weekly, and daily compounding. Determine the future values after 2 years.

**Solution.** We have $F_0 = \$1,000$, $r = 0.07$, $\tau = 2$, and $k = 12$ (monthly), 52 (weekly), and 365 (daily). The respective number of compounding periods is then 24 (monthly), 104 (weekly), and 730 (daily). By (2.13) on page 23, the future values at the end of 2 years are

$$F_{24} = \$1,000 \times 1.14981 = \$1,149.81 \quad \text{(monthly compounding)}$$
$$F_{104} = \$1,000 \times 1.15017 = \$1,150.17 \quad \text{(weekly compounding)}$$
$$F_{730} = \$1,000 \times 1.15026 = \$1,150.26 \quad \text{(daily compounding).}$$

                                                                                 □

### 2.4.2 Compounding: Nonnegative Real Number of Periods

Suppose a principal of $10,000,000 undergoes monthly compounding at 10% per annum over a time span of 15.36 mth. What is the principal's value at the end of the time span? First, view 15.36 mth as 15 mth + 0.36 mth. By Equation (2.13), the value at the end of the first 15 months is

$$F_{15} = \left(1 + \frac{0.1}{12}\right)^{15} \times \$10,000,000 = \$11,325,616.82.$$

To how much will $F_{15}$ grow during the remaining 0.36 mth? For the partial interest period, assume that a bank applies simple interest growth to $F_{15}$, which yields a total of

$$\widetilde{F}_{15.36} = \widetilde{F}_{15+0.36} = \left(1 + 0.36 \times \frac{0.1}{12}\right) \times F_{15} = \$11,359,593.67. \qquad (2.14)$$

However, it may concern the reader that compounding occurs over the first 15 mth, but then stops during the remaining 0.36 mth, and is replaced by simple interest growth. We claim that the latter is actually an approximation of the exact mathematical compounding that should be applied during the partial month. We apply *fractional compounding* to $F_{15}$ during the remaining 0.36 mth, which gives

$$F_{15.36} = F_{15+0.36} = \left(1 + \frac{0.1}{12}\right)^{0.36} \times F_{15} = \$11{,}359{,}503.48. \tag{2.15}$$

In this example, we see that the accrued total in (2.14) is higher by \$90.19 than the total in (2.15) obtained from exact modeling. The bank would be paying more interest if (2.14) is used.

We now present a theoretical basis for (2.15) and the approximation used in (2.14). First, we shall introduce the key defining mathematical property of compound interest as in the treatment by Kellison [10, Sec. 1.5]. For an integral number of interest periods, Equation (2.13) shows that

$$F_{m+n} = \left(1 + \frac{r}{k}\right)^{m+n} F_0 = \left(1 + \frac{r}{k}\right)^{m} \left(1 + \frac{r}{k}\right)^{n} F_0,$$

where $m$ and $n$ are nonnegative integers. We denote the *compound interest growth function over n interest periods* by

$$G(n) = \left(1 + \frac{r}{k}\right)^{n},$$

where

$$G(0) = 1, \qquad G(1) = \left(1 + \frac{r}{k}\right), \qquad G(n) > 1 \quad \text{for } n = 1, 2, \ldots.$$

The inequality $G(n) > 1$ for positive integers $n$ means that the principal will increase for compounding over at least one interest period. The compound interest growth function satisfies:

$$G(m+n) = G(m)G(n). \tag{2.16}$$

In other words, compound interest is such that compounding a principal $F_0$ over $m+n$ interest periods is the same as compounding $F_0$ over $n$ interest periods and then compounding the balance at the end of the $n$th interest period over the remaining $m$ interest periods. Of course, one can interchange $m$ and $n$. Equation (2.16) embodies the core multiplication property of compound interest.

We then extend (2.16) to a more general defining mathematical property of compound interest, one applicable to a nonintegral number of interest periods. Specifically, for any nonnegative real number $x$, a principal $F_0$ is said to grow to the value

$$F_x = G(x)F_0$$

by *k-periodic compounding over x interest periods at interest rate r* if the growth function $G(x)$ satisfies the following properties:

$$G(x+y) = G(x)G(y) \quad \text{for all real numbers } x \geq 0 \text{ and } y \geq 0,$$
$$G(0) = 1,$$
$$G(1) = \left(1 + \frac{r}{k}\right),$$
$$G(x) > 1 \quad \text{for all real numbers } x > 0.$$

$$(2.17)$$

The top equation in (2.17) generalizes (2.16) to a nonintegral number of periods. The same intuition carries over from the integral case: compounding a principal $F_0$ over $x + y$ interest periods to the value

$$F_{x+y} = G(x+y)F_0$$

is identical to compounding $F_0$ for $y$ interest periods to the value $F_y = G(y)F_0$ and then compounding $F_y$ over the remaining $x$ interest periods to the value $G(x)F_y$. The equation $G(0) = 1$ in (2.17) states that no growth occurs when there is no interest period, while $G(1) = \left(1 + \frac{r}{k}\right)$ means that the growth over one interest period is given by simple interest (as we have done all along). Finally, we require the condition $G(x) > 1$ for all $x > 0$ because we assume that compound interest growth increases the principal over a nonzero interest period, even if it is fractional.

Let us now solve for the growth function satisfying (2.17). For mathematical modeling reasons, we shall assume that $G(x)$ is differentiable. Applying a trick similar to the one used in deriving an exponential function, we first consider the derivative of the growth function at $x$. Using the limit definition of a derivative, we find (Exercise 2.30)

$$G'(x) = G(x)G'(0). \qquad (2.18)$$

Dividing by $G(x)$, which is allowed since $G(x) > 0$ for all $x \geq 0$, and recalling that $G'(x)/G(x)$ is the derivative of $\ln G(x)$, we obtain

$$\frac{d\ln G(x)}{dx} = G'(0),$$

or, equivalently,

$$d\ln G(x) = G'(0)\,dx.$$

Integrating the equation from 0 to $x$ yields:

$$\ln G(x) - \ln G(0) = G'(0)\, x.$$

But $\ln G(0) = \ln 1 = 0$. Hence:

$$\ln G(x) = G'(0)\, x. \tag{2.19}$$

Equation (2.19) implies:

$$G'(0) = \ln G(1) = \ln\left(1 + \frac{r}{k}\right).$$

Inserting $G'(0)$ back into (2.19), we find:

$$\ln G(x) = x \ln\left(1 + \frac{r}{k}\right) = \ln\left(1 + \frac{r}{k}\right)^x.$$

The binomial series $\left(1 + \frac{r}{k}\right)^x$ with nonintegral $x$ converges for $0 \le r/k < 1$. Exponentiating both sides of the above equation, we obtain the growth function:

$$G(x) = \left(1 + \frac{r}{k}\right)^x.$$

We summarize the result in the following theorem:

**Theorem 2.1.** *Under k-periodic compound interest at r per annum over a time span of x interest periods, where x is a nonnegative real number, a principal $F_0$ will increase to the following future value at the end of the time span:*

$$F_x = \left(1 + \frac{r}{k}\right)^x F_0, \qquad \left(0 \le \frac{r}{k} < 1, \quad x \ge 0\right), \tag{2.20}$$

*where $k = 1, 2, \ldots$.*

The periodic interest rate $\frac{r}{k}$ in (2.20) is constrained to $0 \le \frac{r}{k} < 1$ to assure convergence of $F_x$ when the nonnegative real $x$ is not an integer. We do not need this requirement when $x$ is a nonnegative integer. In most applications, we consider $0 < r < k$. For example, under monthly compounding $(k = 12)$, the upper-bound condition expressed in percent means that the compounding interest rate satisfies $r < 1200\%$, which will surely be the case in most applications.

We also call $F_x$ the *future value* of $F_0$ at the end of $x$ interest periods from the present and refer to

$$F_0 = \frac{F_x}{\left(1 + \frac{r}{k}\right)^x}, \qquad \left(0 \le \frac{r}{k} < 1, \quad x \ge 0\right), \tag{2.21}$$

as the *present value* of $F_x$. The interest rate $r$ is applied as a growth rate in the future valuing of (2.20) and as a discount rate in the context of (2.21). Since $x$ interest periods is $x/k$ years, the future value $F_x$ occurs $x/k$ years from now, i.e.,

$$F_x = F\left(\frac{x}{k}\right).$$

The number $x$ of interest periods can always be expressed as the sum of an integral number $n$ of interest periods and a fraction $v$ of an interest period:

$$x = n + v,$$

where $n$ is the greatest integer part of $x$ and $0 \le v < 1$. For example, $x = 15.36$ interest periods splits into a sum of $n = 15$ and $v = 0.36$ interest periods.

We can then rewrite (2.20) as

$$F_x = \left(1 + \frac{r}{k}\right)^n F_v = \left(1 + \frac{r}{k}\right)^v F_n, \qquad \left(0 \le \frac{r}{k} < 1, \quad 0 \le v < 1\right). \quad (2.22)$$

Here $F_v$ is the amount to which $F_0$ grows over the fraction $v$ of an interest period, i.e., we have *fractional compounding* during $v$ mth:

$$F_v = \left(1 + \frac{r}{k}\right)^v F_0.$$

For a proper fractional period, i.e., for $0 < v < 1$, the leftmost equality in (2.22) states that the fractionally compounded amount $F_v$ is compounded over $n$ interest periods, and the rightmost equality captures that the accrued amount $F_n$ is compounded over the fraction $v$ of an interest period. The left equality applies to settings where the start of the time span does not coincide with the beginning or end of an interest period, while the right equality is for when the end of the time span is not the beginning or end of an interest period.

The rightmost equality in (2.22) also shows that if the interest rate per interest period $r/k$ is sufficiently small, expanding the binomial series yields:

$$\left(1 + \frac{r}{k}\right)^v \approx 1 + v\frac{r}{k}. \qquad (2.23)$$

The amount accrued at the end of $x$ interest periods can then be approximated as follows:

$$F_x = \left(1 + \frac{r}{k}\right)^v F_n \approx \left(1 + v\frac{r}{k}\right) F_n, \qquad (0 \le v < 1, \quad 0 \le r/k \ll 1). \quad (2.24)$$

**Example 2.5.** Returning to the example from the start of this section (page 24), Equation (2.20) shows that a principal of $\$10,000,000$ compounded monthly at 10% per annum for 15.36 mth will grow to:

$$F_{15.36} = \left(1 + \frac{0.1}{12}\right)^{15.36} \times \$10,000,000 = \$11,359,503.48.$$

Equivalently,

$$F_{15.36} = \$11{,}359{,}503.48 = \left(1 + \frac{0.1}{12}\right)^{0.36} \times F_n,$$

which is the form in (2.22) and the origin of (2.15). Equation (2.14) uses simple interest, rather than fractional compounding, during the remaining 0.36 mth and is justified by (2.24):

$$\tilde{F}_{15.36} = \left(1 + v\frac{r}{k}\right) F_n = \left(1 + 0.36\frac{0.1}{12}\right) \times F_n = \$11{,}359{,}593.67$$

and since $r/k = 0.0083 \ll 1$, we have $\tilde{F}_{15.36} \approx F_{15.36}$.                    □

For $k$-periodic compounding at a constant interest rate $r$ per year, Equation (2.20) yields that a principal $F_0$ will grow to the following future value over $\tau$ years or $k\tau$ periods:

$$F_{k\tau} = \left(1 + \frac{r}{k}\right)^{k\tau} F_0. \tag{2.25}$$

**Example 2.6. (Doubling Your Investment)** Suppose that you invest $F_0$ today in an account with $k$-periodic compounding at $r$ per year. Find a formula for how long it will take you to increase your investment to $x_0 F_0$, where $x_0 > 1$. Does the length of time depend on the initial amount $F_0$? In particular, how long will it take to double an investment of \$1,000 using 6% per annum with daily compounding? What about \$2,000? Compare with the time it would take using simple interest growth at the same interest rate.

**Solution.** We want to find how many years $\tau$ it will take to have $F_0$ grow to $F_{k\tau} = x_0 F_0$. By (2.25),

$$x_0 F_0 = \left(1 + \frac{r}{k}\right)^{k\tau} F_0,$$

which implies that

$$\tau = \frac{\ln x_0}{k \ln\left(1 + \frac{r}{k}\right)} \qquad (r > 0).$$

The time does not depend on the initial $F_0$.

For $F_0 = \$1{,}000$, $x_0 = 2$, $r = 0.06$, and $k = 365$ (daily), we obtain

$$\tau = \frac{\ln 2}{365 \ln\left(1 + \frac{0.06}{365}\right)} \approx 11.55,$$

so it will take 11.55 years. Since the time span will not depend on the initial investment, we obtain the same answer for \$2,000. For simple interest growth, we have $x_0 F_0 = (1 + r\tau)F_0$, which yields $\tau = \dfrac{x_0 - 1}{r} \approx 16.67$. The doubling time is 5.12 years longer.                    □

### *2.4.3 Fractional Compounding Versus Simple Interest*

Compound interest is constructed by applying simple interest over each interest period to the balance at the start of the interest period. This may suggest that simple interest should then be applied over each (proper) fraction of an interest period to the balance at the start of the fractional interest period (since simple interest adds over different time segments). However, if simple interest is applied over a given fraction of a period, then it does not account for the compounding that has to occur over every fraction of the given fractional interest period. Indeed, a new insight from Theorem 2.1 is that *compounding occurs over every portion of an interest period*. For example, start with a balance $F_*$ and have it compound for $\frac{1}{3}$ prd. You should not apply simple interest over the $\frac{1}{3}$ prd because the balance $F_*$ also has to compound over every fraction of the $\frac{1}{3}$ prd. For instance, compounding occurs over the first fourth of the $\frac{1}{3}$ prd and the remaining three-fourths of the period. By (2.20), the correct growth is:

$$F_{1/3} = \left(1 + \frac{r}{k}\right)^{\frac{1}{3}} F_* = \left(1 + \frac{r}{k}\right)^{\frac{3}{4}\left(\frac{1}{3}\right)} \left(1 + \frac{r}{k}\right)^{\frac{1}{4}\left(\frac{1}{3}\right)} F_*.$$

By (2.23), using simple interest over a fraction $v$ of an interest period would only yield an approximation under the following condition:

$$F_v = \left(1 + \frac{r}{k}\right)^v F_* \approx \left(1 + v\frac{r}{k}\right) F_*, \qquad (0 < v < 1, \quad 0 \le r/k \ll 1).$$

If we do not use simple interest over a fraction of an interest period, then why can we apply simple interest over a whole interest period? The reason is that simple interest over one interest period is equivalent to fractional compounding over the interest period. In fact, decompose an interest period into any two fractional periods, say, $v$ prd and $(1 - v)$ prd. Suppose that the balance at the start of the interest period is $F_*$. Then the balance at the end of the period is:

$$F_1 = \left(1 + \frac{r}{k}\right) F_* = \left(1 + \frac{r}{k}\right)^{1-v} \left(1 + \frac{r}{k}\right)^v F_*.$$

In other words, the simple interest growth of $F_*$ over 1 prd is the same as fractional compounding of $F_*$ over $v$ prd followed by fractional compounding of the accrued amount $\left(1 + \frac{r}{k}\right)^v F_*$ over the remaining $(1 - v)$ prd. We could divide an interest period into an arbitrary finite number of subperiods and still obtain that simple interest over one period is fractional compounding over the subperiods:

$$F_1 = \left(1 + \frac{r}{k}\right) F_* = \left[\prod_{j=1}^{m} \left(1 + \frac{r}{k}\right)^{v_j}\right] F_*,$$

where $1 = v_1 + v_2 + \cdots + v_m$ and $0 < v_j < 1$ with $j = 1, \ldots, m$.

### 2.4.4 Continuous Compounding

When the number $k$ of compounding periods per year increases without bound, we have continuous compounding. Applying (2.25), the future value under continuous compounding is

$$F_{cts} = \lim_{k \to \infty} F_{k\tau} = \lim_{k \to \infty} \left(1 + \frac{r}{k}\right)^{k\tau} F_0 = F_0 \left(\lim_{k \to \infty} \left(1 + \frac{r}{k}\right)^{k}\right)^{\tau} = F_0 e^{r\tau}, \quad (2.26)$$

with return rate

$$R(\tau) = e^{r\tau} - 1. \quad (2.27)$$

Under continuous compounding, \$1 will grow to $\$e^{rt}$ over the time interval $[0,t]$. We can apply the same idea to a security (possibly risky) paying a continuous cash dividend at a constant yield rate $q$. Suppose that at time 0, you have 1 unit of a security, and as the cash dividend flows in, you continuously buy more units of the security, i.e., the number of units of the security is continuously compounded at rate $q$. Then 1 unit of the security at time 0 will grow to $e^{qt}$ at time $t$. Consequently, the cum-dividend value of the security at $t$ is

$$S_t^c = e^{qt} S_t, \quad (2.28)$$

where $S_t$ is the ex-dividend price at $t$ of one unit of the security. *We assume that* $S_t$ *models the market price at* $t$ since it discounts the cum-dividend price at the dividend yield rate: $S_t = e^{-qt} S_t^c$. See the discussion on page 18.

## 2.5 Generalized Compound Interest

### 2.5.1 Varying Interest and Varying Compounding Periods

This section extends compound interest from a fixed interest rate over a non-negative real number of compounding periods to discretely varying interest rates across compounding intervals of different lengths.

We begin with some needed notation. Suppose that you put the amount $F_0$ (principal) in an account for a time interval $[t_0, t_f]$, where $t_0 \geq 0$. Assume that each compound interest period is $\frac{1}{k}$ yr. Divide $[t_0, t_f]$ into $n$ subintervals (not necessarily of the same length), say,

$$[t_0, t_1], \quad [t_1, t_2], \quad \ldots, \quad [t_{i-1}, t_i], \quad \ldots, \quad [t_{n-1}, t_n],$$

where $n$ is a positive integer, $t_n = t_f$. Denote the length of the $i$th subinterval $[t_{i-1}, t_i]$ by $\tau_i$, which corresponds to the following number of periods:

$$\tau_i \text{ yr} = k\tau_i \text{ prd}, \qquad\qquad i = 1, \ldots, n.$$

Suppose that $k$-periodic compounding at $r_i$ per annum applies during the $i$th interval $[t_{i-1}, t_i]$ for $i = 1, \ldots, n$.

We now determine a formula for the amount to which the principal $F_0$ will grow at the future time $t_n$.

➤ Over the time interval $[t_0, t_1]$, we have $k$-periodic compounding at interest rate $r_1$ of the principal $F_0$. Applying (2.20) on page 27 with $x = k\tau_1$, the value of $F_0$ grows to the following at time $t_1$:

$$F(t_1) = \left(1 + \frac{r_1}{k}\right)^{k\tau_1} F_0.$$

Reinvest the entire amount $F_1$ in the account.

➤ Over the next time interval $[t_1, t_2]$, the balance $F_1$ at time $t_1$ is $k$-periodically compounded at rate $r_2$. By (2.20) with $x = k\tau_2$, the value of $F(t_1)$ grows to:

$$F(t_2) = \left(1 + \frac{r_2}{k}\right)^{k\tau_2} F(t_1) = \left(1 + \frac{r_2}{k}\right)^{k\tau_2} \left(1 + \frac{r_1}{k}\right)^{k\tau_1} F_0.$$

Reinvest $F_2$ in the account.

➤ Continuing this process, we find that over the final time interval $[t_{n-1}, t_n]$, the balance $F(t_{n-1})$ at time $t_{n-1}$ is $k$-periodically compounded at rate $r_n$. Again (2.20) yields that $F(t_{n-1})$ grows to:

$$F(t_n) = \left(1 + \frac{r_n}{k}\right)^{k\tau_n} F(t_{n-1}).$$

Explicitly:

$$F(t_n) = \left(1 + \frac{r_n}{k}\right)^{k\tau_n} \cdots \left(1 + \frac{r_2}{k}\right)^{k\tau_2} \left(1 + \frac{r_1}{k}\right)^{k\tau_1} F_0. \qquad (2.29)$$

Observe $F(t_n)$ depends on the lengths of the subintervals over which the various interest rates are constant.

We call $F(t_n)$ the generalized compound interest *future value* of $F_0$ at time $t_n$. It is given in product notation as follows:

$$F(t_n) = \left[\prod_{i=1}^{n} \left(1 + \frac{r_i}{k}\right)^{k\tau_i}\right] F_0, \qquad\qquad \left(0 \le \frac{r_i}{k} < 1\right). \qquad (2.30)$$

Here $F_0$ is termed the *present value* of $F(t_n)$.

## Special Case

Assume that each interval $[t_{i-1}, t_i]$, where $i = 1, \ldots, n$, coincides with a compound interest period. Then the generalized future value (2.30) becomes:

$$F(t_n) = \left(1 + \frac{r_n}{k}\right) \cdots \left(1 + \frac{r_2}{k}\right)\left(1 + \frac{r_1}{k}\right) F_0, \qquad (2.31)$$

where $F_0$ is the principal at the initial time $t_0$. For a constant interest rate amount $r_i = r$, where $i = 1, \ldots, n$, we recover the usual $k$-periodic compounding formula over $n$ periods or $\frac{n}{k}$ years:

$$F_n = F\left(\frac{n}{k}\right) = \left(1 + \frac{r}{k}\right)^n F_0 = F_n.$$

**Example 2.7.** How much will $1,000 grow after 1.5 years if it is compounding semiannually with annual interest rate 7% applied at the end of the first 6 months, 8% at the end of the first year, and 9% at the end of 1.5 years?

**Solution.** Use the generalized compound interest formula (2.31) with $k = 2$ (semiannual compounding), $n = 3$ (number of periods), $t_0$ the current time, $t_3 = t_0 + 1.5$ (future time), $r_1 = 0.07$, $r_2 = 0.08$, and $r_3 = 0.09$. We obtain the following future value:

$$F(t_3) = \left(1 + \frac{r_3}{k}\right)\left(1 + \frac{r_2}{k}\right)\left(1 + \frac{r_1}{k}\right) F_0 = 1.045 \times 1.040 \times 1.035 \times \$1,000$$
$$= \$1,124.84.$$

$\square$

## 2.5.2 APR Versus APY

We begin by showing how the interest rate $r$ relates to the return rate in the context of compound interest.

At time $t_0$ invest an amount $F_0 > 0$ (principal) in an account that grows under $k$-periodic compounding at interest rate $r$. Suppose that the account pays no dividend. Let $F(t_f) > 0$ be the value of the principal at a future time $t_f = t_0 + \tau$. Since a time span of $\tau$ years has $k\tau$ periods, Equation (2.20) on page 27 yields that the return rate on the principal $F_0$ is:

$$R_{CI}(\tau) = \frac{F(t_f)}{F_0} - 1 = \left(1 + \frac{r}{k}\right)^{k\tau} - 1, \qquad (2.32)$$

where the subscript CI indicates that the return rate is in the context of compound interest. Note the dependence on the length $\tau$ of the time interval $[t_0, t_f]$. For $n$ periods, the return rate becomes:

$$R_{CI}\left(\frac{n}{k}\right) = \left(1 + \frac{r}{k}\right)^n - 1. \tag{2.33}$$

In addition, the interest rate $r$ can be expressed in terms of $R_{CI}(\tau)$ as follows:

$$r = \frac{(1 + R_{CI}(\tau))^{\frac{1}{k\tau}} - 1}{1/k}. \tag{2.34}$$

Equation (2.32) also shows that *growing the initial amount $V(t_0)$ to the value $V(t_f)$ under compounding at interest rate $r$ is the same as growing $V(t_0)$ to $V(t_f)$ under simple interest using the return rate $R_{CI}(\tau)$ over the time span $\tau$:*

$$V(t_f) = \left(1 + R_{CI}(\tau)\right) V(t_0) = \left(1 + \frac{r}{k}\right)^{k\tau} V(t_0).$$

The return rate $R_{CI}(1)$ over a year is also commonly used. Equation (2.32) yields:

$$R_{CI}(1) = \left(1 + \frac{r}{k}\right)^k - 1, \tag{2.35}$$

which is also called the *annual percentage yield* (APY) or *effective interest rate* and denoted by $R_{CI}(1) = \text{APY}$. The interest rate $r$ corresponding to $R_{CI}(1)$ is called the *annual percentage rate* (APR) or *nominal interest rate* and is given by:

$$\text{APR} = \frac{(1 + \text{APY})^{\frac{1}{k}} - 1}{1/k}.$$

The APR should not be confused with the APY, which involves compounding:

$$\text{APY} = \left(1 + \frac{\text{APR}}{k}\right)^k - 1.$$

For instance, if you are quoted an APR of 12% per annum on a loan, then the APR arises from a monthly interest rate of $\text{APR}/12 = 1\%$. However, since interest on debt typically involves compounding, the APY gives a true reflection of the interest rate a borrower pays. In this case, the 1% per month interest compounds to an annual percentage yield of

$$\text{APY} = (1 + 0.01)^{12} - 1 = 12.68\%,$$

not 12%. The next example further illustrates the difference.

**Example 2.8.** If a credit card company quotes only its APR on the card, say, 10.99%, it can cause a consumer to think that after 1 year, the interest amount on a balance of $2,500 is

$$0.1099 \times \$2,500 = \$274.75.$$

However, this is not correct because it assumes simple interest for the year. Most credit cards compound daily or monthly (and may add fees). The true interest rate for a 365-day year with daily compounding is given by the APY:

$$\text{APY} = \left(1 + \frac{0.1099}{365}\right)^{365} - 1 = 11.6148\%.$$

The actual interest amount for the year is then the (effective) return amount:

$$\text{APY} \times \$2,500 = 0.116148 \times \$2,500 = \$290.37,$$

which is more than the amount \$274.75 naively inferred from an APR of 10.99%. $\square$

### 2.5.3 Geometric Mean Return Versus Arithmetic Mean Return

An argument essentially the same as the one used to derive (2.35) shows that, given any period return rate $R_{\text{prd}}$, the return rate over a year with compounding at rate $R_{\text{prd}}$ per period is given by:

$$R_{\text{ann}} = \left(1 + R_{\text{prd}}\right)^k - 1, \tag{2.36}$$

where (as usual) a year is assumed to have $k$ periods. For example, a weekly return rate of 1% annualizes as follows under weekly compounding:

$$R_{\text{ann}} = (1 + 0.01)^{52} - 1 = 67.8\%.$$

We can generalize (2.36) further. First, the return rate (2.33) extends naturally to compound interest with varying interest rates over a time span of $n$ compounding periods, where each period is $\frac{1}{k}$ yr. Assume that the annual interest rates used for the various $n$ consecutive compounding periods are $r_1, \ldots, r_n$, i.e., the interest over the $i$th period is $\frac{r_i}{k}$. By (2.31) on page 33, the return rate (2.33) generalizes to:

$$R_{\text{CI}}(t_0, t_n) = \frac{F(t_n)}{F_0} - 1 = \left(1 + \frac{r_n}{k}\right)\left(1 + \frac{r_{n-1}}{k}\right) \cdots \left(1 + \frac{r_1}{k}\right) - 1. \tag{2.37}$$

Now, assume that you invest $F_0$ in a nondividend-paying investment that has return rate $R_i$ over the $i$th period, where $i = 1, \ldots, n$. Explicitly, if $V_{i-1}$ and $V_i$ are the respective values of the investment at the start and end of the $i$th period, then return rate is

$$R_j^{\text{prd}} = \frac{V_j - V_{j-1}}{V_{j-1}}.$$

Employing arguments similar to those used to derive (2.37), we can extend (2.37) from an $i$th-period interest rate of $\frac{r_i}{k}$, which is always positive, to the return rate of $R_i^{\text{prd}}$, which is not necessarily positive. In other words, we are generalizing (2.36) to the return rate over $n$ periods by compounding at the respective return rates $R_1^{\text{prd}}, \ldots, R_n^{\text{prd}}$:

$$R_{\text{tot}} \equiv R\left(t_0, t_0 + \frac{n}{k}\right) = \left(1 + R_n^{\text{prd}}\right)\left(1 + R_{n-1}^{\text{prd}}\right) \cdots \left(1 + R_1^{\text{prd}}\right) - 1. \quad (2.38)$$

Note that by (2.1), each factor in the product (2.38) is nonnegative since it is a gross return:

$$1 + R_j^{\text{prd}} = \frac{V_j}{V_{j-1}} \geq 0, \qquad\qquad (j = 1, \ldots, n).$$

Equation (2.38) shows that the initial investment $F_0$ will grow to the following value:

$$\text{F}\left(\frac{n}{k}\right) = (1 + R_{\text{tot}})\, \text{F}_0. \quad (2.39)$$

We remind the reader that in (2.39), we assume you do not withdraw or add any funds to the investment during the $n$ periods. Unless otherwise stated, this is always our assumption when compounding; see Section 2.4.1.

Now, suppose that $n$ periods ago, an investor put $F_0$ into a nondividend-paying fund and her investment grew by the process in (2.39) to the current value $\text{F}(\frac{n}{k})$. She would now like to forecast the behavior of the fund over the next period using a *single* "mean return rate" $x$. In other words, we seek a single rate $x$ such that when compounding $F_0$ using $x$ over each of the past $n$ periods, we obtain the same answer as compounding $F_0$ using the $n$ return rates $R_1^{\text{prd}}, \ldots, R_n^{\text{prd}}$:

$$\text{F}\left(\frac{n}{k}\right) = (1 + x)^n\, \text{F}_0. \quad (2.40)$$

Comparing (2.39) and (2.40), we see that $x$ must be the *geometric mean return* $\overline{R}_{\text{geom}}^{\text{prd}}$ of $R_1^{\text{prd}}, \ldots, R_k^{\text{prd}}$, namely,

$$\overline{R}_{\text{geom}}^{\text{prd}} = \left[\left(1 + R_n^{\text{prd}}\right)\left(1 + R_{n-1}^{\text{prd}}\right) \cdots \left(1 + R_1^{\text{prd}}\right)\right]^{1/n} - 1.$$

We have:

$$\text{F}\left(\frac{n}{k}\right) = \left(1 + \overline{R}_{\text{geom}}^{\text{prd}}\right)^n \text{F}_0.$$

The geometric mean return relates as follows to the total return rate:

$$R_{\text{tot}} = \left(1 + \overline{R}_{\text{geom}}^{\text{prd}}\right)^n - 1. \quad (2.41)$$

In general, the geometric mean return does not equal the *arithmetic mean return*,

$$\overline{R}_{\text{prd}} = \frac{1}{n} \sum_{j=1}^{n} R_j^{\text{prd}}.$$

In fact,

$$\overline{R}_{\text{geom}}^{\text{prd}} \leq \overline{R}_{\text{prd}}.$$

The two means coincide when the period return rates $R_j^{\text{prd}}$ are identical for $j = 1, \ldots, n$. The example below illustrates these two means; see Reilly and Brown [16, Sec. 1.2.2] for more.

**Example 2.9. (Geometric Mean Return Versus the Arithmetic Mean Return)**
Suppose that you initially invest \$3,000 in a fund that pays no dividend. Assume that the investment decreases to \$2,000 at the end of 1 year, decreases from \$2,000 to \$1,000 from the end of year 1 to the end of year 2, and increases from \$1,000 to \$3,000 from the end of year 2 to the end of year 3. Then the total return rate on your investment over the 3 years is zero.

Let us compare what the arithmetic and geometric mean returns forecast for the total return rate. The year-to-year return rates over the 3 years are:

$$R(t_0, t_0 + 1)) = \frac{\$2,000}{\$3,000} - 1 = -\frac{1}{3}, \quad R(t_0 + 1, t_0 + 2)) = \frac{\$1,000}{\$2,000} - 1 = -\frac{1}{2},$$

$$R(t_0 + 2, t_0 + 3)) = \frac{\$3,000}{\$1,000} - 1 = 2.$$

The arithmetic mean return is:

$$\overline{R}_{\text{yr}} = \frac{1}{3} \left( -\frac{1}{3} - \frac{1}{2} + 2 \right) = \frac{7}{18} = 0.3889,$$

which when compounded annually over the 3 years yields a grossly incorrect return rate:

$$(1 + \overline{R}_{\text{yr}})^3 - 1 = (1 + 0.3889)^3 - 1 = 1.6792 = 168\%!$$

On the other hand, the geometric mean return is:

$$\overline{R}_{\text{geom}}^{\text{yr}} = \left[ (1 + 2) \left( 1 - \frac{1}{2} \right) \left( 1 - \frac{1}{3} \right) \right]^{1/3} - 1 = 0,$$

which gives the correct total return rate:

$$R(t_0, t_0 + 3) = \left( 1 + \overline{R}_{\text{geom}}^{\text{yr}} \right)^3 - 1 = 0\%.$$

This is, of course, an illustration of (2.41).

The arithmetic mean return deviated significantly from the geometric mean because of the high volatility in the yearly return rates. The geometric mean return was better able to capture this dispersion and, hence, produced the correct total return rate. The two measures approximate each other when the return rates do not change significantly from period to period. The geometric mean return is usually employed for longer time horizons, where there is more opportunity for higher volatility.                                                                              □

## 2.6 The Net Present Value and Internal Rate of Return

Compound interest can also be applied to develop the notion of a "net present value." This tool helps with deciding whether to partake in a particular investment opportunity. The opportunity can be a project, product line, start-up company, etc. We assume that with an initial capital, the investment opportunity produces net cash flows, i.e., cash inflows minus cash outflows, at different future dates.

> *Unless stated to the contrary, assume that each net cash flow takes taxes into account.*

In addition, when the net cash flow on a particular future date is being estimated, the estimate usually reflects activities over the year leading up to the date. We shall then consider net cash flows on future dates separated by a year. Furthermore, over the time span that an investment opportunity is analyzed for its growth potential, we assume that all the annual net cash flows can be modeled as arising from annual compounding at a constant interest rate. We refer to this constant interest rate as the *compounding growth (annual) rate from investing in the opportunity.*

### 2.6.1 Present Value and NPV of a Sequence of Net Cash Flows

For concreteness, we shall consider a credible, innovative start-up company. Suppose that the start-up forecasts that, with an initial investment of $250,000, it will generate net cash flows of $155,000 1 year from now, $215,000 2 years from now, and $350,000 3 years from now. The entire initial capital is assumed to be invested to produce these cash flows; e.g., none of the money is put aside in an account unrelated to the company's activities.

An important mathematical function we shall employ is the *present value* $PV(r)$ of the sequence of net cash flows at an annual discount rate $r$. Note that in the current context, the present value is expressed as a function of the discount rate $r$ rather than the number of periods $n$ since $n$ will be fixed and $r$ will play a more key role. In our example, we have:

$$PV(r) = \frac{\$155,000}{1+r} + \frac{\$215,000}{(1+r)^2} + \frac{\$350,000}{(1+r)^3}.$$

Equally important will be the spread between $PV(r)$ and the initial capital. This function is called the *net present value* (NPV) at rate $r$ of the sequence of cash flows and is given by:

$$NPV(r) = PV(r) - \$250,000.$$

*For simplicity, we shall write expressions such as* $NPV(r) > 0$, *where it is understood that the "0" represents a zero amount of cash in the currency of the net cash flows.*

Now, an important step in deciding whether to invest in the start-up is to research the marketplace to find the mean compounding growth rate from investing in an alternative opportunity with a similar business profile and risk—e.g., research competitor companies comparable to the start-up in scale, risk, business sector, etc. For illustration, assume that the mean compounding growth rate from investing in an appropriate alternative opportunity is estimated to be:

$$r_{RRR} = 15\%.$$

We then take $r_{RRR}$ as our *required return rate* for investing in the start-up.

The current market value of the start-up's projected stream of future net cash flows is the present value of these net cash flows discounted at the required return rate of 15%:

$$PV(r_{RRR}) = \frac{\$155,000}{1+0.15} + \frac{\$215,000}{(1+0.15)^2} + \frac{\$350,000}{(1+0.15)^3}$$

$$= \$134,782.61 + \$162,570.89 + \$230,130.68$$

$$= \$527,484.18. \tag{2.42}$$

It is important to observe that *when determining the present value in (2.42), no assumption is being made about reinvesting the net cash flows* $\$155,000, \$215,000,$ *and* $\$350,000.$

Equation (2.42) tells us that since the alternative opportunity grows your investment at 15% per annum compounded annually, such an opportunity can generate the start-up's forecasted net cash flows if you invest $527,484.18 in the opportunity today. To see this, separate the required investment of $527,484.18 into three parts as follows:

$$PV(r_{RRR}) = \$527,484.18 = \underbrace{\$134,782.61}_{A_{RRR}} + \underbrace{\$162,570.89}_{B_{RRR}} + \underbrace{\$230,130.68}_{C_{RRR}}.$$

We can then produce the net cash flows by thinking theoretically of the alternative opportunity as growing $A_{RRR}$ to a future value of $FV_{A_{RRR}}(1)$ at 1 year out, $B_{RRR}$ to a future value of $FV_{B_{RRR}}(2)$ at 2 years out, and $C_{RRR}$ to a future value of $FV_{C_{RRR}}(3)$ at 3 years out:

$$FV_{A_{RRR}}(1) = (1 + 0.15) \times \$134{,}782.61 = \$155{,}000$$

$$FV_{B_{RRR}}(2) = (1 + 0.15)^2 \times \$162{,}570.89 = \$215{,}000$$

$$FV_{C_{RRR}}(3) = (1 + 0.15)^3 \times \$230{,}130.68 = \$350{,}000.$$

On the other hand, the credible start-up claims that it can generate the above future net cash flows with an investment today of less than the amount $527,484.18 required by the alternative opportunity, namely, with an initial investment of only

$$C_0 = \$250{,}000.$$

Naturally, investors will favor the start-up since the amount $PV(r_{RRR})$ required by the alternative opportunity is more expensive than the amount $C_0$ required by the start-up. In other words, the start-up appears favorable when the net present value at the market required return rate is positive:

$$NPV(r_{RRR}) = PV(r_{RRR}) - C_0 > 0.$$

*The net present value at the required return rate, namely,*

$$NPV(r_{RRR}) = \$527{,}484.18 - \$250{,}000 = \$277{,}484.18,$$

*then measures how much cheaper (or more expensive, if the difference were negative) it is to invest in the start-up than in the alternative opportunity.* Of course, any final decision to invest in a start-up will not rely solely on the NPV, but will be complemented with a detailed analysis of the start-up's business plan, innovative products/services, market environment, management team, etc.

If we had $NPV(r_{RRR}) = 0$, i.e., the initial capital required by the start-up to produce the given future net cash flows was the same as that required by the alternative opportunity, then there would be no extra value received from investing in the start-up. In this borderline situation, however, some investors may still invest in the start-up if, for example, it has more long-term promise.

The start-up would not be attractive to investors if $NPV(r_{RRR}) < 0$, i.e., if it costs more to receive the same future net cash flows from the start-up than from the alternative opportunity.

## 2.6.2 *The Internal Return Rate*

Indeed, the start-up can achieve these future net cash flows with less initial capital only if it grows the initial capital at a rate greater than the alternative opportunity's compounding annual growth rate of 15%. The start-up's compounding annual growth rate on the initial capital $C_0$ is called the *internal rate of return* (IRR) and denoted $r_{IRR}$. To determine the start-up's IRR, we must find the interest rate $r_{IRR}$ that generates the forecasted net cash flows starting from $C_0 = \$250,000$:

$$
\begin{aligned}
C_1 &= \$155,000 \qquad \text{end of year 1} \\
C_2 &= \$215,000 \qquad \text{end of year 2} \\
C_3 &= \$350,000 \qquad \text{end of year 3.}
\end{aligned}
$$

$$(2.43)$$

First, separate $\$250,000$ into three amounts given by the present values of the net cash flows $\$155,000$, $\$215,000$, and $\$350,000$ at the unknown discount rate $r_{IRR}$. The sum of these individual present values is the present value $PV(r_{IRR})$ of the sequence of net cash flows. Explicitly:

$$
\$250,000 = \underbrace{\frac{\$155,000}{(1+r_{IRR})}}_{A_{IRR}} + \underbrace{\frac{\$215,000}{(1+r_{IRR})^2}}_{B_{IRR}} + \underbrace{\frac{\$350,000}{(1+r_{IRR})^3}}_{C_{IRR}} = PV(r_{IRR}). \quad (2.44)
$$

Then the future values at rate $r_{IRR}$ of the three portions of the $\$250,000$ in (2.44) yield the desired future net cash flows. Specifically, the future value of $A_{IRR}$ at 1 year out is $\$155,000$, of $B_{IRR}$ at 2 years out is $\$215,000$, and of $C_{IRR}$ at 3 years out is $\$350,000$. It suffices then to find the IRR by solving (2.44) for $r_{IRR}$. Note that (2.44) is equivalent to the vanishing of the net present value at the rate $r_{IRR}$:

$$
NPV(r_{IRR}) = PV(r_{IRR}) - \$250,000 = 0. \quad (2.45)
$$

Employing a software, we find that an approximate solution of (2.44) or, equivalently, (2.45) is:

$$
r_{IRR} = 0.652811.
$$

Note that inserting this IRR into (2.44) actually produces $\$250,000.04$, which, of course, is not the exact value $\$250,000$ due to the approximate value of $r_{IRR}$. In other words, decomposing the start-up capital approximately as

$$
\$250,000 \approx \underbrace{\$93,779.63}_{A_{IRR}} + \underbrace{\$78,703.14}_{B_{IRR}} + \underbrace{\$77,517.27}_{C_{IRR}} = PV(0.652811)
$$

and future valuing each term by compounding annually at the rate $r_{IRR}$ will yield the desired stream of net cash flows.

The start-up's IRR of 65.2811% compounded annually exceeds the alternative opportunity's compounding annual growth rate of 15%, which makes the start-up favorable. If it turned out that the IRR were 15%, then the start-up's growth rate would be no better than that of the alternative opportunity in the market (borderline case). If, on the other hand, the IRR were less than 15%, the start-up would be unattractive to investors (start-up not favorable).

In our example, we have: $r_{IRR} > r_{RRR}$ if and only if $NPV(r_{RRR}) > 0$. In other words, the IRR basis for deciding whether to favor the start-up is equivalent, *in this example*, to the choice being based on the NPV. *We have to be careful to not generalize this observation widely.* The example's equivalence of the IRR and NPV criteria is actually based on (2.44) or (2.45) having a unique positive solution and on the net present value being a strictly decreasing function. These two requirements need not hold in general. We address these issues next.

### 2.6.3 NPV and IRR for General Net Cash Flows

Extend the previous example to a general sequence of net cash flows. Suppose that you are considering a new investment opportunity requiring an initial capital of $C_0 > 0$ to generate future net cash flows,

$$C_1, \quad C_2, \quad \ldots, \quad C_n,$$

at respective future years $1, 2, \ldots, n$.

Making no assumptions about reinvesting the net cash flows $C_1, C_2, \ldots, C_n$, we see that the present value of this sequence of cash flows at the compound-interest discount rate of $r$ is:

$$PV(r) = \frac{C_1}{(1+r)} + \frac{C_2}{(1+r)^2} + \cdots + \frac{C_n}{(1+r)^n}, \qquad (r > 0). \qquad (2.46)$$

The net present value of the net cash flows is the cost of the alternative investment opportunity minus the cost of the new investment opportunity:

$$NPV(r) = PV(r) - C_0, \qquad (r > 0, \quad C_0 > 0). \qquad (2.47)$$

As before, denote the required return rate of the new investment opportunity by $r_{RRR}$. Recall that $r_{RRR}$ is the mean compounding (annual) growth rate from investing in an alternative opportunity in the marketplace with business profile and risk similar to the new investment opportunity. *An NPV-based decision-making rule about whether to invest in the new opportunity is as follows:*

➤ If $NPV(r_{RRR}) > 0$, then the new investment opportunity is cheaper than the alternative investment and so is favorable.

➤ If $NPV(r_{RRR}) < 0$, then the new opportunity is more expensive and not favorable.

➤ If $NPV(r_{RRR}) = 0$, then the cost of the new opportunity is the same as the alternative investment and it is borderline whether to invest.

As noted in Section 2.6.1, even with a robust NPV estimate, a real-world business decision about whether to invest in a new opportunity will not use the NPV as the only measure. One has to factor in the business environment, experience of the management team, etc.

An IRR of the new investment is a positive solution, $r = r_{IRR}$, of the following equation:

$$0 = NPV(r) = -C_0 + \frac{C_1}{(1+r)} + \frac{C_2}{(1+r)^2} + \cdots + \frac{C_n}{(1+r)^n}. \tag{2.48}$$

Equation (2.48) is equivalent to a real polynomial, so we are seeking the positive roots of such a real polynomial. Without loss of generality, suppose that the real polynomial has degree $k$ and is of the following form:[5]

$$a_k r^k + a_{k-1} r^{k-1} + \cdots + a_1 r + a_0 = 0, \qquad (a_k \neq 0). \tag{2.49}$$

There is no general formula for the real solutions of (2.49) for all positive integers $k$.

Perhaps the most cited general result about the number of positive solutions of (2.49) is Descartes's Rule of Signs. Before stating this result, we gather some notation. Let $N_+$ denote the number of positive solutions of (2.49), where we count the solutions with multiplicity. For example, the polynomial,

$$r^2 - 10r + 25 = (r - 5)^2 = 0,$$

has $N_+ = 2$, corresponding to two positive solutions $r = 5$ counted with multiplicity. Let $N_{sgn}$ be the number of sign changes in the ordered sequence of the coefficients in (2.49):

$$a_k, a_{k-1}, \ldots, a_1, a_0. \tag{2.50}$$

Since the zero coefficients do not contribute to a sign change, it suffices to consider the sign changes due to the ordered nonzero coefficients.

**Theorem 2.2. (Descartes's Rule of Signs)** *The number $N_+$ of positive solutions of (2.49) is the number $N_{sgn}$ of sign changes of its ordered coefficients in (2.50) or is $N_{sgn}$ minus an even positive integer. Specifically, $N_+$ equals either $N_{sgn}$, $N_{sgn} - 2$, $N_{sgn} - 4, \ldots, N_{sgn} - 2(n-1)$, or $N_{sgn} - 2n$ for some nonnegative integer $n$.*[6]

---

[5] If $a_k = 0$, then simply apply the same discussion to the lower degree polynomial.

[6] Using $N_+ \leq N_{sgn}$, the reason a nonnegative even integer is subtracted from $N_{sgn}$ in the theorem is because $N_+$ and $N_{sgn}$ have the same parity, i.e., $N_+$ is even (odd) if and only if $N_{sgn}$ is even (odd). This

*Proof.* See Meserve [14, p. 156] and Wang [17] for a proof.                    □

For example, the polynomial equation

$$r^5 - r^2 + r - 1 = 0$$

has three sign changes in its ordered nonzero coefficients: $+1, -1, +1, -1$. By Theorem 2.2, this polynomial equation has either 3 or 1 positive solutions.

The IRR Equation (2.48) is equivalent to a polynomial equation of the form (2.49) with ordered coefficients (2.50). By Theorem 2.2, if these ordered coefficients have one sign change, then there is at most one positive solution. If, in addition, you can prove that the polynomial equation has at least one positive solution, then this solution is the unique positive solution and the desired IRR. In the example of the start-up, Equation (2.45) is equivalent to a cubic equation:

$$p(r) = -250{,}000\,r^3 - 595{,}000\,r^2 - 225{,}000\,r + 470{,}000 = 0. \qquad (2.51)$$

There is one sign change, so there is at most one positive solution. Since $p(r) > 0$ at $r = 0$ and $p(r) \to -\infty$ as $r \to \infty$, its graph must cross the positive $r$-axis, which means that $p(r)$ must have at least one positive solution. Hence, the cubic equation has a unique, positive solution, which is the desired $r_{IRR}$. Using a software, we found the approximate positive solution to be $r_{IRR} = 0.652811$.

We also observed that for the required return rate of $r_{RRR} = 15\%$, the IRR criterion to favor the start-up, namely, $r_{IRR} > r_{RRR}$, is equivalent to the NPV criterion of $NPV(r_{RRR}) > 0$. This is not true in general, but holds in the example because the function $NPV(r)$ in (2.45) is strictly decreasing. The next result shows when the situation of the example holds.

**Theorem 2.3.**
*1) Suppose that all the future net cash flows are positive. Then $NPV(r)$ is a strictly decreasing function of $r$ and, if there is an $r = r_{IRR}$, then $r_{IRR}$ is the only IRR.[7]*
*2) If there is an $r_{IRR}$ and $NPV(r)$ is strictly decreasing, then the IRR and NPV decision-making criteria are equivalent:*

*a) $r_{IRR} > r_{RRR}$   if and only if   $NPV(r_{RRR}) > 0$.*
*b) $r_{IRR} = r_{RRR}$   if and only if   $NPV(r_{RRR}) = 0$.*
*c) $r_{IRR} < r_{RRR}$   if and only if   $NPV(r_{RRR}) < 0$.*

*Proof.*

1) Since $C_i > 0$ for $i = 1, \ldots, n$, the derivative of the NPV function satisfies:

---

implies $N_{sgn} - N_+$ is a nonnegative even number, i.e., $N_+ = N_{sgn}$ − even. In particular, $N_+$ is either $N_{sgn}, N_{sgn} - 2, \ldots, N_{sgn} - 2(n-1)$, or $N_{sgn} - 2n$ for some nonnegative integer $n$.
[7] By definition, we assume $r_{IRR} > 0$.

$$\frac{d}{dr}\text{NPV}(r) = -\frac{C_1}{(1+r)^2} - 2\frac{C_2}{(1+r)^3} - \cdots - n\frac{C_n}{(1+r)^{n+1}} < 0.$$

Consequently, the function $\text{NPV}(r)$ is strictly decreasing and, hence, if the graph of $\text{NPV}(r)$ crosses the positive $r$-axis, i.e., there is an IRR, the graph will do so only once, namely, at a unique value $r_{\text{IRR}}$.

2) We are given that an IRR exists: $r_{\text{IRR}} > 0$. Since $\text{NPV}(r)$ is strictly decreasing, we have $r_2 > r_1 > 0$ if and only if $\text{NPV}(r_1) > \text{NPV}(r_2)$. For part a) choose $r_2 = r_{\text{IRR}}$, $r_1 = r_{\text{RRR}}$, and observe that $\text{NPV}(r_{\text{IRR}}) = 0$. For part c) choose $r_2 = r_{\text{RRR}}$, $r_1 = r_{\text{IRR}}$. Part b) holds since $\text{NPV}(r_{\text{IRR}}) = 0$. □

It is important to note that in a real-world decision-making setting, the NPV and IRR criteria have aspects not explicitly spelled out in Theorem 2.3. For example, consider the start-up we have been exploring:

| $C_0$ | $C_1$ | $C_2$ | $C_3$ | $r_{\text{RRR}}$ | $\text{NPV}(r_{\text{RRR}})$ | $r_{\text{IRR}}$ |
|---|---|---|---|---|---|---|
| \$250,000 | \$155,000 | \$215,000 | \$350,000 | 15% | \$277,484.18 | 65.28% |

The table shows that with an initial investment of $C_0$, the start-up is attractive because it has:

➤ Positive future net cash flows $C_1$, $C_2$, and $C_3$ that are nontrivial as a percent of the initial capital and are nontrivially increasing. In particular, $C_1$ is more than half the initial capital, $C_2$ is about 86% of the initial capital, and $C_3$ is 140% of the initial capital. Moreover, the net cash flow increases by about 39% from year 1 to 2 and about 63% from year 2 to 3.

➤ A nontrivially positive NPV value of $\text{NPV}(r_{\text{RRR}}) = \$277,484.18$, which makes the start-up much cheaper to invest in than a comparable alternative opportunity by more than the initial capital.

➤ A quite large compounding growth rate of $r_{\text{IRR}} = 65.28\%$ compared to the required return rate of $r_{\text{RRR}} = 15\%$, i.e., the $r_{\text{IRR}}$ is more than four times $r_{\text{RRR}}$.

## No IRR and Multiple IRRs

The discussion so far assumes a unique IRR. However, complications already arise in the simple case of net cash flows over 2 years ($n = 2$), where the IRR Equation (2.48) becomes a quadratic in $r$:

$$\underbrace{-C_0}_{a} r^2 + \underbrace{(C_1 - 2C_0)}_{b} r + \underbrace{(-C_0 + C_1 + C_2)}_{c} = 0, \qquad (C_0 > 0). \qquad (2.52)$$

For example, if the net cash flows are $C_1 = 2C_0$ and $C_2 = -2C_0$, then the quadratic reduces to one with no real solution: $r^2 + 1 = 0$. *In this case, there is no IRR.*

*It is also possible to have multiple IRRs.* The quadratic (2.52) has two positive solutions, say, $r_1$ and $r_2$, if and only if the following positivity conditions hold:

$$0 < b^2 - 4ac = C_1^2 + 4C_0 C_2$$
$$0 < r_1 + r_2 = -\frac{b}{a} = \frac{C_1 - 2C_0}{C_0}$$
$$0 < r_1 r_2 = \frac{c}{a} = \frac{C_0 - C_1 - C_2}{C_0}.$$

These positivity conditions are equivalent to

$$2C_0 < C_1, \qquad \frac{-C_1^2}{4C_0} < C_2 < C_0 - C_1, \qquad (C_0 > 0).$$

Choosing $C_0 = \$10,000$, $C_1 = \$25,000$, and $C_2 = -\$15,620$, we obtain two IRRs:

$$r_1^{\text{IRR}} = 22.76\%, \qquad r_2^{\text{IRR}} = 27.24\%.$$

In the case of multiple IRRs, it is possible to construct a modified IRR. Nonetheless, in practice it is simplest to work with the NPV when there is no IRR or multiple IRRs. Additionally, in cases where usage of the NPV and/or IRR are unclear, it may be wise to hold off from making an investment decision.

Readers are referred to Bodie, Kane, and Marcus [1, Chaps. 5, 6] for a detailed practical discussion of the uses of the NPV, IRR, and other tools in investment decision-making.

## 2.7  Annuity Theory

In this section, we continue our study of cash flow sequences by considering annuities. An *annuity* is a series of payments made at equal time periods with interest. Examples of annuities are the payment sequences of Social Security funds, pensions, car loans, credit card debt, and mortgages. We shall study annuities with identical payments and a constant interest rate and then generalize them to payments and interest rates that vary discretely in time.

For simplicity, we shall explicitly indicate when the annuity payments vary and, by default, abide by the following:

> *Unless stated to the contrary, assume that each annuity payment is the same amount.*

The *term* of an annuity is the time from the start of the first payment period to the end of the last payment period. For an *ordinary annuity*, payments occur at the end of each time period. When the payments occur at the start of each period, we have an *annuity due*, which will not be treated in the text; see Guthrie and Lemon [8] and Muksian [15] for an introduction.

An ordinary annuity is called *simple* if, at the end of each payment period, both a payment and the simple interest on the balance from the beginning of the payment period are applied. Note that the entire balance from the previous period is reinvested. Hence, for a simple ordinary annuity, the total accrued at the end of a payment period has the following form:

$$(\text{total accrued}) = (\text{payment}) + (\text{previous balance})$$
$$+(\text{simple interest on previous balance}). \quad (2.53)$$

Here "previous balance" refers to the balance from the end of the previous payment period, which recall we treat mathematically the same as the start of the current period. Since the simple interest applied to a previous balance will yield interest on the principal and interest on the interest, we obtain compound interest naturally.

> *Unless stated to the contrary, assume that all loans are simple ordinary annuities.*

### 2.7.1 Future and Present Values of Simple Ordinary Annuities

**Future Value of a Simple Ordinary Annuity**

The *future value* of a simple ordinary annuity is the amount to which the sequence of payments of the annuity will grow, taking into account appreciation due to periodic compounding. We shall see that the annuity's future value is the sum of the end-of-term future values of the individual payments of the annuity.

Consider a simple ordinary annuity based on $k$-periodic compounding at interest rate $r$. This divides each year into $k$ equal-length payment periods. Assume that each payment is the same amount $\mathcal{P}$ and the annuity has a term of $n$ periods, where $n$ is a positive integer. The total accrued at the end of the $i$th period will be denoted by $\mathcal{S}_i$. We shall apply (2.53) to obtain an expression for the total amount $\mathcal{S}_n$ accrued over the $n$ periods:

➢ At the end of the first payment period, a payment $\mathcal{P}$ is made. Since there is no balance from the beginning of this period, the total accrued at the end of the first period is:

$$\mathcal{S}_1 = \mathcal{P}.$$

Reinvest the entire amount $\mathcal{S}_1$ in the annuity.

➤ At the end of the second period, the payment is $P$, the previous balance is $S_1$, and the simple interest earned on the entire reinvested amount $S_1$ is $(r/k)S_1$. The total accrued at the end of the second period is then:

$$S_2 = P + S_1 + \frac{r}{k}S_1 = P + \left(1 + \frac{r}{k}\right)P.$$

Reinvest the entire amount $S_2$ in the annuity.

➤ At the end of the 3rd period, the payment is $P$, the previous balance is $S_2$, and the simple interest earned on $S_2$ is $(r/k)S_2$. The total accrued is:

$$S_3 = P + S_2 + \frac{r}{k}S_2 = P + \left(1 + \frac{r}{k}\right)P + \left(1 + \frac{r}{k}\right)^2 P.$$

Reinvest $S_3$.

➤ Continuing the above process, at the end of the $n$th period, the payment is $P$, the previous balance is $S_{n-1}$, and the simple interest earned on $S_{n-1}$ is $(r/k)S_{n-1}$. The total accrued at the end of the $n$th period is:

$$S_n = P + S_{n-1} + \frac{r}{k}S_{n-1}$$

or

$$S_n = P + \left(1 + \frac{r}{k}\right)P + \left(1 + \frac{r}{k}\right)^2 P + \cdots + \left(1 + \frac{r}{k}\right)^{n-1} P. \qquad (2.54)$$

Equation (2.54) shows that *the future value of a simple ordinary annuity is the sum of each of the payments future valued to the end of the annuity.* To see this, in (2.54) the future values of these payments are shown from right to left. Explicitly, the 1st payment $P$ is at the end of the first period, so its future value at the end of term (i.e., end of the $n$th period) is $(1 + r/k)^{n-1}P$. The 2nd payment $P$ is at the end of the second period, which has a future value at the end of the term of $(1 + r/k)^{n-2}P$. The $(n-2)$nd payment $P$ has an end-of-term future value of $(1 + r/k)^2 P$, and the $(n-1)$st payment $P$ has $(1 + r/k)P$. The $n$th payment $P$ is at the end of the term so it equals its end-of-term future value. By (2.54), the sum of these future values is $S_n$.

The right-hand side of (2.54) has a simpler expression. Applying the geometric sum,

$$a + ax + \cdots + ax^{m-1} = \left(\frac{1 - x^m}{1 - x}\right)a, \qquad (m \geq 1, \quad x \neq 1), \qquad (2.55)$$

with $a = P$, $x = 1 + r/k \neq 1$ (since $r > 0$), and $m = n \geq 1$, we obtain:

$$S_n = \frac{\left[(1 + \frac{r}{k})^n - 1\right]}{r/k}P \qquad (r > 0, \quad n \geq 1).$$

**Remark 2.2.** In actuarial science, the future value $S_n$ is denoted by $s_{\overline{n}|}$ (pronounced "s angle $n$"). □

We summarize the result of the above analysis in the following theorem.

**Theorem 2.4.** *At the end of n periods, the future value of the simple ordinary annuity with payments $\mathcal{P}$ and k-periodic compounding at r per annum is:*

$$S_n = \frac{\left[(1+\frac{r}{k})^n - 1\right]}{r/k}\mathcal{P} \qquad (r > 0, \quad n = 1,2,\dots). \qquad (2.56)$$

Let us consider how $S_n$ behaves as a function of $r \geq 0$. Intuitively, we expect that as the interest rate $r$ increases, the total $S_n$ accumulated after $n$ periods should increase. Of course, this is not true for $n = 1$.[8] However, for $n = 2,3,\dots$, Equation (2.54) readily yields:

$$\frac{dS_n}{dr} = \frac{\mathcal{P}}{k} + 2\left(1+\frac{r}{k}\right)\frac{\mathcal{P}}{k} + \cdots + (n-1)\left(1+\frac{r}{k}\right)^{n-2}\frac{\mathcal{P}}{k} > 0.$$

It follows that *for $n \geq 2$, the total amount $S_n$ accrued over n periods increases as r increases.* Additionally, for $n = 2$ we have

$$\frac{d^2 S_2}{dr^2} = 0.$$

However, if $n = 3,4,\dots$, then

$$\frac{d^2 S_n}{dr^2} = 2\frac{\mathcal{P}}{k^2} + \cdots + (n-1)(n-2)\left(1+\frac{r}{k}\right)^{n-3}\frac{\mathcal{P}}{k^2} > 0.$$

Hence, *for $n \geq 3$, the total amount $S_n$ accumulated over n periods accelerates[9] in value as the interest rate r increases.*

## Present Value of a Simple Ordinary Annuity

The *present value*, denoted by $\mathcal{A}_n$, of a simple ordinary annuity is the amount needed today, taking interest into account, in order to be able to pay the amount $\mathcal{P}$ at the end of each period for a total of $n$ periods. In particular, an interest per period of $r/k$ is applied at the end of each period to the balance from the start of that period. The funds are, of course, exhausted by the end of the last period. For example, if $\mathcal{A}_n$ is a loan, then it would be paid off completely at the end of the $n$th period. The total payout would be $n\mathcal{P}$.

---

[8] If there is only one period, then $S_1 = \mathcal{P}$ (constant) for all $r$ since the principal is added only at the end of the first period, but the first interest payment occurs at the end of the second period.
[9] That is, $S_n$ is concave up as a function of $r$ (it has an increasing slope).

Let us determine a formula for $\mathcal{A}_n$. The first payment $\mathcal{P}$ will be made at the end of the first period. The present value of that future payment $\mathcal{P}$ is then $\dfrac{\mathcal{P}}{(1+r/k)}$. This is the amount needed at the start of the annuity's term in order for the amount to grow to $\mathcal{P}$ after one period. The present value of the second payment $\mathcal{P}$ is $\dfrac{\mathcal{P}}{(1+r/k)^2}$ since after two compoundings, i.e., at the end of the second period, it grows to $\mathcal{P}$. Consequently, at the start of the annuity, the individual present values of the $n$ payments are given in sequential order as follows:

$$\frac{\mathcal{P}}{\left(1+\frac{r}{k}\right)}, \quad \frac{\mathcal{P}}{\left(1+\frac{r}{k}\right)^2}, \quad \cdots, \quad \frac{\mathcal{P}}{\left(1+\frac{r}{k}\right)^n}.$$

Then $\mathcal{A}_n$ is the sum of the present values of all payments:

$$\mathcal{A}_n = \frac{\mathcal{P}}{\left(1+\frac{r}{k}\right)} + \frac{\mathcal{P}}{\left(1+\frac{r}{k}\right)^2} + \cdots + \frac{\mathcal{P}}{\left(1+\frac{r}{k}\right)^n}. \tag{2.57}$$

Equation (2.57) can be expressed as:

$$
\begin{aligned}
\mathcal{A}_n &= \left(1+\frac{r}{k}\right)^{-1}\mathcal{P} + \left(1+\frac{r}{k}\right)^{-2}\mathcal{P} + \cdots + \left(1+\frac{r}{k}\right)^{-n}\mathcal{P} \\
&= \left(1+\frac{r}{k}\right)^{-1}\left[1 + \left(1+\frac{r}{k}\right)^{-1} + \left(1+\frac{r}{k}\right)^{-2} + \cdots \right. \\
&\qquad\qquad\qquad\qquad\qquad\qquad\qquad\left. + \left(1+\frac{r}{k}\right)^{-(n-1)}\right]\mathcal{P} \\
&= \frac{\left(1+\frac{r}{k}\right)^{-1}\left[1 - \left(1+\frac{r}{k}\right)^{-n}\right]\mathcal{P}}{1-(1+r/k)^{-1}}.
\end{aligned}
$$

The last equality above follows from the geometric series (2.55) with $a = \mathcal{P}$ and $x = (1+r/k)^{-1}$ and $m = n$. Further simplification yields:

**Theorem 2.5.** *The present value of a simple ordinary annuity over $n$ periods and with payments $\mathcal{P}$ and $k$-periodic compounding at $r$ per annum is:*

$$\mathcal{A}_n = \frac{\left[1 - \left(1+\frac{r}{k}\right)^{-n}\right]}{\frac{r}{k}}\mathcal{P} \qquad (r > 0, \quad n = 1,2,\ldots). \tag{2.58}$$

Theorem 2.5 gives a formula for the amount needed today at interest rate $r$ in order to be able to pay out the amount $\mathcal{P}$ each period for $n$ periods.

**Remark 2.3.** The present value $\mathcal{A}_n$ is usually denoted by $a_{\overline{n}|}$ and the discount factor $(1+r/k)^{-1}$ by $v$ in actuarial science.                                        □

**The Total Number of Periods as a Function of the Payment per Period**

We know intuitively that increasing the per-period payments of a loan will shorten the time it takes to pay off the loan. In fact, we can solve (2.58) for an exact formula relating the number $n$ of periods to payoff in terms of the inputs $\mathcal{P}$, $\mathcal{A}_n$ (loan amount), $r$, and $k$:

$$n = -\frac{\ln\left(1 - \frac{(r/k)\mathcal{A}_n}{\mathcal{P}}\right)}{\ln\left(1 + \frac{r}{k}\right)}. \tag{2.59}$$

To understand how $n$ varies with $\mathcal{P}$, we can fix $\mathcal{A}_n$, $r$, and $k$ and treat $n$ formally as a function of $\mathcal{P}$ given by (2.59). This treatment, of course, will lead to noninteger values of $n$, which we round off to find the approximate integer value. *For general values of $\mathcal{A}_n > 0$, $r > 0$, and $k$ (nonnegative integer), as $\mathcal{P}$ increases, the total number of periods $n$ strictly decreases and the rate of decrease slows down.* In other words, the quantity $n$ as a function of $\mathcal{P}$ is convex, i.e., $n(\mathcal{P})$ is everywhere concave up. Explicitly, though the function $n(\mathcal{P})$ is a strictly decreasing function, it has an increasing slope:

$$\frac{dn}{d\mathcal{P}} = -\frac{(r/k)\mathcal{A}_n}{\left(1 - \frac{(r/k)\mathcal{A}_n}{\mathcal{P}}\right)\mathcal{P}^2 \ln\left(1 + \frac{r}{k}\right)} < 0 \qquad (r > 0)$$

$$\frac{d^2 n}{d\mathcal{P}^2} = \frac{\left(2 - \frac{(r/k)\mathcal{A}_n}{\mathcal{P}}\right)(r/k)\mathcal{A}_n}{\left(1 - \frac{(r/k)\mathcal{A}_n}{\mathcal{P}}\right)^2 \mathcal{P}^3 \ln\left(1 + \frac{r}{k}\right)} > 0.$$

Here we used $\ln\left(1 + \frac{r}{k}\right) > 0$ (since $r > 0$) and employed (2.58) to conclude that

$$1 - \frac{(r/k)\mathcal{A}_n}{\mathcal{P}} = \left(1 + \frac{r}{k}\right)^{-n} > 0.$$

Note that the quantity $(r/k)\mathcal{A}_n$ is the (simple) interest on the loan at the end of the first period. An example of $n$ as a function of the per-period payment $\mathcal{P}$ is shown in Figure 2.1.

**Present Value of a Perpetuity**

A *perpetuity* is a sequence of cash flows that continues indefinitely. Though a perpetuity has no future value, a simple-ordinary-annuity perpetuity has a present value given by the following geometric series:

$$\mathcal{A}_\infty = \lim_{n\to\infty} \mathcal{A}_n = \lim_{n\to\infty} \frac{\left[1 - \left(1 + \frac{r}{k}\right)^{-n}\right]}{\frac{r}{k}}\mathcal{P} = \frac{\mathcal{P}}{\frac{r}{k}} \qquad (r > 0), \tag{2.60}$$

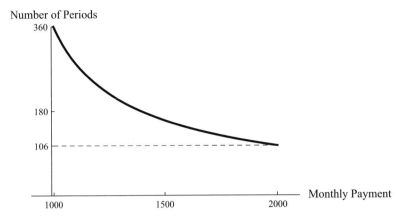

**Fig. 2.1** The graph shows the total number of periods $n$ as a function of monthly payments $\mathcal{P}$ for a loan of $\mathcal{A}_n = \$162,412$ at 6.25% per annum compounded monthly ($k = 12$). The loan is paid off in about 30 years (or 360 months) if the monthly payment is $\$1,000$. Doubling the payments yields a payoff time of 8 years and 10 months (or 106 months), which is much less than half of the time for a $\$1,000$ monthly payment.

which is an immediate consequence of (2.58). See Section 2.9.1 on page 64 for a *growing perpetuity*, i.e., one where the payments $\mathcal{P}$ increase at a certain rate.

**Relating Future and Present Values of a Simple Ordinary Annuity**

If you put aside the amount $\mathcal{A}_n$ today and have it grow by $k$-periodic compounding with interest rate $r$, then after $n$ periods, the initial amount will grow to $\mathcal{S}_n$. In other words, the initial amount $\mathcal{A}_n$ is the present value of the future amount $\mathcal{S}_n$ under periodic compounding. To see this, note that by (2.56) and (2.58), we obtain:

$$
\begin{aligned}
\frac{\mathcal{S}_n}{\left(1 + \frac{r}{k}\right)^n} &= \frac{\left[\left(1 + \frac{r}{k}\right)^n - 1\right] \mathcal{P}}{\frac{r}{k}} \left(1 + \frac{r}{k}\right)^{-n} \\
&= \frac{\left[1 - \left(1 + \frac{r}{k}\right)^{-n}\right]}{\frac{r}{k}} \mathcal{P} \\
&= \mathcal{A}_n,
\end{aligned}
\tag{2.61}
$$

where $r > 0$. Equation (2.61) shows that an equivalent way of determining the future value of a simple ordinary annuity is to take the present value of the sequence of payments and then take the future value of that present value.

### 2.7.2 *Amortization Theory*

*Amortization* is the reducing of a given loan amount (the principal) through a series of payments over a fixed time span whereby one accounts explicitly for the portion of each payment that goes toward the principal and the portion toward the interest owed on the loan. The most common amortization is through a *mortgage*, which is a loan where the borrower (mortgagor) gives the lender (mortgagee) a lien on property as security for the repayment of the loan. The mortgagor has use of the property and the lien is removed when the obligation is fully paid. A mortgage usually involves real estate.[10]

What happens if you are amortizing a debt with equal periodic payments and at some point decide to pay off the remainder of the debt in one lump-sum payment? This occurs each time a house with an outstanding mortgage is sold. How much of each periodic payment is used for interest and how much is used to reduce the unpaid balance of the principal? This issue is also important because the interest part of the payment may be tax deductible (as it is in the USA). In order to answer these questions, we must take a close look at the mathematical structure of amortization.

A loan is paid off with interest through the full sequence of its stipulated minimal payments. We then model the amount of a loan by the present value of the entire sequence of its required future payments. Specifically, we model the loan amount by the present value of a simple ordinary annuity. The initial amount of the loan is $A_n$ (principal balance), the payment at the end of each period is $\mathcal{P}$, and the loan is for $n$ periods. Each period is $(1/k)$th of a year and the annual interest rate is $r$, i.e., the interest applied at the end of each period is $r/k$.

### Unpaid Principal Balances

We determine the unpaid balance on the principal at the end of each period of the loan.

For notational simplicity, define

$$y \equiv 1 + \frac{r}{k}, \qquad (r > 0).$$

Then by (2.58), each end-of-period payment can be expressed as:

$$\mathcal{P} = \frac{(y-1)}{1 - y^{-n}} A_n = \frac{y^n (y-1)}{y^n - 1} A_n. \qquad (2.62)$$

---

[10] While a typical mortgage is a loan used to buy a fixed asset like a house or land, which also secures the loan, a mortgage used to buy movable property such as a mobile home or operational equipment that acts as security for the loan is called a *chattel mortgage* or *secured transaction*.

Denote the initial amount of the loan by:

$$\mathcal{B}_0 = \mathcal{A}_n.$$

Then the unpaid principal balances at the end of the different periods are given as follows:

➤ At the end of the first period, an interest $(r/k)\mathcal{A}_n$ is added to the starting balance $\mathcal{A}_n$ and a payout/withdrawal of $\mathcal{P}$ is made. The unpaid principal balance at the end of the first period is:

$$\mathcal{B}_1 = \mathcal{A}_n + \frac{r}{k}\mathcal{A}_n - \mathcal{P} = y\mathcal{A}_n - \mathcal{P}.$$

➤ At the end of the second period, an interest $(r/k)\mathcal{B}_1$ is added to the balance $\mathcal{B}_1$ from the start of the second period and then a payout/withdrawal of $\mathcal{P}$ is made. The unpaid principal balance at the end of the second period is:

$$\mathcal{B}_2 = \mathcal{B}_1 + \frac{r}{k}\mathcal{B}_1 - \mathcal{P} = y\mathcal{B}_1 - \mathcal{P} = y(y\mathcal{A}_n - \mathcal{P}) - \mathcal{P} = y^2\mathcal{A}_n - (1+y)\mathcal{P}.$$

➤ At the end of the 3rd period, an interest $(r/k)\mathcal{B}_2$ is added to the balance $\mathcal{B}_2$ from the start of the 3rd period and then a payout/withdrawal of $\mathcal{P}$ is made. The unpaid principal balance at the end of the 3rd period is

$$\begin{aligned}
\mathcal{B}_3 &= \mathcal{B}_2 + \frac{r}{k}\mathcal{B}_2 - \mathcal{P} = y\mathcal{B}_2 - \mathcal{P} \\
&= y\left[y^2\mathcal{A}_n - (1+y)\mathcal{P}\right] - \mathcal{P} \\
&= y^3\mathcal{A}_n - (1+y+y^2)\mathcal{P}.
\end{aligned}$$

➤ Continuing the above process, at the end of the $\ell$th period, an interest $(r/k)\mathcal{B}_{\ell-1}$ is added to the balance $\mathcal{B}_{\ell-1}$ from the start of the $\ell$th period and then a payout/withdrawal of $\mathcal{P}$ is made. The unpaid principal balance at the end of the $\ell$th period is:

$$\begin{aligned}
\mathcal{B}_\ell &= \mathcal{B}_{\ell-1} + \frac{r}{k}\mathcal{B}_{\ell-1} - \mathcal{P} \\
&= y^\ell\mathcal{A}_n - (1+y+y^2+\cdots+y^{\ell-1})\mathcal{P} \\
&= y^\ell\mathcal{A}_n - \frac{1-y^\ell}{1-y}\mathcal{P} \\
&= y^\ell\mathcal{A}_n + \frac{(1-y^\ell)}{(y-1)}\frac{y^n(y-1)}{(y^n-1)}\mathcal{A}_n, \qquad [by\ (2.62)] \\
&= \frac{y^\ell(y^n-1)+y^n(1-y^\ell)}{y^n-1}\mathcal{A}_n \\
&= \frac{y^n-y^\ell}{y^n-1}\mathcal{A}_n \qquad (\ell = 1,2,\ldots,n),
\end{aligned}$$

where $\mathcal{B}_0 = \mathcal{A}_n$. Hence:

**Theorem 2.6.** *The unpaid principal balance at the end of the $\ell$th period is given in terms of $\mathcal{A}_n$, $r > 0$, and $k$ by:*

$$\mathcal{B}_\ell = \frac{\left(1 + \frac{r}{k}\right)^n - \left(1 + \frac{r}{k}\right)^\ell}{\left(1 + \frac{r}{k}\right)^n - 1} \mathcal{A}_n, \tag{2.63}$$

*where $n = 1, 2, \ldots$ and $\ell = 0, 1, 2, \ldots, n$.*

The unpaid principal balance at the end of the loan's term is $\mathcal{B}_n = 0$.

### Amount of Per-Period Payment Toward Interest and Unpaid Balance

At the end of each period, a portion of the payment $\mathcal{P}$ is used toward interest on the loan, the other portion toward reduction of the loan's unpaid principal balance.

**Notation.** Let:

$\mathcal{I}_\ell = $ the portion of the payment $\mathcal{P}$ at the end of $\ell$th period that is applied toward interest on the loan (i.e., the interest payment at the end of period $\ell$).

$\mathfrak{P}_\ell = $ the portion of payment $\mathcal{P}$ at the end of the $\ell$th period that is applied toward the unpaid principal balance of the loan.

We now express $\mathcal{I}_\ell$ and $\mathfrak{P}_\ell$ in terms of $\mathcal{A}_n$ and $r$. The interest payment at the end of period $\ell$ is

$$\mathcal{I}_\ell = \left(\frac{r}{k}\right) \mathcal{B}_{\ell-1} = \left(\frac{r}{k}\right) \frac{\left[\left(1 + \frac{r}{k}\right)^n - \left(1 + \frac{r}{k}\right)^{\ell-1}\right]}{\left(1 + \frac{r}{k}\right)^n - 1} \mathcal{A}_n, \tag{2.64}$$

where $r > 0$, $n = 1, 2, \ldots$ and $\ell = 1, 2, \ldots, n$.

For the payment $\mathfrak{P}_\ell$ toward the principal, Equations (2.62) and (2.64) yield:

$$\begin{aligned}
\mathfrak{P}_\ell &= \mathcal{P} - \mathcal{I}_\ell \\
&= \frac{(y-1)y^n}{y^n - 1} \mathcal{A}_n - \frac{(y-1)\left[y^n - y^{\ell-1}\right]}{y^n - 1} \mathcal{A}_n \\
&= \frac{(y-1)y^{\ell-1}}{y^n - 1} \mathcal{A}_n \\
&= \left(\frac{r}{k}\right) \frac{\left(1 + \frac{r}{k}\right)^{\ell-1}}{\left(1 + \frac{r}{k}\right)^n - 1} \mathcal{A}_n. \tag{2.65}
\end{aligned}$$

As a check, we show that the payments, $\mathfrak{P}_1, \mathfrak{P}_2, \ldots, \mathfrak{P}_n$, toward the principal add up to the total loan amount $\mathcal{A}_n$:

$$\sum_{\ell=1}^{n} \mathfrak{P}_\ell = \sum_{\ell=1}^{n} \frac{(y-1)y^{\ell-1}}{y^n - 1} \mathcal{A}_n = \mathcal{A}_n \frac{y-1}{y^n - 1} \sum_{\ell=1}^{n} y^{\ell-1}$$

$$= \mathcal{A}_n \frac{y-1}{y^n - 1} \sum_{\ell=0}^{n-1} y^\ell = \mathcal{A}_n \frac{y-1}{y^n - 1} \frac{y^n - 1}{y - 1}$$

$$= \mathcal{A}_n.$$

Therefore:

**Theorem 2.7.** *The total interest paid during a loan with n periods and k-periodic compounding at interest rate r is then:*

$$\sum_{\ell=1}^{n} \mathcal{I}_\ell = \sum_{\ell=1}^{n} (\mathcal{P} - \mathfrak{P}_\ell) = n\mathcal{P} - \mathcal{A}_n, \tag{2.66}$$

*where* $n = 1, 2, \ldots$.

Note that $n\mathcal{P}$ is the total amount paid into the loan over the life of the loan and $n\mathcal{P} - \mathcal{A}_n$ is the total cost of the loan.

**Remark 2.4.** If you receive a loan today for the amount $\mathcal{A}_n$ at fixed interest rate $r$, fixed payment $\mathcal{P}$ per period, and a term of $n$ periods, then the sum $n\mathcal{P}$ of all your future payments adds money at different future times without present or future valuing them. In fact, the present value of all the future payments is the loan amount $\mathcal{A}_n$ and the future value is $\mathcal{S}_n$, neither of which is $n\mathcal{P}$. The meaning of $n\mathcal{P}$ is the amount you would, in principle, have to pay the lender today if immediately after receiving the loan you want to pay the loan off, but the lender penalizes you by requiring you to pay the principal $\mathcal{A}_n$ plus the total interest for the full term of the loan. Of course, this is merely theoretical since the majority of loans would not have such a drastic penalty.  □

### 2.7.3 Annuities with Varying Payments and Interest Rates

Applying essentially the same arguments used to establish the future value annuity Equation (2.54), we can generalize to a simple ordinary annuity with a sequence of varying payments, $\mathcal{P}_1, \mathcal{P}_2, \ldots, \mathcal{P}_n$, and respective varying interest rates, $r_1, r_2, \ldots, r_n$, over $n$ interest periods that coincide with the payment periods. We assume $k$-periodic compounding. *The payment $\mathcal{P}_\ell$ occurs at the end of the $\ell$th period, and the interest $r_\ell$ is applied at the end of the $\ell$th period to the balance from the start of the $\ell$th interest period, where $\ell = 1, \ldots, n$.* Assume that there is no balance at the start of the first period.

**Future Value of a Generalized Simple Ordinary Annuity**

The pattern for the future value of a simple ordinary annuity generalized to varying payments and varying interest rates emerges as follows:

➤ At the end of the first payment period, a payment $\mathcal{P}_1$ is made. Because there is no balance from the beginning of this period, the total accrued at the end of the first period is:

$$\mathcal{S}_1 = \mathcal{P}_1.$$

Reinvest $\mathcal{S}_1$ in the annuity.

➤ At the end of the second period, the payment is $\mathcal{P}_2$, the previous balance is $\mathcal{S}_1$, and the simple interest earned on $\mathcal{S}_1$ is $(r_2/k)\mathcal{S}_1$. The total accrued is:

$$\mathcal{S}_2 = \mathcal{P}_2 + \mathcal{S}_1 + \frac{r_2}{k}\mathcal{S}_1 = \mathcal{P}_2 + \left(1 + \frac{r_2}{k}\right)\mathcal{P}_1.$$

Reinvest $\mathcal{S}_2$ in the annuity.

➤ At the end of the 3rd period, the payment is $\mathcal{P}_3$, the previous balance is $\mathcal{S}_2$, and the simple interest earned on $\mathcal{S}_2$ is $(r_3/k)\mathcal{S}_2$. The total accrued is:

$$\mathcal{S}_3 = \mathcal{P}_3 + \mathcal{S}_2 + \frac{r_3}{k}\mathcal{S}_2 = \mathcal{P}_3 + \left(1 + \frac{r_3}{k}\right)\mathcal{P}_2 + \left(1 + \frac{r_3}{k}\right)\left(1 + \frac{r_2}{k}\right)\mathcal{P}_1.$$

Reinvest $\mathcal{S}_3$ in the annuity.

➤ Continuing the above process, at the end of the $n$th period, the payment is $\mathcal{P}_n$, the previous balance is $\mathcal{S}_{n-1}$, and the simple interest earned on $\mathcal{S}_{n-1}$ is $(r_n/k)\mathcal{S}_{n-1}$. The total accrued is

$$\mathcal{S}_n = \mathcal{P}_n + \mathcal{S}_{n-1} + \frac{r_n}{k}\mathcal{S}_{n-1}$$

or

$$\mathcal{S}_n = \mathcal{P}_n + \left(1 + \frac{r_n}{k}\right)\mathcal{P}_{n-1} + \left(1 + \frac{r_n}{k}\right)\left(1 + \frac{r_{n-1}}{k}\right)\mathcal{P}_{n-2} +$$
$$\cdots + \left(1 + \frac{r_n}{k}\right)\left(1 + \frac{r_{n-1}}{k}\right)\cdots\left(1 + \frac{r_2}{k}\right)\mathcal{P}_1.$$

$$(2.67)$$

Observe that, by letting $r_{n+1} = 0$ and rewriting (2.67) as

$$\mathcal{S}_n = \left(1 + \frac{r_{n+1}}{k}\right)\mathcal{P}_n + \left(1 + \frac{r_{n+1}}{k}\right)\left(1 + \frac{r_n}{k}\right)\mathcal{P}_{n-1}$$
$$+ \left(1 + \frac{r_{n+1}}{k}\right)\left(1 + \frac{r_n}{k}\right)\left(1 + \frac{r_{n-1}}{k}\right)\mathcal{P}_{n-2}$$
$$\cdots + \left(1 + \frac{r_{n+1}}{k}\right)\left(1 + \frac{r_n}{k}\right)\left(1 + \frac{r_{n-1}}{k}\right)\cdots\left(1 + \frac{r_2}{k}\right)\mathcal{P}_1 ,$$

we see that (2.67) can be expressed more compactly as follows:

**Theorem 2.8.** *The future value at the end of $n$ payment periods, which coincide with the interest periods, of the simple ordinary annuity with payments $\mathcal{P}_1,\ldots,\mathcal{P}_n$ and $k$-periodic compounding at respective interest rates $r_2,\ldots,r_n$ during the consecutive interest periods is:*

$$\mathcal{S}_n = \sum_{\ell=0}^{n-1} \left[ \prod_{j=0}^{\ell} \left( 1 + \frac{r_{n+1-j}}{k} \right) \right] \mathcal{P}_{n-\ell}, \qquad (n = 1,2,\ldots), \qquad (2.68)$$

*where $r_i > 0$ for $i = 2,\ldots,n$ and $r_{n+1} = 0$.*

When $r_i = r$ for $i = 2,\ldots,n$, and $\mathcal{P}_i = \mathcal{P}$ for $i = 1,\ldots,n$, Equation (2.68) recovers (2.54) on page 48.

An application of (2.68) to sinking funds is given in Section 2.8.3.

### Present Value of a Generalized Simple Ordinary Annuity

Similarly, the present value Equation (2.57) generalizes naturally to the case of a sequence of payments, $\mathcal{P}_1,\ldots,\mathcal{P}_n$, and interest rates $r_1,\ldots,r_n$. Here the amount $\mathcal{P}_i$ is paid at the end of the $i$th period, and the interest $r_i$ is applied at the end of the $i$th period to the balance from the end of the $(i-1)$st period.

When simple interest at rate $r_1$ is applied at the end of the first period to the initial amount $\mathcal{P}_1 (1 + r_1/k)^{-1}$, we obtain the first payment $\mathcal{P}_1$. Applying compound interest with rates $r_1$ and $r_2$ at the end of the first and second periods, respectively, to the initial amount $\mathcal{P}_2 (1 + r_1/k)^{-1}(1 + r_2/k)^{-1}$ yields the second payment $\mathcal{P}_2$. Continuing this process gives the initial amount that will grow to the $n$th payment $\mathcal{P}_n$. These initial amounts are the present values of the sequence of payments under compound interest at different rates. Summing all the present values gives the following *present value* for the generalized annuity:

$$\mathcal{A}_n = \frac{\mathcal{P}_1}{\left(1 + \frac{r_1}{k}\right)} + \frac{\mathcal{P}_2}{\left(1 + \frac{r_1}{k}\right)\left(1 + \frac{r_2}{k}\right)}$$
$$+ \cdots + \frac{\mathcal{P}_n}{\left(1 + \frac{r_1}{k}\right)\left(1 + \frac{r_2}{k}\right)\cdots\left(1 + \frac{r_n}{k}\right)},$$
$$(2.69)$$

or, more compactly,

$$\mathcal{A}_n = \sum_{\ell=1}^{n} \left[ \frac{\mathcal{P}_\ell}{\prod_{j=1}^{\ell} \left( 1 + \frac{r_j}{k} \right)} \right], \qquad (2.70)$$

where $n = 1,2,\ldots$ and $r_i > 0$ for $i = 1,\ldots,n$. In the special case where $r_i = r$ and $\mathcal{P}_i = \mathcal{P}$ for $i = 1,\ldots,n$, Equation (2.70) yields (2.57) on page 50.

Applications of (2.69) to the dividend discount model and bond pricing are given, respectively, in Section 2.9.1 and 2.10 (see page 68).

## Relating Future and Present Values of a Generalized Simple Ordinary Annuity

We now show that $\mathcal{S}_n$ in (2.67) is the generalized future value of $\mathcal{A}_n$ in (2.69) under the generalized periodic compounding in (2.31) (see page 33). Using Equation (2.67), direct calculation shows that

$$
\frac{\mathcal{S}_n}{\left(1+\frac{r_1}{k}\right)\left(1+\frac{r_2}{k}\right)\cdots\left(1+\frac{r_{n-1}}{k}\right)\left(1+\frac{r_n}{k}\right)}
$$
$$
= \frac{\mathcal{P}_n}{\left(1+\frac{r_1}{k}\right)\left(1+\frac{r_2}{k}\right)\cdots\left(1+\frac{r_n}{k}\right)} + \cdots + \frac{\mathcal{P}_2}{\left(1+\frac{r_1}{k}\right)\left(1+\frac{r_2}{k}\right)} + \frac{\mathcal{P}_1}{\left(1+\frac{r_1}{k}\right)}
$$
$$
= \mathcal{A}_n.
$$

It immediately follows that the relationship between the future and present values in Equation (2.61) generalizes to

$$
\mathcal{A}_n = \frac{\mathcal{S}_n}{\prod_{j=1}^{n}\left(1+\frac{r_j}{k}\right)}, \tag{2.71}
$$

where $n = 1, 2, \ldots$ and $r_i > 0$ for $i = 1, \ldots, n$.

## 2.8 Applications of Annuities

### 2.8.1 Saving, Borrowing, and Spending

**Example 2.10. (Saving During College)** A prospective college student plans to deposit $25 every month in an "untouchable" savings account, starting the first of July of the year she enters college until the last deposit on the thirtieth of June of her graduating year. Assume that she secured a fixed interest rate of 2.25% annually. Assume that the account compounds monthly.

a) Using this average interest rate, estimate how much she would have on July 1st of her graduating year.

   **Solution.** Use the future value $\mathcal{S}_n$ in (2.56). We have $k = 12$ for monthly compounding and since the period is 4 years, we have $n = 4 \times 12 = 48$ periods, $r = 0.0225$, and $\mathcal{P} = \$25$. By (2.56),

$$S_{48} = \frac{\left[\left(1 + \frac{0.0225}{12}\right)^{48} - 1\right]}{\frac{0.0225}{12}} \times \$25 = \$1,254.43.$$

b) If her target is to have at least \$1,300 on July 1st of her graduating year, determine the minimum required interest rate.

**Solution.** Solving the equation,

$$\$1,300 = S_{48} = \frac{\left[\left(1 + \frac{r}{12}\right)^{48} - 1\right]}{\frac{r}{12}} \times \$25,$$

implicitly for $r$ (use a software package), we obtain the smallest interest rate to be $r = 4.04\%$. Note that this is the smallest value of $r$ that works since $S_n$ is a strictly increasing function of $r$ for natural numbers $n \geq 2$ (see page 49). □

**Example 2.11. (Saving for Retirement)** Suppose that you open a retirement fund at the start of a month and you deposit \$200 at the end of each month. If the fund pays 4% per annum compounded monthly, how much would you accumulate at the end of 25 years?

**Solution.** This problem deals with the future value $S_n$ in (2.56). For monthly compounding ($k = 12$), we have $n = 25 \times 12 = 300$ periods, $r = 0.04$, and $\mathcal{P} = \$200$. Equation (2.56) then yields the following future value:

$$S_{300} = \frac{\left[\left(1 + \frac{0.04}{12}\right)^{300} - 1\right]}{\frac{0.04}{12}} \times \$200 = 514.13 \times \$200 = \$102,826.$$

□

**Example 2.12. (Total Paid on Loan)** A relative is considering a 20-year loan of \$150,000 with an interest rate of 8% compounded monthly. Assuming you hold the loan the entire term and make the minimum payment at the end of each month, what is the total amount you pay into the loan?

**Solution.** We have $\mathcal{A}_n = \$150,000, k = 12, r = 0.08$, and $n = 20 \times 12 = 240$ periods, so by the present value annuity formula, we obtain the minimum monthly payment:

$$\mathcal{P} = \frac{(r/k)\,\mathcal{A}_n}{1 - \left(1 + \frac{r}{k}\right)^{-n}} = \$1,254.66.$$

Since there are 240 months, the total paid is: $240 \times \$1,254.66 = \$301,118.40.$

□

**Example 2.13. (Paying Off Debt)** Suppose that you borrow $100,000 at an annual interest rate of 6% with monthly compounding. For an ordinary annuity based on this compounding, what is your minimum payment per month to pay off the loan in 10 years?

**Solution.** The problem requires the present value $\mathcal{A}_n$. Since $k = 12$, there are $10 \times 12 = 120$ periods. Using $\mathcal{A}_{120} = \$100,000$ and $\frac{r}{k} = \frac{0.06}{12} = 0.005$, we get:

$$P = \frac{(0.005) \times \$100,000}{1 - (1.005)^{-120}} = \$1,110.21.$$

□

**Example 2.14. (How Much Loan Can You Afford)** Suppose that you can pay $1,495 per month for the next 15 years. What is the largest loan you can afford at 6.25% per annum with monthly compounding?

**Solution.** Assume that the first payment is made 1 month from now. We have $n = 15 \times 12 = 180$ periods (months), $P = \$1,495$, and $r = 0.0625$. The maximum loan you can afford is:

$$\mathcal{A}_n = \frac{\left[1 - \left(1 + \frac{0.0625}{12}\right)^{-180}\right]}{\frac{0.0625}{12}} \times \$1,495 = \$174,359.71.$$

□

**Example 2.15. (Living Off a Lump Sum)** Suppose that you inherited $300,000 and invested it in an account with an annual interest rate of 7% compounded monthly. For an ordinary annuity based on this compounding, if you want your inheritance to last 20 years, what is the maximum fixed amount you can spend from the account per month?

**Solution.** Using the present value annuity formula with $\mathcal{A}_n = \$300,000, k = 12$, $r = 0.07, n = 20 \times 12 = 240$ periods, we obtain:

$$P = \frac{(r/k)\,\mathcal{A}_n}{1 - \left(1 + \frac{r}{k}\right)^{-n}} = \frac{(0.07/12) \times \$300,000}{1 - \left(1 + \frac{0.07}{12}\right)^{-240}} = \$2,325.90.$$

□

### 2.8.2 Equity in a House

**Example 2.16. (House Equity)** A couple bought their house 11 years ago for $225,000 and put down 10% on the house. On the balance, they took out a

15-year mortgage at 5.75% per annum with monthly compounding. The current *net market value* of the house is its current market value minus all costs in selling the house today. Suppose that the current net market value is now $350,000 and the couple wants to sell their house.

a) How much equity (to the nearest dollar) is in the house today? *Equity* in a house is defined as:

$$\text{equity} = (\text{current net market value}) - (\text{unpaid loan balance}).$$

**Solution.** The couple puts down 10% or $22,500 at the start, so the mortgage is for $\mathcal{A}_n = \$225,000 - \$22,500 = \$202,500$. Since $n = 15 \times 12 = 180$, $r = 0.0575$, $k = 12$, and $\ell = 132$, Equation (2.63) yields the unpaid balance at the end of the 132nd month:

$$\mathcal{B}_{132} = \frac{\left(1 + \frac{r}{k}\right)^n - \left(1 + \frac{r}{k}\right)^\ell}{\left(1 + \frac{r}{k}\right)^n - 1} \mathcal{A}_n = \$71,952.87.$$

Hence, the equity is: $\$350,000 - \mathcal{B}_{132} = \$278,047.13$.

b) What are the 1st and 132nd interest payments?

**Solution.** We have $\mathcal{I}_1 = \frac{r}{k}\mathcal{B}_0 = \$970.31$ and $\mathcal{I}_{132} = \frac{r}{k}\mathcal{B}_{131} = \$351.15$.

□

### 2.8.3 Sinking Funds

A *sinking fund* is an account into which one (individual or company) regularly deposits money in order to cover an obligation or debt that will come due at a known future date.

**Example 2.17. (Saving for College Tuition)** When a child was born in 2011, her parents decided to invest in her college education. This was motivated by a forecast that 4 years of in-state tuition at an average public college will be about $96,000 when she will attend college. Suppose that the parents want to accumulate that amount by their child's 17th birthday. They open a sinking fund into which they make a deposit on each birthday of the child up to the 17th birthday. Assume that the first deposit is for the amount $\mathcal{P}$ and thereafter the parents increase the deposited amount by 4% annually. Suppose that the bank where they have the sinking fund pays a fixed 5.5% per annum compounded annually. What should the minimum annual deposits be in order for the amount in the fund to reach at least $96,000 after her 17th deposit?

**Solution.** This example applies generalized future value annuity formula in (2.68), i.e.,

$$S_n = \sum_{\ell=0}^{n-1} \left[ \prod_{j=0}^{\ell} \left( 1 + \frac{r_{n+1-j}}{k} \right) \right] P_{n-\ell}, \qquad n = 1, 2, 3 \dots,$$

where $r_{n+1} = 0$. Note that $r_1$ does not appear in the formula since no interest is paid at the end of the first period (because the first deposit is not made at the start of the first period, but at the end of the first period).

We have $n = 17$, $S_{17} = \$96{,}000$, $r_2 = \cdots = r_{17} = 0.055$, and $k = 1$. The product in the sum above then becomes:

$$\prod_{j=0}^{\ell} \left( 1 + \frac{r_{n+1-j}}{k} \right) = \left( 1 + \frac{r_n}{k} \right) \left( 1 + \frac{r_{n-1}}{k} \right) \cdots \left( 1 + \frac{r_{n-(\ell-1)}}{k} \right) = (1.055)^{\ell},$$

where $\ell = 0, 1, 2, \dots, n-1$.

Let us determine the deposits $P_1, \dots, P_n$. The deposit $P_1 = P$ is made on the first birthday. On the second birthday, it is increased by 4% to $P_2 = P_1 + 0.04 P_1 = (1.04) P$. On the 3rd birthday, the deposit is $P_3 = P_2 + 0.04 P_2 = (1.04)^2 P$. For $j = 1, \dots, n$, the deposit on the $j$th birthday is then $P_j = (1.04)^{j-1} P$. It follows: $P_{n-\ell} = (1.04)^{16-\ell} P$.

The target amount for the sinking fund can then be expressed as:

$$\$96{,}000 = P \times \sum_{\ell=0}^{16} (1.055)^{\ell} (1.04)^{16-\ell} = P \times (1.04)^{16} \times \sum_{\ell=0}^{16} \left( \frac{1.055}{1.04} \right)^{\ell}$$
$$= P \times 1.87298 \times 19.1104 = 35.7934 P.$$

This yields $P = \$2{,}682.06$, which is the first deposit. Hence, for $j = 1, \dots, 17$, the minimum deposit on the $j$th birthday must be: $P_j = (1.04)^{j-1} \times \$2{,}682.06$, which has values

$$P_1 = \$2{,}682.06, \quad P_2 = \$2{,}789.34, \quad \dots, \quad P_{16} = \$4{,}830.24, \quad P_{17} = \$5{,}023.45.$$

$\square$

## 2.9 Applications to Stock Valuation

This section applies the theory of annuities to determining the present values of preferred and common stocks. The main tool is the dividend discount model. A stochastic model for the future value of a stock will be taken up in a later chapter.

### 2.9.1 The Dividend Discount Model

The *dividend discount model* (DDM) was pioneered by Williams [18] (1938) and Gordon [7] (1959). *The fundamental hypothesis of the DDM is that, if a stock is held for n years, then its current value is the present value of the sequence of its expected future cash dividends through n years plus the present value of the stock's expected price in n years.*

A stock has no maturity date and so is a security in perpetuity. Suppose that the stock pays a dividend and the (annual) required return rate of the stock is $\hbar$.[11] Assume that you will hold the stock for $n$ years. Let $\mathcal{D}_0$ be the current cash dividend, i.e., the total cash dividend per share over the previous year. Suppose that all future cash dividends are expected to grow at a constant annual rate $g$, which we assume is less than the required return rate ($\hbar > g$). Let $\mathcal{D}(i)$ denote the expected cash dividend per share for the interval from the present time to $i$ years out, where $i = 1, \ldots, n$. Then the expected sequence of future cash dividends per share for years 1 through $n$ is:

$$\mathcal{D}(1) = (1+g)\,\mathcal{D}_0, \qquad \mathcal{D}(2) = (1+g)^2\,\mathcal{D}_0, \qquad \ldots, \qquad \mathcal{D}(n) = (1+g)^n\,\mathcal{D}_0.$$

The share value $S_0^{(n)}$ of the stock today is the present value of the expected dividend cash flows and the expected price of the stock $n$ years from now:

$$S_0^{(n)} = \frac{\mathcal{D}(1)}{1+\hbar} + \frac{\mathcal{D}(2)}{(1+\hbar)^2} + \cdots + \frac{\mathcal{D}(n)}{(1+\hbar)^n} + \frac{\mathcal{T}_n}{(1+\hbar)^n}, \qquad (\hbar > 0), \quad (2.72)$$

where $\mathcal{T}_n$ is the *terminal price*, i.e., the expected price of the stock in $n$ years. Note that (2.72) is a special case of the generalized present value equation (2.69) (page 58) with $k = 1$, $r_i = \hbar$, $\mathcal{P}_i = \mathcal{D}(i)$ for $i = 1, \ldots, n-1$, and $\mathcal{P}_n = \mathcal{D}(n) + \mathcal{T}_n$.

Now, if you hold the stock in perpetuity (indefinitely) rather than for $n$ years, then there is no terminal price, and the stock's present share price becomes:

$$S_0 = \lim_{n \to \infty} S_0^{(n)} = \mathcal{D}_0 \sum_{\ell=1}^{\infty} \left( \frac{1+g}{1+\hbar} \right)^{\ell}, \qquad (\hbar > g > 0).$$

Since $\frac{1+g}{1+\hbar} < 1$ due to $\hbar > g$, the geometric series yields

$$\frac{1}{1 - \left( \frac{1+g}{1+\hbar} \right)} = \sum_{\ell=0}^{\infty} \left( \frac{1+g}{1+\hbar} \right)^{\ell} = 1 + \sum_{\ell=1}^{\infty} \left( \frac{1+g}{1+\hbar} \right)^{\ell}.$$

Hence, the present share price of the stock becomes:

---

[11] Recall that the marketplace is assumed to be in equilibrium, which allows for the required return rate of the stock to be estimated using the CAPM model; see Chapter 4 for an introduction.

$$S_0 = \frac{(1+g)\mathcal{D}_0}{\hbar - g} = \frac{\mathcal{D}(1)}{\hbar - g}, \qquad (\hbar > g > 0). \qquad (2.73)$$

Equation (2.73) is called the *Gordon growth model*. This is an example of a *growing perpetuity*, i.e., a perpetuity with payments that increase each period.

The Gordon growth model generalizes naturally to allow for $k$ compoundings per year through the replacements $g \to g/k$ and $\hbar \to \hbar/k$:

$$S_0 = \frac{(1+\frac{g}{k})\mathcal{D}_0}{\frac{\hbar}{k} - \frac{g}{k}} = \frac{\mathcal{D}(1)}{\frac{\hbar}{k} - \frac{g}{k}} \qquad (\hbar > g > 0),$$

where $\mathcal{D}(j) = (1+\frac{g}{k})^j \mathcal{D}_0$.

## 2.9.2 Present Value of Preferred and Common Stocks

A preferred stock grants its holder ownership in a corporation, but no voting rights, and a claim on assets in the event of bankruptcy that comes before any claim of the common stock holders. Preferred stocks are considered fixed-income securities because they promise to pay a fixed cash dividend, which has priority over any cash dividends paid to common stock holders. Since the future cash dividends are fixed and expected to be paid indefinitely, the value of a preferred stock is then obtained from Equation (2.73) using $g = 0$:

$$S_0 = \frac{\mathcal{D}_0}{\hbar} \qquad (\hbar > 0) \qquad (2.74)$$

**Example 2.18. (Preferred Stocks)** Suppose that a preferred stock has a fixed total annual cash dividend per share of $2.50. Assume an annual required return rate of 13% for the stock. How much should you pay for the preferred stock?

**Solution.** We apply Equation (2.74) with $\mathcal{D}_0 = \$2.50$ and $\hbar = 0.13$. The current share price of the preferred stock is: $S_0 = \frac{\mathcal{D}_0}{\hbar} = \frac{\$2.50}{0.13} = \$19.23$.

□

Common stocks do not have a promise to pay cash dividends. Nonetheless, if a common stock currently pays no cash dividends, there is still investor expectation that earnings are being reinvested in the company to create growth which will lead to cash dividends in the future. Due to the uncertainty of future cash dividends for common stocks, we shall model their valuation under certain assumptions about the expected cash dividends.

**Example 2.19. (Common Stocks)** Suppose that the total cash dividend of a stock last year was $2.75 per share and dividends are expected to increase at 3% per annum. If the annual required return rate is 10%, find the share price of the stock today.

**Solution.** By (2.73), the price is: $S_0 = \frac{(1+g)\mathcal{D}_0}{\hbar - g} = \frac{(1+0.03) \times \$2.75}{(0.1-0.03)} = \$40.46.$

$\square$

## 2.10 Applications to Bond Valuation

The US bond market is vast—much bigger than its stock market. As measured at the end of 2012 in terms of capitalizations, the US bond market was twice as big as the US stock market for domestic companies.[12] As with other fixed-income financial investments, the price of a bond is the present value of its cash flow. We shall explore how to value bonds.

### 2.10.1 Bond Terminologies

A *bond* is a contract between an issuer (bond seller) and a lender (bondholder) legally binding the issuer to repay the lender a specified fixed amount at maturity and a series of interest payments during the life of the bond. In essence, a bond is an IOU.[13] The specific terms for a bond's duration, interest payment, etc. are described fully in the contract (indenture). The funds raised by bond issues are used for capital expenditures, operations, corporate takeovers, public projects, etc.

Bonds are usually redeemed on the maturity date. However, some bonds have the option to be *callable*,[14] i.e., such bonds give the issuer the right, but not the obligation, to redeem (call) the bond prior to the maturity date. There are also bonds with the option to be *convertible*, i.e., the bondholder has the right to exchange the bond for a different security (e.g., shares of common stock), and have a prescribed variable interest rate or even deferred interest. To avoid confusion about which types of bonds are intended, we assume

> *Unless stated to the contrary, all bonds are without options, i.e., they are noncallable, nonconvertible, etc., and have a fixed interest paid every 6 months.*

---

[12] http://www.learnbonds.com/how-big-is-the-bond-market/

[13] IOU is an abbreviation for "I owe you."

[14] Most corporate bonds are callable. Also, the US Treasury has not issued callable bonds since 1985.

Although both bonds and stocks are securities of a company, they are different in the sense that bondholders are creditors of the company, whereas stockholders are owners of the company. The cash flows from a company's bonds are more reliable than those from its stocks since the company has a legal obligation to repay its bondholders. Sometimes even when a company becomes insolvent, its bondholders may still get back some compensation, while compensation is not guaranteed for its common stockholders.

We now list and discuss some basic terminologies and features of bonds:

➤ The *issue date* of a bond is the date on which the bond issuer receives the loan from the lender and from which the lender is entitled to receive interest from the issuer.

➤ The *maturity value* $M$ (also known as the *par value, face value, principal*) of a bond is the unit of the amount borrowed at the time it was issued. It is traditionally in units of $1,000, but municipal bonds are usually sold in units of $5,000.

➤ There are two main markets for bonds: *primary market*, where bonds are sold for the first time to institutional investors, and *secondary market*, where the resale of bonds taking place after their initial offering is open to the public, though individual investors will need to have a brokerage account to transact trades. Bonds selling at their maturity value are called *par bonds*. In the secondary market, bonds are traded at prices that are typically different from the maturity value. If a bond sells at a market price above (respectively, below) its maturity value, then it is called a *premium bond* (respectively, *discount bond*).

**Remark 2.5.** The primary bond market is essentially an institutional market. In practice, the US Treasury uses an auction process to sell treasury bills, notes, and bonds in the primary market, whereas the pricing of newly issued corporate bonds is negotiated between the corporation (or its representative) on the one hand and investment bankers and large institutional investors on the other.                                                        □

➤ The *maturity date* is the date on which the bond issuer must repay the lender the bond's maturity value. Note that callable bonds have features which allow for the principal to be repaid before the maturity date.

➤ The *term to maturity*, or simply *maturity*, of a bond is the length of the time interval between the issue date and the maturity date.

Bonds can be classified into three groups: *short term*, *intermediate term* and *long term* according to maturities of, respectively, 1–5 years, 5–12 years, and greater than 12 years.

➤ A *coupon payment* $C$, or simply *coupon*, is an interest payment of a bond.

Most bonds have a fixed coupon that does not change during the life of the bond and is paid at regular time intervals, usually semi-annually.

If a bond does not pay a coupon during its life, it is called a *zero-coupon bond*. To compensate for no coupon payments, such bonds are issued at a deep discount from their value at maturity.[15] For this reason, they are also called *discount bonds* or *deep discount bonds*. Zero-coupon bonds are similar to US government savings bonds[16] in concept and have significant theoretical value.

➤ The *current yield* indicates the yield of a security based on its current market value. The *current yield of a bond*, denoted by $r$, is determined by the formula below:

$$r = \frac{\text{annual coupon amount}}{\text{current bond price}}. \tag{2.75}$$

➤ The *coupon rate* or *interest rate*, denoted by $r_C$, is defined by the current yield when the bond price is equal to its maturity value. That is:

$$r_C = \frac{\text{annual coupon amount}}{\text{maturity value}}.$$

➤ A bond's *yield to maturity* (YTM), denoted by $r_Y$, is the marketplace's annual required return rate of the bond held to maturity and whose future coupon payments are reinvested at the same rate. Equivalently, a bond's YTM equates the present value of the bond's future cash flows to the bond's current market price:

$$\text{current bond price} = \sum_{\ell=1}^{n-1} \frac{C}{(1 + \frac{r_Y}{k})^\ell} + \frac{C + M}{(1 + \frac{r_Y}{k})^n}, \tag{2.76}$$

where $n$ is the number of coupon payments remaining on the bond and $k$ is the number of coupon payments per annum (typically, $k = 2$). Equation (2.76) is obtained by applying Equation (2.69) on page 58 with $\mathcal{P}_i = C$ for $i = 1, 2, \ldots, n-1$, $\mathcal{P}_n = C + M$, and $r_i = r_Y$ for $i = 1, 2, \ldots, n$.

Denote the current bond price and $\frac{r_Y}{k}$ in (2.76) by $B(n)$ and $\hat{r}_Y$, respectively. Then:

$$B(n) = \frac{M}{(1 + \hat{r}_Y)^n} + \sum_{\ell=1}^{n} \frac{C}{(1 + \hat{r}_Y)^\ell}. \tag{2.77}$$

---

[15] For example, such a bond might be issued at a 50% discount from its maturity value.
[16] A savings bond offers a fixed rate of interest over a fixed period of time, but cannot be traded after being purchased.

By (2.58) on page 50, Equation (2.77) is equivalent to

$$B(n) = \frac{M}{(1+\hat{r}_Y)^n} + \frac{((1+\hat{r}_Y)^n - 1)\, C}{(1+\hat{r}_Y)^n\, \hat{r}_Y} \qquad (\hat{r}_Y > 0). \qquad (2.78)$$

➤ Finally, the following relationships among yield to maturity ($r_Y$), current yield ($r$), and coupon rate ($r_C$) hold (Exercise 2.35):

1. A bond trades at par iff $r_Y = r = r_C$ (see Proposition 2.1).
2. A bond trades as a discount bond iff $r_Y > r > r_C$.
3. A bond trades as a premium bond iff $r_Y < r < r_C$.

**Remark 2.6.** Generally, evaluating financial investment performance can be a complicated task as there are different measures to be applied to serve different purposes. Yield is a measure of an investment income that an investor receives annually. As a bond investor, if you just want to hold on to your bond until its maturity, the coupon rate is the only measure that matters. However, if you need to sell your bond before maturity, you have to adopt the current yield as a measure. Yield to maturity is a measure that enables you to compare different bonds by taking the effect of compound interest into consideration under the assumption that all the coupon payments are reinvested at the same rate and you hold the bond to maturity.[17]                                                        □

### 2.10.2 Bond Prices Versus Interest Rates and Yield to Maturity

**Bond Price with YTM at the Coupon Rate**

For a bond being traded after it was originally issued, we expect intuitively that when the YTM is at the coupon rate, then the market value of the bond should be its maturity value. The following proposition confirms that intuitive result and its converse:

**Proposition 2.1.** *Suppose that a bond has n coupon payments remaining. The market price of the bond equals its maturity value exactly when its coupon rate is the YTM:*

$$r_Y = r_C \quad \text{if and only if} \quad B(n) = M. \qquad (2.79)$$

---

[17] It is worth noting that comparing different bonds by their percentage change in price is often mis-leading since the significance is not the same for an identical percentage price change of bonds with different interest rates. Also, it is important to realize that reinvesting all the coupon payments at the same rate is rather difficult if not impossible in practice.

*Proof.* If $\hat{r}_C = \hat{r}_Y$, where $\hat{r}_C = r_C/2$ and $\hat{r}_Y = r_Y/2$, then $\mathcal{C} = \hat{r}_C \mathcal{M} = \hat{r}_Y \mathcal{M}$ and the bond valuation Equation (2.78) becomes:

$$B(n) = \frac{\mathcal{M}}{(1+\hat{r}_Y)^n} + \frac{\left((1+\hat{r}_Y)^n - 1\right)\mathcal{M}}{(1+\hat{r}_Y)^n} = \frac{\mathcal{M}}{(1+\hat{r}_Y)^n}\left(1 + (1+\hat{r}_Y)^n - 1\right)$$
$$= \mathcal{M}.$$

Conversely, if $B(n) = \mathcal{M}$, then (since $\mathcal{C} = \hat{r}_C \mathcal{M}$) Equation (2.78) reduces to:

$$1 = \frac{1}{(1+\hat{r}_Y)^n} + \frac{\left((1+\hat{r}_Y)^n - 1\right)\hat{r}_C}{(1+\hat{r}_Y)^n \hat{r}_Y}.$$

Multiplying through by $(1+\hat{r}_Y)^n$ yields: $(1+\hat{r}_Y)^n - 1 = \frac{\hat{r}_C\left((1+\hat{r}_Y)^n - 1\right)}{\hat{r}_Y}$. Hence, $\hat{r}_C = \hat{r}_Y$.                                                            □

**Bond Prices Move in Opposite Direction to Interest Rates**

The interest rate probably has the single largest impact on the prices of all bonds. The following three examples are related to each other and illustrate the relationship between bond prices on the one hand and interest rates and YTM on the other.

**Example 2.20.** Suppose that a 30-year bond with an annual 3% coupon rate payable semiannually was issued by the US Treasury on the first trading day of 2013. If the maturity value is $1,000, what is the semiannual coupon amount?

**Solution.** Solve for the semiannual coupon amount $\mathcal{C}$ from the equation

$$3\% = \frac{2\mathcal{C}}{\$1,000}$$

to obtain $\mathcal{C} = \$15$. Assume, for simplicity, that the bond was sold in the primary market at its maturity value. Then by Proposition 2.1, the YTM equals 3%.
                                                                                   □

In the next two examples, the bond in Example 2.20 will be referred to as "the first bond."

**Example 2.21.** Since the Feds kept interest rates artificially low in 2013, doubling the interest rate in 10 years from 2013 is not an unreasonable speculation. Suppose that another 30-year bond with an annual 6% coupon rate payable semiannually will be issued by the Treasury on the first trading day of 2023. What will be the price of the first bond at the time of the second bond initial offering?

**Solution.** For the simplicity of our argument, we assume almost no intraday bond price fluctuations on the issue date of the second bond. Let

$$t_0 = \text{the issue date of the second bond,}$$
$$r_1 = \text{the (current) yield of the first bond at } t_0,$$
$$B_1 = \text{the price of the first bond at } t_0.$$

Since no investors will buy a bond with 3% annual yield when they have the choice to purchase a bond of the same type with 6% annual yield, we have $r_1 = 6\%$. In other words, the current yield of the first bond will be forced to approach 6% on the issue date of the second bond under the law of supply and demand. To speculate on the price of the first bond, we apply (2.75) and solve for $B_1$ from the equation, $r_1 = 6\% = \frac{2 \times \$15}{B_1}$, to obtain $B_1 = \$500$. *Observe that when the interest rate rises from 3% to 6%, the first bond's price will fall from* $\$1,000$ *to* $\$500$.

□

**Example 2.22.** Suppose that you will purchase the first bond on the first trading day of 2023 at the price $500 and hold it to the maturity date of the first trading day of 2043. What will be the yield to maturity?

**Solution.** We need to solve the bond Equation (2.78) for $r_Y$, which in our setting is[18]

$$B_1 = \frac{((1 + \frac{r_Y}{k})^n - 1)\,C}{\frac{r_Y}{k}(1 + \frac{r_Y}{k})^n} + \frac{M}{(1 + \frac{r_Y}{k})^n}.$$

Using $B_1 = \$500$, $C = \$15$, $M = \$1,000$, $k = 2$ (semiannual compounding), and $n = 40$ (number of coupon payments remaining), we obtain $r_Y = 8.084\%$. Indeed, *when the price of a bond drops from* $\$1,000$ *to* $\$500$, *the yield to maturity rises from* 3% *to* 8.084%. Note that we assumed the bond was sold in the primary market at its maturity value.

□

We now establish, in general, the observation at the end of the solution of Example 2.22. Take the first and second derivatives of the bond's present value (2.77):

$$\frac{dB(n)}{d\hat{r}_Y} = -n\frac{M}{(1 + \hat{r}_Y)^{n+1}} - C\sum_{\ell=1}^{n}\frac{\ell}{(1 + \hat{r}_Y)^{\ell+1}} < 0$$

and

---

[18] As before, there is no general analytical solution $r_Y$ for every $n$. In most applications, we can only estimate $r_Y$ numerically using a software.

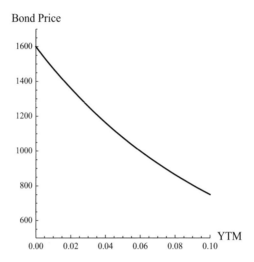

Bond Price

**Fig. 2.2** The price of a bond is a strictly decreasing, concave-up function of the bond's YTM. The graph illustrates this for a bond with $1,000 maturity value and 6% coupon rate. Note that when the YTM is 6%, the bond's price is its maturity value.

$$\frac{d^2 B(n)}{d\hat{r}_Y^2} = n(n+1)\frac{M}{(1+\hat{r}_Y)^{n+2}} + C \sum_{\ell=1}^{n} \frac{\ell(\ell+1)}{(1+\hat{r}_Y)^{\ell+2}} > 0.$$

In other words, the bond's price is not only strictly decreasing as the yield increases, but has a convex graph, i.e., the graph is everywhere concave up (increasing slope). Figure 2.2 depicts this property for a bond with $1,000 maturity value and 6% annual coupon rate.

## 2.11 Exercises

### 2.11.1 Conceptual Exercises

**2.1.** A physicist summed up the growth rate of an initial sum of money held over a fixed time span as follows: "If simple interest is applied during the time span, then the initial sum will grow with uniform (constant) velocity as the interest rate increases. If periodic compound interest is applied, then the growth of the initial sum will accelerate as interest increases." Do you agree with this interpretation? Justify your answer.

**2.2.** Theorem 2.1 on page 27 yields that $k$-periodic compounding of a principal $\mathcal{F}_0$ at $r$ per annum over a time span of $\tau$ years consisting of $x$ interest periods gives a future value,

$$\mathcal{F}_x = \left(1 + \frac{r}{k}\right)^x \mathcal{F}_0,$$

where $x$ is a nonnegative real number and $0 \leq \frac{r}{k} < 1$ for $k = 1,2,\dots$. Does $\mathcal{F}_x$ increase or decrease as $k$ increases indefinitely? Justify your answer.

**2.3.** Suppose you purchase a lottery ticket for $2. What is your return rate if you lose? What if you win $200 million? Express your answer as a percentage.

**2.4.** Consider an investment that promises a fixed sequence of future cash dividends. Briefly explain why an increase in the required return rate on the investment would decrease the current value of the investment.

**2.5.** Explain what the following is stating financially about the start-up: "A start-up's NPV at 30% is $35,000."

**2.6.** A friend borrows $1,000 from a lender that gives him the loan as a simple ordinary annuity at a fixed interest rate over 2 years with a payment of $100 per month. If your friend carries the loan to its full term, then he will have to pay more than the amount of the loan in just interest. True or false? Justify your answer.

**2.7.** A loan with a fixed payment of $1,000 per month for 5 years has the stipulation that you will have to pay all the interest due on the loan even if you pay the loan off early. If immediately after you receive the loan, you want to pay it off, how much do you have to pay the lender?

**2.8.** How would you modify the interpretation of the noncallable bond pricing formula (2.78) on page 69 to obtain the current price of a callable bond, i.e., a bond where the issuer has the right, but not the obligation, to redeem (in practice, cancel) the bond before maturity? Use a single call date, i.e., a date when the issuer can redeem the bond before maturity. Compare the price of a callable bond with a noncallable one. Corporations issue callable bonds because if interest rates go down, they can call their bonds and refinance their debt at a lower interest rate.

**2.9.** How would you modify the interpretation of the noncallable bond pricing formula (2.78) on page 69 to obtain the current price of a puttable bond, i.e., a bond where the investor has the right, but not the obligation, to redeem the bond before maturity? Use a single put date, i.e., a date on which the investor can redeem the bond before maturity. Compare the price of a puttable bond with a noncallable one. Investors buy puttable bonds because if interest rates increase, they can sell back their original bonds at the put value and invest the proceeds in a higher interest rate bond.

### 2.11.2  Application Exercises

**2.10.** Consider a principal $\mathcal{F}_0$ that is held for $n_{exact}$ days during a non-leap year at the simple interest rate $r$. By what percent is the simple interest amount using Banker's Rule greater than the simple interest amount employing exact time and exact interest?

**2.11. (Selling or Buying a Loan)** On November 12, 2007, a borrower closes on a loan for $176,000 at 6.25% per annum compounded daily. Repayment of the loan's maturity value (principal plus interest) is due in full on April 15, 2008. Suppose that the fine print of the original loan stipulated that the lender can sell the loan on the condition that the interest rate and maturity date remain the same. The lender sells the loan to another lender on January 5, 2008. The new lender agrees to purchase the debt for the present value of the maturity value at 10% per annum compounded daily. Assume that interest compounds daily and the borrower does not default on the loan. Use Banker's Rule when solving the following:

a) What is the maturity value of the loan?
b) What will the first lender receive for selling the loan? Is any profit made by the first lender?
c) What profit will the second lender make on the loan's maturity date if the conditions of the original loan are unchanged?
d) Though the original interest rate and maturity date are unchanged, the second lender is not prevented from reissuing the loan with a new start date set as the loan's purchase date and with the new loan's principal set as the value of the loan on the purchase date. Does the second lender make more profit by resetting the loan in this way? Explain.

**2.12.** For an interest rate of 4% per year, compare the future value 2 years from now to which $10,000 increases under daily compounding versus continuous compounding. Assume 365 days per year and express your answer as a fractional-difference percentage of the daily compounding case.

**2.13.** Suppose that at the start of college, you have $1,000 to invest and would like for it to grow to $1,250 at the end of your senior year through monthly compounding. Determine the general formula for the interest rate required for the growth and then compute the interest rate.

**2.14.** Assume that college tuition is currently 30 times its cost 15 years ago. Assuming annual compounding, what is the interest rate $r$ that gives the rate of increase in tuition?

**2.15.** How much should you have today in an account with monthly compounding and annual interest rate of 4% to receive $1,000 per month forever?

**2.16. (Equity in a House)** A couple purchased a house 7 years ago for $375,000. The house was financed by paying 20% down and signing a 30-year mortgage at 6.5% on the unpaid balance. The net market value of the house is now $400,000. Assume that the couple wishes to sell the house.

a) How much equity (to the nearest dollar) does the family have in the house now, after making 84 monthly payments?

b) Find the first interest payment $\mathcal{I}_1$ and the 84th interest payment $\mathcal{I}_{84}$.

**2.17. (Social Security Benefits)** We present a simplified problem to illustrate Social Security benefits. A college graduate begins work at age 22. She has an annual income of $70,000 until retirement (a simplification), pays 12.4% of this income into Social Security each year, and retires at age 65 with Social Security benefits of $20,000 annually. How long must she live before the present value of these benefits equals the present value of her annual contributions? In other words, how long must she live after retirement to get back the full value of her contributions to Social Security? Will she get the entire value? Assume a discount rate of 4% per year, no change in her salary, and that all payments and benefits occur at the end of each year.

**2.18. (Worker's Compensation)** The usual legal settlement for an industrial accident is the present value of the employee's lifetime earnings. If you expect to work for 10 more years, make $70,000 a year in the next 2 years, and get a raise of $5,000 every 2 years, what would be your settlement? Assume an annual discount rate of 4% in the first 5 years and 6% in the second 5 years, and that your paycheck is received at the end of each year.

**2.19. (Bonds)** Suppose that you bought a 30-year bond with 4% annual coupon rate. You wish to sell that bond at a later date when the remaining life of the bond is 2.5 years and the current YTM of your bond has declined to 2%.

a) What is the fair value, as determined by the present value method, of the bond at the time of your sale?

b) How much would you earn if you purchased the bond for $1,000, sold it at the fair value, and did not reinvest the coupon payments?

**2.20. (Bonds)** Bonds are generally quoted as a percentage of their face value. A bond selling at 99.2% of its face value is quoted as 99.2. The following information for a treasury bond was provided by the WSJ market data center on December 4, 2013:

| Maturity | Coupon | Current price | Previous price | Change | Yield |
|----------|--------|---------------|----------------|--------|-------|
| 11/30/20 | 2.000  | 99.20         | 99.00          | 0.203  | 2.123 |

The coupon column refers to the annual coupon rate. Verify that the last column indicates YTM.

## Purchasing a House

The remaining Application Exercises deal with purchasing a house. Assume that you are currently renting an apartment for $1,040 per month and you have been considering buying a house. You have saved $10,000 toward a down payment for the house.

A salesperson informs you that he has a new house for sale, where the house and land were independently appraised at $200,000, but are being sold by the builder at a discount price of $185,000. The builder wants to get rid of the property quickly because the house is the last one to be sold in the development and the builder is moving on to construction of a new development.

The salesperson connects you with his in-house lender, to whom you give details about your income and grant permission to review your credit and eligibility for a loan. You inform her that you are prepared to make a down payment of $10,000 toward the house if necessary. She gets back to you with good news that, if you put $8,100 toward the house, then they can give you a 30-year loan for the balance of $176,900 at 6.25% per annum (compounded monthly). Note that lenders require the house to appraise at or above the purchase price; otherwise, they may reject the loan or require more down payment. The lender computes the monthly mortgage payment at $1,089.20. She informs you that the remaining $1,900 of your $10,000 can be used toward costs associated with the final evaluation of the physical property and the closing of the purchase (property inspector fee, termite inspector fee, official survey, attorney fees, etc.). The builder agrees to pay for costs beyond your $1,900 and make necessary repairs you identify during the period you have to inspect the property (the due diligence period).

Hearing the news about your qualification for the loan, the salesperson asks you how much rent you are now paying. When you inform him that you pay $1,040 per month, he quickly points out that it would be a mere extra $50 per month for you to meet the mortgage payments. He emphasizes that it is better to own than to rent, especially if the mortgage is just a bit more than your current rent.

You are thrilled! After the excitement subsides, however, you decide to run the numbers yourself to make sure you get a clear understanding of what you are getting into financially.[19] The problems in this project help guide you through some of this analysis.

**2.21.** Show that the monthly loan payment on the unpaid principal balance of $176,900 is $1,089.20.

**2.22.** In addition to closing fees paid to settle the loan, there are expenses beyond the monthly mortgage payments.

First, since your deposit was less than 20% of the purchase price, you are required to take out a private mortgage insurance (PMI) to protect the lender

---

[19] Mortgages on a house are generally modeled as simple ordinary annuities by lenders.

if you default on the loan. The PMI typically lasts until the unpaid principal balance of the mortgage is paid down to 80% of the original value of the house, where the house's original value is the lesser of the purchase price and the official appraised value of the house used in closing the sale. Note that the bank may also require your payment history to be in good standing (e.g., no late payments in the past year or two) before removing PMI. Of course, if the value of the house increases nontrivially, you may be able to remove the PMI earlier. Suppose that the PMI is $141.52 per month.

Second, along with PMI, you have to pay for hazard insurance to cover un-planned damages to the house due to fire, smoke, wind, etc. Assume that the hazard insurance is $36.50 per month.

Third, you have to pay property taxes to the tax district (e.g., county and city) where the house is located. The property (i.e., house and land) will be valued within your tax district, which is a valuation that is separate from the appraisal done when purchasing the house. The resulting tax district's valua-tion is the taxable value of the house and is the amount to which the property tax rate will be applied. Suppose that the annual property tax rate is 1.3% and the taxable value of the property is $189,986. For this project, the taxable prop-erty value is less than the appraised value (i.e., $200,000) used for the purchase. Sometimes, however, the taxable value can be higher which was not uncom-mon in the aftermath of the 2008 mortgage crisis.

The PMI, hazard insurance, and property tax payments are in addition to the monthly loan payment, and all together they form a single payment you make to the lender. The lender or a company hired by the lender manages these payments by taking out the portion for the loan payment (principal plus interest) and depositing the rest into an escrow account, which is used to pay the annual insurance premiums and property taxes on behalf of the borrower.

Finally, assume that the property is in a housing development that comes with a mandatory Homeowners Association (HOA) fee. The HOA fee is used to maintain the grounds, roads, etc. in the development. If you do not pay the fee, the HOA can foreclose on your property. Assume an HOA fee of $100 per month.

a) What is the estimated total monthly PITI, i.e., the minimum monthly pay-ment covering the principal, interest, taxes, and (hazard) insurance?
b) Identify two other mandatory house expenses that are outside of the PITI payment and other basic house costs like utilities and repairs. Do exclude costs like groceries, tuition, medical expenses, etc., which are more associ-ated with running a home. What is the minimum monthly cost of the house during the first year if you now include these two mandatory house ex-penses and PITI? Which of these housing costs will likely increase in the future?

c) What is your opinion about the salesperson's pitch about the cost of renting versus buying a house?

**2.23.** Fill out the amortization schedule below, which is for the first 5 months of the loan.

| Payment # | Payment ($\mathcal{P}$) | Principal ($\mathfrak{P}_\ell$) | Interest ($\mathcal{I}_\ell$) | Bal. ($\mathcal{B}_\ell$) |
|---|---|---|---|---|
| 1 | 1,089.20 | 167.85 | 921.35 | 176,732.15 |
| 2 | 1,089.20 | | | |
| 3 | 1,089.20 | | | |
| 4 | 1,089.20 | | | |
| 5 | 1,089.20 | | | |

**2.24.** Are there discrepancies in the above amortization table? If so, explain how to remove them mathematically.

For the remaining problems, note that only the payments toward principal and interest (PI) are relevant to the loan's balance. Costs associated with property taxes, hazard insurance, PMI, HOA, etc. are separate expenses and do not impact the balance of the loan. Such costs are typically not included in the loan's cost.

**2.25.** Using a software, compute the numbered payment at which the unpaid balance on the loan will first dip below 80% of the original value of the house. Roughly how many years and months does it take to reach that balance? If the value of the house has not decreased below its original value at that point in time, you would stop paying PMI henceforth.

**2.26.** Determine the total amount you would pay into the mortgage, excluding escrow payments, if you make only the minimum payment over the full 30 years. What is the total cost of the mortgage? Is it more than the mortgage?

**2.27.** Estimate the number of years and months it would take to pay off the mortgage if you double your monthly payments.

**2.28.** Estimate the total you would pay into the mortgage if you double your monthly payments. What is the total cost of the mortgage for doubled payments? Is it more than the mortgage?

### 2.11.3 Theoretical Exercises

**2.29.** Suppose that an initial capital $\mathcal{F}_0$ grows to an amount $\mathcal{F}(\tau)$ over a time span $\tau$. A mathematician modeling the growth observes that for all time spans

$x$ and $y$, the accumulated amount $\mathcal{F}(x)$ is a differentiable function satisfying the following:

$$\mathcal{F}(x+h) = \mathcal{F}(x) + \mathcal{F}(h) - \mathcal{F}_0, \qquad \mathcal{F}(0) = \mathcal{F}_0, \qquad \frac{d\mathcal{F}}{dx}(0) = r\,\mathcal{F}_0,$$

where $r \geq 0$. Determine the type of growth model, i.e., find $\mathcal{F}(x)$.

**2.30.** Derive Equation (2.18) on page 26: $G'(x) = G(x)\,G'(0)$.

**2.31. (Capital After Spending, Inflation, and Interest)** Consider the following setup:

- Begin with an initial capital $C(0)$ in an interest-bearing account and let $C(n)$ be the remaining capital at the end of the $n$th year.
- Assume an interest rate $r$ is applied at the end of each year to the capital remaining on that date.
- At the end of the first year, assume that an amount $S$ was spent from $C(0)$ on goods and services, and money will be spent on similar goods and services in each of the subsequent years.
- Suppose that the amount spent at the end of any specific year is the total amount spent by the end of the first year increased in subsequent years at the annual inflation rate i compounding annually until the end of the specified year. Assume that $r > $ i since investors are not interested in a market interest rate that is below the inflation rate.

a) Show that the total capital at the end of the $(n+1)$st year can be expressed recursively as follows in terms of the capital at the end of the previous year, taking into account spending, inflation, and interest growth:

$$C(n+1) = (1+r)\left[C(n) - (1+\text{i})^n S\right]. \qquad (2.80)$$

b) Use induction to show that

$$C(n) = (1+r)^n \left[C_0 - \frac{1+r}{r-\text{i}}S\right] + \left(\frac{1+r}{r-\text{i}}\right)(1+\text{i})^n S.$$

**2.32.** Suppose that after this year, your grandmother will receive regular payments from a retirement fund, but she has to choose between two options for how to receive the payments during $n+1$ years. She does not plan to spend any of the money until after the $n+1$ years. Assume that she will save all the disbursements in an account that accrues the payments as a simple ordinary annuity with $k$-periodic compounding at interest rate $r$ (e.g., each payment date coincides with an interest date).

The payment start date will differ for the two plans, but both payment options will have the last payment at the start of the last interest period during the $(n+1)$st year. Your job is to help her choose between the two options.

a) **(A General Future Value Formula)** The current problem determines a general formula that incorporates the future value of the payments into your grandmother's account. Suppose that regular payments of $P$ into an interest-bearing account form a simple ordinary annuity with $k$-periodic compounding at interest rate $r$. Assume that the account receives the first payment at the end of the first interest period of a certain year and the last payment at the end of the $N$th interest period going forward, with no payment at the end of the $(N+1)$st interest period. Show that the amount accrued in the account at the end of the $(N+1)$st period is:

$$FV \equiv \left[ \frac{(1+r/k)^{N+1} - (1+r/k)}{r/k} \right] P, \tag{2.81}$$

where $N$ is the total number of payments into the account.

b) We now explore the future values associated with the following two plans for receiving payment.

   i. *Plan A.* Assume that Plan A begins officially at the start of next year with payments of $A$ starting at the end of the first interest period of next year. Show that the total amount she would accrue at the end of the $(n+1)$st year is:

$$FV_A \equiv \left[ \frac{(1+r/k)^{(n+1)k} - (1+r/k)}{r/k} \right] A.$$

   ii. *Plan B.* Under Plan B, your grandmother receives payments of $B$ with the choice of officially starting at the beginning of the $(q+1)$st year after Plan A starts and the first payment disbursing at the end of the first interest period of the official starting year. Show that the total amount she would accrue by the end of the $(n+1)$st year is

$$FV_B \equiv \left[ \frac{(1+r/k)^{[(n+1)-q]k} - (1+r/k)}{r/k} \right] B,$$

   where $q = 1, 2, \ldots$. Note that for $q = 0$, the two options coincide.

c) **(Choosing Between Plans A and B)** Naturally, since Plan B starts out later than Plan A and both have the same last-payment date, the payment amount of Plan B has to be higher than that of Plan A, i.e., $B > A$. Suppose that the account's interest rate exceeds a threshold as follows:

$$r > k \left[ (B/A)^{1/q} - 1 \right].$$

   i. Show that there is no $n$ such that the amounts accrued under both options are equal by the end of the $(n+1)$st year.
   ii. Show that Plan A is superior to Plan B, i.e., prove $FV_A > FV_B$.

**2.33. (Relating Present and Future Values of a Generalized Annuity)** Using

$$S_n = \sum_{\ell=0}^{n-1} \left[ \prod_{j=0}^{\ell} \left( 1 + \frac{r_{n+1-j}}{k} \right) \right] \mathcal{P}_{n-\ell},$$

verify the formula

$$\mathcal{A}_n = \frac{S_n}{\prod_{j=1}^{n} \left( 1 + \frac{r_j}{k} \right)},$$

where $n = 1, 2, \ldots, r_j > 0$ for $j = 1, \ldots, n$, and $r_{n+1} = 0$.

**2.34. (Bonds)** Given a coupon bond described by Equation (2.76) on page 68, find the future value at maturity of the bond's cash flow.

**2.35. (Bonds)** Show that for a coupon bond, its yield to maturity ($r_Y$), current yield ($r$), and coupon rate ($r_C$) have the following relationships:

a) A bond trades at a discount if and only if $r_Y > r > r_C$.
b) A bond trades at a premium if and only if $r_Y < r < r_C$.

# References

[1] Bodie, Z., Kane, A., Marcus, A.: Investments, 9th edn. McGraw-Hill Irwin, New York (2011)

[2] Brealey, R., Myers, S., and Allen, F.: Principles of Corporate Finance. McGraw-Hill Irwin, New York (2011)

[3] Brown, S., Kritzman, M.: Quantitative Methods for Financial Analysis. Dow Jones-Irwin, Homewood (1990)

[4] Chaplinsky, S., Doherty, P., Schill, M.: Methods of evaluating mergers and acquisitions. Note Number UVA-F-1274. University of Virginia Darden Business Publishing (2000)

[5] Choudhry, M.: Fixed-Income Securities and Derivatives Handbook. Bloomberg Press, New York (2005)

[6] Davis, M.: The Math of Money. Copernicus Books, New York (2001)

[7] Gordon, M.J.: Dividends, earnings and stock prices. Rev. Econ. Stat. **41**, 99 (1959)

[8] Guthrie, G, Lemon, L.: Mathematics of Interest Rates and Finance. Prentice Hall, Upper Saddle River (2004)

[9] Hull, J.: Options, Futures, and Other Derivatives, 7th edn. Pearson Prentice Hall, Upper Saddle River (2009)

[10] Kellison, S.: The Theory of Interest, 2nd edn. Irwin McGraw-Hill, Boston (1991)

[11] Koller, T., Goedhart, M., Wessels, D.: Valuation: Measuring and Managing the Value of Companies. Wiley, Hoboken (2010)

[12] L.E.K. Consulting, LLC: Discounted Cash Flow Valuation Primer. L.E.K. Consulting, Chicago (2003)

[13] Lovelock, D., Mendel, M., Wright, A.: An Introduction to the Mathematics of Money. Springer, New York (2007)

[14] Meserve, B.: Fundamental Concepts of Algebra. Dover, New York (1981)

[15] Muksian, R.: Mathematics of Interest Rates, Insurance, Social Security, and Pensions. Prentice Hall, Upper Saddle River (2003)

[16] Reilly, F., Brown, K.: Investment Analysis and Portfolio Management. South-Western Cengage Learning, Mason (2009)

[17] Wang, X.: A simple proof of Descartes's Rule of Signs. Am. Math. Mon. **111**, 525 (2004)

[18] Williams, J.B.: The Theory of Investment Value. Harvard University Press, Cambridge (1938). Reprinted in 1997 by Fraser Publishing

# Chapter 3
# Markowitz Portfolio Theory

In 1952 Harry Markowitz [17] pioneered a Nobel Prize-winning[1] mathematical model that showed how to distribute an initial capital across a collection of risky securities to create an efficient portfolio, namely, one with the least risk given an expected return and largest expected return given a level of portfolio risk.

We shall first present the assumptions of the Markowitz portfolio model and the model's formulas for the expected return rate and risk of a portfolio (Section 3.1). These formulas are then applied in detail over a single period to two-security portfolios (Section 3.2) and extended to $N$ securities to show how to obtain efficient portfolios (Sections 3.3–3.4). This is followed by an application of the theory to determining the global minimum-variance portfolio and the diversified portfolio, resulting in the Mutual Fund Theorem (Section 3.5). In Section 3.6 we introduce an investor's utility function. Given an infinite collection of efficient portfolios, we illustrate how optimizing an investor's expected utility function leads to the selection of an efficient portfolio that maximizes the investor's satisfaction relative to the investor's risk tolerance. The chapter then concludes with portfolio diversification and the issue of systematic and unsystematic risk (Section 3.7). We shall see that the major contributor to the risk of a portfolio with a sufficiently large number of securities from across the marketplace is not the risks of the securities, but the movements of the securities' returns relative to each other (their covariances).

## 3.1 Markowitz Portfolio Model: The Setup

In this section, we overview the key concepts, assumptions, and quantities of the Markowitz portfolio model and introduce some needed notation.

---

[1] Harry Markowitz, Merton Miller, and William F. Sharpe shared the 1990 Nobel Prize in Economic Sciences. Markowitz won for his work on portfolio selection (see Press Release at Novelprize.org).

© Arlie O. Petters and Xiaoying Dong 2016

A.O. Petters, X. Dong, *An Introduction to Mathematical Finance with Applications*, Springer Undergraduate Texts in Mathematics and Technology, DOI 10.1007/978-1-4939-3783-7_3

We begin with the market assumptions assumed throughout the book:

*Unless stated to the contrary, assume a US market environment and the following ideal market conditions hold:*

➤ *Investors:* all investors are rational, i.e., they make financial decisions that maximize their expected satisfaction with possible wealth gains in the face of the risks these possible gains require.

➤ *Equilibrium:* supply equals demand.

➤ *No arbitrage:* no-arbitrage opportunities exist, which means intuitively that there is no opportunity to make a costless, riskless profit. For an arbitrage, none of your funds is required, loans would be settled with interest, and a profit is still guaranteed. However, for mathematical modeling purposes, a broader and more precise definition of arbitrage is used later; see Definition 7.1 (page 334).

➤ *Access to information:* rapid availability of accurate information on securities exists.

➤ *Efficiency:* a security's price adjusts quickly to new information, so its current price reflects all known information impacting the security, which includes information about the past and expected future behavior of the security.

➤ *Liquidity:* any number of units of a security can be bought and sold quickly.

➤ *No transaction costs:* transaction costs are assumed to be negligible compared to the value of the trades and so are ignored.

➤ *No taxes:* transactions occur without taxation.

➤ *Borrowing/lending:* borrowing and lending are at the risk-free rate r.[2]

We now turn to the setup of the Markowitz model. First, a *portfolio* is a collection of different securities, each of whose value is a specific percentage of the entire portfolio's value.

*Throughout the chapter, all portfolios consist only of risky securities, i.e., securities whose future return rates cannot be predicted with certainty, and each portfolio has at least two securities.*

Chapter 4 will explore the consequences of adding a risk-free asset to a portfolio with a sufficiently large number of risky securities. Furthermore, given the vast array of different risky securities, it may be helpful to the reader to keep stocks or ETFs in mind specifically.

---

[2] No lending or borrowing of money will be done in the current chapter, but it will be part of the modeling in Chapter 4, which generalizes the Markowitz model.

### 3.1.1 Security Return Rates

Let $N \geq 2$ be an integer and consider a portfolio of $N$ securities with unit prices $S_1(t),\ldots,S_N(t)$ at time $t$. Assume that the liability of each security is limited to its value, i.e., for $i = 1,\ldots,N$, we have $S_i(t) \geq 0$ with $S_i(t) = 0$ corresponding to bankruptcy. We shall exclude the latter situation:

> *Unless stated to the contrary, assume that each security has a positive price at every moment of time: $S_i(t) > 0$ for $i = 1,\ldots,N$.*

We shall fix an investment time interval $[t_0,t_f]$, where $t_0 \geq 0$ is the present time, $t_f$ is a future time, and $\tau = t_f - t_0$ is the length of the investment interval. Intuitively, the return rate of a security (or portfolio) for the given investment interval measures what comes back to you beyond your initial investment. By (2.2) on page 19, the return rate from $t_0$ to $t_f$ of the $i$th risky security is the following random percentage:

$$R_i(t_0,t_f) \quad = \quad \underbrace{\frac{S_i(t_f) - S_i(t_0)}{S_i(t_0)}}_{\text{capital-gain return}} \quad + \quad \underbrace{\frac{D_i(t_0,t_f)}{S_i(t_0)}}_{\text{dividend yield}}, \quad (3.1)$$

where $i = 1,\ldots,N$. Here $D_i(t_0,t_f) \geq 0$ is the per-unit total cash dividend from the $i$th security during the time interval $[t_0,t_f)$. Furthermore, the return rate is unchanged when the total number of units of the $i$th security is included; see Section 2.2.3 for more. The amount that comes back to you for the investment interval is then a percentage $R_i(t_0,t_f)$ of the initial investment $S_i(t_0)$:

$$R_i(t_0,t_f)\, S_i(t_0), \qquad\qquad (i = 1,\ldots,N).$$

We shall see that the return rates of the securities are the core quantities from which all the other Markowitz portfolio inputs are calculated.

**Notation.** In most of this chapter, we shall consider security return rates over a fixed investment time interval $[t_0,t_f]$ and so the following simpler notation will be used:

$$R_i = R_i(t_0,t_f), \qquad i = 1,\ldots,N.$$

In the formula (3.1) for $R_i$, the futures price $S_i(t_f)$ is a discrete random variable in models like binomial trees (Chapter 5), while in some continuous-time models, it is lognormal (Chapter 6). Additionally, the future dividend $D_i(t_0,t_f)$ is also random since it is typically unknown. However, in most applications in the book, we shall model the dividend as a known percentage of the security's unit price at $t_0$:

$$D_i(t_0,t_f) = q_i\,\tau\, S_i(t_0), \qquad i = 1,\ldots,N,$$

where $q_i$ is the (assumed known) annual dividend yield rate of the $i$th security.

*A basic assumption about the securities' returns is that the covariance matrix $V$ of the return rates $R_1, R_2, \ldots, R_N$ is invertible:* the financial implication is that *there is no redundant security in the portfolio*, i.e., no security with a return rate that is a linear combination of the others. To see this, suppose for illustration that

$$R_1 = a_2 R_2 + \cdots a_N R_N.$$

Write the covariance matrix as

$$V = \begin{bmatrix} \sigma_{11} & \sigma_{12} & \cdots & \sigma_{1N} \\ \sigma_{21} & \sigma_{22} & \cdots & \sigma_{2N} \\ \vdots & \vdots & \ddots & \vdots \\ \sigma_{N1} & \sigma_{N2} & \cdots & \sigma_{NN} \end{bmatrix},$$

where $\sigma_{ij} = \text{Cov}(R_i, R_j) = \sigma_{ji}$, and consider the first column of $V$. The top entry in the first column expands to

$$\begin{aligned} \sigma_{11} = \text{Cov}(R_1, R_1) &= \text{Cov}(R_1, a_2 R_2 + \cdots + a_N R_N) \\ &= a_2 \, \text{Cov}(R_1, R_2) + \cdots + a_N \, \text{Cov}(R_1, R_N) \\ &= a_2 \sigma_{12} + \cdots + a_N \sigma_{1N}. \end{aligned}$$

Similarly, for the other entries $\sigma_{21}, \ldots, \sigma_{N1}$, we obtain

$$\begin{bmatrix} \sigma_{11} \\ \sigma_{21} \\ \vdots \\ \sigma_{N1} \end{bmatrix} = \begin{bmatrix} a_2 \sigma_{12} + \cdots + a_N \sigma_{1N} \\ a_2 \sigma_{22} + \cdots + a_N \sigma_{2N} \\ \vdots \\ a_2 \sigma_{N2} + \cdots + a_N \sigma_{NN} \end{bmatrix}.$$

In other words, the first column of $V$ is a linear combination of the other columns:

$$c_1 = a_2 c_2 + \cdots + a_N c_N,$$

where $c_i$ is the $i$th column of $V$. This contradicts $V$ being invertible.

Now, a covariance matrix $V$ is always semi-positive definite since for all $x = (x_1, \ldots, x_N)$ in $\mathbb{R}^N$, we have

$$x^T V x = \text{Var}(x_1 R_1 + \cdots + x_N R_N) \geq 0.$$

Additionally, the eigenvalues of a semi-positive definite matrix are nonnegative. Since the determinant of $V$ is the product of its eigenvalues and because $V$ is invertible (so $\det V > 0$), the eigenvalues of $V$ are all positive. Hence, *the covariance matrix $V$ of security return rates is positive definite.*[3]

---

[3] The theoretical importance of the positive definite property will be seen in Section 3.3.

## 3.1.2 *What About Multivariate Normality of Security Return Rates?*

An assumption about the Markowitz model that is often encountered in text-books (in part, maybe because students are very familiar with the normal distribution) is the multivariate normal condition on the securities' return rates:

> Multivariate normality condition: The $N$-tuple of return rates $(R_1, \ldots, R_N)$ of the risky securities in a portfolio has a joint multivariate normal density with a positive definite covariance matrix. In other words, any linear combination $\sum_{i=1}^{N} a_i R_i$ of the return rates is a normal random variable; hence, each $R_i$ is normal.

The validity of this assumption has been widely debated. One immediate criticism is as follows: since $S_i(t_f) + D_i(t_0, t_f) > 0$, the return rate satisfies

$$R_i = \frac{S_i(t_f) + D_i(t_0, t_f)}{S_i(t_0)} - 1 > -1.$$

But, if $R_i$ is normal, then it has a nonzero probability of satisfying $R_i \leq -1$, which is inconsistent with a positive security price and nonnegative dividend. However, the issue depends on the length $\tau$ of the time interval for which one is considering the return rate. It is typically assumed that the normality of security and portfolio return rates holds for a sufficiently short time span $\tau$. See Bodie, Kane, and Marcus [1, pp. 139–153] for an elementary introduction as well as the research paper [15] by Levy and Duchin.

*The multivariate normality condition is actually not necessary for Markowitz mean-variance analysis*—see Markowitz [18] and Markowitz and Blay [19]—and so will not be enforced. Readers are also referred to the insightful reviews by Goldberg [6] and Levy [14], which provide excellent guides to the book [19].

### 3.1.3 *Investors and the Efficient Frontier*

Assume the following about investors:

➤ Investors assess a portfolio only through its *expected return rate* and *risk* and agree on the joint distribution of the securities' return rates from $t_0$ to $t_f$.

➤ Investors are *risk averse*, i.e., for a portfolio with a given risk, investors demand the largest possible expected return rate *and* for a portfolio with a given expected return rate, they demand the least possible risk.

By construction, risk-averse investors are interested only in portfolios that are efficient. An *efficient portfolio* is a portfolio having simultaneously the smallest possible risk for its given level of expected return rate *and* the largest possible expected return rate for its given level of risk. The collection of all efficient portfolios is called the *efficient frontier*. The efficient frontier contains infinitely many efficient portfolios and each represents a different risk-return tradeoff. We shall show how the Markowitz model is applied to determine the efficient frontier of a two-security portfolio (Section 3.2.2) and then the general case of a portfolio with $N$ securities. We shall show how the Markowitz model yields that the more one spreads a portfolio's capital across different risky securities, the more the portfolio's risk is reduced (Sections 3.2.3 and 3.7), lending theoretical support to "don't put all your eggs in one basket."

Finally, the place where an investor positions her portfolio on the efficient frontier will have to do with her *utility function*, which indicates her satisfaction with the risk-return tradeoff. In other words, she will rank potential returns in the face of the potential risks it takes to realize those returns in such a way as to maximize her expected satisfaction or happiness (utility). In other words, we assume that an investor will seek an *optimal portfolio*, i.e., an efficient portfolio that maximizes her expected utility function—see Section 3.6 for more.

### 3.1.4 The One-Period Assumption, Weights, and Short Selling

We now turn to the setup of a portfolio in the Markowitz model.

**One-Period Assumption**

Assume that today, denoted by $t_0$, we use an initial investment capital

$$\mathcal{V}_P(t_0) > 0$$

to create a portfolio by distributing the entire capital among $N$ different preselected risky securities. We shall show how the Markowitz model addresses the issue of how to allocate the initial capital among different securities. However, the details of how to screen the marketplace to preselect the $N$ different risky securities for the portfolio are beyond the scope of this book. For the practicalities of portfolio management, readers are referred to, for example, Grinold and Kahn [9] and Reiley and Brown [22]. Furthermore, it is possible that not all the preselected securities will be used. Some securities can have a zero percent assigned to them, which we refer to as having *no position* or being *flat* in the security.

Now suppose that we will hold the portfolio until a future end date $t_f$. During the investment period $[t_0, t_f]$, we abide by

*The one-period assumption: from the current time $t_0$ to the final time $t_f$, we make no change to the total number of securities, the type of securities, or the number of units of any security in the portfolio.*

## Weights and Number of Units of the Security

Assume that the percentage of the initial capital $V_P(t_0)$ invested in the $i$th security is $w_i$, which is called the *weight* of the $i$th security. Since the entire initial capital $V_P(t_0)$ is distributed across the $N$ securities, the weights of all the securities add up to 100%:

$$w_1 + \cdots + w_N = 1. \tag{3.2}$$

The column vector

$$w = [w_1 \ w_2 \ \ldots \ w_N]^T$$

is called the *portfolio weight vector* at $t_0$. Note that $w \neq \mathbf{0}$. The amount $V_P(t_0)$ invested in the $i$th security is then $w_i V_P(t_0)$ and the total investment spreads among the securities are follows:

$$V_P(t_0) = w_1 V_P(t_0) + \cdots + w_N V_P(t_0).$$

Because the amount of the initial capital $V_P(t_0)$ to be used to purchase units of the $i$th security at $t_0$ is $w_i V_P(t_0)$, the number of units the money buys is

$$n_i = \frac{w_i V_P(t_0)}{S_i(t_0)}, \qquad i = 1,\ldots,N, \tag{3.3}$$

where $S_i(t_0)$ is the price of the $i$th security at $t_0$. Equivalently, the cost of $n_i$ units of the $i$th security is $n_i S_i(t_0)$. Note that *a non-integer number of units of a security is allowed.*

The initial value of the portfolio can then be expressed as the sum of the costs of the various securities, where a security's cost is a product of cost per unit and the number of units:

$$V_P(t_0) = n_1 S_1(t_0) + \cdots + n_N S_N(t_0). \tag{3.4}$$

Since the portfolio is constructed at $t_0$ by obtaining the following specific numbers of units of the $N$ securities,

$$n(t_0) = (n_1,\ldots,n_N), \tag{3.5}$$

this vector (3.5) is called the *trading strategy* of the portfolio at $t_0$. The *value of the trading strategy* at $t_0$ is defined to be the initial capital $V_P(t_0)$ as expressed in (3.4). By the one-period assumption, the number of units of each security is held fixed to the end date $t_f$, i.e., the trading strategy is held constant during the period.

**Example 3.1.** Assume that today an initial capital of $5,000 is used to create a three-security portfolio with 20% of the money in stock 1, 30% in stock 2, and 50% in stock 3. Suppose that the current share prices of the stocks are, respectively, $40, $70, and $10. Then the current trading strategy of the portfolio is to buy the following numbers of shares of stocks 1, 2, and 3, respectively:

$$n_1 = \frac{0.2 \times \$5,000}{\$40} = 25, \quad n_2 = \frac{0.3 \times \$5,000}{\$70} = 21.43, \quad n_3 = \frac{0.5 \times \$5,000}{\$10} = 250.$$

$\square$

### Short Selling

At this stage, you may be implicitly assuming that the number of units and weight of a security are nonnegative real numbers. However, we apply no such restriction. This is because *we assume that each security in the portfolio is obtained by either buying, short selling, or taking no trading position (being flat).* When you buy $n_i$ units of a security, you are adding redundant securities to your portfolio and so we represent this position by $n_i > 0$. When you sell $n_i$ units of a security, your portfolio has $n_i$ fewer units of the security and we express this position by $n_i < 0$. When you do not hold a security, we represent that position by $n_i = 0$. In general, to close or liquidate a buy (respectively, sell) position in $n_i$ units of a security, you must sell (respectively, buy) $n_i$ units of the security. The weight $w_i$ corresponding to a position of $n_i$ units in a security will have the same sign as $n_i$.

*Short selling* securities is a selling of securities that varies in its details depending on the type of security. The most common example is *short selling a stock*, where you sell a certain number of shares of a stock borrowed from a broker. You will almost definitely need to have a margin (a certain amount of required funds) in your account in case you are unable to return the borrowed shares. You close the stock short sale by buying back the shares of the given stock. The rationale behind short selling a stock is that you hope to make a profit from a nontrivial decrease in the stock's price. If you sell the borrowed shares of stock for $50 per share and the share price drops to $45 a month later, then you can use your proceeds to buy back the shares and still have a payoff of $5 per share (excluding transaction fees). When you buy back and return the borrowed shares, you are said to have *closed* the short position.

Next, consider an *option*, i.e., a legal contract between two parties whereby one party (the *issuer/writer*) sells to the other (the *holder*) the right, but not the obligation, to buy from or sell to the issuer a fixed amount of a security (e.g., stock) at a preagreed price (called the *strike price* or *exercise price*) on or by a preagreed date (called the *expiration date*). In particular, a *call option* is a legal contract between a buyer (holder) and seller (issuer) granting the holder the right, but not the obligation, to *buy* a stipulated amount of the asset from the issuer at the strike price on or by the expiration date.

Short selling an option contract is *issuing* an option contract. For example, short selling an *equity call option*, i.e., a call option contract on a stock means that you are obligated to sell 100 shares of the stock at the strike price if the option is exercised and you are assigned the exercise. Specifically, when an equity option is exercised by a holder, the exercise is randomly assigned (in the USA, by the Options Clearing Corporation) to a market participant who short sold the same call option (i.e., same underlying stock, strike price, and expiration). If you are assigned, then you close the short position by obtaining 100 shares of the stock (if you do not already have them) and sell each share at the strike price. If you were not assigned yet, then you can close the position by buying back the exact call contract (same stock, strike price, and expiration). In general, you close a buy position in an option by selling the exact option and close a short-sell position by buying back the exact option.

In our Markowitz context, for any proper subset of securities in a portfolio that are short sold, we always use the proceeds along with the initial capital to purchase the units of the remaining securities. *We shall not consider a portfolio where all its securities are short sold.*

Now, at $t_0$ our portfolio has $n_1,\ldots,n_N$ units, respectively, of securities 1 through $N$. When $n_i > 0$, it means that $n_i$ units of the $i$th security are bought at time $t_0$, while for $n_i < 0$, the interpretation is that $n_i$ units of the security are short sold at $t_0$. When no action is taken on the $i$th security, we write $n_i = 0$.

Furthermore, by (3.3) the $i$th weight can be written as

$$w_i = \frac{n_i\,S_i(t_0)}{V_P(t_0)}, \qquad i = 1,\ldots,N. \tag{3.6}$$

Since $S_i(t_0) > 0$ and $V_P(t_0) > 0$, we see from (3.6) that the weight $w_i$ has the same sign as $n_i$. We then interpret the sign of the weights as:

➤ $w_i > 0$ means buy $n_i$ units of the $i$th security at time $t_0$ (long position).

➤ $w_i = 0$ means neither buy nor sell the ith security at time $t_0$ (flat position).

➤ $w_i < 0$ means short sell $n_i$ units of the $i$th security at time $t_0$ (short position).

Equation (3.4) also shows that the sum of the weights is still unity as in (3.2), even if a proper subset of weights is negative:

$$w_1 + \cdots + w_N = 1.$$

When no short selling is used to construct the portfolio, the weights satisfy

$$w_i \geq 0 \quad \text{(no short selling)}, \qquad\qquad i = 1, \ldots, N.$$

Consequently, in the absence of short selling, since the weights are nonnegative and sum to unity, we always have

$$0 \leq w_i \leq 1, \qquad\qquad i = 1, \ldots, N.$$

**Example 3.2.** Suppose that you identified a stock and a call option contract on the stock to create a portfolio at $t_0$. Assume that the positions you take in these securities are to long (buy) $\Delta(t_0)$ shares of the stock priced at $S(t_0)$ per share and to short (sell) the call option contract on the same stock, where the call is sold at price $C(t_0)$ per share of the stock. In practice, call option contracts are typically based on 100 units of the underlier, but for modeling purposes, it is simpler to quote the call price per unit of the underlier.

Since long positions are an inflow of securities into the portfolio and short positions are an outflow, they are represented by positive and negative signs, respectively, when tallying the total value of a portfolio. The portfolio's value at $t_0$ is then

$$V_P(t_0) = S(t_0)\Delta(t_0) \; - \; C(t_0). \tag{3.7}$$

Equivalently, you can obtain (3.7) if you unwind the two positions. Specifically, liquidate (sell) the $\Delta(t_0)$ shares of the stock at $t_0$, which yields a cash inflow of $S(t_0)\Delta(t_0)$, and close your short position on the call (assuming it was not assigned at $t_0$). The latter means that you buy back a call contract on the same stock and with the exact strike price and expiration date. The latter yields a cash outflow of $C(t_0)$, which is represented mathematically as $-C(t_0)$. The total value of the portfolio at $t_0$ is then the net sum of these cash flows, which gives (3.7).

The trading strategy that created this portfolio is

$$\boldsymbol{n}(t_0) = (\Delta(t_0), -1).$$

In other words, short selling the call contract brings in proceeds of $C(t_0)$, which when added to the initial capital $V_P(t_0)$ in (3.7) yields the funds $S(t_0)\Delta(t_0)$ to buy $\Delta(t_0)$ shares of the stock.                                        □

**Example 3.3.** Suppose that we selected stocks 1, 2, and 3 to create a portfolio. Assume that the current prices of the stocks are:

$40  per share for stock 1

$70  per share for stock 2

$10  per share for stock 3

Starting with an initial capital of $5,000, we create a portfolio by distributing the money across the three stocks using the following weights:

$$w_1 = -20\%, \qquad w_2 = 50\%, \qquad w_3 = 70\%.$$

Note that $w_1 + w_2 + w_3 = 1$.

Let us interpret the meaning of the above weight assignment. First, the weights state that we use the following trading strategy to form the portfolio:

$$n_1 = \frac{-0.2 \times \$5,000}{\$40} = -25 \qquad \text{(short sell 25 shares of stock 1)}$$

$$n_2 = \frac{0.5 \times \$5,000}{\$70} = 35.7143 \qquad \text{(buy 35.7143 shares of stock 2)}$$

$$n_3 = \frac{0.7 \times \$5,000}{\$10} = 350 \qquad \text{(buy 350 shares of stock 3)}.$$

Specifically, we create the portfolio by first short selling 25 shares of stock 1 to obtain $1,000, which is $-20\%$ of the initial capital (since the proceeds are from a short position). Adding these proceeds to the initial capital, we then have $6,000 to invest in stocks 2 and 3.

Weight $w_2$ tells us that we take 50% of the initial capital $5,000 to buy 35.7143 shares of stock 2. This reduces the initial capital to $2,500. Weight $w_3$ indicates that we use 70% of the initial capital, i.e., $3,500, to purchase 350 shares of stock 3. Though the cost of the purchase exceeds the $2,500 remaining from the initial capital, we have an extra $1,000 from the short sale to cover the purchase. Finally, observe that (3.4) holds (to two decimal places)

$$\$5,000.00 = -25 \times \$40 + 35.7143 \times \$70 + 350 \times \$10.$$

Note that using four decimal places in 35.7143 gives the desired accuracy, while 35.71 would yield $4,999.70$.                                                                    □

To avoid confusion about whether short selling is or is not allowed in a portfolio, we shall abide by the following convention:

> *Unless stated to the contrary, assume that short selling is allowed in constructing a portfolio.*

### 3.1.5 Expected Portfolio Return Rate

Consider the investment interval $[t_0, t_f]$. As with a security, Equation (2.2) on page 19 yields that the return rate (the percentage you get back) of the portfolio for the given investment interval comes from the percentage change in its market value and from the total cash dividend it pays:

$$R_P(t_0, t_f) = \frac{\mathcal{V}_P(t_f) - \mathcal{V}_P(t_0)}{\mathcal{V}_P(t_0)} + \frac{D_P(t_0, t_f)}{\mathcal{V}_P(t_0)}, \tag{3.8}$$

where $\mathcal{V}_P(t_0)$ and $\mathcal{V}_P(t_f)$ are the values of the portfolio at the start and end of the investment period, and $D_P(t_0, t_f)$ is the total cash dividend during $[t_0, t_f)$ from all the securities in the portfolio. For the investment interval, the amount you get back beyond the initial investment is the percentage $R_P(t_0, t_f)$ of the initial investment $\mathcal{V}_P(t_0)$:

$$R_i(t_0, t_f)\, S_i(t_0), \qquad\qquad (i = 1, \dots, N).$$

Since the investment interval $[t_0, t_f]$ is fixed, we use the following simpler notation:

$$R_P = R_P(t_0, t_f).$$

By (3.4), the initial portfolio market value is

$$\mathcal{V}_P(t_0) = n_1\, S_1(t_0) + \cdots + n_N\, S_N(t_0)$$

and the final portfolio market value is

$$\mathcal{V}_P(t_f) = n_1\, S_1(t_f) + \cdots + n_N\, S_N(t_f). \tag{3.9}$$

The total cash dividend received from the securities during $[t_0, t_f)$ is

$$D_P(t_0, t_f) = n_1\, D_1(t_0, t_f) + \cdots + n_N\, D_N(t_0, t_f). \tag{3.10}$$

Using (3.9), the portfolio return rate (3.8) from $t_0$ to $t_f$ becomes

$$
\begin{aligned}
R_P &= \sum_{i=1}^{N} \frac{n_i}{\mathcal{V}_P(t_0)} \left( S_i(t_f) - S_i(t_0) + D_i(t_0, t_f) \right) \\
&= \sum_{i=1}^{N} \frac{n_i\, S_i(t_0)}{\mathcal{V}_P(t_0)} \left( \frac{S_i(t_f) - S_i(t_0) + D_i(t_0, t_f)}{S_i(t_0)} \right).
\end{aligned}
$$

By (3.1) and (3.6), the portfolio return rate is the weighted sum of the securities' return rates:

$$R_P = \sum_{i=1}^{N} w_i\, R_i. \tag{3.11}$$

The expected (or mean) portfolio return rate for the period $[t_0, t_f]$ is then

$$\mu_P = \mathbb{E}(R_P) = \sum_{i=1}^{N} w_i \, \mu_i, \tag{3.12}$$

where

$$\mu_i = \mathbb{E}(R_i) \tag{3.13}$$

is the expected return rate of the $i$th security. In (3.12), *every weight $w_i$ is assumed nonrandom, unless stated otherwise, and each expected return rate $\mu_i$ is assumed finite and known.* The weights are to be determined in the search for an efficient portfolio, while the expected return rates $\mu_1, \ldots, \mu_N$ are typically estimated using samples of historical return rates.

**Example 3.4.** Consider a portfolio with two stocks over a time interval $[t_0, t_f]$ corresponding to the next month. Denote the expected return rate over the next month of a stock in the portfolio by

$$\mu_{\text{monthly}} = \mathbb{E}\left( R(t_0, t_f) \right).$$

Suppose that $t_0 = t'_0 > t'_1 > \cdots > t'_n$ denotes a sample of end-of-month to end-of-month trading dates for $m$ consecutive months from the present date $t_0$ into the past. Denote the corresponding historical return rates as follows:[4]

$$\widehat{R}(t'_n, t'_{n-1}), \quad \widehat{R}(t'_{n-1}, t'_{n-2}), \quad \ldots, \quad \widehat{R}(t'_1, t'_0).$$

The stock's theoretical ensemble expected monthly return rate $\mu_{\text{monthly}}$ is estimated using the time average of the monthly return data (see Exercise 3.12):

$$\overline{R}_{\text{monthly}} = \frac{1}{n} \sum_{j=1}^{n} \widehat{R}(t'_j, t'_{j-1}).$$

$\square$

**Remark 3.1.** The question of whether a data-sample time average $\overline{R}_{\text{monthly}}$ is an accurate approximation of the theoretical expected value $\mu_{\text{monthly}}$ is a rather thorny issue. Should we have used weekly or daily data rather than monthly? Or, how many return rates should we have used? In general, the sampling frequency (daily, monthly, quarterly, etc.) and sample size used to estimate statistical financial quantities will depend on the context and typically become a debatable matter. See Graham, Smart, and Megginson [8, p. 212] for a discussion. $\square$

---

[4] Note that the returns are notationally the reverse of the case for future times: for past times $t'_j < t'_{j-1}$, we use $\widehat{R}(t'_j, t'_{j-1})$ instead of $R(t_{j-1}, t_j)$, which is for future times $t_{j-1} < t_j$.

### 3.1.6 Portfolio Risk

We saw that since the securities in the portfolio have random futures prices and (in general) random future dividends, the portfolio return rate $R_P$ is also random. This uncertainty in the return rate gives rise to the portfolio's risk. In other words, portfolio risk is determined by how much the possible values of the random portfolio return rate $R_P$ can spread away from the expected return rate $\mu_P$. More precisely, the *risk of a portfolio* is modeled in Markowitz theory by the standard deviation of its return rate:

$$\sigma_P = \sqrt{\text{Var}(R_P)} = \sqrt{\mathbb{E}\left((R_P - \mu_P)^2\right)}. \tag{3.14}$$

Sometimes the standard deviation of a random variable $X$ will be called the *volatility* of $X$. Note that some authors refer to the variance $\sigma_P^2$, rather than the volatility $\sigma_P$, as the risk of the portfolio, but we shall not abide by that usage. The larger the portfolio risk $\sigma_P$, the more the portfolio return rate $R_P$ can spread away from the expected return rate $\mu_P$, while the smaller $\sigma_P$ becomes, the closer $R_P$ concentrates to $\mu_P$.

The portfolio variance $\sigma_P^2$ is given as follows explicitly in terms of the volatilities of the individual securities' return rates and the covariances of these return rates:

$$\sigma_P^2(w_1, \ldots, w_N) = \sum_{i=1}^{N} w_i^2 \sigma_i^2 + 2 \sum_{1 \leq i < j \leq N} w_i w_j \sigma_{ij}, \tag{3.15}$$

where

$$\sigma_i = \sqrt{\text{Var}(R_i)} = \sqrt{\mathbb{E}\left((R_i - \mu_i)^2\right)} \tag{3.16}$$

is the volatility of $R_i$ and

$$\sigma_{ij} = \text{Cov}(R_i, R_j) = \mathbb{E}\left((R_i - \mu_i)(R_j - \mu_j)\right) = \sigma_{ji}$$

is the covariance of $R_i$ and $R_j$, with the variance of $R_i$ denoted by

$$\sigma_{ii} = \sigma_i^2.$$

In (3.15), *the volatilities $\sigma_i$'s and covariances $\sigma_{ij}$'s are assumed to be finite and known.*

### 3.1.7 Risks and Covariances of the Portfolio's Securities

The Markowitz model (3.15) of portfolio risk implies that the risk of a portfolio comes from two sources: the weighted contributions of the variances $\sigma_i^2$, where

$i = 1, \ldots, N$, of the individual securities's returns and the covariances $\sigma_{ij}$, where $1 \leq i < j \leq N$, between the returns of all pairs of the securities. Putting this informally in terms of the portfolio variance, we have

$$\left\{ \text{portfolio variance} \right\} = \left\{ \begin{array}{c} \text{weighted sum of} \\ \text{the} \\ \textit{securities' variances} \end{array} \right\} + \left\{ \begin{array}{c} \text{weighted sum of} \\ \text{the} \\ \textit{securities' covariances} \end{array} \right\}.$$

In the current section, we review some basic results about the volatility $\sigma_i$ and covariances $\sigma_{ij}$ of the securities in a portfolio.

## Risk of a Security

At the start of the Section 3.1 we informally defined a "risky security" as one whose return rates cannot be predicted with certainty. In Markowitz setting, risk is modeled more narrowly using volatility. Specifically, the *risk of the $i$th security* is modeled by $\sigma_i$ for $i = 1, \ldots, N$. In other words, the $i$th security's risk is a measure of how much the random return rate $R_i$ spreads about the security's expected return rate $\mu_i$. The risks $\sigma_1, \ldots, \sigma_N$, are usually estimated from historical data (see Exercise 3.12).

**Remark 3.2.** The portfolio risk (3.14) and security risk (3.16) measure how the random return rates $R_P$ and $R_i$ disperse above and below the expected return rates $\mu_P$ and $\mu_i$, respectively. Some have argued that risk should instead be modeled by how much the return rates spread below the mean (downside risk) or the probability of the return rates being below some threshold (shortfall probability). Later we shall explore three portfolio risk measures: the Sortino ratio (Section 4.2.2), the maximum drawdown (Section 4.2.3), and the value-at-risk (Section 4.2.5). See Grinold and Kahn [9, pp. 41–46] for a critique of these risk measures relative to the standard deviation. Throughout our text, however, the primary measure of risk shall be the standard deviation. □

**Example 3.5.** (We continue with Example 3.4 on page 95.) Let us consider the variance of a stock in the portfolio using $n$ historical monthly consecutive return rates over times $t_0 = t_0' > t_1' > t_2' > \cdots > t_{n-1}' > t_n'$. The data runs from the past time $t_n'$ to the present time $t_0'$: $\quad \widehat{R}(t_n', t_{n-1}'), \quad \ldots, \quad \widehat{R}(t_2', t_1'), \quad \widehat{R}(t_1', t_0')$. The theoretical variance $\sigma_{\text{month}}^2$ of the stock for the next month is estimated using the sample monthly variance $\widehat{\sigma}_{\text{monthly}}^2$, which can be expressed as

$$\widehat{\sigma}_{\text{monthly}}^2 = \frac{1}{n-1} \left[ \left( \sum_{j=1}^n \widehat{R}^2(t_j', t_{j-1}') \right) - n \overline{R}_{\text{monthly}}^2 \right]. \tag{3.17}$$

Exercise 3.12 involves a computation of $\widehat{\sigma}_{\text{monthly}}^2$. □

Mathematically, the risk of a security is nonnegative. What happens when the risk vanishes? Intuitively, the more tightly concentrated the return rates are about their mean, the smaller the risk and vice versa. The following proposition expresses this insight for discrete random variables, which is the context when dealing with data:

**Proposition 3.1.** *Let $X$ be a discrete random variable with finite mean $\mu = \mathbb{E}(X)$ and finite variance $\sigma^2 = \mathrm{Var}(X)$. Then the volatility $\sigma$ of $X$ vanishes if and only if $X = \mu$ almost surely, i.e., with probability 1.*

Motivated by Proposition 3.1, we define a *risky security* to be one whose return rate has a positive volatility.

*Unless stated to the contrary, assume $\sigma_i > 0$ for $i = 1, \ldots, N$.*

### Covariance Between Two Securities

The other contributor to portfolio risk is the weighted collection of the covariances of the random return rates of the securities in the portfolio. We shall see in Section 3.7 that, for a portfolio with a sufficiently large number of different securities, the weighted sum of the covariances of the securities dominates the weighted sum of the securities' volatilities. In this section, we instead review some basic insights into the covariance of a pair of return rates of risky securities using the associated correlation coefficient. The correlation coefficient of the return rates $R_i$ and $R_j$ of the $i$th and $j$th securities will be written as follows:

$$\rho(R_i, R_j) = \frac{\sigma_{ij}}{\sigma_i \sigma_j} = \rho_{ij},$$

The respective risks of the two securities are $\sigma_i$ and $\sigma_j$, while the covariance is $\sigma_{ij} = \mathrm{Cov}(R_i, R_j)$.

In general, the *(Pearson) correlation coefficient* of random variables $X$ and $Y$ with nonzero volatilities $\sigma_X$ and $\sigma_Y$, respectively, is defined by

$$\rho(X, Y) = \frac{\mathrm{Cov}(X, Y)}{\sigma_X \sigma_Y}.$$

A basic property is

$$-1 \leq \rho(X, Y) \leq 1.$$

The correlation coefficient is a unit-independent measure of how $X$ and $Y$ vary relative to each other, which is not the case for the covariance $\mathrm{Cov}(X, Y)$, where we assume the units carry a positive sign. This is a special case of the following general property showing how the covariance and correlation coefficient behave under affine transformations of $X$ and $Y$ (see Exercise 3.16 on page 146):

$$\text{Cov}(aX + b, cY + d) = ac \, \text{Cov}(X, Y),$$

and for $ac \neq 0$,

$$\rho(aX + b, cY + d) = \pm\rho(X, Y) \qquad \text{with } +1 \text{ if } ac > 0 \text{ and } -1 \text{ if } ac < 0.$$

The securities' correlation coefficients $\rho_{ij}$, where $i, j = 1, \ldots, N$, are assumed known and are estimated using historical data. In the discrete context of estimates from data, we can think intuitively of $\sigma_{ij}$ as measuring the degree to which the random return rates of the $i$th and $j$th securities move together along a straight line. The following proposition captures this insight more generally:

**Proposition 3.2.** *Let $X$ and $Y$ be discrete random variables. Then $\rho(X, Y) = \pm 1$ if and only if $Y = aX + b$ with probability 1, where $a \neq 0$ and $b$ are real numbers with $a > 0$ corresponding to $\rho(X, Y) = +1$ and $a < 0$ to $\rho(X, Y) = -1$.*

In other words, for any two discrete random variables $X$ and $Y$, which is the setting when working with data, the closer the random variables are to being *perfectly positively correlated*, i.e., $\rho(X, Y) = 1$, the more likely the values of $X$ and $Y$ are close to a positively sloped line. Similarly, the closer the random variables are to having a *perfectly negative correlation*, $\rho(X, Y) = -1$, the more the values of $X$ and $Y$ concentrate near a negatively sloped line. For $-1 < \rho(X, Y) < 1$, we then have varying degrees of how much the values of $X$ and $Y$ spread away from a straight line.

A pair of random variables $X$ and $Y$ is called *uncorrelated* if $\rho(X, Y) = 0$. This means that a scatter plot of possible values of the two random variables has no linear relationship and so may appear as a cluster of independent points or points showing an overall nonlinear relationship. Indeed, if $X$ and $Y$ are independent, then they are uncorrelated. The converse is not true since it is possible for two uncorrelated random variables to be dependent, though their dependence will be nonlinear.

As noted earlier, the covariances $\sigma_{ij}$ and correlation coefficients $\rho_{ij}$ are estimated using historical data. There are several Web resources that compute the correlation coefficients of pairs of stocks.[5]

**Example 3.6.** (We continue with Example 3.4 on page 95.) Let $R_{\text{month}}^A$ and $R_{\text{month}}^B$ be the random return rates over the next month of the two stocks in the portfolio. Write their covariance and correlation coefficient as follows:

$$\sigma_{\text{month}}^{AB} = \text{Cov}\left(R_{\text{month}}^A, R_{\text{month}}^B\right), \qquad \rho_{\text{month}}^{AB} = \frac{\sigma_{\text{month}}^{AB}}{\sigma_{\text{month}}^A \, \sigma_{\text{month}}^B}.$$

---

[5] For example, see the Correlation Tracker at: http://www.sectorspdr.com/correlation.

The theoretical covariance $\sigma^{AB}_{\text{month}}$ and correlation coefficient $\rho^{AB}_{\text{month}}$ of the stocks for the next month are estimated using the sample data as follows:

$$\widehat{\sigma}^{AB}_{\text{monthly}} = \frac{1}{n-1} \sum_{j=1}^{n} \left( \widehat{R}^A(t'_j, t'_{j-1}) - \overline{R}^A_{\text{monthly}} \right) \left( \widehat{R}^B(t'_j, t'_{j-1}) - \overline{R}^B_{\text{monthly}} \right)$$

and

$$\widehat{\rho}^{AB}_{\text{monthly}} = \frac{\widehat{\sigma}^{AB}_{\text{monthly}}}{\widehat{\sigma}^A_{\text{monthly}} \, \widehat{\sigma}^B_{\text{monthly}}},$$

respectively. Here $\widehat{\sigma}^A_{\text{monthly}}$ and $\widehat{\sigma}^B_{\text{monthly}}$ are the estimated volatilities of the stocks for the next month. Note that for historical data, we have $t'_j < t'_{j-1}$. Exercise 3.12 illustrates these estimates.

### 3.1.8 Expectation and Volatility of Portfolio Log Return

Throughout this section assume that the portfolio pays no dividend.

**Portfolio Log Return for a Single Time Horizon**

Suppose that a nondividend-paying asset has a current value of \$500,000 and its value 1 month later is \$505,000. At what interest rate[6] was \$500,000 continuously compounded to reach the value \$505,000 after 1 month? In other words, we want an interest rate $r$ such that

$$\$505,000 = e^{r \times \left(\frac{1}{12}\right)} \times \$500,000,$$

which yields an annual interest rate of

$$r = 12 \ln \left( \frac{\$505,000}{\$500,000} \right).$$

Expressing the interest as a monthly rate, we obtain

$$r \times \frac{1}{12} = \ln \left( \frac{\$505,000}{\$500,000} \right) = 0.995\% \quad \text{(per month)}.$$

The continuously compounded interest rate,

$$\ln \left( \frac{\$505,000}{\$500,000} \right) = 0.995\%,$$

is called the *log return of the asset over a month*.

---

[6] Recall from Chapter 2 that interest rate is per annum, unless otherwise stated.

Using the logarithmic return is advantageous not only for continuous compounding but also for discrete compounding problems. In general, for a nondividend-paying portfolio with initial value $V_P(t_0)$ and end-of-period value $V_P(t_f)$, we define the *portfolio log return* from $t_0$ to $t_f$ to be

$$r^L_{P,\text{span}} = \ln\left(\frac{V_P(t_f)}{V_P(t_0)}\right).$$

Intuitively, the log return $r^L_{P,\text{span}}$ is the continuously compounded rate across $[t_0, t_f]$ that transforms $V_P(t_0)$ to $V_P(t_f)$, namely,

$$V_P(t_f) = V_P(t_0)\, e^{r^L_{P,\text{span}}}.$$

In the example above, $t_0$ is the current time, $t_f$ is 1 month from now, and the log return over $[t_0, t_f]$ is

$$r^L_{P,\text{span}} = 0.995\%.$$

For the typical situations we shall consider, the initial value $V_P(t_0)$ is known and $V_P(t_f)$ is a random future value, so $r^L_{P,\text{span}}$ is random. In this case, the expectation and volatility of $r^L_{P,\text{span}}$ over $[t_0, t_f]$ are denoted by

$$\mu^L_{P,\text{span}} = \mathbb{E}\left(r^L_{P,\text{span}}\right), \qquad \sigma^L_{P,\text{span}} = \sqrt{\text{Var}\left(r^L_{P,\text{span}}\right)}.$$

**Portfolio Log Return for Different Time Horizons**

Now, let us consider how the portfolio log returns, as well as their expectation and volatility, behave under different time horizons. Divide $[t_0, t_f]$ into $n$ equal-length subintervals:

$$[t_0, t_1], \quad [t_1, t_2], \quad \cdots, \quad [t_{n-1}, t_n], \qquad \left(t_n = t_f, \quad t_j - t_{j-1} = \frac{\tau}{n} \equiv h_n\right).$$

Let $V_P(t_j)$ be the value of the portfolio at time $t_j$, where $j = 1, \ldots, n$. The log return from $t_{j-1}$ to $t_j$ is given by

$$r^L_{P,\text{prd}}(t_j) = \ln\left(\frac{V_P(t_j)}{V_P(t_{j-1})}\right).$$

These log returns over the subintervals relate to the log return over the entire interval as follows:

$$r_{P,\text{span}}^{L} = \ln\left(\frac{\mathcal{V}_P(t_n)}{\mathcal{V}_P(t_0)}\right) = \ln\left(\frac{\mathcal{V}_P(t_1)}{\mathcal{V}_P(t_0)}\frac{\mathcal{V}_P(t_2)}{\mathcal{V}_P(t_1)}\cdots\frac{\mathcal{V}_P(t_n)}{\mathcal{V}_P(t_{n-1})}\right) = \sum_{j=1}^{n}\ln\left(\frac{\mathcal{V}_P(t_j)}{\mathcal{V}_P(t_{j-1})}\right)$$

$$= \sum_{j=1}^{n} r_{P,\text{prd}}^{L}(t_j). \tag{3.18}$$

Before considering the expectation and volatility over different time horizons, we need two assumptions:

➤ For the first assumption, recall that the present value $\mathcal{V}_P(t_0)$ of the portfolio is known and reflects all available information about the asset at the current time $t_0$.[7] Suppose that the value of the portfolio $\mathcal{V}_P(t_1)$ at the future date $t_1$ is dependent on information that is not known today. Similarly, the value $\mathcal{V}_P(t_2)$ at date $t_2$ is assumed to be based on information not known on date $t_1$. For this reason, *we assume that the random log returns*

$$r_{P,\text{prd}}^{L}(t_1), \quad \ldots, \quad r_{P,\text{prd}}^{L}(t_n)$$

*are uncorrelated random variables across the n successive subintervals:*

$$\text{Cov}\left(r_{P,\text{prd}}^{L}(t_k), r_{P,\text{prd}}^{L}(t_j)\right) = 0,$$

where $k \neq j$ and $k,j = 1,\ldots,n$.

➤ For the second assumption, recall that by the one-period assumption, we do not make changes to the portfolio during the time interval $[t_0, t_f]$. Consequently, *we assume that the probability distributions of the log returns* $r_{P,\text{prd}}^{L}(t_j)$ *across the future subintervals are identically distributed*. Write the expectation and volatility of the log returns for each subinterval as

$$\mu_{P,\text{prd}}^{L} = \mathbb{E}\left(r_{P,\text{prd}}^{L}(t_j)\right), \qquad \sigma_{P,\text{prd}}^{L} = \sqrt{\text{Var}\left(r_{P,\text{prd}}^{L}(t_j)\right)},$$

where $j = 1,\ldots,n$.

We can now relate the expectation of the log return for the full time interval $[t_0, t_f]$, where $t_f = t_n$, to the expected log return over the $n$ subintervals. Using (3.18), we obtain

$$\mu_{P,\text{span}}^{L} = \mathbb{E}\left(\ln\left(\frac{\mathcal{V}_P(t_n)}{\mathcal{V}_P(t_0)}\right)\right) = \mathbb{E}\left(\sum_{j=1}^{n} r_{P,\text{prd}}^{L}(t_j)\right) = n\,\mu_{P,\text{prd}}^{L}. \tag{3.19}$$

Hence, by (3.19) the expected portfolio log return over the time span $\tau$ is $n$ times the expected portfolio log return over a period. For instance, an annual

---

[7] This issue pertains to the Weak Efficient Market Hypothesis.

expected portfolio log return is 12 times the monthly expected portfolio log return. In other words, *longer time horizons have a higher expected log return compared to shorter ones.*

For the variance of the log return over $[t_0, t_f]$ with uncorrelated and identically distributed log returns across consecutive subintervals, we obtain

$$\left(\sigma_{P,\text{span}}^L\right)^2 = n \left(\sigma_{P,\text{prd}}^L\right)^2,$$

which yields the volatility:

$$\sigma_{P,\text{span}}^L = \sqrt{n}\, \sigma_{P,\text{prd}}^L. \tag{3.20}$$

The volatility of the portfolio log return increases as the square root of the number of periods in the time horizon increases. *Longer time horizons will have higher volatility than shorter time horizons.* In particular, the volatility of an annual portfolio log return is $\sqrt{12}$ times the volatility of a monthly portfolio log return.

### Relating the Portfolio Log Return and Portfolio Return Rate

Lastly, the portfolio log return $r_{P,\text{span}}$ relates to the portfolio return rate $R_P$ as follows:

$$r_{P,\text{span}} = \ln\left(\frac{V_P(t_n)}{V_P(t_0)}\right) = \ln\left(1 + R_P\right),$$

where $t_n = t_f$. Taylor expanding the log return yields

$$r_{P,\text{span}} = R_P - \frac{R_P^2}{2} + \frac{R_P^3}{3} - \frac{R_P^4}{4} + \cdots \qquad \text{for } |R_P| < 1. \tag{3.21}$$

Consequently, if $|R_P|$ is sufficiently small, then the log return and the return rate are approximately equal:

$$r_{P,\text{span}} \approx R_P, \qquad\qquad |R_P| \ll 1.$$

However, over a sufficiently long time span, there is no guarantee that the risk of a portfolio will not increase or the magnitude of its return rate will be sufficiently small. In this case, we cannot treat the portfolio log return and portfolio return rate as approximately equal.

### Looking Ahead

The remainder of the chapter will show how to apply the infrastructure of the Markowitz model to selecting the weights that produce efficient portfolios as well as optimal portfolios, which is an efficient portfolio determined by maximizing an investor's expected utility function (Section 3.6). It will also be shown how diversification reduces portfolio risk.

We shall first study portfolios with two risky securities and then extend to $N$ risky securities. The original Markowitz treatment (e.g., [18]) employs a geometric approach using iso-mean and iso-variance curves and surfaces. This approach, though, gets extremely intricate in higher dimensions. Instead, we shall carry out the efficient portfolio selection using the optimization approach by Merton [20].

## 3.2 Two-Security Portfolio Theory

Suppose that you have an amount of money $\mathcal{V}_P(t_0)$ to create a portfolio with two risky securities. After an appropriate amount of due diligence, you have identified two risky securities to buy today $t_0$ and hold in the portfolio until future date $t_f$. The fundamental portfolio question we shall address is

*What percentage of the money $\mathcal{V}_P(t_0)$ should you allocate today to each security to create an efficient portfolio?*

In other words, find the weights that give rise to an efficient portfolio.

Let us first collect some quantities needed for Markowitz model in a setting of two securities, say, securities 1 and 2. The respective random return rates are $R_1$ and $R_2$. The securities' expected return rates $\mu_1$ and $\mu_2$, risks $\sigma_1 > 0$ and $\sigma_2 > 0$, and correlation coefficient $\rho_{12} = \rho$ are assumed to have been estimated (either by you or by a company offering such a service). For real data the two securities will typically *not* have identical expected return rates ($\mu_1 = \mu_2$), identical risks ($\sigma_1 = \sigma_2$), or return rates with either a perfectly positive correlation ($\rho = 1$) or perfectly negative correlation ($\rho = -1$). These mathematical idealizations will not be assumed by default and will explicitly be identified when considered:

*Unless stated to the contrary, assume that $\mu_1 \neq \mu_2$, $\sigma_1 \neq \sigma_2$, and $-1 < \rho < 1$. Without loss of generality, we assume $\sigma_2 > \sigma_1 > 0$.*

Though it may not be obvious at this stage, the assumption on $\rho$ is actually already included in the Markowitz setup. We show this and more in Section 3.2.1.

### 3.2.1 Preliminaries

Before addressing how to determine the weights that produce an efficient port-folio, we shall present the needed quantities conveniently in matrix form.

For two securities, the weights are given by a vector

$$w = \begin{bmatrix} w_1 \\ w_2 \end{bmatrix}.$$

Since $w_1 + w_2 = 1$, we have $w \neq 0$. Furthermore, letting

$$e = \begin{bmatrix} 1 \\ 1 \end{bmatrix},$$

the sum-of-weight condition becomes

$$1 = w_1 + w_2 = w^T e.$$

The random return rates $R_1$ and $R_2$, the expected return rates $\mu_1$ and $\mu_2$, and the covariances of $R_1$ and $R_2$ can also be compiled conveniently in matrix form:

$$R = \begin{bmatrix} R_1 \\ R_2 \end{bmatrix}, \qquad \mu = \mathbb{E}(R) = \begin{bmatrix} \mu_1 \\ \mu_2 \end{bmatrix}, \qquad V = \begin{bmatrix} \sigma_1^2 & \rho\sigma_1\sigma_2 \\ \rho\sigma_1\sigma_2 & \sigma_2^2 \end{bmatrix},$$

where $\rho\sigma_1\sigma_2 = \sigma_{12} = \mathrm{Cov}(R_1, R_2)$. Here $w$, $R$, $\mu$, and $V$ are the *weight vector, return rate vector, expected return rate vector*, and *covariance matrix*, respectively, of the portfolio. The portfolio expected return (i.e., $\mu_P(w) = \mathbb{E}(R_P)$) and portfolio risk (i.e., $\sigma_P(w) = \sqrt{\mathrm{Var}(R_P)}$) are

$$\mu_P(w) = w_1\mu_1 + w_2\mu_2 = \mu^T w \tag{3.22}$$

$$\sigma_P(w) = \sqrt{w_1^2\sigma_1^2 + w_2^2\sigma_2^2 + 2w_1w_2\rho\sigma_1\sigma_2} = \sqrt{w^T V w}. \tag{3.23}$$

The assumption of distinct expected return rates is captured by

$$\mu_1 \neq \mu_2 \qquad \Longleftrightarrow \qquad \mu \neq ce \text{ for any constant } c.$$

In particular, we have $\mu \neq 0$ (i.e., for $c = 0$) and $\mu \neq e$. In general, $\mu$ and $e$ are linearly independent vectors.

We saw that the covariance matrix $V$ of security returns is positive definite (page 86). Recall that an $n \times n$ real symmetric matrix is defined to be *positive definite* if

$$x^T A x > 0 \quad \text{for all} \quad x \neq 0 \quad \text{in} \quad \mathbb{R}^n. \tag{3.24}$$

This is equivalent to each leading principal submatrix having a positive determinant (Sylvester's criterion). In particular, the positive definiteness of $V$ is equivalent to

$$\sigma_1^2 > 0 \quad \text{and} \quad \det V = (1 - \rho^2)(\sigma_1 \sigma_2)^2 > 0.$$

Since $\sigma_2 > \sigma_1 > 0$ and $V$ is positive definite within our setup of the Markowitz model, we also have $-1 < \rho < 1$.

We now collect some basic consequences of the positive definiteness of the covariance matrix $V$ and introduce additional notation:

➤ Since $w \neq 0$, by (3.24) the positive definiteness of $V$ implies that *the portfolio always has risk*, i.e.,

$$\sigma_P(w) = \sqrt{w^T V w} > 0 \tag{3.25}$$

for all portfolio weight vectors $w$.

➤ The positive definiteness of $V$ implies that $V$ is invertible and its symmetric inverse

$$V^{-1} = \frac{1}{\det V} \begin{bmatrix} \sigma_2^2 & -\rho \sigma_1 \sigma_2 \\ -\rho \sigma_1 \sigma_2 & \sigma_1^2 \end{bmatrix}, \tag{3.26}$$

is positive definite.

➤ We introduce the following quantities which are needed later:

$$A \equiv e^T V^{-1} e = \frac{\sigma_1^2 + \sigma_2^2 - 2\rho \sigma_1 \sigma_2}{\det V} \tag{3.27}$$

$$B \equiv \mu^T V^{-1} e = \frac{(\sigma_2^2 - \rho \sigma_1 \sigma_2) \mu_1 + (\sigma_1^2 - \rho \sigma_1 \sigma_2) \mu_2}{\det V} \tag{3.28}$$

$$C \equiv \mu^T V^{-1} \mu = \frac{\sigma_1^2 \mu_2^2 + \sigma_2^2 \mu_1^2 - 2\rho \sigma_1 \sigma_2 \mu_1 \mu_2}{\det V} \tag{3.29}$$

$$AC - B^2 = \frac{(\mu_1 - \mu_2)^2}{\det V}. \tag{3.30}$$

Let us consider their signs. Since $e \neq 0$ and $\mu \neq 0$, the positive definiteness of $V^{-1}$ gives

$$A > 0 \tag{3.31}$$

$$C > 0. \tag{3.32}$$

For the cross term $B$, we cannot draw any conclusion about its sign at this stage. However, Equation (3.32) and the linear independence of $\mu$ and $e$ yield[8]

---

[8] If $B\mu - Ce = 0$, then the linear independence of $\mu$ and $e$ implies $B = C = 0$. This contradicts $C > 0$. Hence, $B\mu - Ce \neq 0$.

$$B\mu - Ce \neq 0.$$

Consequently, the positive definiteness of $V^{-1}$ gives

$$(B\mu - Ce)^T V^{-1} (B\mu - Ce) > 0.$$

This implies (Exercise 3.19)

$$AC - B^2 > 0. \tag{3.33}$$

Note that we could have obtained (3.33) directly from (3.30) since $\mu_1 \neq \mu_2$ and $\det V > 0$. However, our previous arguments for the signs of $A$, $C$, and $AC - B^2$ were independent of the detailed expression for these quantities. This will allow us to carry these results over to the $N$-security case.

### 3.2.2 Efficient Frontier of a Two-Security Portfolio

To determine an efficient portfolio for two securities, we must find a weight vector for the two securities such that the resulting portfolio has the minimum risk for a given portfolio expected return rate *and* the maximum expected return rate for a given portfolio risk. The set of all such portfolio weight vectors will determine the two-security efficient frontier.

We shall approach this problem by first determining a weight that gives the smallest portfolio risk given an expected portfolio return rate $\mu$. More precisely, we seek a portfolio weight vector $w = w_\mu$ that solves the following optimization problem:

$$\text{minimize} \quad \sigma_P(w) = \sqrt{w^T V w} \tag{3.34}$$

$$\text{subject to} \quad w^T e = 1 \quad \text{and} \quad w^T \mu = \mu, \tag{3.35}$$

where $-\infty < \mu < \infty$. There is no constraint $w_i \geq 0$ on the weights, i.e., unlimited short selling is allowed. The constraints (3.35) are two equations in the two unknowns $(w_1, w_2)$ and have a unique solution $w = w_\mu$. In fact, Equation (3.35) is given explicitly by

$$w_1 + w_2 = 1, \qquad w_1\mu_1 + w_2\mu_2 = \mu,$$

or, equivalently, in matrix form by

$$\underbrace{\begin{bmatrix} 1 & 1 \\ \mu_1 & \mu_2 \end{bmatrix}}_{\mathbb{K}} \begin{bmatrix} w_1 \\ w_2 \end{bmatrix} = \begin{bmatrix} 1 \\ \mu \end{bmatrix}. \tag{3.36}$$

Because $\mu \neq e$, which is equivalent to $\mu_1 \neq \mu_2$, we see that

$$\det \mathbb{K} = \mu_2 - \mu_1 \neq 0.$$

Consequently, Equation (3.36) has a unique solution:

$$\boldsymbol{w}_\mu \equiv \begin{bmatrix} w_{1,\mu} \\ w_{2,\mu} \end{bmatrix} = \mathbb{K}^{-1} \begin{bmatrix} 1 \\ \mu \end{bmatrix} = \frac{1}{\mu_2 - \mu_1} \begin{bmatrix} \mu_2 - \mu \\ \mu - \mu_1 \end{bmatrix}.$$

Now, since $w_1 + w_2 = 1$, we can write the weights in terms of a single parameter $w_1 = w$:

$$\boldsymbol{w} = \begin{bmatrix} w_1 \\ w_2 \end{bmatrix} = \begin{bmatrix} w \\ 1-w \end{bmatrix}.$$

The portfolio weight vector $\boldsymbol{w}_\mu$ corresponding to $\mu$ is then written simply as

$$\boldsymbol{w}_\mu = \begin{bmatrix} w_\mu \\ 1 - w_\mu \end{bmatrix} = \frac{1}{\mu_1 - \mu_2} \begin{bmatrix} \mu - \mu_2 \\ \mu_1 - \mu \end{bmatrix}. \tag{3.37}$$

Note that since unlimited short selling is permitted, we have $-\infty < w_\mu < \infty$ and so $-\infty < \mu < \infty$. If short selling is forbidden (i.e., $0 \leq w_\mu \leq 1$) and if we assume $\mu_1 < \mu_2$, then $\mu_1 \leq \mu \leq \mu_2$.

Though the expression of $\boldsymbol{w}_\mu$ in (3.37) is specific to the two-security case, we can actually cast $\boldsymbol{w}_\mu$ in the following form, which carries over to $N$-securities:

$$\boldsymbol{w}_\mu = \left( \frac{C - \mu B}{AC - B^2} \right) V^{-1} e + \left( \frac{\mu A - B}{AC - B^2} \right) V^{-1} \mu. \tag{3.38}$$

Establishing (3.38) is a rather lengthy computation (Exercise 3.20). Moreover, the form (3.38) is more complicated than (3.37) and its origin seems mysterious at this stage. However, it will appear naturally during the $N$-security efficient frontier analysis in Section 3.3.2 and allow us to link back readily to the two-security efficient frontier (see page 123).

Given that the constraint equations in (3.35) have a unique solution $\boldsymbol{w}_\mu$, this is the only portfolio weight vector available to solve (3.34). In other words, the solution $\boldsymbol{w}_\mu$ yields a unique portfolio risk $\sigma_P(\boldsymbol{w}_\mu)$ given a portfolio expected return rate $\mu_P(\boldsymbol{w}_\mu) = \mu$. The quantity $\sigma_P(\boldsymbol{w}_\mu)$ is then the minimum possible portfolio risk, being the only risk associated with $\mu$. However, this is not enough to decide whether $\boldsymbol{w}_\mu$ determines an efficient portfolio because we do not know whether $\mu_P(\boldsymbol{w}_\mu) = \mu$ is the maximum possible expected return rate given $\sigma_P(\boldsymbol{w}_\mu)$.

We shall show that $\mu$ will have to lie in a restricted range in order for $\boldsymbol{w}_\mu$ to give an efficient portfolio. Allowing the expected portfolio return $\mu$ to vary over $\mathbb{R}$, we shall see that the corresponding portfolio risk $\sigma_P(\boldsymbol{w}_\mu)$ traces out a branch of a hyperbola. The turning point on this branch will be the global minimum portfolio risk. The efficient frontier will be the curve segment from the turning point to the upper part of the given branch of the hyperbola. We now detail the computation of the efficient frontier.

First, to emphasize that the portfolio risk is a function of $\mu$, we also write

$$\sigma_P(\boldsymbol{w}_\mu) = \sigma_P(\mu).$$

Next, we express the portfolio variance as a quadratic in $\mu$. Since

$$\sigma_P^2(\mu) = \boldsymbol{w}_\mu^T \boldsymbol{V} \boldsymbol{w}_\mu = \left(\sigma_1^2 + \sigma_2^2 - 2\rho\sigma_1\sigma_2\right) w_\mu^2 + 2\left(\rho\sigma_1\sigma_2 - \sigma_2^2\right) w_\mu + \sigma_2^2,$$

substituting (using (3.37))

$$w_\mu = \frac{\mu - \mu_2}{\mu_1 - \mu_2}$$

yields

$$\sigma_P^2(\mu) = \mathbb{A}\mu^2 + \mathbb{B}\mu + \mathbb{C}, \tag{3.39}$$

where

$$\mathbb{A} = \frac{\sigma_1^2 + \sigma_2^2 - 2\rho\sigma_1\sigma_2}{(\mu_1 - \mu_2)^2}$$

$$\mathbb{B} = -2\frac{(\sigma_2^2 - \rho\sigma_1\sigma_2)\,\mu_1 + (\sigma_1^2 - \rho\sigma_1\sigma_2)\,\mu_2}{(\mu_1 - \mu_2)^2}$$

$$\mathbb{C} = \frac{\sigma_1^2\mu_2^2 + \sigma_2^2\mu_1^2 - 2\rho\sigma_1\sigma_2\mu_1\mu_2}{(\mu_1 - \mu_2)^2}.[-1pt]$$

Using Equations (3.27), (3.28), (3.29), and (3.30), we obtain

$$\mathbb{A} = \frac{A}{AC - B^2}, \qquad \mathbb{B} = -\frac{2B}{AC - B^2}, \qquad \mathbb{C} = \frac{C}{AC - B^2}.$$

Hence, the portfolio variance (3.39) becomes

$$\sigma_P^2(\mu) = \frac{A\mu^2 - 2B\mu + C}{AC - B^2}. \tag{3.40}$$

To identify the graph of (3.40), complete the square of the numerator to get

$$\sigma_P^2(\mu) = \frac{A}{AC - B^2}\left(\mu - \frac{B}{A}\right)^2 + \frac{1}{A}, \tag{3.41}$$

and introduce the following (upright) variables:

$$\sigma_P = \sigma_P(\mu), \qquad \mu_P = \mu.$$

Then (3.41) is equivalent to the equation of a hyperbola in the $(\sigma_P, \mu_P)$-plane:

$$\frac{\sigma_P^2}{\frac{1}{A}} - \frac{\left(\mu_P - \frac{B}{A}\right)^2}{\frac{AC-B^2}{A^2}} = 1, \tag{3.42}$$

Because $\sigma_P > 0$, the graph is the branch of a hyperbola opening to the right along the $\sigma_P$-axis. Each point on the graph is a portfolio risk-mean pair $(\sigma_P(\mu), \mu)$. Figure 3.1 illustrates a portion of the graph, which consists of solid and dotted curves. The turning point or vertex of the graph is called the *global minimum-variance portfolio*, denoted $(\sigma_G, \mu_G)$, and given by

$$\text{turning point} = (\sigma_G, \mu_G) = \left( \frac{1}{\sqrt{A}}, \frac{B}{A} \right). \tag{3.43}$$

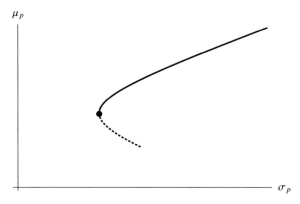

**Fig. 3.1** The solid curve (including ●) shows the Markowitz efficient frontier $M_{E,2}$ for a two-security portfolio. The turning point ● of the curve is the global minimum-variance portfolio $(\sigma_G, \mu_G)$. The efficient frontier curve and the dotted curve extend to infinity when unlimited short selling is allowed (since $-\infty < \mu < \infty$). If there is no short selling and $\mu_1 < \mu_G < \mu_2$, then $\mu_1 \le \mu \le \mu_2$ and the upper portion of the efficient frontier ends at security 2, while the dotted curve ends at security 1. Here $-1 < \rho < 1$, $\sigma_1 \ne \sigma_2$, and $\mu_1 \ne \mu_2$

Since the turning point is the furthest point to the left on the graph, the portfolio risk $\sigma_G$ is indeed the global minimum value of the portfolio risk function $\sigma_P(\mu)$ as $\mu$ varies over $\mathbb{R}$. Furthermore

*Unless stated to the contrary, assume that $\mu_G > 0$, so $B > 0$.*

This is a financially reasonable assumption because it requires the global minimum-variance portfolio to have a positive expected return rate; otherwise, no investor will be interested in the portfolio. Note that by (3.27) and (3.28), we obtain

$$\sigma_G = \frac{1}{\sqrt{A}} = \sqrt{\frac{(1 - \rho^2)(\sigma_1 \sigma_2)^2}{\sigma_1^2 + \sigma_2^2 - 2\rho\sigma_1\sigma_2}} \tag{3.44}$$

$$\mu_G = \frac{B}{A} = \frac{(\sigma_2^2 - \rho\sigma_1\sigma_2)\,\mu_1 + (\sigma_1^2 - \rho\sigma_1\sigma_2)\,\mu_2}{\sigma_1^2 + \sigma_2^2 - 2\rho\sigma_1\sigma_2}. \tag{3.45}$$

The portfolio weight vector producing $(\sigma_G, \mu_G)$ is given by (3.38)

$$w_G \equiv w_{\mu_G} = \left(\frac{C - \mu_G B}{AC - B^2}\right) V^{-1} e + \left(\frac{\mu_G A - B}{AC - B^2}\right) V^{-1}\mu = \frac{V^{-1}e}{A}. \quad (3.46)$$

Explicitly

$$w_G = \begin{bmatrix} w_G \\ 1 - w_G \end{bmatrix},$$

where

$$w_G = \frac{\sigma_2^2 - \rho\sigma_1\sigma_2}{\sigma_1^2 + \sigma_2^2 - 2\rho\sigma_1\sigma_2}, \qquad 1 - w_G = \frac{\sigma_1^2 - \rho\sigma_1\sigma_2}{\sigma_1^2 + \sigma_2^2 - 2\rho\sigma_1\sigma_2}. \quad (3.47)$$

Let us now determine the efficient frontier, which is the set of all efficient portfolios. Recall that an efficient portfolio is determined by a risk-mean pair $(\sigma_P, \mu_P)$, where $\sigma_P$ is the smallest possible portfolio risk for the expected return rate $\mu_P$ *and* $\mu_P$ is the largest possible expected return rate for the portfolio risk $\sigma_P$. Inspection of Figure 3.1 shows that the efficient frontier for the given plot consists of the solid curve, including the turning point.

More generally, given a portfolio expected return rate $\mu_P = \mu$, there is a unique portfolio risk $\sigma_P = \sigma_P(\mu)$ determined by (3.40) with $\sigma_P$ the minimum possible portfolio risk associated with $\mu_P$. However, given the risk $\sigma_P$, for the pair $(\sigma_P, \mu_P)$ to be efficient, we also need the expected return rate $\mu_P$ to be the maximum possible. Solving (3.40) for the expected return rate $\mu = \mu_P$ yields either no solution, one solution, or two solutions depending on whether $\sigma_P < \sigma_G$, $\sigma_P = \sigma_G$, or $\sigma_P > \sigma_G$, respectively. This is captured by Figure 3.1. In fact, the expected return rate solutions correspond in Figure 3.1 to the inter-section points of the vertical lines, $\sigma_P = $ constant, with the right branch of the hyperbola. For $\sigma_P = \sigma_G$, the unique intersection point determines the portfolio expected return $\mu_P = \mu_G$. Consequently, the turning point is an efficient port-folio. For $\sigma_P = $ constant $> \sigma_G$, there are two intersection points with the largest possible portfolio expected return rate determined by the upper intersection point. Indeed, the two-security efficient portfolios are given by the turning point and the upper part of the hyperbola.

Equations (3.40) and (3.43) and the discussion above show that the *Markowitz efficient frontier* $M_{E,2}$, i.e., the collection of all efficient two-security portfolios, is given by

$$M_{E,2} = \left\{ (\sigma_P, \mu_P) : \sigma_P = \sqrt{\frac{A\mu_P^2 - 2B\mu_P + C}{AC - B^2}}, \ \mu_P \geq \frac{B}{A} \right\}. \quad (3.48)$$

The set of portfolio weight vectors that give rise to $M_{E,2}$ is

$$W_{E,2} = \left\{ w : \ w = \left( \frac{C - \mu_P B}{AC - B^2} \right) V^{-1} e + \left( \frac{\mu_P A - B}{AC - B^2} \right) V^{-1} \mu \ , \ \mu_P \geq \frac{B}{A} \right\}.$$

$$(3.49)$$

Finally, observe from Figure 3.1 that the efficient frontier indicates theoretically that *to obtain a higher expected return rate the portfolio has to take on more risk.* Note, however, in a real-world setting, portfolio management involves more complexity due to transaction costs, taxes, the trading platform, etc.

**Example 3.7.** Suppose that you have $2,000 to invest in two stocks, say, stocks 1 and 2, which have current share prices of $40.25 and $35.10, respectively. From an analysis of historical data of the two stocks, suppose that

$$\mu_1 = 8\%, \quad \mu_2 = 12\%, \quad \sigma_1 = 9\%, \quad \sigma_2 = 15\%, \quad \rho = -0.5,$$

where these are annualized percentages. Using these two stocks, create an efficient portfolio that has an expected annual return rate of 20%.

**Solution.** The goal is to find a trading strategy[9] $(n_1, n_2)$ such that $n_1$ and $n_2$ are the respective number of shares of stocks 1 and 2 needed to build an efficient portfolio with $\mu_P = 0.2$.

The initial capital is $V_P(t_0) = \$2,000$. Let us now collect the quantities from Section 3.2.1 that are used in determining the portfolio weight vector. We shall employ the expressions of these quantities that involve vectors and matrices since that form carries over to the $N$-security analysis. First,

$$e = \begin{bmatrix} 1 \\ 1 \end{bmatrix}, \quad \mu = \begin{bmatrix} 0.08 \\ 0.12 \end{bmatrix}, \quad V = \begin{bmatrix} 0.0081 & -0.00675 \\ -0.00675 & 0.0225 \end{bmatrix},$$

and

$$V^{-1} = \begin{bmatrix} 164.609 & 49.3827 \\ 49.3827 & 59.2593 \end{bmatrix}.$$

Second

$$A = e^T V^{-1} e = 322.6337449$$

$$B = \mu^T V^{-1} e = 30.1563786$$

$$C = \mu^T V^{-1} \mu = 2.8549794$$

$$AC - B^2 = 11.7055327.$$

_____
[9] See (3.5) on page 89.

Third

$$\frac{C - \mu_P B}{AC - B^2} = -0.27135, \qquad \frac{\mu_P A - B}{AC - B^2} = 2.93625.$$

Note that the global minimum-variance portfolio,

$$\mu_G = \frac{B}{A} = 9.3\%, \qquad \sigma_G = \frac{1}{\sqrt{A}} = 5.6\%,$$

does not meet the requirement of the portfolio we are trying to create since $\mu_P = 20\% > \mu_G$.

The portfolio weight vector is actually given by

$$w = \left( \frac{C - \mu_P B}{AC - B^2} \right) V^{-1} e + \left( \frac{\mu_P A - B}{AC - B^2} \right) V^{-1} \mu = \begin{bmatrix} -2 \\ 3 \end{bmatrix}.$$

Consequently, we create the desired efficient portfolio using the following trading strategy:

$$n_1 = \frac{w_1 V_P(t_0)}{S_1(t_0)} = \frac{(-2) \times \$2,000}{\$40.25} = -99.379 \quad \text{(short sell 99.379 shares of stock 1)}$$

$$n_2 = \frac{w_2 V_P(t_0)}{S_2(t_0)} = \frac{3 \times \$2,000}{\$35.10} = 170.9402 \qquad \text{(buy 170.9402 shares of stock 2).}$$

In other words, to create the portfolio, short selling $n_1$ shares of stock 1 brings in

$$n_1 \times \$40.25 = 99.379 \times \$40.25 = \$4,000.$$

Adding this amount to the initial \$2,000 then allows one to buy $n_2$ shares of stock 2, which costs

$$n_2 \times \$35.10 = 170.9402 \times \$35.10 = \$6,000.$$

The decimal places are maintained simply for mathematical consistency with the amounts received from shorting and needed for purchasing. In an actual trading setting, an integer number of shares is traded. Also, note that, to obtain the required high expected portfolio return rate of 20%, the constructed efficient portfolio ends up with a risk much higher than that of the individual stocks:

$$\sigma_P = \sqrt{w^T V w} = 56.2\% \quad \gg \quad \max\{\sigma_1, \sigma_2\} = 15\%.$$

$\square$

### 3.2.3 Reducing Risk Through Diversification

The discussion so far makes no mention about the diversification of the two securities in the portfolio. We first discuss how diversification relates to the correlation coefficient between the return rates of two stocks. Using two separate online correlation coefficient Calculators,[10] we see that for the time span from August 10, 2010, to August 10, 2013, using adjusted closing prices, the for-profit stocks Apollo Group, which owns the University of Phoenix, and Strayer Education, Inc., which owns Strayer University, have a positive correlation of $\rho \approx 0.77$. Indeed, we would expect intuitively that on average two stocks from the same business sector have a nontrivial positive correlation. On the other hand, the correlation calculators output a negative correlation of $\rho \approx -0.5$ for American Airlines[11] and Exxon-Mobil. We also expect this intuitively since increases in oil prices benefit oil companies, but hurt airlines due to the resulting higher fuel cost. Finally, the correlation between the Apollo Group and American airlines was quite weak, namely, $\rho \approx 0.15$, which is expected intuitively since the online education sector and the airline industry do not compete with each other.

**Remark 3.3.** Bear in mind that there are always exceptions to the above simplistic intuition. Historical-data estimates of correlation coefficients (and other quantities) are part art and part science. The degree to which two stocks vary relative to each other is influenced by overall market movements during the time span of the data—e.g., a period of an overall market rise lifts the return rates of most stocks, creating positive correlations between them. Additionally, correlation estimates are affected by the data's sample size, sample frequency, etc.                                                                              □

Overall, as the correlation coefficient $\rho$ decreases from 1 down to $-1$, we are considering pairs of securities whose return rates covary less and less in the same direction, yielding more and more diversification in their return rates. The two-security efficient frontier in Figure 3.1 was plotted for varying weights $w$ and fixed values of $\mu_1, \mu_2, \sigma_1, \sigma_2$, and $\rho$. Suppose, instead, we consider various pairs of securities with the same $\mu_1, \mu_2, \sigma_1$, and $\sigma_2$, but different correlation coefficients, i.e., we allow $\rho$ to vary. To identify the associated efficient frontier, each $\rho$ in $(-1, 1)$, we shall vary the weight $w$ assuming no short selling, i.e., $0 \le w \le 1$. The portfolio risk is given by

---

[10] We used the free online Correlation Tracker tool at (www.sectorspdr.com) and Stock Correlation Calculator at Buyupside (www.buyupside.com/calculators).

[11] Note that the ticker symbol of American Airlines at the time was AAMRQ, which changed after the merger with US Airways.

$$\sigma_P(w) = \sqrt{w^2\sigma_1^2 + (1-w)^2\sigma_2^2 + 2w(1-w)\rho\sigma_1\sigma_2}.$$

Since $w(1-w) \geq 0$, we see that the portfolio risk decreases as $\rho$ decreases from 1 to $-1$ (the securities covary less and less in the same direction). In other words, *portfolio risk decreases as the diversification increases*.

Figure 3.2 illustrates the above situation. The curves from right to left depict the different efficient frontier curves for increasing diversification, i.e., as $\rho$ decreases from positive to negative values over the interval $(-1,1)$. Figure 3.2 then captures the benefit of diversification in the two-security setting, namely, increasing diversification creates efficient-frontier curves that push to the left, reducing the overall portfolio risk. Diversification in the $N$-security setting will be explored in Section 3.7.

**Is There a Riskless Portfolio with Two Risky Securities?**

The answer is affirmative if we drop the requirement that $\rho^2 < 1$ and consider a portfolio with two risky securities having perfect negative correlation:

$$\rho = -1,$$

which is an idealization. The portfolio variance becomes

$$\sigma_P^2(w) = w^2\sigma_1^2 + (1-w)^2\sigma_2^2 - 2w(1-w)\sigma_1\sigma_2 = (w\sigma_1 - (1-w)\sigma_2)^2,$$

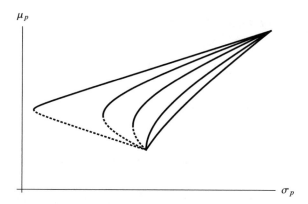

**Fig. 3.2** From left to right, the solid curves show typical efficient frontiers for a two-security portfolio as the correlation coefficient $\rho$ varies over $(-1,1)$ from negative to positive values. Each associated set of feasible portfolios for a given $\rho$ has risk and expected return rates determined by the union of the solid and dashed curves. The turning point of each curve identifies the efficient portfolio with the lowest risk for the given curve

which yields the portfolio risk for $w$ in $\mathbb{R}$: $\sigma_P(w) = |w\sigma_1 - (1-w)\sigma_2| \geq 0$. Hence, the smallest value of the portfolio risk is zero:

$$\sigma_P(w_*) = 0 \qquad \text{at} \qquad w_* = \frac{\sigma_2}{\sigma_1 + \sigma_2}.$$

*We now resume our assumption that $-1 < \rho < 1$.*

### Diversification: Two Securities Versus One Security

Is it better to put all your money in two risky securities versus one risky security? Intuitively, it would seem better to spread your money between the two risky securities to lower your risk, i.e., not to put all your eggs in one basket.

*We shall construct a portfolio with two uncorrelated risky securities having less risk than a portfolio consisting of either one of the securities.* In particular, we consider securities 1 and 2 with risks $\sigma_1 > 0$ and $\sigma_2 > 0$ and correlation coefficient

$$\rho = 0.$$

Then the two-security portfolio has variance given by

$$\sigma_P^2(w) = (\sigma_1^2 + \sigma_2^2)w^2 - 2\sigma_2^2 w + \sigma_2^2.$$

Moreover, Equation (3.44) yields the global minimum-variance portfolio:

$$\sigma_G = \frac{\sigma_1 \sigma_2}{\sqrt{\sigma_1^2 + \sigma_2^2}}, \qquad \mu_G = \frac{\sigma_2^2 \mu_1 + \sigma_1^2 \mu_2}{\sigma_1^2 + \sigma_2^2}.$$

Furthermore, by (3.47) the global minimum portfolio has weight

$$w_G = \frac{\sigma_2^2}{\sigma_1^2 + \sigma_2^2}, \qquad 1 - w_G = \frac{\sigma_1^2}{\sigma_1^2 + \sigma_2^2}. \tag{3.50}$$

Note that since $\sigma_1 > 0$ and $\sigma_2 > 0$, we have

$$0 < w_G < 1.$$

Consequently, the weight $w_G$ is not a short selling position.

Since

$$\frac{\sigma_1}{\sqrt{\sigma_1^2 + \sigma_2^2}} = \sqrt{\frac{\sigma_1^2}{\sigma_1^2 + \sigma_2^2}} < 1 \qquad \text{and} \qquad \frac{\sigma_2}{\sqrt{\sigma_1^2 + \sigma_2^2}} = \sqrt{\frac{\sigma_2^2}{\sigma_1^2 + \sigma_2^2}} < 1,$$

we see that the smallest value of the portfolio risk satisfies

$$0 < \sigma_G < \sigma_1 \qquad \text{and} \qquad 0 < \sigma_G < \sigma_2,$$

or, more compactly,

$$0 < \sigma_G < \min\{\sigma_1, \sigma_2\}.$$

Therefore

> *A portfolio of uncorrelated risky securities with its fraction $w_G$ of the initial capital invested in one security and $1 - w_G$ in the other will have less risk than either of the two securities. Moreover, this portfolio involves no short selling since $0 < w_G < 1$.*

In other words, spreading the investment capital strategically as above between the two uncorrelated securities yields a portfolio with less risk than a portfolio consisting of just one of the two securities.

## 3.3 Efficient Frontier for *N* Securities with Short Selling

Many of the ideas and quantities introduced for two securities (Section 3.2) carry over naturally to *N* securities in determining the efficient frontier. In this section, we shall compute the efficient frontier for an *N*-security portfolio without putting any restrictions on short selling. The only restriction on the weights will then be that they sum to unity. When there is no short selling, the efficient frontier is analytically more complex and usually presented by using numerical plots; see Section 3.4.

Suppose that, after researching a collection of different risky securities, you have identified *N* of them for which you are confident in your estimates of their expected return rates $\mu_i$, of their risks $\sigma_i$, and of the correlations $\rho_{ij}$ of the return rates of each pair of the securities. It would also be unrealistic for all of the securities to have the same expected return and risk and to have a perfect correlation between any pair of return rates. For these reasons

> *Unless stated to the contrary, for any N-security portfolio, assume:*
>
> - *The securities are risky: $\sigma_i > 0$ for $i = 1, \ldots, N$.*
>
> - *The expected returns $\mu_1, \ldots, \mu_N$, risks $\sigma_1, \ldots, \sigma_N$, and correlation coefficients $\rho_{ij}$, where $1 \leq i, j \leq N$ and $i \neq j$, are known and fixed.*
>
> - *None of the following occurs: identical expected returns*
>
> $$\mu_1 = \mu_2 = \cdots = \mu_N,$$
>
> *identical risks*
>
> $$\sigma_1 = \sigma_2 = \cdots = \sigma_N,$$
>
> *or perfect correlation $\rho_{ij} = \pm 1$ for any distinct pair $i, j$.*
>
> - *Unlimited short sales are allowed, i.e., $-\infty < w_i < \infty$.*

At an initial time $t_0$, you want to use an amount of money $\mathcal{V}_P(t_0)$ to purchase the $N$ risky securities. Our goal is to form an *efficient portfolio* out of these securities, i.e., a portfolio whose risk is the least for a given expected return and whose expected return is the most for a given level of risk. In other words, we shall determine *the percentage of $\mathcal{V}_P(t_0)$ that should be allocated today to each security to obtain an efficient portfolio.*

### 3.3.1 N-Security Portfolio Quantities in Matrix Notation

The portfolio weight vector and column of 1's are

$$
w = \begin{bmatrix} w_1 \\ w_2 \\ \vdots \\ w_N \end{bmatrix}, \qquad e = \begin{bmatrix} 1 \\ 1 \\ \vdots \\ 1 \end{bmatrix}.
$$

The summing of the weights to unity is expressed by

$$
1 = w_1 + \cdots + w_N = w^T e.
$$

The *weight space* of an $N$-security portfolio that allows for unlimited short selling is

$$
W_N = \left\{ w \in \mathbb{R}^N : w^T e = 1 \right\}. \tag{3.51}
$$

Note that $W_N$ is a line for $N = 2$ and plane for $N = 3$. In general, the space $W_N$ is an $(N-1)$-dimensional plane in $\mathbb{R}^N = \{(w_1, \ldots, w_N)\}$ passing through the $N$ standard unit basis vectors:

$$
e_1 = [1\,0\,0\ldots 0\,0]^T, \quad e_2 = [0\,1\,0\ldots 0\,0]^T, \quad \ldots, \quad e_N = [0\,0\,0\ldots 0\,1]^T.
$$

Of course, because a realistic weight space will not extend to infinity, it will be a proper subset of the mathematical space $W_N$.

The random return rates and expected return rates of the securities are

$$
R = \begin{bmatrix} R_1 \\ R_2 \\ \vdots \\ R_N \end{bmatrix}, \qquad \mu = \mathbb{E}(R) = \begin{bmatrix} \mu_1 \\ \mu_2 \\ \vdots \\ \mu_N \end{bmatrix}.
$$

The portfolio return rate and expected portfolio return rate can then be expressed as

$$R_P(w) = w^T R = \sum_{i=1}^{N} w_i R_i, \qquad \mu_P(w) = \mathbb{E}(w^T R) = w^T \mu = \sum_{i=1}^{N} w_i \mu_i.$$

The assumption that the securities do not all have the same expected return means that $\mu \neq ce$ for some constant $c$.

The covariance matrix of the return rates $R_1, \ldots, R_N$ is

$$V = \begin{bmatrix} \sigma_1^2 & \sigma_{12} & \cdots & \sigma_{1N} \\ & \sigma_2^2 & \cdots & \sigma_{2N} \\ & & \ddots & \vdots \\ & & & \sigma_N^2 \end{bmatrix},$$

where the entries below the diagonal are not shown since the covariance matrix is symmetric, $\sigma_{ij} = \text{Cov}(R_i, R_j) = \sigma_{ji}$. The matrix $V$ is invertible and so positive definite (see page 86). It follows that $V^{-1}$ is symmetric and positive definite. Estimating the covariance matrix using historical data then requires

$$N + \frac{N(N-1)}{2} \text{ estimates,}$$

which correspond to $N$ variances and $\frac{N(N-1)}{2}$ correlation coefficients. For example, one hundred stocks require $5,050$ estimates to determine $V$.

The *portfolio risk* is given by

$$\sigma_P(w) = \sqrt{w^T V w} > 0.$$

Note that the positivity follows since $V$ is positive definite and $w \neq 0$.

The quantities $A$, $B$, and $C$ that were introduced in the two-security case generalize naturally to an $N$-security portfolio. Essentially the exact arguments used for the two-security setting give the following (see (3.31), (3.32), and (3.33); page 106):

$$A = e^T V^{-1} e > 0 \tag{3.52}$$
$$B = \mu^T V^{-1} e \tag{3.53}$$
$$C = \mu^T V^{-1} \mu > 0 \tag{3.54}$$
$$AC - B^2 > 0. \tag{3.55}$$

It will be argued in the next section that it is financially reasonable to assume $B > 0$ (as we did in the two-security case).

Finally, we shall employ the following notation for the gradient and Hessian of a twice continuously differentiable function $g(w)$:

$$\frac{\partial g}{\partial w} = \begin{bmatrix} \frac{\partial g}{\partial w_1} \\ \vdots \\ \frac{\partial g}{\partial w_N} \end{bmatrix}, \qquad \frac{\partial^2 g}{\partial w \partial w^T} = \begin{bmatrix} \frac{\partial^2 g}{\partial w_1^2} & \frac{\partial^2 g}{\partial w_1 \partial w_2} & \cdots & \frac{\partial^2 g}{\partial w_1 \partial w_N} \\ & \frac{\partial^2 g}{\partial w_2^2} & \cdots & \frac{\partial^2 g}{\partial w_2 \partial w_N} \\ & & \ddots & \vdots \\ & & & \frac{\partial^2 g}{\partial w_N^2} \end{bmatrix},$$

where since the Hessian matrix is symmetric, we do not show the entries below its diagonal. Two useful properties (Exercise 3.25) to keep in mind are that for an $n \times n$ real matrix $A$, the gradient and Hessian of $x^T A x$ are given by

$$\frac{\partial(x^T A x)}{\partial x} = (A + A^T)x, \qquad \frac{\partial^2(x^T A x)}{\partial x \partial x^T} = A + A^T, \qquad x \in \mathbb{R}^n. \qquad (3.56)$$

### 3.3.2 Derivation of the N-Security Efficient Frontier

We first find the portfolio with the smallest risk given an expected portfolio return rate. We then vary through all possible expected portfolio return rates to obtain the corresponding set of minimum-risk portfolios. The set of all such pairs of portfolio risks and expected return rates forms the right branch of a horizontal hyperbola. We shall argue that the portion of the branch from the turning point to the upper part of the hyperbola forms the desired efficient frontier.

Let us now detail the above. We wish to find a portfolio weight vector $w$ that solves the following:

$$\text{Problem I:} \qquad \text{minimize} \qquad \sigma_P(w) = \sqrt{w^T V w} \qquad (3.57)$$

$$\text{subject to} \qquad w^T e = 1 \quad \text{and} \quad w^T \mu = \mu. \qquad (3.58)$$

Note that unrestricted short selling is allowed. A major difference between the above optimization problem and the two-security problem on page 107 is that the constraints (3.58) are now two equations in $N$ unknowns $w_1, w_2, \ldots, w_N$. In other words, for $N \geq 3$, there is not a unique portfolio weight vector $w$ satisfying (3.58). In fact, generically there are infinitely many solutions $w$ of (3.58) and, hence, infinitely many portfolio risks corresponding to $\mu$ for $N \geq 3$. However, we shall show that there is a unique portfolio weight vector $w = w_\mu$ yielding the smallest portfolio risk associated with the given $\mu$. That is, we find a unique solution of (3.57) and (3.58) together.

It is analytically simpler to solve instead the following optimization problem:

*Problem II:*      minimize      $f_c(w) = \dfrac{w^T V w}{c}$,    where $c > 0$,    (3.59)

subject to      $w^T e = 1$   and   $w^T \mu = \mu$.      (3.60)

Problems I and II have the same solution set (Exercise 3.6).[12]

**Remark 3.4.** The equivalence of Problems I and II is a simple example of obtaining a desired solution by first identifying the mathematical equivalence of the two optimization problems and then solving the simpler of the two problems, which in this case is Problem II (since it does not involve the differentiation of square roots). For a deeper study of how to exploit the equivalence of portfolio optimization problems to develop an iterative scheme to find a solution, see Korn [12, p. 8] (compare with Korn and Korn [13, pp. 3–5]). Students interested in the programming aspects of optimization problems are referred to Vanderbei's text on linear programming [25].

It suffices to consider Problem II with $c = 2$, which is a convenient choice since $f_c(w)$ is a quadratic:

minimize      $f(w) = \dfrac{w^T V w}{2}$

subject to      $w^T e = 1$   and   $w^T \mu = \mu$.

Since the constraints are equalities, the tool for solving this problem is the method of Lagrange multipliers.[13] The Lagrange function for Problem II is

$$\mathcal{L}(w, \lambda) = f(w) + \lambda_1(1 - w^T e) + \lambda_2(\mu - w^T \mu) = f(w) + \lambda^T h(w),$$

with multipliers and constraints given, respectively, by

$$\lambda = \begin{bmatrix} \lambda_1 \\ \lambda_2 \end{bmatrix}, \quad h(w) = \begin{bmatrix} h_1 \\ h_2 \end{bmatrix} = \begin{bmatrix} 1 - w^T e \\ \mu - w^T \mu \end{bmatrix}.$$

By (3.56) and the fact that $V^T = V$, we have the following:

$$\frac{\partial \mathcal{L}}{\partial w}(w, \lambda) = V w - \lambda_1 e - \lambda_2 \mu \qquad (3.61)$$

---

[12] Do not attempt to show the equivalence via the Lagrange conditions (i.e., (3.64) to (3.66)). Simply use the statement of the problems and logically imply one from the other.

[13] For optimization problems with inequality constraints, the Karush-Kuhn-Tucker Theorem is employed.

$$\frac{\partial \mathcal{L}}{\partial \lambda}(w, \lambda) = h(w) \tag{3.62}$$

$$\frac{\partial^2 \mathcal{L}}{\partial w \, \partial w^T}(w, \lambda) = V. \tag{3.63}$$

The Lagrange Multiplier Theorem yields that $w_\mu$ is a solution of Problem II if and only if there is a pair $(w_\mu, \lambda_\mu)$, where $\lambda_\mu$ is unique to $w_\mu$, satisfying

$$\frac{\partial \mathcal{L}}{\partial w}(w_\mu, \lambda_\mu) = 0 \tag{3.64}$$

$$\frac{\partial \mathcal{L}}{\partial \lambda}(w_\mu, \lambda_\mu) = 0 \tag{3.65}$$

$$x^T \left( \frac{\partial^2 \mathcal{L}}{\partial w \partial w^T}(w_\mu, \lambda_\mu) \right) x \geq 0 \quad \text{for every } x \neq 0 \text{ such that} \tag{3.66}$$

both of the following hold

$$0 = \left( \frac{\partial h_1}{\partial w}(w_\mu) \right)^T x = -e^T x = -(x_1 + \cdots + x_N)$$

$$0 = \left( \frac{\partial h_2}{\partial w}(w_\mu) \right)^T x = -\mu^T x = -(\mu_1 x_1 + \cdots + \mu_N x_N).$$

In other words, the set of solutions of Problem II is in 1-1 correspondence with the set of solutions of (3.64)–(3.66). We shall show that the latter set of equations has a unique solution $(w_\mu, \lambda_\mu)$, which, by the Lagrange Multiplier Theorem, gives a unique solution $w_\mu$ to Problem II, which in turn is a unique solution to Problem I. Without loss of generality, we set $c = 2$ for convenience.

To determine solutions $(w_\mu, \lambda_\mu)$ of (3.64)–(3.66), first observe that condition (3.66) holds automatically because by (3.63) the Hessian matrix,

$$\frac{\partial^2 \mathcal{L}}{\partial w \partial w^T}(w_\mu, \lambda_\mu) = V,$$

is positive definite. We next search for the pairs $(w_\mu, \lambda_\mu)$ satisfying (3.64). Let

$$\lambda_\mu = \begin{bmatrix} \lambda_{1,\mu} \\ \lambda_{2,\mu} \end{bmatrix}.$$

Equation (3.61) shows that (3.64) is equivalent to

$$V w_\mu = \lambda_{1,\mu}\, e + \lambda_{2,\mu}\, \mu.$$

Multiplying through by $V^{-1}$, we get

$$w_\mu = \lambda_{1,\mu}\, V^{-1} e + \lambda_{2,\mu}\, V^{-1} \mu \tag{3.67}$$

or

$$\left(w_\mu\right)^T = \lambda_{1,\mu}\, e^T V^{-1} + \lambda_{2,\mu}\, \mu^T V^{-1}.$$

Though $(w_\mu, \lambda_\mu)$ solves (3.64), we also need the pair to satisfy (3.65). But Equation (3.65) is equivalent to the two constraint equations, which fortunately are linear in the weight vector and so, by (3.67), linear in the multipliers:

$$1 = w_\mu^T e = \lambda_{1,\mu}\, (e^T V^{-1} e) + \lambda_{2,\mu}\, (\mu^T V^{-1} e)$$
$$\mu = w_\mu^T \mu = \lambda_{1,\mu}\, (e^T V^{-1} \mu) + \lambda_{2,\mu}\, (\mu^T V^{-1} \mu).$$

Employing the quantities $A$, $B$, and $C$, the above equations simplify to

$$1 = \lambda_{1,\mu}\, A + \lambda_{2,\mu}\, B \tag{3.68}$$
$$\mu = \lambda_{1,\mu}\, B + \lambda_{2,\mu}\, C. \tag{3.69}$$

Note that $B = \mu^T V^{-1} e = e^T V^{-1} \mu$ (since $B$ is a scalar). The constraints in Equations (3.68) and (3.69) can be expressed more compactly in the matrix form:

$$\begin{bmatrix} A & B \\ B & C \end{bmatrix} \begin{bmatrix} \lambda_{1,\mu} \\ \lambda_{2,\mu} \end{bmatrix} = \begin{bmatrix} 1 \\ \mu \end{bmatrix}. \tag{3.70}$$

Setting $K = \begin{bmatrix} A & B \\ B & C \end{bmatrix}$, Equation (3.55) yields $\det K = AC - B^2 > 0$. Because $K$ is invertible, we obtain a unique solution for the Lagrange multipliers:

$$\lambda_\mu = \begin{bmatrix} \lambda_{1,\mu} \\ \lambda_{2,\mu} \end{bmatrix} = K^{-1} \begin{bmatrix} 1 \\ \mu \end{bmatrix} = \frac{1}{AC - B^2} \begin{bmatrix} C & -B \\ -B & A \end{bmatrix} \begin{bmatrix} 1 \\ \mu \end{bmatrix},$$

where

$$\lambda_{1,\mu} = \frac{C - \mu B}{AC - B^2}, \qquad \lambda_{2,\mu} = \frac{\mu A - B}{AC - B^2}. \tag{3.71}$$

Then the uniqueness of $\lambda_\mu$ yields by (3.67) a unique portfolio weight vector:

$$w_\mu = \left( \frac{C - \mu B}{AC - B^2} \right) V^{-1} e + \left( \frac{\mu A - B}{AC - B^2} \right) V^{-1} \mu. \tag{3.72}$$

We call $w_\mu$ the minimum-variance portfolio weight vector with expected portfolio return rate $\mu$. Observe that, for a two-security portfolio, Equation (3.72) coincides with (3.38) (page 108).

To summarize, we found a unique pair $(w_\mu, \lambda_\mu)$ satisfying (3.64) and (3.65), and automatically satisfying (3.66) by the positive definiteness of $V$. The Lagrange Multiplier Theorem then implies that the weight vector $w_\mu$ is a unique solution of Problem II or, equivalently, Problem I. Note that our derivation of $(w_\mu, \lambda_\mu)$ drew fundamentally upon the linearity of the constraint equations and the positive definiteness of $V$.

With $w_\mu$ available, let us determine the efficient frontier. The portfolio variance associated with the expected portfolio return $\mu$ is

$$\sigma_P^2(w_\mu) = w_\mu^T V w_\mu = w_\mu^T V (\lambda_{1,\mu} V^{-1} e + \lambda_{2,\mu} V^{-1} \mu)$$
$$= \lambda_{1,\mu} (w_\mu^T e) + \lambda_{2,\mu} (w_\mu^T \mu)$$
$$= \lambda_{1,\mu} + \lambda_{2,\mu} \mu$$

since $w_\mu^T e = 1$ and $w_\mu^T \mu = \mu$. Using (3.71), we get

$$\sigma_P^2(\mu) = \frac{A\mu^2 - 2B\mu + C}{AC - B^2}. \tag{3.73}$$

Equation (3.73) has the exact form as the portfolio variance (3.40) for two securities (page 109). Indeed, the methodology for the two-security case in Section 3.2.2 carries over almost verbatim to determining the $N$-security efficient frontier. Equation (3.73) is equivalent to the right branch of the following hyperbola in the $(\sigma_P, \mu_P)$-plane:

$$\frac{\sigma_P^2}{\frac{1}{A}} - \frac{\left(\mu_P - \frac{B}{A}\right)^2}{\frac{AC-B^2}{A^2}} = 1. \tag{3.74}$$

For our portfolios, we have $\sigma_P = \sigma_P(\mu)$ and $\mu_P = \mu$ and consider the right branch since $\sigma_P(\mu) > 0$. The turning point of the right branch of the hyperbola yields the *global minimum-variance portfolio*, which has risk $\sigma_G$, expected return rate $\mu_G$, and weight vector $w_G$:

$$(\sigma_G, \mu_G) = \left(\frac{1}{\sqrt{A}}, \frac{B}{A}\right), \qquad w_G = \frac{V^{-1} e}{A}. \tag{3.75}$$

Since investors are not interested in a portfolio with negative expected return rate, *we assume $\mu_G > 0$, which yields $B > 0$.*

The portion of the branch of the hyperbola from the turning point to along the upper portion of the branch then forms the efficient frontier, which we call the *Markowitz efficient frontier for N securities*

$$M_{E,N} = \left\{ (\sigma_P, \mu_P) : \sigma_P = \sqrt{\frac{A\mu_P^2 - 2B\mu_P + C}{AC - B^2}}, \ \mu_P \geq \frac{B}{A} \right\}$$
$$= \left\{ (\sigma_P, \mu_P) : \frac{\sigma_P^2}{(1/A)} - \frac{(\mu_P - (B/A))^2}{((AC - B^2)/A^2)} = 1, \ \sigma_P > 0, \ \mu_P \geq \frac{B}{A} \right\}. \tag{3.76}$$

The set of portfolio weight vectors that produce the efficient frontier $M_{E,N}$ is:

$$W_{E,N} = \left\{ w : w = \left( \frac{C - \mu_P B}{AC - B^2} \right) V^{-1} e + \left( \frac{\mu_P A - B}{AC - B^2} \right) V^{-1} \mu , \mu_P \geq \frac{B}{A} \right\}.$$

$$(3.77)$$

Equations (3.76) and (3.77) coincide, respectively, with (3.48) and (3.49) (see page 111) when $N = 2$.

Finally, since not all portfolios are efficient, we consider how the efficient frontier sits in the *set of feasible portfolios*, i.e., the collection of all possible pairs of portfolio risk and expected return rate. The discussion will highlight a difference between the N-security portfolio case with $N \geq 3$ and the two-security case. Introduce a mapping $f_{P,N}$ from the weight space $W_N$ into the $(\sigma_P, \mu_P)$-plane by

$$f_{P,N}(w) = (\sigma_P(w), \mu_P(w)).$$

The *set of feasible portfolios with no short selling* is defined as the range of $f_{P,N}$:

$$F_{P,N} = f_{P,N}[W_N] = \left\{ (\sigma_P(w), \mu_P(w)) : w^T e = 1 \right\}.$$

For two securities, the weight space $W_2$ is a line and the transformation $f_{P,2}$ maps it into the right branch of a hyperbola. In fact, by (3.37) and (3.40) (see pages 108 and 109), we can express $w$ and $\sigma_P(w)$ as functions of the expected portfolio return rate $\mu$:

$$\mu_P(w) = \mu, \qquad w = \frac{1}{\mu_1 - \mu_2} \begin{bmatrix} \mu - \mu_2 \\ \mu_1 - \mu \end{bmatrix}, \qquad \sigma_P(w) = \sqrt{\frac{A\mu^2 - 2B\mu + C}{AC - B^2}}.$$

Employing the upright variables $\mu_P$ and $\sigma_P$ defined by

$$\mu_P = \mu_P(w) = \mu, \qquad \sigma_P = \sigma_P(w)$$

and using Equation (3.42) on page 109, we see that the feasible set $F_{P,N}$ is the right branch of a hyperbola:

$$F_{P,2} = \left\{ (\sigma_P, \mu_P) : \sigma_P = \sqrt{(A\mu_P^2 - 2B\mu_P + C)/(AC - B^2)}, -\infty < \mu_P < \infty \right\}$$

$$= \left\{ (\sigma_P, \mu_P) : \frac{\sigma_P^2}{(1/A)} - \frac{(\mu_P - (B/A))^2}{((AC - B^2)/A^2)} = 1, \sigma_P > 0, -\infty < \mu_P < \infty \right\}.$$

The efficient frontier $M_{P,2}$ is the top half, including the turning point, of the curve $F_{P,2}$, i.e., the solid curve in Figure 3.1 on page 110.

For $N \geq 3$, the set of feasible portfolios is different. For each expected portfolio return rate $\mu_P(\boldsymbol{w}) = \mu_P$, an efficient portfolio has the least risk, which by (3.76) is

$$\sigma_P(\boldsymbol{w}) = \sqrt{\frac{A\mu_P^2 - 2B\mu_P + C}{AC - B^2}}, \qquad (N \geq 3).$$

In other words, a portfolio with $\mu_P(\boldsymbol{w}) = \mu_P$ is inefficient if and only if it has risk greater than the above amount. The set of feasible portfolios, i.e., the locus of all efficient and inefficient portfolios, is then given as follows for $N \geq 3$:

$$F_{P,N\geq3} = \left\{ (\sigma_P, \mu_P) : \sigma_P \geq \sqrt{\frac{A\mu_P^2 - 2B\mu_P + C}{AC - B^2}}, \ -\infty < \mu_P < \infty \right\}$$

$$= \left\{ (\sigma_P, \mu_P) : \frac{\sigma_P^2}{(1/A)} - \frac{(\mu_P - (B/A))^2}{((AC - B^2)/A^2)} \geq 1, \ \sigma_P > 0, \ -\infty < \mu_P < \infty \right\}.$$

$$(3.78)$$

Consequently, the feasible set $F_{P,N}$, where $N \geq 3$, forms an infinite (unbounded) region whose boundary is the right branch of a hyperbola. In this case, we call $F_{P,N}$ a *Markowitz bullet*. The efficient frontier $M_{E,N}$ is then a proper subset of $F_{P,N}$ and forms the upper half of the feasible set's boundary curve, including the turning point. Consequently, *any portfolio in the feasible region that is not in $M_{E,N}$ is inefficient*.

Figure 3.3 illustrates the feasible region $F_{P,3}$ and its efficient frontier.

## 3.4 $N$-Security Efficient Frontier Without Short Selling

By default, we assume that short selling is allowed with restriction. In this section, we briefly treat the case of no short selling. Unfortunately, in the latter case the study of the efficient frontier is much more complex mathematically and so is usually treated numerically using quadratic programming software. For our purpose, since short selling will be part of most models considered, we shall simply illustrate the no-short-selling case graphically. Interestingly, when short selling is forbidden, many of the efficient portfolios have several vanishing weights (e.g., Luenberger [16, p. 161]).

When there is no short selling, the space of weights for $N$ securities is

$$W_N^* = \left\{ \boldsymbol{w} : \boldsymbol{w}^T \boldsymbol{e} = 1, \quad w_i \geq 0 \right\}. \qquad (3.79)$$

The space $W_N^*$ is a standard $(N-1)$-simplex in $\mathbb{R}^N$. Figure 3.4 depicts the weight spaces for portfolios with two and three securities. The two-security case is a line segment, while the three-security case is an equilateral triangle.

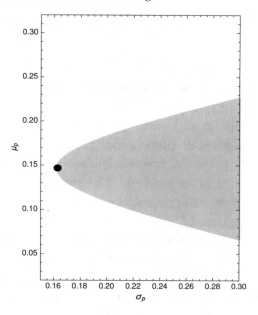

**Fig. 3.3** The feasible region $F_{P,3}$ for three securities with unlimited short selling. The outer boundary curve is a branch of a hyperbola. The efficient frontier $M_{E,3}$ is the portion of the hyperbola starting from the turning point ● and going along the upper segment of the hyperbola

Note that $W_N^*$ is the intersection of the weight space $W_N$, where short selling is allowed, with the positive orthant $\{w : w_i \geq 0,\ i = 1, \ldots, N\}$ in $\mathbb{R}^N$. For example, in Figure 3.4 the line segment and equilateral-triangle weight spaces arise, respectively, from the intersection of the straight line $W_1$ with the first quadrant and from the plane with the first octant.

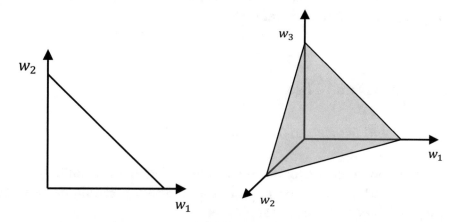

**Fig. 3.4** The weight spaces for two-security (left) and three-security (right) portfolios

Define a mapping $f_{P,N}^*$ from the weight space $W_N^*$ into the $(\sigma_P, \mu_P)$-plane $\mathbb{R}^2$ by

$$f_{P,N}^*(w) = (\sigma_P(w), \mu_P(w)), \qquad w \in W_N^*.$$

The range of $f_{P,N}^*$ is defined to be the set $F_{P,N}^*$ of feasible portfolios with no short selling:

$$f_{P,N}^*[W_N^*] = F_{P,N}^*.$$

For a two-security portfolio, the no-short-selling feasible set $F_{P,2}^*$ is a segment of the right branch of a hyperbola. The end points of the segment correspond to the risk-mean points $(\sigma_1, \mu_1)$ and $(\sigma_2, \mu_2)$ due to the individual securities. Figure 3.1 on page 110 gives an illustration.

For a three-security portfolio, the feasible set $F_{P,3}^*$ forms a region. The left panel in Figure 3.5 depicts an example. The three cusp points in the figure correspond to the pair of risk and expected return rate due to the securities, namely, $(\sigma_1, \mu_1)$, $(\sigma_2, \mu_2)$, and $(\sigma_3, \mu_3)$. As in the short selling case, the turning point (shown as •) designates the portfolio with the global minimum risk and expected return rate. We can then identify the efficient frontier as the outer boundary curve segment from the turning point to the uppermost cusp point. Each bold curve joins a pair of cusp points corresponding to two securities, so each point on such a curve is a two-security portfolio.

The right panel in Figure 3.5 shows how the three-security feasible set with no short selling fits inside the feasible set in Figure 3.3 with short selling. The two figures are generated using the same three-security inputs, except that the weights are nonnegative in the case of no short selling. The efficient frontier for short selling in Figure 3.3 extends higher up than its analog for no short selling in the left panel of Figure 3.5.

**Remark 3.5.** Portfolio theory with no short selling and with bounds on the portfolio expectation and portfolio variance is usually treated with optimization software. For a mathematical treatment, see, for example, Korn and Korn [13, Chap. 1] and references therein.                                                    □

## 3.5 The Mutual Fund Theorem

### 3.5.1 The Global Minimum-Variance Portfolio

Let us recall how to characterize an efficient portfolio when unlimited short selling is allowed. By (3.77) on page 125, an efficient portfolio has a weight vector $w_\mu$ and expected return $\mu$ satisfying

$$w_\mu = \underbrace{\left(\frac{C - \mu B}{AC - B^2}\right)}_{\lambda_{1,\mu}} V^{-1}e + \underbrace{\left(\frac{\mu A - B}{AC - B^2}\right)}_{\lambda_{2,\mu}} V^{-1}\mu, \qquad \mu \geq \mu_G = \frac{B}{A}. \quad (3.80)$$

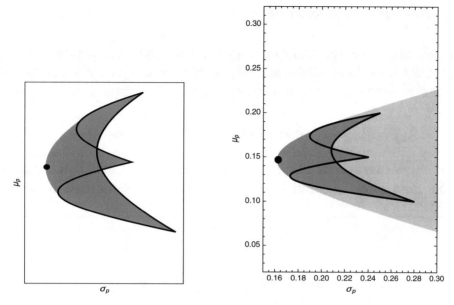

**Fig. 3.5** Left: set $F_{P,3}^*$ of feasible portfolios for three securities without short selling (shaded region). We use the same input parameters as in Figure 3.3, except the weights are restricted to being nonnegative. Each cusp point corresponds to a pair of risk and expected return rate due to one of the three securities. The efficient frontier is the outer curve segment from the turning point • to the uppermost cusp point. The three solid curves joining pairs of cusp points consist of two-security portfolios containing only the two joined securities (cusp points). Right: superposition of the feasible sets for no short selling (left) and short selling (Figure 3.3). The efficient frontier for the short selling example extends higher up than the efficient frontier with no short selling

The global minimum-variance portfolio in (3.75) on page 124 arises when the second Lagrange multiplier $\lambda_{2,\mu}$ vanishes:

$$\lambda_{2,\mu} = 0 \quad \Longleftrightarrow \quad \mu = \mu_G \quad \Longleftrightarrow \quad w_\mu = w_G = \frac{V^{-1}e}{A}.$$

To see

$$w_\mu = \frac{V^{-1}e}{A} \quad \Longrightarrow \quad \mu = \mu_G,$$

note that since

$$w_\mu = \frac{V^{-1}e}{A} \quad \Longleftrightarrow \quad \left(\lambda_{1,\mu} - \frac{1}{A}\right) V^{-1}e + \lambda_{2,\mu} V^{-1}\mu = 0$$

and because $V^{-1}e$ and $V^{-1}\mu$ are linear independent (due $e$ and $\mu$ being linearly independent), we get

$$\lambda_{1,\mu} = \frac{1}{A}, \qquad \lambda_{2,\mu} = 0,$$

which yields $\mu = \mu_G$.

### 3.5.2 The Diversified Portfolio

When the first Lagrange multiplier $\lambda_{1,\mu}$ vanishes, the corresponding portfolio is called the *diversified portfolio*, and its weight vector, expected return rate, and risk are denoted by $w_D$, $\mu_D$, and $\sigma_D$, respectively. Similar to the above, we find

$$\lambda_{1,\mu} = 0 \quad \Longleftrightarrow \quad \mu = \mu_D = \frac{C}{B} \quad \Longleftrightarrow \quad w_\mu = w_D = \frac{V^{-1}\mu}{B}.$$

However, we must check one more property, i.e., to conclude that the diversified portfolio is efficient, Equation (3.80) shows that we still need to establish

$$\mu = \mu_D \geq \mu_G.$$

Since

$$\mu_D - \mu_G = \frac{C}{B} - \frac{B}{A} = \frac{AC - B^2}{AB},$$

we have

$$\mu_D = \mu_G + \frac{AC - B^2}{AB}.$$

But $AB > 0$ and $AC - B^2 > 0$. Consequently

$$\mu_D > \mu_G.$$

Hence, the diversified portfolio is at a point $(\sigma_D, \mu_D)$ on the efficient frontier $M_{P,N}$ that is higher up than the global minimum-variance portfolio $(\sigma_G, \mu_G)$.

The variance of the diversified portfolio is

$$\sigma_P^2(w_D) = \frac{C}{B^2}.$$

Because we assume that $\mu_G > 0$ and so $B > 0$, the risk of the diversified portfolio is

$$\sigma_P(w_D) = \frac{\sqrt{C}}{B} \equiv \sigma_D.$$

Note that because $\sigma_G$ is the smallest possible variance for the $N$ securities, we have

$$\sigma_D > \sigma_G.$$

### 3.5.3 The Mutual Fund Theorem

It will now readily follow that every minimum-variance portfolio, i.e., a portfolio with weight vector that solves Problem II, can be expressed as an appropriate combination of the global minimum-variance portfolio $w_G$ and the diversified portfolio $w_D$. In fact, for a minimum-variance portfolio, we have

$$w_\mu = \lambda_{1,\mu}(V^{-1}e) + \lambda_{2,\mu}(V^{-1}\mu) = (\lambda_{1,\mu}A)w_G + (\lambda_{2,\mu}B)w_D,$$

where by (3.68) on page 123, the coefficients sum to unity:

$$\lambda_{1,\mu}A + \lambda_{2,\mu}B = 1.$$

Equation (3.71) (page 123) yields

$$a_\mu \equiv \lambda_{1,\mu}A = \frac{A(C - \mu B)}{AC - B^2}, \qquad 1 - a_\mu = \lambda_{2,\mu}B = \frac{B(\mu A - B)}{AC - B^2}.$$

Hence:

$$w_\mu = a_\mu w_G + (1 - a_\mu)w_D. \tag{3.81}$$

The result (3.81) is called a *Mutual Fund Theorem* or *Separation Theorem*. It states that *an N-security minimum-variance portfolio with expected return μ has the same risk-mean as the two-security portfolio with the percentage $a_\mu$ invested in portfolio $w_G$ and percentage $1 - a_\mu$ invested in portfolio $w_D$.* Proxies for the portfolios $w_G$ and $w_D$ could be mutual funds. The idea would then be to purchase two such mutual funds in the proportions $a_\mu$ and $1 - a_\mu$ to create a two-security portfolio that replicates the risk-reward profile of the N-security portfolio.

In general, there are many minimum-variance portfolio separations available:

**Theorem 3.1. (Mutual Fund Theorem)** *Any minimum-variance portfolio $w$ can be expressed in terms of any two distinct minimum-variance portfolios:*

$$w = s_1 w_a + s_2 w_b,$$

*where $w_a \neq w_b$ (distinct portfolios) and the scalars $s_i$ obey $s_1 + s_2 = 1$.*

Readers are referred to Ingersoll [10] for more on the Mutual Fund Theorem.

## 3.6 Investor Utility Function

### 3.6.1 Utility Functions and Expected Utility Maximization

A *utility function* is sometimes informally called a "happiness function." Intuitively speaking, it is a one-to-one function that represents the level of happiness of the investor as a function of return (or wealth). The investor applies her utility function to all investments and chooses the investment that maximizes her expected utility. We shall also see in Section 3.6.2 that the shape of the graph of an investor's utility function carries information about the investor's risk tolerance.

An investor's utility function, denoted $u(x)$, is assumed to be deterministic. It becomes a random variable $u(X)$ after substituting in a random variable $X$ representing a random future return rate (or random future wealth). The possible values of $X$ are denoted by $x$. In particular, if an investment, such as a portfolio, has an array of possible return rates $x_1, x_2, \ldots x_n$, we assume that an investor will assign numerical values $u(x_1), u(x_2), \ldots u(x_n)$ that indicate her preferences for those opportunities based on her satisfaction with the possible returns. The actual numerical values $u(x_i)$, where $i = 1, \ldots, n$, are inessential in the sense that it is how two investment choices are ranked relative to each other that will matter. This brings us to the optimization of an investor's expected utility.

*We assume that investors maximize their expected utility when choosing portfolios.* In other words, when faced with a choice between two investments, an investor will choose the one with higher expected utility. For instance, if the investor owns security A and discovers that A has less expected utility than that of security B, then the investor will sell A and buy B. The investor will favor the security that is expected to bring higher satisfaction. In general, since we assume that an investor always maximizes her expected utility when distributing her wealth across different investment choices, there is no trade that can further increase her expected utility. Hence, if an investor chooses investments with returns $X$ and $Y$, then the investor's expected utilities for these choices are the same:

$$\mathbb{E}\big(u(X)\big) = \mathbb{E}\big(u(Y)\big). \tag{3.82}$$

An *optimal portfolio* for an investor is the one for which the investor's expected utility is maximized.

The derivative $u'(x)$ is called the investor's *marginal utility function.* We shall *assume that utility functions are analytic (i.e., they can be Taylor expanded) and have positive marginal utility,* i.e., $u$ is strictly increasing:

$$x_2 > x_1 \iff u(x_2) > u(x_1). \tag{3.83}$$

In other words, a positive incremental change $x_2 - x_1 > 0$ in returns implies a positive change $u(x_2) - u(x_1) > 0$ in utility and vice versa. This underscores that a higher return ($x_2 > x_1$) leads to higher satisfaction ($u(x_2) > u(x_1)$) and, hence, requires a higher degree of risk tolerance for the potential higher return. There is no free ride: a higher return requires higher risk.

Taylor expanding the utility function about the expectation $\mathbb{E}(X)$, we get

$$u(X) = u(\mathbb{E}(X)) + u'(\mathbb{E}(X))\,(X - \mathbb{E}(X)) + \frac{1}{2!}u''(\mathbb{E}(X))\,(X - \mathbb{E}(X))^2$$
$$+ \frac{1}{3!}u^{(3)}(\mathbb{E}(X))\,(X - \mathbb{E}(X))^3 + \cdots,$$

where $u^{(k)}(x)$ is the $k$th derivative of $u$. Since $u(\mathbb{E}(X))$ and $u^{(k)}(\mathbb{E}(X))$ are constants and because $\mathbb{E}(X - \mathbb{E}(X)) = 0$, it follows

$$\mathbb{E}\big(u(X)\big) = u(\mathbb{E}(X)) + \frac{1}{2!}u''(\mathbb{E}(X))\ \mathrm{Var}(X)$$

$$+ \frac{1}{3!}u^{(3)}(\mathbb{E}(X))\ \mathbb{E}\left(\big(X - \mathbb{E}(X)\big)^3\right) + \cdots.$$

A key assumption of Markowitz portfolio theory is that investors assess an investment only through the expectation and variance of its return. *Unless otherwise stated, let us assume that investors evaluate an investment exclusively in terms of its expected return $\mathbb{E}(X)$ and variance $\mathrm{Var}(X)$ and, for simplicity, treat all higher-order derivatives of the utility function as zero, even if those terms in the Taylor expansion are a function of $\mathbb{E}(X)$ and $\mathrm{Var}(X)$:*

$$u^{(k)}(\mathbb{E}(X)) = 0 \qquad \text{for } k = 3, 4, \ldots.$$

Our simplifying assumption implies

$$\mathbb{E}\big(u(X)\big) = u(\mathbb{E}(X)) + \frac{1}{2}u''(\mathbb{E}(X))\ \mathrm{Var}(X). \qquad (3.84)$$

Examples of utility functions satisfying (3.84) are (Exercise 3.32)

$$u(x) = ax + b \qquad\qquad (a > 0)$$

$$u(x) = ax - \frac{b}{2}x^2 \qquad\qquad \left(b \neq 0, \quad x < \frac{a}{b}\right).$$

### 3.6.2 Risk-Averse, Risk-Neutral, and Risk-Seeking Investors

Investors can be divided into three broad categories. To illustrate, let 0 be the current time and suppose that an investor is considering a choice of one of the following portfolios:

➤ **Portfolio A**: riskless with a 100% guaranteed return rate of $R_t^A = 10\%$ from the present time 0 to a time $t$ in the future. This sure investment has an expected utility given by

$$\mathbb{E}(u(R_t^A)) = \mathbb{E}(u(10\%)) = u(10\%).$$

➤ **Portfolio B**: risky with a random return rate from 0 to a future time $t$ given by

$$R_t^B = \begin{cases} 5\% & \text{with probability of 50\%} \\ 15\% & \text{with probability of 50\%,} \end{cases}$$

which has an expected return $\mathbb{E}(R_t^B) = 10\%$ identical to that of Portfolio A.

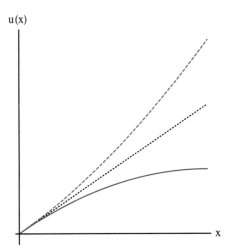

**Fig. 3.6** Depiction of three different utility functions, where $x$ represents possible return rates on an investment and $u(x)$ assigns a number designating the investor's utility for the given possible return rate. Lower, middle, and upper curves show graphs of the utility functions of a risk-averse, risk-neutral, and risk-seeking investor, respectively

We now consider choosing between these two portfolios through the eyes of risk-averse, risk-seeking, and risk-neutral investors.

### Risk-Averse Investors

A *risk-averse investor* is one with a utility function satisfying $u'(x) > 0$ and $u''(x) < 0$. In other words, the utility function is strictly increasing and strictly concave, i.e., it has a decreasing slope; see the lower curve in Figure 3.6.

How would a risk-averse investor evaluate Portfolio B? The increasing, concave shape of the graph shows that the magnitude of the decrease of the investor's utility (happiness) due to a 5% drop in return from 10% to 5% is greater than the increase in utility from a 5% increase in return from 10% to 15%. Such an investor is affected more by a 5% drop in wealth than a 5% increase in wealth and so she assigns a stiffer penalty to losses. This implies that a risk-averse investor assigns a lower expected utility to Portfolio B than to portfolio A:[14]

$$\mathbb{E}(u(R_t^A)) = u(R_t^A) = u\left(\mathbb{E}(R_t^B)\right) > \frac{1}{2}u(0.05) + \frac{1}{2}u(0.15) = \mathbb{E}(u(R_t^B)).$$

In other words, a risk-averse investor will prefer Portfolio A over B. Intuitively, for the investor to be just as happy with risky Portfolio B as she is with riskless

---

[14] A strictly concave function $f$ on an interval $I$ satisfies $f(xa + (1-x)b) > xf(a) + (1-x)f(b)$ for all $0 < x < 1$, and $a,b$ in $I$. In the example, the strict inequality follows using $f = u$ with $xa + (1-x)b = \mathbb{E}(R_B)$, where $x = \mathbb{P}(R_B = 0.05)$, $1 - x = \mathbb{P}(R_B = 0.15)$, $a = 0.05$, and $b = 0.15$.

Portfolio A, the expected return of Portfolio B must exceed that of Portfolio A, which will be shown below. The extra return beyond the 10% of Portfolio A is the *risk premium* required by a risk-averse investor for bearing risk.

Risk premiums show up more generally for risk-averse investors. Let $R_f$ be the return rate across $[0,t]$, where 0 is the current time, of a risk-free security that continuously compounds at the risk-free rate $r$. Then

$$R_f = e^{rt} - 1. \tag{3.85}$$

Note that since $R_f$ is a constant, the value $u(R_f)$ is not random. Let $R_S$ be the random return rate on $[0,t]$ of a risky security that pays a continuous cash dividend at yield rate $q$, which is continuously reinvested to buy more units of the security. Then the cum-dividend price of the risky security at $t$ is given by $S_t^c = e^{qt} S_t$.[15] The return rate becomes

$$R_S = \frac{S_t^c}{S_0} - 1 \qquad (S_t^c = e^{qt} S_t), \tag{3.86}$$

where $S_0$ is the (known) current price and $S_t$ is the random market price at the future time $t$. Since $u(R_f)$ is a constant, Equation (3.82) implies

$$u(R_f) = \mathbb{E}\big(u(R_f)\big) = \mathbb{E}\big(u(R_S)\big).$$

Employing the assumption that investors evaluate their expected utility of a security only through its expectation and variance and assuming that they treat all higher moments as zero, Equation (3.84) yields

$$\mathbb{E}\big(u(R_S)\big) = u\big(\mathbb{E}(R_S)\big) + \frac{1}{2}u''\big(\mathbb{E}(R_S)\big)\ \mathrm{Var}(R_S). \tag{3.87}$$

Because $u''(x) < 0$ for a risk-averse investor and $\mathrm{Var}(R_S) > 0$ (since the security is risky), it follows

$$u(R_f) = u\big(\mathbb{E}(R_S)\big) + \frac{1}{2}u''\big(\mathbb{E}(R_S)\big)\ \mathrm{Var}(R_S)\ < \ u\big(\mathbb{E}(R_S)\big).$$

By (3.83), we see that on $[0,t]$ the expected return rate of the risky security is higher than the risk-free return rate: $R_f < \mathbb{E}(R_S)$. The positive quantity $\mathbb{E}(R_S) - R_f$ is the risk premium required for a risk-averse investor to be equally happy or indifferent between the risky investment and the risk-free one. Using (3.85) and (3.86), we also get $\frac{\mathbb{E}(S_t^c)}{S_0} > e^{rt}$. In other words, a risk-averse investor chooses risky securities with

$$\mathbb{E}(S_t) > S_0\, e^{(r-q)t}. \tag{3.88}$$

Most investors are risk averse.

---

[15] See (2.28) on page 31.

Finally, the reader may wonder how our current definition of a risk-averse investor, i.e., one with utility function where $u'(x) > 0$ and $u''(x) < 0$, relates to our original definition given in the Markowitz theory. Recall that an investor is termed risk averse in the Markowitz model if, for a portfolio with a given level of risk, the investor requires the maximum expected return and if, for a portfolio with a given expected return, the investor requires the minimum risk. To see the link with utility functions, for a portfolio return rate $R_P$, Equation (3.84) yields

$$\mathbb{E}\big(u(R_P)\big) = u\big(\mathbb{E}(R_P)\big) \; + \; \frac{1}{2}u''\big(\mathbb{E}(R_P)\big) \; \text{Var}(R_P).$$

Since $u''\big(\mathbb{E}(R_P)\big) < 0$, we see that maximizing the expected utility $\mathbb{E}\big(u(R_P)\big)$ given a fixed expected return $\mathbb{E}(R_P)$ implies that the variance $\text{Var}(R_P)$ of the portfolio is minimized. Conversely, if we are given a fixed variance $\text{Var}(R_P)$, then maximizing $\mathbb{E}\big(u(R_P)\big)$ implies $u\big(\mathbb{E}(R_P)\big)$ is maximized, which yields that the expected portfolio return $\mathbb{E}(R_P)$ is also maximized (since $u$ is strictly increasing).

**Risk-Seeking Investor**

A *risk-seeking investor* is one whose utility function satisfies $u'(x) > 0$ and $u''(x) > 0$. The graph is an increasing function with an increasing slope; see the upper curve in Figure 3.6. The strict convexity of the function shows that the increase in the investor's utility due to a 5% rise in return from 10% to 15% is greater than the decrease in utility from a 5% drop in return from 10% to 5%. Such investors are less impacted from a 5% drop in wealth than from a 5% increase in wealth. They gravitate toward risk. A risk-seeking investor then assigns a higher expected utility to Portfolio B than to A:[16]

$$\mathbb{E}(u(R_t^A)) = u(R_t^A) = u\left(\mathbb{E}(R_t^B)\right) \; < \; \frac{1}{2}u(0.05) \; + \; \frac{1}{2}u(0.15) \; = \; \mathbb{E}(u(R_B)).$$

A risk-seeking investor will select Portfolio B.

Analogous to the risk-averse case, given a risk-free security and risky security with returns $R_f$ and $R_S$, respectively, and since $u''(x) > 0$ for a risk-seeking investor, we obtain

$$u(R_f) = u\big(\mathbb{E}(R_S)\big) \; + \; \frac{1}{2}u''\big(\mathbb{E}(R_S)\big) \; \text{Var}(R_S) \; > \; u\big(\mathbb{E}(R_S)\big).$$

---

[16] A strictly convex function $f$ on an interval $I$ satisfies $f(xa + (1-x)b) < xf(a) + (1-x)f(b)$ for all $0 < x < 1$ and $a, b$ in $I$. In the example, the strict inequality follows using $f = u$ with $xa + (1-x)b = \mathbb{E}(R_B)$, where $x = \mathbb{P}(R_B = 0.05)$, $1 - x = \mathbb{P}(R_B = 0.15)$, $a = 0.05$, and $b = 0.15$. Also, see Jensen's inequality.

Equation (3.83) yields $R_f > \mathbb{E}(R_S)$, i.e., a risk-seeking investor has a negative risk premium $\mathbb{E}(R_S) - R_f$. By (3.85) and (3.86), we obtain $e^{rt} > \frac{\mathbb{E}(S_t^c)}{S_0}$. Hence, a risk-seeking investor prefers risky securities with

$$\mathbb{E}(S_t) < S_0 e^{(r-q)t}. \tag{3.89}$$

### Risk-Neutral Investors

A *risk-neutral investor* is one with a utility function satisfying $u'(x) > 0$ and $u''(x) = 0$. This means $u(x) = ax + b$, where $a > 0$. The graph is a straight line with positive slope, which is shown in Figure 3.6 for the case $u(x) = ax$, where $a > 0$. The magnitude of the decrease of the investor's utility due to a 5% drop in return from 10% to 5% is the same as the increase in utility from a 5% increase in return from 10% to 15%. Such an investor is equally affected by a 5% drop in wealth compared with a 5% increase in wealth. A risk-neutral investor is then neutral toward or indifferent between Portfolios A and B:

$$\mathbb{E}\big(u(R_t^A)\big) = \mathbb{E}\big(u(R_t^B)\big).$$

Considering the risk-free security with return $R_f$ and risky one with return $R_S$, we have

$$u(R_f) = u\big(\mathbb{E}(R_S)\big) + \frac{1}{2}u''\big(\mathbb{E}(R_S)\big)\ \text{Var}(R_S) = u\big(\mathbb{E}(R_S)\big).$$

Since $u$ strictly increases, we must have $e^{rt} = \frac{\mathbb{E}(S_t^c)}{S_0}$. Thus, a risk-neutral investor prefers risky securities with

$$\mathbb{E}(S_t) = S_0 e^{(r-q)t}. \tag{3.90}$$

### Remark 3.6.

1. Though investors tend to be risk averse versus risk seeking, they are unlikely to be risk neutral. Despite this, risk-neutral investors will play an important role in the pricing of derivatives (Chapter 8).

2. Portfolio theory can also be presented starting from the theory of utility functions; see Pennacchi [21, Chap. 2] for more. This approach, though, would take us far deeper into utility functions than is appropriate for this introductory text.  □

*We now return to our original assumption that all investors are risk averse.*

## 3.7 Diversification and Randomly Selected Securities

*Diversification* is a risk management technique that mixes a variety of assets in a portfolio. Its effect is to reduce portfolio risk by distributing a given investment capital across more and more securities to spread the risk around. We first explored diversification in Section 3.2.3 (on page 116) for a two-security portfolio. We showed how a portfolio's risk can be reduced by spreading an initial capital in a certain way across two uncorrelated securities as opposed to investing all of the capital in one of the securities. This section explores how the mean portfolio risk is impacted when we create diversification by randomly and uniformly selecting more and more securities and weights. We shall see that the Markowitz portfolio model provides a mathematical basis for the statement, "Don't put all your eggs in one basket."

Consider a market of all possible risky securities. By default we focus on securities in the USA and one may choose say, the NASDAQ, as a proxy. To create an $N$-security portfolio, we pick a random pair of $N$ risky securities and $N$ weights. Specifically, we simultaneously pick $N$ random securities uniformly from across the market and uniformly choose $N$ nonnegative random weights for these securities. For each such random pair, there is a portfolio variance, and it is the mean of this variance for many random pairs that we shall explore as $N$ increases.

*It is important to emphasize that since we shall randomly draw securities from the marketplace and randomly assign weights, the resulting portfolios will, in general, not be efficient. Also, as $N \to \infty$ we assume that all the securities in the marketplace are included in the portfolio.*

We now model these ideas mathematically under the assumption of no short selling.

### 3.7.1 Mean Portfolio Variance and the Uniform Dirichlet Distribution

We determine a formula for the expectation of the portfolio variance of $N$ random securities.

First, consider an $N \times N$ random covariance matrix $V$ resulting from $N$ randomly and uniformly selected securities, where $N \geq 2$. Since $V$ is symmetric, fix an ordering of the entries of the diagonal and upper triangular portion of $V$ as follows:

$$(\sigma_{11}, \sigma_{12}, \ldots, \sigma_{1N},\ \sigma_{22}, \sigma_{23}, \ldots, \sigma_{2N},\ \sigma_{33}, \sigma_{34}, \ldots, \sigma_{3N},\ \ldots,\ \sigma_{NN}),$$

where $\sigma_{ii} = \sigma_i^2$. Denote the joint p.d.f. of these $N + \frac{N(N-1)}{2}$ entries by $f_V$ and the marginal p.d.f.s of the entries $\sigma_{ij}$ by $f_{ij}$, respectively. Expectations with respect to $f_V$ and $f_{ij}$ are denoted by $\mathbb{E}_V$ and $\mathbb{E}_{ij}$, respectively. We shall set $\mathbb{E}_{ii} = \mathbb{E}_i$. The expectation of the sample mean of the random variances $\sigma_i^2$ is

$$\bar{\sigma}_{\text{Var}}(N) \equiv \mathbb{E}_w \left( \frac{1}{N} \sum_{i=1}^{N} \sigma_i^2 \right) = \frac{1}{N} \sum_{i=1}^{N} \mathbb{E}_i(\sigma_i^2), \tag{3.91}$$

which is the average of the expected individual security variances. The expectation of the sample mean of the covariances $\sigma_{ij}$, where $i \neq j$, is

$$\bar{\sigma}_{\text{Cov}}(N) \equiv \mathbb{E}_V \left[ \frac{1}{\frac{N(N-1)}{2}} \left( \sum_{1 \leq i < j \leq N} \sigma_{ij} \right) \right] = \frac{1}{\frac{N(N-1)}{2}} \left( \sum_{1 \leq i < j \leq N} \mathbb{E}_{ij}[\sigma_{ij}] \right),$$
$$\tag{3.92}$$

which is the average of the expected individual security covariances. *Assume that as $N \to \infty$, the quantities $\bar{\sigma}_{\text{Var}}(N)$ and $\bar{\sigma}_{\text{Cov}}(N)$ converge to finite values, which we denote simply as $\bar{\sigma}_{\text{Var}}$ and $\bar{\sigma}_{\text{Cov}}$, respectively.*

Second, uniformly and randomly choose weight vectors $w$ from the weight space $W_N^*$, which excludes short selling:

$$W_N^* = \left\{ (x_1, \ldots, x_{N-1}, x_N) \in \mathbb{R}^N : x_i \geq 0,\ x_1 + \cdots + x_{N-1} + x_N = 1 \right\}.$$

Since $W_N^*$ is a standard $(N-1)$-simplex, *the uniform Dirichlet density is used to choose randomly the elements of $W_N^*$.* The joint Dirichlet p.d.f of

$$w = [w_1\ \ldots\ w_{N-1}\ w_N]^T,$$

where $w_i \geq 0$ and $w_N = 1 - (w_1 + \cdots + w_{N-1})$, is then given relative to the first $N-1$ weights:

$$f_w(x_1, \ldots, x_N) = \begin{cases} (N-1)! & (x_1, \ldots, x_N) \in W_N^* \\ 0 & (x_1, \ldots, x_N) \notin W_N^*, \end{cases}$$

where $x_i$ is the variable representing the possible values of $w_i$ for $i = 1, \ldots, N$. See, for example, Fine [4, Secs 7.4.3, 7.7] and Frigyik, Kapila, and Gupta [5] for introductions to the Dirichlet distribution.

Figure 3.7 illustrates the uniform random selection from the weight spaces for two and three securities. The associated marginal density of each weight is identical and given by

$$f_w(x_i) = (N-1)(1 - x_i)^{N-2}, \qquad i = 1, \ldots, N \quad (N \geq 2).$$

Expectations, variances, and covariances relative to the Dirichlet density (joint or marginal) will be indicated with $w$ as a subscript. The following basic properties hold, where $i, j = 1, \ldots, N$ and $N \geq 2$:

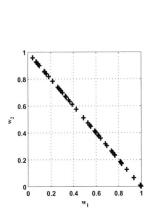
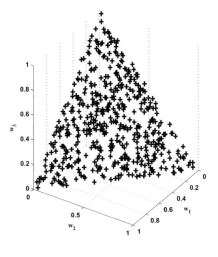

**Fig. 3.7** Randomly and uniformly chosen weights from the weight spaces with no short selling for two securities (left) and three securities (right). The weights were drawn using a uniform Dirichlet distribution

$$\mathbb{E}_w(w_i) = \frac{1}{N}$$
$$\mathrm{Var}_w(w_i) = \frac{N-1}{N^2(N+1)}$$
$$\mathrm{Cov}_w(w_i, w_j) = \frac{-1}{N^2(N+1)}, \quad i \neq j.$$

Moreover:

$$\mathbb{E}_w(w_i^2) = \mathrm{Var}_w(w_i) + (\mathbb{E}_w(w_i))^2 = \frac{2}{N(N+1)} \tag{3.93}$$

and

$$\mathbb{E}_w(w_i w_j) = \mathrm{Cov}_w(w_i, w_j) + \mathbb{E}_w(w_i)\,\mathbb{E}_w(w_j) = \frac{1}{N(N+1)} \qquad (i \neq j), \tag{3.94}$$

where $i, j = 1, \dots, N$ and $N \geq 2$.

Third, let the pair $(w, V)$ represent the following string of random variables:

$$(w_1, \dots, w_N, \ \sigma_{11}, \sigma_{12}, \dots, \sigma_{1N}, \ \sigma_{22}, \sigma_{23}, \dots, \sigma_{2N}, \ \sigma_{33}, \sigma_{34}, \dots, \sigma_{3N}, \ \dots, \ \sigma_{NN}).$$

Denote the joint p.d.f. of $(w, V)$ by $f_{(w,V)}$ and expectations relative to $f_{w,V}$ by $\mathbb{E}_{(w,V)}$. With respect to the degree of independence between $w$ and $V$, *we assume*:

$$\mathbb{E}_w(w_i w_j \sigma_{ij}) = \mathbb{E}_w(w_i w_j)\,\mathbb{E}_{ij}(\sigma_{ij}). \tag{3.95}$$

Now, turning to the portfolio variance, we have the random variable:

$$\sigma_{P,N}^2 = w^T V w = \sum_{i=1}^{N} w_i^2 \sigma_i^2 + 2 \sum_{1 \leq i < j \leq N} w_i w_j \sigma_{ij}.$$

Using (3.91)—(3.95), we compute the expectation of the variance:

$$\mathbb{E}_{(w,V)}\left(\sigma_{P,N}^2\right) = \mathbb{E}_{(w,V)}\left(\sum_{i=1}^N w_i^2 \sigma_i^2\right) + 2\,\mathbb{E}_{(w,V)}\left(\sum_{1\le i<j\le N} w_i w_j \sigma_{ij}\right)$$

$$= \sum_{i=1}^N \mathbb{E}_w(w_i^2)\,\mathbb{E}_i(\sigma_i^2) + 2\sum_{1\le i<j\le N}\mathbb{E}_w(w_i w_j)\,\mathbb{E}_{ij}(\sigma_{ij})$$

$$= \frac{2}{N+1}\sum_{i=1}^N\frac{\mathbb{E}_i(\sigma_i^2)}{N} + \frac{N-1}{N+1}\sum_{1\le i<j\le N}\frac{\mathbb{E}_{ij}(\sigma_{ij})}{(N(N-1)/2)}.$$

Consequently:

$$\mathbb{E}_{(w,V)}\left(\sigma_{P,N}^2\right) = \frac{1}{N+1}\left(2\bar\sigma_{\mathrm{Var}}(N) - \bar\sigma_{\mathrm{Cov}}(N)\right) + \frac{1}{1+\frac{1}{N}}\bar\sigma_{\mathrm{Cov}}(N)$$

$$(3.96)$$

Since $\bar\sigma_{\mathrm{Cov}}(N) \to \bar\sigma_{\mathrm{Cov}}$ and $\bar\sigma_{\mathrm{Var}}(N) \to \bar\sigma_{\mathrm{Var}}$ as $N\to\infty$, Equation (3.96) yields

$$\mathbb{E}_{(w,V)}\left(\sigma_{P,N}^2\right) \longrightarrow \bar\sigma_{\mathrm{Cov}} \qquad \text{as } N \longrightarrow \infty. \qquad (3.97)$$

Hence, *the limiting value of the mean portfolio variance as N increases shows that the mean sample variance of the individual securities' return rates is dominated by the mean sample covariance of these returns.* That is, for a sufficiently large number of securities, the covariances between securities have a greater impact on a typical portfolio's variance than the variances of the individual securities. The next section illustrates these ideas using data from the NASDAQ.

**Remark 3.7.**

1. The above is consistent with our findings in Section 3.2.3 for two securities with no short selling. By decreasing the correlation coefficient $\rho$ between the two securities from 1 toward $-1$, we increased the diversification and reduced the portfolio's risk. See Figure 3.2 on page 115.
2. For the case when the weights are nonrandom and equal, namely,

$$w_i = \frac{1}{N}, \qquad i = 1,\dots,N,$$

but the securities are randomly chosen (i.e., the covariance matrix is random), the expected portfolio variance becomes

$$\mathbb{E}_{(w,V)}\left(\sigma_{P,N}^2\right) = \frac{1}{N}\left(\bar\sigma_{\mathrm{Var}}(N) - \bar\sigma_{\mathrm{Cov}}(N)\right) + \bar\sigma_{\mathrm{Cov}}(N).$$

By increasing the number of securities indefinitely, we obtain the same result as in (3.97):

$$\mathbb{E}_{(w,V)}\left(\sigma^2_{P,N}\right) \longrightarrow \overline{\sigma}_{\mathrm{Cov}} \qquad \text{as} \qquad N \to \infty.$$

$\square$

### 3.7.2 Mean Portfolio Variance using the NASDAQ

**Example 3.8.** We estimated the mean portfolio variance for 2,385 NASDAQ stocks using 503 adjusted closing daily prices for each stock over the time span from March 15, 2011, to March 15, 2013. The data was obtained from finance.yahoo.com. Figure 3.8 depicts the results generated by 100,000 pairs of random stock picksand random weights for each $N$. The portfolio variance for

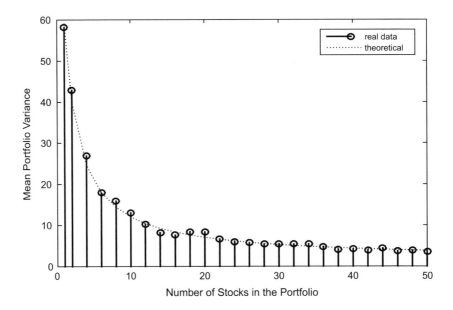

**Fig. 3.8** Mean portfolio variance as a function of the number of randomly chosen stocks on the NAS-DAQ (see text). The theoretical line is the mean portfolio variance based on Equation (3.96), which employs a uniform Dirichlet distribution of weights with no short selling. The real data line is based on random draws from the NASDAQ. Courtesy of Li Li

all 100,000 pairs was computed and averaged to estimate the mean portfolio variance for the given $N$. The figure also depicts the theoretical mean portfolio variance determined using the uniform Dirichlet distribution.                                    $\square$

Equation (3.97) and Figure 3.8 illustrate that *even after diversification by randomly and uniformly choosing a large number of securities across the marketplace and across weights, portfolio risk arising from the mean sample covariance remains.*

This type of risk, i.e., risk that cannot be eliminated by diversification, is called *undiversifiable risk* or *systematic risk*. It will play an important role in Chapter 4. The portion of portfolio risk that can be removed by diversification, namely, the risk that contributes to the mean portfolio variance values above the convergence value in Figure 3.8, is called *unsystematic risk* or *diversifiable risk*.

> *Unless stated to the contrary, assume that a portfolio's risk is systematic, i.e., the portfolio is sufficiently diversified that all other risks are negligible.*

**Remark 3.8.** Readers interested in learning more about Markowitz portfolio theory should consult the book *Risk-Return Analysis: The Theory and Practice of Rational Investing* by Markowitz and Blay[19].

## 3.8 Exercises

### 3.8.1 Conceptual Exercises

**3.1.** An investor plans to create a portfolio of ten stocks by shorting all of them. Can he use the Markowitz theory presented in this chapter? Explain your answer.

**3.2.** Can you find five examples of pairs of stocks in the USA that are negatively correlated? Are such occurrences common?

**3.3.** The volatility of the log-return rate of a portfolio over 120 days is roughly 11 times the volatility over a day. Agree or disagree? Explain your answer.

**3.4.** Decide whether you agree or disagree with the statements below. Justify your answer.

a) "Investors who are not risk averse are irrational."
b) "When the risk of a portfolio vanishes, the risk of each security has to vanish."

**3.5.** A clever wealth manager constructed a portfolio of stocks such that the portfolio has no risk *and* has an expected return of 25%. What is the probability that the portfolio return rate will actually be 25%?

**3.6.** Show that Problem I on page 120 is equivalent to Problem II on page 121, i.e., show that these two optimization problems have the same set of solutions for all $c > 0$.

**3.7.** Explain the financial meaning of minimizing the function $f_P(w) = \frac{\sigma_P^2(w)}{\mu_P(w)}$, where $\mu_P(w)$ is the portfolio expected return.

**3.8.** It can be shown that the covariance of the return of the global minimum-variance $N$-security portfolio with the return of any other efficient $N$-security portfolio is always $1/A$ (Exercise 3.28). Interpret this result.

**3.9.** If your initial capital increases by \$100, then we would expect an increase in your utility. To which scenario would you assign a higher utility, assuming the same risk for both? Scenario A: initial capital of \$1,000. Scenario B: initial capital of \$10,000.

**3.10.** Is $u(x) = a$ a utility function for $a$ a constant? Justify your answer.

**3.11.** Is $u(x) = 1 - e^{-bx}$, where $b > 0$, a risk-averse, risk-neutral, or risk-seeking utility function? Justify your answer.

### 3.8.2 Application Exercises

**3.12. (Two Securities)** The table below gives a sample of artificial historical data for an energy company (Stock A) and a phone company (Stock B). Each indicated return rate is end of month to end of month and is based on adjusted closing prices. For example, $R^A_{\text{Jan}-2012}$ is the return rate from the last trading day in December 2011 to the last trading day in January 2012. Using the table, express your answers in percentages where appropriate.

| Date | $R^A_{monthly}$ | $R^B_{monthly}$ |
|------|-----------------|-----------------|
| Jan-2012 | 2.45% | 2.69% |
| Feb-2012 | 3.35% | 1.81% |
| Mar-2012 | 3.24% | 4.94% |
| Apr-2012 | 2.93% | 5.88% |
| May-2012 | 6.13% | 2.51% |
| Jun-2012 | 6.19% | -0.35% |
| Jul-2012 | 0.78% | 1.59% |
| Aug-2012 | -0.19% | -3.83% |
| Sep-2012 | 4.65% | 5.24% |
| Oct-2012 | 3.53% | 4.85% |
| Nov-2012 | 5.03% | 2.48% |
| Dec-2012 | -1.71% | 4.03% |

a) Sketch the graph of the monthly total return rates of each stock as a function of time during 2012. Briefly discuss the movement of the stocks during two equal-length-time periods in 2012.

b) Estimate the expected monthly total return rate of each stock in the table. What is your answer if you use data only from Dec-2011 to Jun-2012? Compute the monthly volatility of each stock for the year 2012.

c) Estimate the monthly variance of each stock and determine the covariance and correlation coefficient between the monthly total return rates for the two stocks during 2012. Is the result what you expected? Briefly discuss.

d) For a portfolio consisting of stocks A and B, use the data in the table to determine the portfolio's expected monthly total return rate and the portfolio's monthly risk under selection I, where funds are split evenly between the two stocks, and selection II, where two-thirds of your funds are in stock A and one-third in stock B. Annualize the portfolio expected returns and risks.

e) Would you recommend portfolio selection I or II to an investor? Briefly discuss your answer.

f) Critique how the data in the table is being applied to the theoretical framework used in these exercises. Include two drawbacks with using historical data to estimate the expected returns and risks.

**3.13. (Three Securities)** Suppose that you have $5,000 to invest in stocks 1, 2, and 3 with current prices $\begin{bmatrix} S_1(t_0) \\ S_2(t_0) \\ S_3(t_0) \end{bmatrix} = \begin{bmatrix} \$10.20 \\ \$53.75 \\ \$30.45 \end{bmatrix}$, covariance matrix

$$V = \begin{bmatrix} 0.03 & -0.04 & 0.02 \\ -0.04 & 0.08 & -0.04 \\ 0.02 & -0.04 & 0.04 \end{bmatrix},$$

and expected return vector

$$\mu = \begin{bmatrix} 0.10 \\ 0.15 \\ 0.075 \end{bmatrix}.$$

For example, stock 3 has a volatility of $\sigma_3 = 20\%$ and expected return rate of $\mu_3 = 7.5\%$. The values in $V$ and $\mu$ are pure numbers (not percentages). Answer the following using an appropriate software.

a) Determine the weights needed to create the global minimum-variance portfolio of these three stocks.
b) Create an efficient portfolio with an expected return rate of 18%. Explicitly state the number of shares one must hold for each stock and how you fund each position. State the portfolio risk and compare it with maximum risk among the stocks.

**3.14. (N Securities)** Consider managing a portfolio with 500 risky securities, and assume that the variances of the securities are robustly estimated from reliable historical data. If 1% of the remaining independent covariances of the securities are poorly estimated due to inaccuracies in the data, then determine the number of poorly estimated covariances.

**3.15.** Suppose a client just inherited $1,000,000 and has come to you seeking advice on how to split the money between two of his favorite securities so as to maximize return. Security A has expected rate of return $r_A = 0.13$ and standard deviation of $\sigma_A = 0.15$. Security B has expected rate of return $r_B = 0.14$ and standard deviation $\sigma_B = 0.20$. The correlation coefficient between their rates of return is $\rho = -0.3$. If the investor has a utility function $U(x) = \sqrt[3]{x}$, how should he invest in each stock to maximize his overall rate of return?

### 3.8.3 Theoretical Exercises

**3.16.** Let $X$ and $Y$ be two random variables. Let $a_i, i = 1,2,3,4$ be real numbers with $a_2 a_4 \neq 0$. Let $\rho$ be the correlation coefficient. Prove that

$$\rho(a_1 + a_2 X, a_3 + a_4 Y) = \begin{cases} \rho(X,Y) & \text{if } a_2 a_4 > 0 \\ -\rho(X,Y) & \text{if } a_2 a_4 < 0. \end{cases}$$

The significance of the result is to allow a change of scale to one convenient for computation.

**3.17.** Prove (3.17) on page 97.

**3.18.** Verify Equation (3.20) on page 103, where the portfolio log-return rates are assumed uncorrelated and identically distributed.

**3.19. (Two Securities)** Let $A = e^T V^{-1} e$, $B = \mu^T V^{-1} e$, and $C = \mu^T V^{-1} \mu$. Show that if

$$(B\mu - Ce)^T V^{-1} (B\mu - Ce) > 0,$$

then $AC - B^2 > 0$. See (3.33) on page 107.

**3.20. (Two Securities)** Verify Equation (3.38) on page 108, i.e., show that

$$w_\mu = \left( \frac{C - \mu B}{AC - B^2} \right) V^{-1} e + \left( \frac{\mu A - B}{AC - B^2} \right) V^{-1} \mu,$$

where

$$w_\mu = \begin{bmatrix} w_\mu \\ 1 - w_\mu \end{bmatrix} = \frac{1}{\mu_1 - \mu_2} \begin{bmatrix} \mu - \mu_2 \\ \mu_1 - \mu \end{bmatrix}.$$

**3.21.** Let $f(x) = ax^2 + bx + c$, where $a$, $b$, and $c$ are real numbers with $a \neq 0$.

a) Show that if $b^2 - 4ac < 0$ and $f(x) \geq 0$ for all $x \in \mathbb{R}$, then $f(x) > 0$ for all $x \in \mathbb{R}$.

b) Show that if $a > 0$ and $f(x) > 0$ for all $x \in \mathbb{R}$, then the global minimum point of $\sqrt{f}$ is

$$x_* = -\frac{b}{2a}$$

and the corresponding global minimum value is

$$\sqrt{f(x_*)} = \sqrt{-\frac{(b^2 - 4ac)}{4a}}.$$

c) Use the above results to give an alternative proof that the two-security portfolio variance,

$$\sigma_P^2(w) = w^2 \sigma_1^2 + (1 - w)^2 \sigma_2^2 + 2w(1 - w)\rho \sigma_1 \sigma_2,$$

is strictly positive. Compute the global minimum point $w_*$ of the portfolio risk $\sigma_P(w)$ and find $\sigma_P(w_*)$. Compare with (3.44) and (3.47); see page 110.

**3.22. (Two Securities)** Given two securities $S_i$, $i = 1,2$. Let $R_i$, $i = 1,2$, be their return rates, respectively. Assuming that $R_i$, $i = 1,2$, are independent and identically distributed continuous random variables, determine the portfolio that a risk-averse investor would select.

**3.23. (Two Securities)** Consider a portfolio with two securities having returns $\mu_1$ and $\mu_2$, risks $\sigma_1$ and $\sigma_2$, and a correlation coefficient $\rho$ that vanishes. To minimize this portfolio's risk-to-reward ratio, a natural quantity to minimize is

$$f(w) = \frac{\sigma_P^2(w)}{\mu_P w)},$$

where $w$ is the fraction of the total investment in the security with expected return $\mu_1$.

a) Determine an equation that any critical point of $f$ must satisfy. What type of equation is it?
b) Show that if $\mu_1 = \mu_2$, then we obtain a linear equation for $w$ with solution

$$w = \frac{\sigma_2^2}{\sigma_1^2 + \sigma_2^2}.$$

This critical point coincides with the global minimum $w$ we found for the two-security portfolio when minimizing the variance $\sigma_P^2$ with $\rho = 0$. Why are the two critical points identical?

**3.24. (Three Securities)** Consider a portfolio with three securities having risks $\sigma_1$, $\sigma_2$, and $\sigma_3$ and correlation coefficients $\rho_{12}$, $\rho_{13}$, and $\rho_{23}$. Let $V$ be the covariance matrix of the security returns. Show that $V$ is positive definite if and only if the following hold:

a) $\sigma_1 > 0$, $\sigma_2 > 0$, and $\sigma_3 > 0$,
b) $|\rho_{12}| < 1$ and $\rho_{12}^2 + \rho_{13}^2 + \rho_{23}^2 - 2\rho_{12}\rho_{13}\rho_{23} < 1$.

**3.25.** Let $A$ be an $n \times n$ matrix. Show that the gradient and Hessian of the quadratic $x^T A x$ are

$$\frac{\partial(x^T A x)}{\partial x} = (A + A^T)x, \qquad \frac{\partial^2(x^T A x)}{\partial x \partial x^T} = A + A^T, \qquad x \in \mathbb{R}^n,$$

where $\left(\frac{\partial f}{\partial x}\right) = \left[\frac{\partial f}{\partial x_1} \cdots \frac{\partial f}{\partial x_n}\right]^T$ and $\frac{\partial^2(x^T A x)}{\partial x \partial x^T} = \left[\frac{\partial^2 f}{\partial x_i \partial x_j}\right]_{n \times n}$.

**3.26. (N Securities)** For an $N$ security portfolio, show that the portfolio vector $w$ which minimizes the variance $\sigma_P^2(w) = w^T V w$, subject to $w^T e = 1$, is the global minimum-variance portfolio vector. Explain why this is expected.

**3.27. (N Securities)** Determine the equations for the lines asymptotic to the set of all minimum-variance $N$-security portfolios.

**3.28.** Show that the covariance of the return of the global minimum-variance $N$-security portfolio with the return of any other efficient $N$-security portfolio is always $1/A$.

**3.29.** Where does the tangent line at the diversified portfolio on the Markowitz $N$-security efficient frontier intersect the $\mu_P$-axis?

**3.30.** Determine the equation of a line asymptotic to the Markowitz $N$-security efficient frontier in the $(\sigma_P, \mu_P)$-plane.

**3.31.** Let $w_a$ and $w_b$ be any two distinct minimum-variance portfolio vectors. For suitable constants $a$ and $b$, these vectors can be expressed in the following form:

$$w_a = (1-a)w_G + aw_D, \qquad w_b = (1-b)w_G + bw_D,$$

where $w_G$ and $w_D$ are the global minimum-variance and diversified portfolio vectors, respectively.

a) Show that any minimum-variance portfolio vector $w$ can be expressed as

$$w = \left(\frac{\lambda_1 A + b - 1}{b - a}\right) w_a + \left(\frac{1 - a - \lambda_1 A}{b - a}\right) w_b.$$

b) Show that the covariance is given as follows:

$$\mathrm{Cov}(R_P(w_a), R_P(w_b)) = \frac{1}{A} + ab\,\frac{AC - B^2}{AB^2},$$

where $A = e^T V^{-1} e$, $B = \mu^T V^{-1} e$, and $C = \mu^T V^{-1} \mu$.

**3.32.** Let $a > 0$ and $b \neq 0$. Show that the utility functions $u(x) = ax + b$ and $u(x) = ax - \frac{b}{2}x^2$, where $x < \frac{a}{b}$, obey

$$\mathbb{E}(u(X)) = u(\mathbb{E}(X)) + \frac{1}{2}u''(\mathbb{E}(X))\,\mathrm{Var}(X).$$

**3.33.** A power utility function refers to one of the form $u(x) = x^a$. When is $u$ a risk-averse utility function?

# References

[1] Bodie, Z., Kane, A., Marcus, A.: Investments, 9th edn. McGraw-Hill, New York (2011)
[2] Capiński, M., Zastawniak, T.: Mathematics forFinance. Springer, New York (2003)
[3] Durrett, R.: Probability: Theory and Examples, 4th edn. University Press, Cambridge (2010)
[4] Fine, T.: Probability and Probabilistic Reasoning for Electrical Engineering. Pearson Prentice Hall, Upper Saddle River (2006)

[5] Frigyik, B., Kapila, A., Gupta, M.: Introduction to the Dirichlet distribution and related processes. University of Washington Electrical Engineering Technical Report, Number UWEETR-2010-0006 (2010)

[6] Goldberg, L.: A review of "Risk-Return Analysis: The Theory and Practice of Rational Investing (Volume I)" by Harry Markowitz and Kenneth Blay. http://www.cfapubs.org/doi/full/10.2469/br.v9.n1.9%40faj.2014.70.issue-3 (2014)

[7] Goodman, J.: Statistical Optics. Wiley Classics Library Edition. Wiley, New York (2000)

[8] Graham, J., Smart, S., Megginson, W.: Corporate Finance. South-Western Cengage Learning, Mason (2010)

[9] Grinold, R., Kahn, R.: Active Portfolio Management. McGraw-Hill, New York (2000)

[10] Ingersoll, J.: Theory of Financial Decision Making. Rowman and Littlefield, Savage (1987)

[11] Jorion, P.: Value at Risk, 3rd edn. McGraw-Hill, New York (2007)

[12] Korn, R.: Optimal Portfolios. World Scientific, River Edge (1997)

[13] Korn, R., Korn, E.: Option Pricing and Portfolio Optimization. American Mathematical Society, Providence (2001)

[14] Levy, H.: A review of "Risk-Return Analysis: The Theory and Practice of Rational Investing (Volume I)" by Harry Markowitz and Kenneth Blay (2014). Quant. Finance **14**(7), 1141 (2014)

[15] Levy, H., Duchin, R.: Asset return distributions and the investment horizon. J. Portf. Manag. **30**(3), 47 (2004)

[16] Luenberger, D.: Investment Science. Oxford University Press, New York (1998)

[17] Markowitz, H.: Portfolio selection. J. Financ. Res. **7**(1), 77 (1952)

[18] Markowitz, H.: Portfolio Selection. Blackwell, Cambridge (1959)

[19] Markowitz, H., Blay, K.: Risk-Return Analysis: The Theory and Practice of Rational Investing, vol. I. McGraw-Hill, New York (2014)

[20] Merton, R.: An analytic derivation of the efficient portfolio frontier. J. Financ. Quant. Anal. **7**(3), 1851 (1972)

[21] Pennacchi, G.: Theory of Asset Pricing. Pearson Addison Wesley, Boston (2008)

[22] Reiley, F., Brown, K.: Investment Analysis and Portfolio Management. Dryden Press, Fort Worth (1997)

[23] Roman, S.: Introduction to the Mathematics of Finance. Springer, New York (2004)

[24] Ross, S.: An Elementary Introduction to Mathematical Finance. Cambridge University Press, Cambridge (2011)

[25] Vanderbei, R.: Linear Programming: Foundations and Extensions, 4th edn. Springer, New York (2014)

# Chapter 4
# Capital Market Theory and Portfolio Risk Measures

The process of dividing a portfolio among major asset categories such as stocks, bonds, real estate[1], and cash is generally referred to as *asset allocation*. A key aspect of asset allocation is portfolio risk management.

The risk that the entire financial system bears is called the *systematic risk*, whereas the risk that a portfolio bears is called *portfolio risk*.

➤ *Systematic risk* can be characterized by the potential of financial system disruption with substantial and adverse effects on the economy. The most recent example of such risks was exhibited by 2007–2008 financial crisis.

Portfolio risk can be classified into two broad categories: market risk and idiosyncratic risk.

➤ *Market risk* is the risk that is correlated with price fluctuations of the general market. Since an essential feature of the market is cross-sectional, it is also called *undiversifiable risk* or *aggregated risk*. Sources of macroeconomic factors such as inflation and changes in exchange rates are often considered to be such a risk since the likelihood they will cause adverse market price fluctuations is high.

➤ *Idiosyncratic risk* is company-specific risk that is uncorrelated with price fluctuations of the general market. As the Markowitz portfolio theory showed in the last chapter, it can be greatly reduced through diversification. Thus, it is also called *diversifiable risk*. Sources such as strikes, slumping sales, and unexpected poor earning reports or forecasts are examples of such risks.

Though in the current chapter we shall extend Markowitz's portfolio theory, risk will still be measured by employing the variance and covariance of returns. We shall see that the sensitivity to the market return plays an important role. But later, we will introduce other measures of risk and tools used to determine the risk-reward profile of a portfolio.

---

[1] For example, REITs, which is an acronym for Real Estate Investment Trusts.

A.O. Petters, X. Dong, *An Introduction to Mathematical Finance with Applications*, Springer
Undergraduate Texts in Mathematics and Technology, DOI 10.1007/978-1-4939-3783-7_4

**Remark 4.1.**

1. Unless stated otherwise, in this chapter, $R_P$ and $r_P$ denote the return[2] and logarithmic return of a portfolio $P$ on a general time interval $[t_0, t_f]$, respectively. On the other hand, by default the risk-free rate r is a percent quoted on an annual basis. To keep the mathematical expressions simple, in the current chapter, both $R_P$ and r are for the same period; consequently, so are $\mu_P = \mathbb{E}(R_P)$ and $\sigma_P = \sqrt{\mathrm{Var}(R_P)}$. For example, if one of them is monthly, so are the rest.

2. Although we will provide all the concepts in this chapter using only total returns, some concepts such as the Sharpe ratios and linear factor models can be defined using logarithmic returns as well. The latter has advantages in studying the properties of individual securities since logarithmic returns are more tractable than ordinary returns. Furthermore, statistical tools can be more conveniently applied with logarithmic returns, particularly under the lognormal assumption of security prices.

$\square$

## 4.1 The Capital Market Theory

Capital market theory, which includes the Capital Asset Pricing Model (CAPM[3]), was developed by William Sharpe,[4] John Lintner, and Jan Mossin. This theory naturally generalizes the Markowitz mean-variance portfolio model by introducing both a new efficient frontier that extends beyond the Markowitz efficient frontier and a model for pricing individual securities. The new efficient frontier is formed by adding a risk-free borrowing or lending consideration, which turns the old efficient frontier into a half line.

Recall that the one-period Markowitz model is interested in risk-averse investors selecting portfolios at time $t_0$ that produce stochastic returns at time $t_f$. Besides all the assumptions we made about the Markowitz model in the last chapter,[5] we also assume that:

➢ all investors have equal access to borrowing and lending, which occur at the same risk-free rate r, and lenders bear no risk of not being repaid;

➢ the inflation rate is no more than the risk-free rate r.[6]

---

[2] The return can be applied either in a simple or compounded context.

[3] Pronounced "CAP-M."

[4] Harry Markowitz, Merton Miller, and William F. Sharpe shared the 1990 Nobel Prize in Economic Sciences. Sharpe won for his contributions to the Capital Asset Pricing Model. See the Press Release at Novelprize.org.

[5] See Section 3.1.

[6] Under normal circumstances, inflation constitutes a major portion of the risk-free rate. The only problem with this is when the inflation is far above the risk-free rate due to a central bank intervention.

## 4.1.1 The Capital Market Line (CML)

Let $A$ be a risk-free security with risk-free rate r.[7] Consider a set of $N$ risky securities,[8] and assume that the initial investment at time $t_0$ is the amount $V_0$. We shall investigate the best risk-return trade-off portfolio design that is based only on allocating the initial investment between $A$ and a portfolio $B$ consisting of the given $N$ risky securities.

As in Markowitz portfolio theory, we represent each portfolio $P$ by a point in the $(\sigma_P, \mu_P)$-plane. To avoid possible confusion about the time basis for these quantities and r, we clarify our usage in the remark below:

**Remark 4.2.** The definitions of $R_P$, $\mu_P$, and $\sigma_P$ used in this section are the same as that in (3.8) (page 94), (3.13) (page 95), and (3.14) (page 96), respectively, which are all relative to a time interval $[t_0, t_f]$. As noted in Remark 4.1, we allow in this chapter for r to be not necessarily on an annual basis, but to have the same period as $R_P$. The same applies to $R_M$, $R_i$, etc., which will be introduced later. $\qquad\square$

Now, if we let $V_0$ be the amount of the initial investment at time $t_0$, then $P$ with coordinates $(0, r)$ on the $\mu$-axis means that the investor puts all the money $V_0$ into the risk-free security A. Otherwise, $P$ is a point in the right half plane $\sigma_P > 0$ and with $\mu_P > r$, since no investor would want to take on any risk to get an expected return below or equal to r.

Let $A$ and $B$ be two points in the $(\sigma_P, \mu_P)$-plane described by

$$A = (0, r) \quad \text{and} \quad B = (\sigma_B, \mu_B) \quad \text{with } \sigma_B > 0 \text{ and } \mu_B > r.$$

First, we present a more intuitive discussion. Any point $(x, y)$ on the line determined by the points A and B can be represented by

$$(x, y) = w_0(0, r) + (1 - w_0)(\sigma_B, \mu_B)$$

for some real number $w_0$, and (see Figure 4.1)

if $w_0 \in [0, 1]$, then $(x, y)$ is between A and B;
if $w_0 < 0$, then $(x, y)$ is on the right side of B;
if $w_0 > 1$, then $(x, y)$ is on the left side of A.

---

[7] A *risk-free security* is, of course, a theoretical concept. In reality, any investment carries a certain amount of risk. In this context, by risk-free securities we mean US T-bonds or FDIC insured bank accounts to which we can lend out our money and obtain a sufficient credit line from which we can borrow money. Under our assumption in the last section, the interest rate for lending is equal to that for borrowing and occurs at the *risk-free rate*.

[8] At this stage, $N$ is arbitrary. Eventually we need to consider only a sufficiently large $N$ for which we have a diversified portfolio.

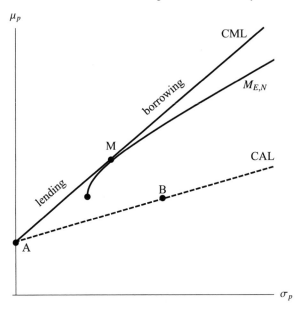

**Fig. 4.1** The points $M$, $B$, and $A$ have coordinates $(\sigma_M, \mu_M)$, $(\sigma_B, \mu_B)$, and $(0, r)$, respectively. The curve $M_{E,N}$ is the efficient frontier of $F_{P,N}$. The line CAL is the graph of equation (4.1), while the CML is the graph of equation (4.3).

In our context, obviously, the case $w_0 > 1$ is not applicable as no investor would want to get an expected return below r. That is to say, a possible portfolio design in terms of the portfolio's expected return rate $\mu_P$ and variance $\sigma_P^2$ is given by

$$(\sigma_P, \mu_P) = w_0(0, r) + (1 - w_0)(\sigma_B, \mu_B) \quad \text{where } w_0 \leq 1. \tag{4.1}$$

The fact that

$$\text{risk with lending} < \sigma_B < \text{risk with borrowing}$$

implies that

if $w_0 \in [0, 1]$, then the investor lends amount $w_0 V_0$ at time $t_0$;

if $w_0 < 0$, then the investor borrows amount $- w_0 V_0$ at time $t_0$.

The graph of (4.1) in the $(\sigma_P, \mu_P)$-plane as $w_0$ varies is shown in Figure 4.1 and is called a *Capital Allocation Line (CAL)*.

Second, we take a more theoretical approach. Let $F_{P,N}$ denote the set of feasible portfolios (the Markowitz bullet) that contains only the given $N$ risky securities, and let $F_{P,N+1}$ denote the set of all feasible portfolios that contain the given risk-free security and the $N$ risky securities. We have

$$F_{P,N} \subset W_N = \{w = [w_1, w_2, \cdots, w_N]^\top : \sum_{i=1}^N w_i = 1\},$$

$$F_{P,N+1} \subset W_{N+1} = \{w = [w_0, w_1, w_2, \cdots, w_N]^\top : \sum_{i=0}^N w_i = 1\},$$

where $w_0$ is the weight of the risk-free security A, and $w_i$ is the weight of the $i$th risky security for $i = 1, 2, \ldots, N$, and $W_N$ and $W_{N+1}$ are defined as indicated. *Since we shall need N sufficiently large, it suffices to assume at this stage that $N \geq 3$.*[9]

We are interested in finding the efficient frontier of $F_{P,N+1}$.[10] Recall that the Markowitz efficient frontier of $F_{P,N}$ is denoted by $M_{E,N}$; see Figure 4.1. *For ease of presentation, we shall slightly abuse our notation and denote the efficient frontier of $F_{P,N+1}$ by $M_{E,N+1}$* (though all the securities are not risky). For each $P \in F_{P,N+1}$, there exists vector $w_P \in W_{N+1}$ such that $P$ can be expressed by

$$w_P = [w_0, w_1, w_2, \cdots, w_N]^\top \quad \text{with} \sum_{i=0}^N w_i = 1. \tag{4.2}$$

Let

$$w_N = [w_1, w_2, \cdots, w_N]^\top,$$

where $w_i, i = 1, 2, \ldots, N$ remain the same as in (4.2). Notice that $w_N \notin W_N$ unless $w_0 = 0$ for $\sum_{i=1}^N w_i = 1 - w_0$. We have

$$\mu_P = w_0 r + \sum_{i=1}^N w_i \mu_i, \quad \sigma_P^2 = w_N^\top V w_N,$$

where $V$ is the covariance matrix of the random vector $[R_1, R_2, \cdots, R_N]^\top$ and here $R_i$ is the return from the $i$th risky security over the time period $[t_0, t_f]$ for $i = 1, 2, \ldots, N$.

Let $\tilde{w}_i = \frac{w_i}{1 - w_0}$, and $\tilde{w} = [\tilde{w}_1, \tilde{w}_2, \cdots, \tilde{w}_N]^\top$. Then $\tilde{w} \in W_N$ as $\sum_{i=1}^N \tilde{w}_i = 1$. Since $w_N = (1 - w_0)\tilde{w}$, we have

$$\mu_P = w_0 r + (1 - w_0) \sum_{i=1}^N \tilde{w}_i \mu_i = w_0 r + (1 - w_0)\tilde{\mu},$$

$$\sigma_P^2 = w_N^\top V w_N = (1 - w_0)^2 \tilde{w}^\top V \tilde{w} = (1 - w_0)^2 \tilde{\sigma}^2,$$

where $(\tilde{\sigma}, \tilde{\mu})$ represents a portfolio containing no risk-free securities. Thus, for $w_0 < 1$,

$$P = (\sigma_P, \mu_P) \in M_{E,N+1} \quad \text{iff} \quad (\tilde{\sigma}, \tilde{\mu}) \in M_{E,N}.$$

---

[9] Recall that the feasible set for $N = 2$ is a curve, while for $N \geq 3$, it is a region (see page 126).

[10] The proof to be provided here is for a one-period portfolio. The proof for $n$-period self-rebalancing portfolios is similar. The details are left as an exercise for the reader.

Obviously, the graph of $M_{E,N+1}$ in the $(\sigma_P, \mu_P)$-plane is the (half) tangent line to the graph of $M_{E,N}$ at the point $(\tilde{\sigma}, \tilde{\mu})$ and with the left end point $(0, r)$. Otherwise, either the graph of $M_{E,N+1}$ is a secant line to that of $M_{E,N}$ or is above that of $M_{E,N}$. In either case, a contradiction can be produced. The graph of $M_{E,N+1}$ in the $(\sigma_P, \mu_P)$-plane is called a *Capital Market Line (CML)*.

Finally, combining our above intuitive and theoretical discussions, we have the equation for the *CML*:

$$(\sigma_P, \mu_P) = w_0(0, r) + (1 - w_0)(\tilde{\sigma}, \tilde{\mu}),$$

where $(\tilde{\sigma}, \tilde{\mu}) \in M_{E,N}$.

It follows from the Markowitz portfolio theory of Chapter 3 that the efficient portfolio $(\tilde{\sigma}, \tilde{\mu})$ offers the best risk-return trade-off. Everyone will then want to invest in the portfolio $(\tilde{\sigma}, \tilde{\mu})$ and will want a risk-free asset to be on a CML. If a security is not part of this portfolio, then there is no interest in the security, which will cause it to drop out the marketplace. Consequently, *the tangent point portfolio $(\tilde{\sigma}, \tilde{\mu})$ then consists of all securities in the marketplace and so N is now the total number of securities in the market. It also follows that the weight of each security in $(\tilde{\sigma}, \tilde{\mu})$ is the percent of the marketplace the security occupies, i.e., the weight is the security's market capitalization.* For this reason, $(\tilde{\sigma}, \tilde{\mu})$ is called the *market portfolio* and denoted by $(\sigma_M, \mu_M)$. Therefore, our desired portfolio should be designed by putting the percentage $w_0$ in the risk-free security and $1 - w_0$ in the market portfolio:

$$(\sigma_P, \mu_P) = w_0(0, r) + (1 - w_0)(\sigma_M, \mu_M). \tag{4.3}$$

Again, the graph of (4.3) in the $(\sigma_P, \mu_P)$-plane is called a *capital market line (CML)*. It is depicted in Figure 4.1.

Using the parametric equations of the CML(obtained from the equation (4.3)),

$$\begin{cases} \mu_P = w_0 r + (1 - w_0)\mu_M \\ \sigma_P = (1 - w_0)\sigma_M \end{cases},$$

to express $\mu_P$ in terms of $\sigma_P$, we obtain the slope-intercept form of the CML:

$$\mu_P(\sigma_P) = \frac{\mu_M - r}{\sigma_M}\sigma_P + r. \tag{4.4}$$

**Example 4.1. (No Leverage Versus Leverage)** Suppose that you have $2,000 to invest. Assume a risk-free rate of 6% and market expected return of 12%.

a) If you invest $1,500 in a risk-free security at the rate of 6% and put the rest in the market portfolio, then what are the expected portfolio return rate and portfolio risk?

**Solution.** Since $w_0 = \frac{\$1,500}{\$2,000} = 0.75$ and $r = 0.06$, we have

$$\mu_P = w_0 r + (1 - w_0)\mu_M = 0.75 \times 0.06 + 0.25 \times 0.12 = 0.075 = 7.5\%$$
$$\sigma_P = (1 - w_0)\sigma_M = (1 - 0.75)\sigma_M = 0.25\sigma_M.$$

The portfolio has less expected return than the market's, but the portfolio risk is only 25% of the market risk.

b) If you add leverage to your portfolio by borrowing $1,500 at the risk-free rate, then what are the expected portfolio return rate and portfolio risk? Compare with the previous case.

**Solution.** In this case, $w_0 = -0.75$. Consequently,

$$\mu_P = w_0 r + (1 - w_0)\mu_M = -0.75 \times 0.06 + 1.75 \times 0.12 = 0.165 = 16.5\%$$
$$\sigma_P = (1 - w_0)\sigma_M = (1 - (-0.75))\sigma_M = 1.75\sigma_M.$$

This means that after paying back the loan, the expected return is 16.5%, which is the sum of the loss in returns due to paying the loan and the gain in returns from investing the loan and the initial capital. The expected return is more than double that for the previous case. However, such higher expected returns expose you to risks that are much more volatile than the market, about 1.75 times the market volatility. The leveraged portfolio is seven times more risky than the portfolio in the previous case.                     □

### 4.1.2 Expected Return and Risk of the Market Portfolio

The CML shows that the market portfolio is the best efficient frontier portfolio to combine with a risk-free security to exceed the expected returns of the Markowitz model.

Though in practical applications we can employ a proxy for the market portfolio to position a portfolio on the CML, the proxy does not give us any quantitative understanding of where the market portfolio is on the Markowitz efficient frontier. Moreover, one may be interested in investing in a limited number of stocks, for which there may be no obvious proxy, and wish to construct a CML-type tangent to the Markowitz efficient frontier of those stocks. For these reasons, we compute, for a general $N$-security risky portfolio, the point of tangency of the CML to the Markowitz efficient frontier of the $N$ securities. We shall obtain explicit expressions for the expected return and risk of the point of tangency, which we call the *market portfolio for the $N$ securities*, even if the securities form a subset of the true market portfolio. The mathematical framework will, therefore, be general enough to incorporate the cases ranging from two securities to any finite number.

Given a risk-free rate $r$ and $N$ risky securities, we apply the definition of $\mu$ and formula for $\sigma$ in the Markowitz portfolio theory:

$$\mu_M = w_M^\top \mu, \qquad \sigma_M(\mu_M) = \sqrt{w_M^\top V w_M} = \sqrt{\frac{A\mu_M^2 - 2B\mu_M + C}{AC - B^2}},$$

where $w_M^\top e = 1$, to obtain the expressions of $\mu_M$, $\sigma_M$ and $w_M$ as follows:

$$\mu_M = \frac{C - Br}{B - Ar}, \quad \sigma_M^2 = \frac{Ar^2 - 2Br + C}{(B - Ar)^2}, \quad w_M = \frac{V^{-1}(\mu_M - re)}{B - Ar}. \qquad (4.5)$$

The computational details are left as an exercise for the reader (Exercise 4.23).

The reader may recall the diversified portfolio that was introduced in the Markowitz model as part of establishing the Mutual Fund Theorem (see (3.81) on page 131). The market portfolio is a generalization of the diversified portfolio $(\sigma_D, \mu_D)$. In fact, if $r = 0$, then we obtain the diversified portfolio from the market portfolio:

$$\mu_M = \frac{C}{B} = \mu_D, \qquad \sigma_M = \frac{\sqrt{C}}{B} = \sigma_D, \qquad w_M = \frac{V^{-1}\mu}{B} = w_D,$$

where we use $B > 0$. In this case, the CML runs from the origin to the diversified portfolio $(\sigma_D, \mu_D)$.

### 4.1.3 The Capital Asset Pricing Model (CAPM)

Consider a security with return $R_i$. The security's *beta*, denoted $\beta_i$, measures the degree to which the security's return moves in step with the market's return:

$$\beta_i \equiv \frac{\text{Cov}(R_i, R_M)}{\sigma_M^2} = \rho_{iM}\frac{\sigma_i}{\sigma_M},$$

where $R_M$ is the return of the market portfolio and $\sigma_M$ is the market portfolio's risk.

The risk of the security is bounded below in terms of beta and the market risk as follows:

$$\sigma_i \geq |\beta_i|\sigma_M.$$

We shall see that the result can be improved to an equality relating the risk premium of the security in terms of beta and the market risk premium.

The *risk premium of a security* is defined to be

$$\mu_i - r,$$

which is how much the security's expected return is above/below the risk-free rate r. The *risk premium of the market portfolio* is

$$\mu_M - r.$$

This is how much the market's expected return is expected to differ from the risk-free rate. The Capital Asset Pricing Theorem, which is due to Sharpe, Lintner, and Mossin, relates the risk premiums of the security and market via beta.

**Theorem 4.1. (Capital Asset Pricing Theorem)** *Assume that the covariance matrix $V$ of all securities is positive definite and $\mu$ and $e$ are linearly independent. Let $\mu_i = e_i^\top \mu$, where $e_i^\top = [0 \cdots 0\, 1\, 0 \cdots 0]$ with 1 in the ith slot, be the expected return on the ith security. Then*

$$\mu_i - r = \beta_i(\mu_M - r). \tag{4.6}$$

*Proof.* The idea is to show that

$$\beta_i = \frac{\text{Cov}(R_i, R_M)}{\sigma_M^2} = \frac{e_i^\top V w_M}{w_M^\top V w_M} = \frac{\mu_i - r}{\mu_M - r}.$$

Let $w_\ell^M$ be the weight of the $\ell$th security in the marketplace. Then

$$\text{Cov}(R_i, R_M) = \text{Cov}(R_i, w_M^\top R_M) = \text{Cov}(R_i, \sum_{\ell=1}^N w_\ell^M R_\ell) = \sum_{\ell=1}^N w_\ell^M \text{Cov}(R_i, R_\ell)$$

$$= e_i^\top V w_M.$$

Furthermore, since

$$e_i^\top V w_M = [0 \cdots 1 \cdots 0] \begin{bmatrix} \sigma_{11} & \cdots & \sigma_{1N} \\ \vdots & & \vdots \\ \sigma_{i1} & & \sigma_{iN} \\ \vdots & & \vdots \\ \sigma_{N1} & \cdots & \sigma_{NN} \end{bmatrix} \begin{bmatrix} w_1^M \\ \vdots \\ w_N^M \end{bmatrix} = \sum_{\ell=1}^N w_\ell^M \sigma_{i\ell},$$

it follows

$$\beta_i = \frac{e_i^\top V w_M}{w_M^\top V w_M}.$$

Now, we saw that the market portfolio vector is given by

$$w_M = \frac{V^{-1}(\mu_M - re)}{B - Ar},$$

or equivalently,

$$V w_M = \frac{\mu_M - re}{B - Ar}.$$

Because $e_i^\top \mu_M = \mu_i$, $e_i^\top e = 1$, $w_M^\top \mu_M = \mu_M$, and $w_M^\top e = 1$, we get

$$\beta_i = \frac{e_i^\top V w_M}{w_M^\top V w_M} = \left(\frac{e_i^\top \mu_M - r e_i^\top e}{B - Ar}\right)\left(\frac{B - Ar}{w_M^\top \mu_M - r w_M^\top e}\right)$$

$$= \frac{\mu_i - r}{\mu_M - r}.$$

□

Although beta is a relatively stable measure of a stock's relative volatility (a stock's risk relative to the market), a stock's beta may change over time. Additionally, (4.6) carries the following messages:

1. Since $\beta$ as the slope of the regression line (see (4.6)) relates the excess returns of the stock to that of the market, $\beta$ measures the sensitivity of the stock's return to the fluctuations in the market.

2. $\beta(\mu_M - r)$ is a risk premium, where $\mu_M - r$ is the premium in units of $\beta$.

3. The sign of the beta of a stock indicates the direction of the movement of the stock price with respect to that of the market portfolio, and moreover, statistically speaking,

$$\beta \begin{cases} < 0 & \text{indicates that the stock is losing money while the market} \\ & \text{as a whole is gaining and vice versa;} \\ = 0 & \text{indicates that the stock fluctuates uncorrelatedly} \\ & \text{with the market;} \\ \in (0,1) & \text{indicates that the stock fluctuates less than the market;} \\ = 1 & \text{indicates that the stock fluctuates the same as the market} \\ & \text{(i.e., same direction and magnitude as the market);} \\ > 1 & \text{indicates that the stock price fluctuates more than the market.} \end{cases}$$

**Example 4.2. (Security Pricing via CAPM)** Let $D(t_0, t_f)$ be the stock cash dividend issued in the time period $[t_0, t_f)$. Let $S(t)$ be the stock's price at time $t$. Recall the one-period return (see (3.1) on page 85):

$$R(t_0, t_f) = \frac{S(t_f) + D(t_0, t_f) - S(t_0)}{S(t_0)},$$

where $S(t_f)$ and $D(t_0, t_f)$ are random variables, whereas $S(t_0)$ is deterministic. Taking expectations on both sides of the last equation yields

$$\mathbb{E}\left(R(t_0, t_f)\right) = \frac{\mathbb{E}\left(S(t_f)\right) + \mathbb{E}\left(D(t_0, t_f)\right) - S(t_0)}{S(t_0)}.$$

Solving for $S(t_0)$ from the last equation produces

$$S(t_0) = \frac{\mathbb{E}\left(S(t_f)\right) + \mathbb{E}\left(D(t_0, t_f)\right)}{1 + \mathbb{E}\left(R(t_0, t_f)\right)}. \tag{4.7}$$

Applying the CAPM to valuation of the stock price at time $t_0$, we obtain the asset pricing formula based on the CAPM

$$S(t_0) = \frac{\mu_{S(t_f)} + \mu_{D(t_0, t_f)}}{1 + r + \beta(\mu_M - r)},$$

where $\mu_{S(t_f)} = \mathbb{E}\left(S(t_f)\right)$ and $\mu_{D(t_0, t_f)} = \mathbb{E}\left(D(t_0, t_f)\right)$. $\quad\square$

**Remark 4.3.**

1. Given a time interval $[t_0, t_f]$, (4.7) provides a relation between an asset price at time $t_0$ and the expectation of its return over the interval. This is to say that every asset return model corresponds to an asset pricing model. This is why authors use terminologies like "modeling asset prices" and "modeling asset returns" interchangeably.

2. It is worth noting that the CAPM shows how the market must price an individual security in relation to its asset class index, which we call beta, a risk measure. Thus, the CAPM as an asset pricing model also shows how the set of all securities can be classified by one risk measure, which is the beta in the case of the CAPM. Looking ahead from these two points, the latter sections of the chapter will branch into risk measures and linear factor models which are a class of much more practically useful asset pricing models.

$\quad\square$

**Example 4.3.** The *hurdle rate* refers to the minimum acceptable (rate of an) investment return. This concept plays an important role in decision-making when a project is under consideration.

Here is a table of project specifications:

| Project | A | B |
|---|---|---|
| Project beta | 1.5 | 1.4 |
| Initial Investment | $10,000 | $10,000 |
| Expected Payoffs | $8,000 in 2 years | $9,000 in 2 years |
| | $16,000 in 5 years | $9,000 in 5 years |
| | | $9,000 in 8 years |

Assuming that the hurdle rate is calculated by the formula,

$$\text{hurdle rate} = \mathbb{E}(R_{project}) = r + \beta_{project}(\mathbb{E}(R_M) - r),$$

where $r = 2\%$ and $\mathbb{E}(R_M) = 20\%$, determine if any of the projects is worth pursuing.

**Solution.** We have

$$\mathbb{E}(R_{project\,A}) = 0.02 + 1.5 \times (0.2 - 0.02) = 0.290,$$
$$\mathbb{E}(R_{project\,B}) = 0.02 + 1.4 \times (0.2 - 0.02) = 0.272.$$

Since

$$\text{NPV}_A(0.290) = -\$10{,}000 + \frac{\$8{,}000}{1.29^2} + \frac{\$16{,}000}{1.29^5} = -\$713.70,$$

$$\text{NPV}_B(0.272) = -\$10{,}000 + \$9{,}000 \times \left( \frac{1}{1.272^2} + \frac{1}{1.272^5} + \frac{1}{1.272^8} \right)$$
$$= -\$421.52,$$

neither project is worth pursuing.                                                     □

Finally, *the CAPM Theorem also applies to portfolios.* First, the concept of beta extends naturally to portfolios. If $P'$ represents a portfolio of $n$ risky securities with portfolio weight vector $w'$ given by

$$w' = \begin{bmatrix} w'_1 \\ \vdots \\ w'_n \end{bmatrix}, \qquad \sum_{i=1}^{n} w'_i = 1, \tag{4.8}$$

the *portfolio beta* is defined to be the weighted average of the individual risky security betas:

$$\beta_{P'} = w'_1 \beta'_1 + w'_2 \beta'_2 + \cdots + w'_n \beta'_n, \tag{4.9}$$

where $\beta'_i$ is the beta of the $i$th risky security in $P'$ for $i = 1, 2, \ldots n$. Second, for portfolio $P'$ we denote by $\mu_{P'}$ and $\mu'_i$ the expected portfolio return and expected return of the $i$th security in $P'$, respectively. Then by applying (4.6) on page 159 to the $i$th risky security in $P'$ and employing (4.8) and (4.9), we obtain

$$\mu_{P'} - r = \sum_{i=1}^{n} w'_i(\mu'_i - r) = \sum_{i=1}^{n} w'_i \beta'_i(\mu_M - r).$$

Hence, we obtain the CAPM for portfolios:

$$\mu_{P'} - r = \beta_{P'}(\mu_M - r). \tag{4.10}$$

**Remark 4.4.**

1. The CAPM is a theoretically significant equilibrium pricing model. Without a model of market equilibrium, the efficient market hypothesis cannot be tested (see Fama [16]).

2. Although the CAPM has a beautiful simplicity in theory, the empirical evidence shows its weaknesses in practice (see Fama and French [19] for a detailed and comprehensive discussion).

$\square$

### 4.1.4 The Security Market Line (SML)

The Capital Asset Pricing Theorem can be viewed as expressing the expected return of a security as an affine function of the security's beta, i.e., it defines a straight line:

$$\mu_i(\beta_i) = (\mu_M - r)\beta_i + r.$$

This line is called the *security market line (SML)*. That is, the SML is a graphical representation of the CAPM on the $(\beta, \mu)$-plane.

An illustration of the SML is shown in Figure 4.2. A security with $\beta_i = 0$ has an expected return at the risk-free rate r—the security has no risk premium. For $\beta_i = 1$, we have a security with $\mu_i = \mu_M$, i.e., the risk premium of the security coincides with the market risk premium. If $\beta_i = 1.5$, then the security's risk premium is 1.5 times the market risk premium or 50% larger than the market risk premium. For $\beta_i = -1$, the security's risk premium is minus the market risk premium, which means that the expected security return is less than the expected market return by twice the market premium. Indeed, if $\beta_i = -b < 0$, then

$$\mu_i = \mu_M - (1 + b)(\mu_M - r).$$

In theory, the SML provides an equilibrium in the sense that each stock on the line is fairly valued. Otherwise, it is located off the line:

1. A stock is overvalued if it is below the SML because such a stock offers too low a risk premium. In other words, the level of expected return is not adequate for the given level of risk measured by $\beta$.

2. A similar argument can be made to show that a stock is undervalued if it is above the SML.

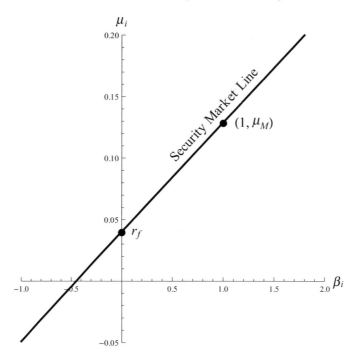

**Fig. 4.2** An illustration of the security market line (SML) for risky securities identified by $(\beta, \mu)$. A security with its risk measured by $\beta_i$ has expected return $\mu_i$. In particular, for $\beta_i = 0$, the security earns the risk-free rate ($\mu_i = r$), so the security has no risk premium. A security with $\beta_i = 1$ has an expected return at the market expected return, $\mu_i = \mu_M \approx 13\%$. For $\beta_i = -1$, the security has $\mu_i \approx -5\%$, i.e., the expected security return is less than the expected market return by twice the market risk premium ($\mu_i = \mu_M - 2(\mu_M - r) \approx -5\%$)

### 4.1.5 CAPM Security Risk Decomposition

By the Capital Asset Pricing Model, we have

$$\mathbb{E}\left(R_i - r\right) = \mathbb{E}\left(\beta_i(R_M - r)\right),$$

which yields the following model for the return of a risky security:[11]

$$R_i = r + \beta_i(R_M - r) + \varepsilon_i,$$

where $\varepsilon_i$ is a random variable with mean zero. Assume that $\varepsilon_i$ is normal with variance denoted by $\sigma_{\varepsilon_i}^2$ and suppose that $\varepsilon_i$ is independent of $R_M$ and $\varepsilon_j$ for $j \neq i$.

The security's risk can be found from

$$\sigma_i^2 = \text{Var}(R_i) = \text{Var}(\beta_i R_M) + \text{Var}(\varepsilon_i) = \beta_i^2 \sigma_M^2 + \sigma_{\varepsilon_i}^2. \qquad (4.11)$$

---

[11] If $\mathbb{E}(X) = \mathbb{E}(Y)$, then $X = Y + \varepsilon$, where $\varepsilon$ is a random variable with $\mathbb{E}(\varepsilon) = 0$, called an *error term*. Typically, an error term $\varepsilon$ is assumed to be normal (with mean zero).

The term

$$\mathrm{Var}(\beta_i R_M) = \beta_i^2 \sigma_M^2$$

is the market risk (also known as *systematic risk*), while

$$\mathrm{Var}(\varepsilon_i) = \sigma_{\varepsilon_i}^2$$

is the idiosyncratic risk (also known as *unsystematic risk*). Expression (4.11) asserts that the total risk of a security can be decomposed into two orthogonal components (under the inner product of covariance). In this sense we write

$$\text{Total Risk} = \text{Systematic Risk} + \text{Unsystematic Risk.}$$

Under the CAPM for security returns, we see that the covariance between the returns of any two securities is determined by the betas of the securities and market risk:

$$
\begin{aligned}
\mathrm{Cov}(R_i, R_j) &= \mathrm{Cov}\left(r + \beta_i(R_M - r) + \varepsilon_i, r + \beta_j(R_M - r) + \varepsilon_j\right) \\
&= \beta_i \beta_j \, \mathrm{Cov}(R_M, R_M) \\
&= \beta_i \beta_j \sigma_M^2.
\end{aligned}
$$

## 4.2 Portfolio Risk Measures

Risk measures are a challenging topic as the notion of risk itself is hard to conceptualize. The most popular measure of risk is volatility, which by definition measures the dispersion of the investment return from its mean, regardless of the direction of an investment price's movement. However, for investors in the real world, risks are often associated with the adverse movement of the market only. This is to suggest we also take a different perspective in the further development of risk measures: if a risk measure is about the sustainability of losing money, then maximum drawdown is employed (Section 4.2.3). If the risk measure is about the odds of losing money, then VaR and CVaR are used (Sections 4.2.5 and 4.2.6). Especially in a leveraged investment, the performance is measured by the return on unit risk, such as the Sharpe ratio (Section 4.2.1), the Sortino ratio (Section 4.2.2), or the ratio of return and maximum drawdown. It is worth pointing out, nevertheless, that in spite of the intuitiveness of the latter two ratios, the system developer often prefers to optimize the in sample Sharpe ratio as it is intimately connected to the statistical t-test, which is a measurement of the reliability of the system in out sample period.

In short, this section addresses several approaches to risk measures, which provide a variety of portfolio evaluation techniques.

**Remark 4.5.**

1. Before formally introducing any mathematical terminologies about risk, we encourage the readers to think about how they themselves as (individual or institutional) investors interpret "risk." For insight into how some of the world's greatest minds have viewed risk, we refer the reader to Peter Bernstein's book [5].

2. The lack of statistics in the prerequisite presents us with a great challenge in this section and the next. To circumvent this difficulty, one of our pedagogical approaches is to focus on basic understanding of concepts and avoid statistical tests.

□

### 4.2.1 The Sharpe Ratio

Consider a portfolio $P$ over a time period $[t_0, t_f]$. Let $R_P$ be the portfolio return over $[t_0, t_f]$ and let r be the best available risk-free rate corresponding to the same period (e.g., T-bills).[12] The portfolio's *Sharpe ratio*,[13] denoted by $S(P)$, is defined by

$$S(P) = \frac{\mathbb{E}(R_P - r)}{\sigma_P(R_P - r)}.$$

The Sharpe ratio was originally developed as a forecasting tool with the expected return to calculate the forward-looking ratio (see Sharpe [36]). But with the historical returns, which can be of any frequency, e.g., hourly, daily, monthly, and so on, it is used to evaluate the risk-reward trade-off of an investment over that period.

**Example 4.4.** Suppose that we have the data set of monthly returns of a portfolio P and that of monthly rates of 90-day T-bills over the past 54 months. This is to say that each data set consists of 54 data points. Using the sample mean and sample standard deviation, we can approximate the Sharpe ratio of the portfolio over the past 4.5 years (54 months). A Sharpe ratio of

$$S(P) = 0.0401$$

is interpreted as the portfolio $P$ earned an average excess return of 4% per unit risk over the time period under consideration.

---

[12] Recall that US Treasuries can be classified into *bills, notes,* and *bonds* according to their initial maturities (in years) in terms of time intervals: $(0, 1], (1, 10],$ and $(10, \infty),$ respectively. We consider only T-bills here since, the longer the maturities, the bigger the risk of inflation, and consequently, the less reliable.

[13] Named after William Sharpe.

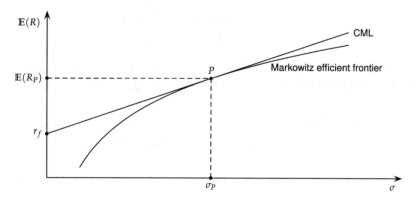

**Fig. 4.3** The slope of the CML equals the Sharpe ratio

Statistically speaking, data sets need to be sufficiently large. The more data points we use (the shorter the sub-sample period), the more accurate our approximation of $S(P)$ becomes. □

For a constant risk-free rate, we have

$$\mathbb{E}(R_P - r) = \mathbb{E}(R_P) - r, \qquad \text{Var}(R_P - r) = \text{Var}(R_P),$$

so

$$S(P) = \frac{\mathbb{E}(R_P) - r}{\sigma_P(R_P)}.$$

The following explains the significance of the Sharpe ratio:

1. Since $\mathbb{E}(R_P - r)$ and $\sigma_P$ represent the expected excess return and risk, respectively, the Sharpe ratio is a measure of the excess return per unit of risk of the portfolio. In other words, the ratio describes how much risk premium you are receiving for the extra volatility that you endure for holding a riskier portfolio. In short, the Sharpe ratio measures risk-adjusted performance of the portfolio.

2. Recall that if you invest part of your money in a risk-free security (e.g., T-bills) and the remainder in an efficient portfolio, the capital market line (CML) can help you find the portfolio $P$ that offers the "most favorable" risk-return trade-off. In fact, assuming that $R_P$ and $r$ are the same as we used in the discussion of the CAPM, the slope of this CML is equal to the Sharpe ratio of $P$ (see Figure 4.3). This observation provides a method for finding the best possible portfolio from the given collection of securities.

3. The Sharpe ratio is a leverage-environment[14] measure of performance in the sense that if $r$ is omitted, then approximately

---

[14] Two basic ways of achieving leverage are (a) to borrow money for investment and (b) to use financial instruments such as futures and options (see Chapter 7).

$$S_P = \frac{\mathbb{E}(R_P)}{\sigma_P(R_P)} = \frac{b\mathbb{E}(R_P)}{b\sigma_P(R_P)},$$

where b denotes the leverage factor. For instance, if we double every investment, then the return is doubled, but the risk (standard deviation $\sigma$) is also
doubled. If a hedge fund predetermines the risk level, then the Sharpe ratio
determines the return from which the leverage level can be determined for
products allowing high leverage (e.g., futures, commodities, currencies, and
options). In this sense, the Sharpe ratio provides a method of optimizing a
portfolio. Otherwise, the Sharpe ratio does not provide a portfolio of highest
return possible.

4.  A negative correlation can considerably reduce the standard deviation (com-
    pare page 115). Even in the case that the portfolio return is reduced, the
    Sharpe ratio may still increase.

**Remark 4.6.**

1. Sharpe ratios can be defined by using logarithmic returns as well. Keeping
   this in mind, we replace $R_P$ by $r_P$ in the definition of the Sharpe ratio to
   obtain

$$S(P) = \frac{\mathbb{E}(r_P - r)}{\sigma_P(r_P - r)}.$$

For a constant risk-free rate, we have

$$S(P) = \frac{\mathbb{E}(r_P) - r}{\sigma_P(r_P)}.$$

2. For the reader who has had an introductory statistical background and
   wishes to delve further into the significance of the Sharpe ratio, the
   t-statistic is useful. In testing the null hypothesis $\mu = \mu_0$, where $\mu$ denotes
   the population mean, one uses

$$t = \frac{\bar{x} - \mu_0}{\hat{\sigma}(x)/\sqrt{n}},$$

where $\bar{x}$ is the sample mean of the data, $\hat{\sigma}(x)$ is the sample standard deviation of the $x$-data, and $n$ is the sample size. The $\hat{\ }$ and $\bar{\ }$ are used to indicate
that a sample-data estimate is being carried out.

   In connection to our interest, we rewrite for constant r,

$$t = \frac{\overline{R_P} - r}{\hat{\sigma}_P/\sqrt{n}} = \sqrt{n}\,\hat{S}_P,$$

where

$$\hat{\sigma}_P = \hat{\sigma}(R_P - r) = \hat{\sigma}(R_P)$$

and $n$ is the number of returns used in the calculation of $\widehat{S}_P$. This relation shows us how reliable the portfolio strategy is in obtaining excess returns.

□

**Example 4.5.** Consider a portfolio with the following annual returns:

| Year | Year 1 | Year 2 | Year 3 | Year 4 | Year 5 | Year 6 |
|------|--------|--------|--------|--------|--------|--------|
| Return Rate | 21% | 7.8% | -13% | 59.4% | 0.2% | -1.2% |

Suppose that the average T-bill return for those 6 years is 4%. Find the Sharpe ratio for the portfolio in the period of these 6 years.

**Solution.** In this case, model the risk-free rate $r$ over the total time span as a constant given by 4% per year. The Sharpe ratio is then

$$\widehat{S}_P = \frac{\overline{R}_P - r}{\widehat{\sigma}_P}.$$

Applying the formula for sample mean

$$\overline{X} = \frac{1}{n}\sum_{i=1}^{n} X_i,$$

we obtain

$$\overline{R}_P = \frac{1}{6} \times \frac{21 + 7.8 - 13 + 59.4 + 0.2 - 1.2}{100} = \frac{74.2}{6 \times 100} = \frac{12.3667}{100}.$$

Note that we converted the percentages to pure fractions. The sample variance formula,

$$\widehat{\sigma}_P^2 = \frac{\sum_{i=1}^{n} X_i^2 - n\overline{X}^2}{n-1},$$

yields

$$\widehat{\sigma}_P^2 = \frac{1}{5} \times \frac{21^2 + 7.8^2 + 13^2 + 59.4^2 + 0.2^2 + 1.2^2 - 6 \times 12.3667^2}{100^2} = \frac{656.6137}{100^2}.$$

Thus,

$$\widehat{S}_P = \frac{\overline{R}_P - r}{\widehat{\sigma}_P} = \frac{100\,\overline{R}_P - 100\,r}{100\,\widehat{\sigma}_P} = \frac{12.3667 - 4}{\sqrt{656.6137}} = \frac{8.3667}{25.6245} = 0.3265.$$

□

**Example 4.6.** Suppose that the risk-free rate is 4% and that portfolio $i$ is operated under strategy $i$, where $i = 1,2$. Consider the following information:

| | $\mathbb{E}(R)$ | $\sigma$ | $S_P$ |
|---|---|---|---|
| Portfolio 1 | 17% | 9% | 1.44 |
| Portfolio 2 | 15% | 5% | 2.2 |

Which portfolio is more attractive to investors?

**Solution.** Assuming that $S_{P_1}$ and $S_{P_2}$ are equally statistically significant, portfolio 2 is preferable. Although strategy 1 generates a higher expected return, strategy 2 is able to generate a higher return on a risk-adjusted basis. Otherwise, since $t = \sqrt{n}\, S_P$, without information on $n$, one cannot tell the reliability of $S_{P_i}$ for $i = 1, 2$.

$\square$

**Remark 4.7.**

1. For long-term investments, a Sharpe ratio of $S_P > 1$ is typically considered desirable. However, some short-term traders may consider only $S_P \geq 3$ good enough, while other fund managers may consider $S_P > 2$ to be an appropriate target level.

2. Under the lognormal assumption, the unit conversions of volatility can easily be made by applying the property that the variance of its increments is linear in the observation interval.

3. One should always remember that the Sharpe ratio is calculated based on the historical returns and that past returns might be an indicator of future performance, but they are certainly not a guarantee.

4. For a complex trading or investing system, the Sharpe ratio may provide false information.

$\square$

### 4.2.2 The Sortino Ratio

The standard deviation of the return of an investment does not distinguish between up (good) and down (bad) volatility. A variation of the Sharpe ratio is the Sortino ratio, which uses semivariance to differentiate between good and bad volatility. Thus, defining the concept of semivariance is a preparation for understanding the Sortino ratio. We begin with a brief review of variance. Let $X$ be a random variable with the density function $f(x)$. The variance of $X$ is a measure of the dispersion of $X$ from its mean $u = \mathbb{E}(X)$ and can be computed by

$$\sigma^2 = \text{Var}(X) = \mathbb{E}((X - u)^2) = \int_{-\infty}^{\infty} (x - u)^2 f(x)\, dx.$$

Using the same idea, we can compute the dispersion of $X$ from a given number $a$ by

$$\sigma_a^2 = \int_{-\infty}^{\infty} (x - a)^2 f(x)\, dx.$$

In a similar fashion, if we are interested only in the dispersion of $X$ from one side of number $a$, say, from the downside $(-\infty, a]$, then the formula below serves that purpose:

$$\int_{-\infty}^{a} (x - a)^2 f(x)\, dx.$$

This leads to a natural way of defining sample semivariance. Let $X_1, \ldots, X_n$ be a random sample of size $n$ drawn from a population $X$. Let $a$ be a number. Let

$$Y_i = \begin{cases} a & \text{if } X_i \geq a \\ X_i & \text{if } X_i < a, \end{cases}$$

where $i = 1, \ldots, n$. The *downside sample semivariance* $\sigma_{a-}^2$ of $X$ is defined by

$$\sigma_{a-}^2 = \frac{1}{n} \sum_{i=1}^{n} (Y_i - a)^2. \tag{4.12}$$

Note that (4.12) is equivalent to

$$\sigma_{a-}^2 = \frac{1}{n} \sum_{i=1}^{n} (\min\{0, X_i - a\})^2. \tag{4.13}$$

We are now ready for the definition of the Sortino ratio. Let

$P$ represent a portfolio over a time period $[t_0, t_f]$,

$R_P$ be the portfolio return over $[t_0, t_f]$,

$r_0$ be the target or required rate of return for the investment strategy under consideration.[15]

The *Sortino ratio*, denoted by $S_D(P)$, is defined by

$$S_D(P) = \frac{\mathbb{E}(R_P) - r_0}{\sigma_{r_0-}},$$

where $\sigma_{r_0-} = \sqrt{\sigma_{r_0-}^2}$ and is the downside deviation of the portfolio equity.

**Example 4.7.** Let us use the same information given in Example 4.5 and find the Sortino ratio in the period of those 6 years.

To apply (4.13), we first compute

$$\min\{0, 21 - 4\} = 0, \quad \min\{0, 7.8 - 4\} = 0, \quad \min\{0, -13 - 4\} = -17,$$
$$\min\{0, 59.4 - 4\} = 0, \quad \min\{0, 0.2 - 4\} = -3.8, \quad \min\{0, -1.2 - 4\} = -5.2.$$

---

[15] The quantity $r_0$ was originally known as the *minimum acceptable return* (or hurdle rate) and is often taken to be r.

Thus,

$$\sigma_{r_0-}^2 = \frac{1}{6} \times \frac{17^2 + 3.8^2 + 5.2^2}{100^2} = \frac{330.48}{6 \times 100^2} = \frac{55.08}{100^2}.$$

We obtain

$$S_D(P) = \frac{\overline{R}_P - r}{\sigma_{r_0-}} = \frac{8.3667}{\sqrt{55.08}} = 1.1273.$$

$\Box$

Since the Sortino ratio captures the downside risk only, one can use the Sortino ratio as a measure to rank the performance of a portfolio or an investment strategy.

### 4.2.3 The Maximum Drawdown

A *drawdown* is a portfolio equity retracement. The *maximum drawdown* is the maximum equity retracement of a portfolio over a period of time. In other words, it is the peak-to-trough portfolio equity decline during a specific time period. The maximum drawdown measures how sustained one's losses can be and is usually quoted as the percentage between the peak and the trough. (For example, let the peak be 1, and the trough be 0.8. Then the maximum drawdown is $1 - 0.8 = 0.2 = 20\%$.) To express this concept in a rigorous fashion we introduce the following notation:

$V(t)$ = the value of the portfolio at time $t$,

$M(t) = \max_{u \in [0,t]} V(u)$, the maximum value of the portfolio over time period $[0,t]$.

Hence, our description of the maximum drawdown given above can be expressed below:

**Definition 4.1.** Given a time period $[0,T]$ and a portfolio, the *maximum drawdown* of portfolio equity over $[0,T]$, denoted by $\mathrm{MDD}(T)$, or simply $\mathrm{MDD}$, is defined by

$$\mathrm{MDD}(T) = \max_{0 \le u \le v \le T} (V(u) - V(v)). \tag{4.14}$$

However, in practice, the formula

$$\mathrm{MDD}(T) = \max_{0 \le t \le T} (M(t) - V(t)) \tag{4.15}$$

may be easier to use in computations as there is only one parameter involved.

For the definition to be meaningful, we must show that (4.14) and (4.15) are equivalent. Let

$$M_{(1)} = \max_{0 \le u \le v \le T} (V(u) - V(v)),$$

$$M_{(2)} = \max_{0 \le t \le T} (M(t) - V(t)).$$

Since $M(v) = \max_{u \in [0,v]} V(u)$ given $v \in [0, T]$, we have

$$\max_{0 \le u \le v \le T} (V(u) - V(v)) \ge M(v) - V(v) \quad \text{for each fixed } v.$$

This in turn implies that

$$\max_{0 \le u \le v \le T} (V(u) - V(v)) \ge \max_{0 \le v \le T} (M(v) - V(v)).$$

That is, $M_{(1)} \ge M_{(2)}$.

On the other hand, since

$$V(u) \le M(u) \le M(v) \quad \text{for} \quad u \le v,$$

$$V(u) - V(v) \le M(u) - V(v) \le M(v) - V(v) \quad \text{for} \quad u \le v,$$

we have

$$\max_{0 \le u \le v \le T} (V(u) - V(v)) \le \max_{0 \le v \le T} (M(v) - V(v)).$$

That is, $M_{(1)} \le M_{(2)}$. Therefore, $M_{(1)} = M_{(2)}$.

We conclude that the definition of $\text{MDD}(T)$ can be given by either of the following equivalent statements:

**Statement 1:** $\text{MDD}(T) = \max_{0 \le u \le v \le T} (V(u) - V(v))$.

**Statement 2:** $\text{MDD}(T) = \max_{0 \le t \le T} (M(t) - V(t))$.

Statement 1 is more intuitive in understanding the concept. Statement 2 is easier to use in computation.

**Example 4.8.** Consider a portfolio with these given daily values:

| Date | Day 1 | Day 2 | Day 3 | Day 4 | Day 5 | Day 6 |
|------|-------|-------|-------|-------|-------|-------|
| \$ ($V_i$) | 1141.95 | 1143.73 | 1147.73 | 1142.24 | 1140.81 | 1149.88 |

By the definition of $M(v)$,

$$M(1) = V_1, \quad M(2) = \max\{V_1, V_2\} = V_2,$$

$$M(3) = \max\{V_1, V_2, V_3\} = \max\{V_2, V_3\} = V_3,$$

and the rest of values of $M$ are

$$M(4) = V_3, \quad M(5) = V_3, \quad M(6) = V_6.$$

Since

$$\max_{0 \leq t \leq 6} \{M(t) - V(t)\} = \max\{0, 0, 0, V_3 - V_4, V_3 - V_5, 0\} = V_3 - V_5,$$

the maximum drawdown of the portfolio equity over the time period of these 6 days is

$$\text{MDD} = 1,147.73 - 1,140.81 = 6.92.$$

Since a drawdown is usually quoted as the percentage between the peak and trough, in terms of percentage, MDD is $6.92/1147.73 \approx 0.00603 \approx 0.6\%$.  $\square$

### 4.2.4 Quantile Functions

The quantile function is one of the basic statistical concepts. We are primarily interested in its fundamental role in defining distortion risk measures.[16]

The materials in quantiles covered in this section mainly serve as a preparation for introduction to the concepts of value-at-risk and conditional value-at-risk in the next two sections.

**Definition 4.2.** Let $X$ be a random variable with c.d.f. $F_X$. Given $p \in (0,1)$, a *p-quantile* of $X$ (or its distribution) is a number $a$ satisfying the properties

$$\mathbb{P}(X < a) \leq p \quad \text{and} \quad \mathbb{P}(X \leq a) \geq p. \tag{4.16}$$

The number $a$ is also referred to as a *quantile of order $p$* for the distribution of $X$.

Let $F_X(a^-)$ denote $\mathbb{P}(X < a)$, (4.16) is equivalent to

$$F_X(a^-) \leq p \quad \text{and} \quad F_X(a) \geq p.$$

➤ Keep in mind that a c.d.f. is a right-continuous and monotonically increasing (i.e., nondecreasing) function whose domain is the entire real line.

**Example 4.9.** Consider a random variable $X$ taking on the values 1, 2, 3, 4, 5, 6, 7, 8, 9, and 10 with p.d.f. given by $\mathbb{P}(X = i) = \frac{1}{11}$ for $i \neq 7$ and $\mathbb{P}(X = 7) = \frac{2}{11}$. The fact that

$$\mathbb{P}(X < 7) = \frac{6}{11} \quad \text{and} \quad \mathbb{P}(X \leq 7) = \frac{8}{11},$$

$$\mathbb{P}(X < 8) = \frac{8}{11} \quad \text{and} \quad \mathbb{P}(X \leq 8) = \frac{9}{11}$$

shows that both 7 and 8 are $\frac{8}{11}$-quantile of $X$, and so is each $a \in [7,8]$.  $\square$

---

[16] A distortion risk measure is a type of risk measure related to the cumulative distribution function of a financial portfolio return. CVaR (to be introduced shortly) is an example of a distortion risk measures.

**Definition 4.3.** The *quantile function* of a random variable $X$ is denoted by $F_X^{-1}$ and defined by

$$F_X^{-1}(p) = \min\{x \in \mathbb{R} | F_X(x) \geq p\}, \quad p \in (0,1). \tag{4.17}$$

In words, the value of a quantile function of $X$ at a point $p$ is defined by the lowest $p$-quantile of $X$.

In a situation where there is no ambiguity of the random variable, to ease the notation, the *quantile function is also denoted by $Q(p)$.*

**Example 4.10.** Let $X$ be an exponential random variable with parameter $\lambda$. Find the quantile function of $X$.

**Solution.** The probability density function of $X$ is

$$f(x) = \begin{cases} 0 & \text{if } x < 0 \\ \lambda e^{-\lambda x} & \text{if } x \geq 0. \end{cases}$$

Let $F$ be the cumulative distribution function of $X$. Then

$$F(x) = \mathbb{P}(X \leq x) = \int_{-\infty}^{x} f(t)\,dt = \begin{cases} 0 & \text{if } x < 0 \\ \int_0^x \lambda e^{-\lambda t}\,dt & \text{if } x \geq 0 \end{cases}$$

$$= \begin{cases} 0 & \text{if } x < 0 \\ 1 - e^{-\lambda x} & \text{if } x \geq 0. \end{cases}$$

By definition, $Q(p) = F^{-1}(p)$, where $p \in (0,1)$. Thus

$$F(x) = p \quad \text{if and only if} \quad 1 - e^{-\lambda x} = p.$$

Solving for $x$, we obtain $Q(p) = -\frac{\ln(1-p)}{\lambda}$.

Observe that $Q(p)$ is a continuously increasing function on $(0,1)$. □

**Example 4.11.** We revisit random variable $X$ taking on the values $1, 2, 3, 4, 5, 6,$ $7, 8, 9,$ and $10$ with p.d.f. defined by $\mathbb{P}(X = i) = \frac{1}{11}$ for $i \neq 7$ and $\mathbb{P}(X = 7) = \frac{2}{11}$. A straightforward verification shows that

$$Q(p) = \begin{cases} i & \text{if } p \in (\frac{i-1}{11}, \frac{i}{11}], \ i = 1,2,3,4,5,6 \\ 7 & \text{if } p \in (\frac{6}{11}, \frac{8}{11}] \\ i & \text{if } p \in (\frac{i}{11}, \frac{i+1}{11}], \ i = 8,9,10. \end{cases}$$

Observe that $Q(p)$ is nondecreasing and left-continuous on $(0,1)$. Shortly we will show that these properties of $Q$ are not accidental. □

The next property helps the reader to visualize graphs of quantile functions.

*Property 4.1.*

1. A quantile function is monotonically increasing on $(0,1)$.
2. A quantile function is left-continuous on $(0,1)$.

*Proof.* Let $X$ be a random variable. As usual, $F$ and $Q$ represent the c.d.f. and quantile function of $X$, respectively.

To show item 1, given $p_1, p_2 \in (0,1)$ with $p_1 < p_2$, we let

$$A_i = \{x \in \mathbb{R} | F(x) \geq p_i\}, \quad i = 1,2.$$

Clearly, $x \in A_2$ implies $x \in A_1$. Thus $A_2 \subseteq A_1$. Consequently, $Q(p_2) \geq Q(p_1)$.

To show item 2, we need to show that $\forall \varepsilon > 0, \exists \delta > 0$, such that

$$0 < p_0 - p < \delta \quad \text{implies} \quad 0 \leq Q(p_0) - Q(p) < \varepsilon.$$

In fact, for a fixed $p_0$, let $x = Q(p_0)$. Given $\varepsilon > 0$, let $x' = x - \varepsilon$ and $p' = F(x')$. Clearly, $p' < p_0$ by the definition of $Q$ and property of $F$. It follows from item 1 that for $\delta = p_0 - p' > 0$,

$$x' < Q(p) \leq Q(p_0) \quad \text{whenever} \quad p \in (p_0 - \delta, p_0).$$

This implies that $0 \leq Q(p_0) - Q(p) < x - x' = \varepsilon$. $\qquad\qquad\square$

Now we summarize the above intuition of $F_X$ and $F_X^{-1}$ as follows:

➤ A distribution function is nondecreasing and continuous from the *right*.
➤ A quantile function is nondecreasing and continuous from the *left*.

**Example 4.12.** In statistics, if $x = (x_1, x_2, \ldots, x_n)$ represents the observed values of a sample of size $n$ corresponding to a random variable $X$, the *order statistic of rank k*, denoted by $x_{(k)}$, is the $k$-th smallest value in the data set $x$. That is,

$$x_{(1)} \leq x_{(2)} \leq \cdots \leq x_{(n)}.$$

The (sample) quantile of order $p = \frac{1}{2}$ is known as the (sample) *median* of $X$.

The (sample) quantile of order $p = \frac{1}{4}$ is known as the *first (sample) quartile* of $X$. Similarly, the (sample) quantile of order $p = \frac{3}{4}$ is known as the *third (sample) quartile* of $X$.

The quantile function is a useful tool in dividing ordered (from lowest to highest) data (i.e., order statistics) into finitely many essentially probability-wise equally sized data subsets (see Exercise 4.7 on page 200). In this sense, the term quantile is synonymous with percentile.

As the data size $n$ increases, a natural question to ask is what are asymptotic behaviors of sample quantile functions. We refer the reader with sufficient statistic backgrounds to the literature (e.g., Ma, Genton, and Parzen [29]). $\qquad\square$

**Example 4.13.** In modern financial theory, the concept of quantile can be used as a measure of the downside portfolio risk. If $X$ represents the possible loss on a portfolio, this measure is determined by a prescribed $p$-quantile (e.g., $p = 1\%$) of $X$ such that the likelihood of $X$ (i.e., loss in dollar amount) to take on a value larger than that $p$-quantile is less than probability $p$ (i.e., 1% chance).

Such a measure of the downside risk is called VaR, an abbreviation for value-at-risk, which will be introduced shortly. □

The next remark is for the reader interested in statistical importance and applications of quantile functions.

**Remark 4.8.**

1. The inverse relation between the quantile function and the cumulative distribution function makes the quantile function one of basic concepts used to describe the probability distribution of a random variable. Since the quantile function also plays an essential role in the concept of mid-distribution, which is important for discrete distributions, the quantile function is especially important for sample distribution functions (therefore for statistical data modeling).

2. Regression analysis is a way to determine whether or not there is a correlation between two or more variables and how strong any correlation may be. Quantile regression is a type of regression analysis used in statistics and econometrics (e.g., detection of heteroscedasticity). Both statistics and econometrics are employed in mathematical finance. Just as the method of least squares enables one to estimate models for conditional means, methods of quantile regression enable one to estimate models for conditional quantile functions (e.g., conditional median function).

   For extensive discussions on the utility of quantile functions in statistical applications, we refer the reader to the literature (e.g., Gilchrist [21]). □

### 4.2.5 *Value-at-Risk*

Value-at-risk (VaR) and conditional value-at-risk (CVaR) are two risk measures widely used by financial institutions and financial regulators. The latter may be viewed as an extension of, or complement to, the former.

We focus on a basic understanding of the concepts of these two measures in Sections 4.2.5 and 4.2.6 and explore a deeper issue behind these concepts in Section 4.2.7.

For the sake of convenience and clarity of notation, we begin with the following notational remarks:

➤ Let $X$ represent data of (random) returns of a portfolio. By saying that $X$ is in the form of profit/loss (P/L), we mean that the data assign positive values to profits and negative values to losses. In a similar fashion, by saying that $X$ is in the form of loss/profit (L/P), we mean that the data assign positive values to losses and negative values to profits.

➤ In studying VaR and CVaR in relation to financial risk management, it is more convenient to define $X$ in L/P form because VaR and CVaR as measures of portfolio risk are denominated in loss terms.

**Definition 4.4.** Given $p \in (0,1)$, the *value-at-risk* of a random variable $X$ for the level of probability $p$ is denoted by $\mathrm{VaR}_p(X)$ and defined by

$$\mathrm{VaR}_p(X) = F_X^{-1}(p). \tag{4.18}$$

That is, $\mathrm{VaR}_p(X) = \min\{x | F_X(x) \geq p\} = Q(p)$, the lowest $p$-quantile of $X$.

Depending on the interpretation of $X$, there are different versions[17] of the notion *VaR* of $X$. In the above definition we interpret $X$ as the possible loss in dollar amount of a portfolio of securities (i.e., $X$ in L/P form).

Assuming normal financial market conditions, under mark-to-market accounting[18] and single-period framework (see Chapter 3), for a given portfolio, a given holding period (time horizon) and a given probability $p \in (0,1)$, the VaR (often referred to the $p$-$VaR$) is a threshold loss value such that the probability that the loss on the portfolio over the given holding period exceeds this value is $p$.

**Example 4.14.** A stock portfolio with a one-day $p$-VaR = 20,000, where $p = 1\%$, is interpreted as that there is a 0.01 probability that the portfolio will lose more than \$20,000 on a day if there is no trading during the day. □

**Example 4.15.** Consider a portfolio of a single asset. Suppose that the return of the asset is normally distributed with mean return of 14% per annum and annual standard deviation of 35%. The value of the portfolio today is \$100,000. With 1% probability (equivalently, a 99% level of confidence), what is the maximum loss at the end of the year (equivalently, what is the annual VaR)?

**Solution.** Let $X$ be the annual return of the portfolio (in P/L form). Then $X$ is a normal random variable with

---

[17] One of those versions is given as follows:

Let $X$ be (random) returns of a portfolio in P/L form, the VaR of $X$ at confidence level $(1-p)100\%$ is defined by

$$\mathrm{VaR}_p(X) = -\min\{x | P(X \leq x) \geq p\}.$$

That is, the VaR at tail probability $p$ is the negative of the lower $p$-quantile of the return distribution.

[18] Mark-to-market accounting is an accounting process by which the price of an asset held in an account is valued each day to reflect the daily closing price of the asset.

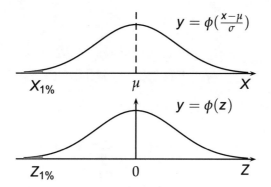

**Fig. 4.4** The probability distributions for $X_{1\%}$ and $Z_{1\%}$ are given by the left corners in the top and bottom plots, respectively

$$\mathbb{E}(X) = 0.14 \times \$100,000 = \$14,000,$$

$$\sqrt{\mathrm{Var}(X)} = 0.35 \times \$100,000 = \$35,000,$$

and

$$Z = \frac{X - \$14,000}{\$35,000}, \qquad \Phi(Z_{1\%}) = 1\%.$$

We illustrate $X$ and $Z$ in Figure 4.4.

A software can be used to obtain $Z_{1\%}$. However, let's use a linear interpolation as in Figure 4.5, which is a plot of the line segment,

$$y - y_1 = \frac{y_2 - y_1}{Z_2 - Z_1}(Z - Z_1).$$

We need to find the value $Z_{1\%}$ satisfying $\Phi(Z_{1\%}) = 0.01$. Since

$$\Phi(-2.33) = 0.0099, \qquad \Phi(-2.32) = 0.0102,$$

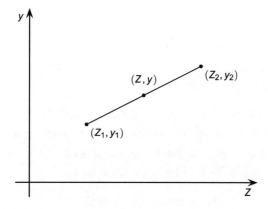

**Fig. 4.5** Linear interpolation

we have

$$Z_1 = -2.33, \quad Z_2 = -2.32,$$
$$y_1 = 0.0099, \quad y_2 = 0.0102.$$

Solving for $Z$ when $y = 0.01$ yields the desired value:

$$Z_{1\%} = Z = Z_1 + \frac{Z_2 - Z_1}{y_2 - y_1}(y - y_1)$$
$$= -2.33 + \frac{0.01}{0.0003} \times 0.0001$$
$$= -2.32667.$$

From

$$X = \$35,000 \, Z + \$14,000,$$

we then obtain

$$\text{VaR}_{1\%}(X) = \$35,000 \times (-2.32667) + \$14,000 = -\$67,433.45.$$

$\square$

In the next example, the historical performance of a portfolio return is described by a frequency distribution.

**Example 4.16.** Suppose that the historical data of a portfolio shows the weekly returns over the past 750 weeks in the table below.

| Percentage Gain/Loss | Number of weeks (Frequency) | Other information |
|---|---|---|
| $< -5$ | 2 | |
| $[-5, -4.5)$ | 2 | Suppose the |
| $[-4.5, -4)$ | 0 | seventh-highest |
| $[-4, -3.5)$ | 3 | weekly loss is 3.6% |
| $[-3.5, -3)$ | 1 | |
| $\vdots$ | $\vdots$ | |

Find the weekly loss that will not be exceeded in 99% of cases. Assume that the initial investment is $100,000.

**Solution.** One percent of observations is $750 \times 1\% = 7.5$. "Not exceeded in 99% of cases" implies that we need to consider only the seven lowest observations. From the table, the seventh-highest weekly loss is 3.6% of the initial investment. Hence, the weekly VaR of the portfolio is

$$0.036 \times \$100,000 = \$3,600$$

Equivalently, $3,600 is the weekly loss that will not be exceeded in 99% of cases.

$\square$

The next property provides a relation between VaRs under two different interpretations of random variable $X$ (e.g., P/L vs L/P). The proof of it provides strategy and tactics in proving (4.20).

*Property 4.2.*

$$\text{VaR}_p(X) = -\text{VaR}_{1-p}(-X), \quad p \in (0,1). \tag{4.19}$$

*Proof.* Notice that (4.19) is equivalent to

$$F_X^{-1}(p) = -F_{-X}^{-1}(1-p), \quad p \in (0,1),$$

and (4.16) is equivalent to

$$\mathbb{P}(X < a) \le p \le \mathbb{P}(X \le a).$$

Let $Y = -X$, $x \in \mathbb{R}$ and $y = -x$. It is sufficient to show that for each fixed $p \in (0,1)$, $x$ is a $p$-quantile of $X$ if and only if $y$ is a $(1-p)$-quantile of $Y$. In fact,

$$X < x \quad \text{iff} \quad -X > -x$$

implies

$$\mathbb{P}(X < x) = \mathbb{P}(-X > -x) = P(Y > y) = 1 - \mathbb{P}(Y \le y).$$

Similarly,

$$X \le x \quad \text{iff} \quad -X \ge -x$$

implies

$$\mathbb{P}(X \le x) = \mathbb{P}(-X \ge -x) = P(Y \ge y) = 1 - \mathbb{P}(Y < y).$$

Therefore,

$$\mathbb{P}(X < x) \le p \le \mathbb{P}(X \le x) \quad \text{iff} \quad 1 - \mathbb{P}(Y \le y) \le p \le 1 - \mathbb{P}(Y < y),$$

which is equivalent to

$$\mathbb{P}(X < x) \le p \le \mathbb{P}(X \le x) \quad \text{iff} \quad \mathbb{P}(Y < y) \le 1 - p \le \mathbb{P}(Y \le y),$$

which implies the desired result. □

By applying arguments similar to that in the proof of Property 4.2, we can show that for any constant $c \neq 0$ and $b$,

$$\text{VaR}_p(cX + b) = \begin{cases} c\,\text{VaR}_p(X) + b & \text{if } c > 0 \\ c\,\text{VaR}_{1-p}(X) + b & \text{if } c < 0. \end{cases} \tag{4.20}$$

Verification of (4.20) is left as an exercise to the reader. (*Hint*: Let $Y = cX + b$ and $y = cx + b$, and consider case 1, $c > 0$, and case 2, $c < 0$ separately.)

**Remark 4.9.**

1. There are three basic methods for calculating VaR:

   a. The historical method.
   b. The variance-covariance method.
   c. The Monte Carlo simulation method.

   Without getting into technical details, we note that the historical method involves historical simulation and is a nonparametric approach, the variance-covariance method involves estimation of the standard deviation and is a parametric approach, and the Monte Carlo simulation involves generation of time series such as (sample) paths of security prices, etc. For sophisticated examples and detailed and comprehensive treatments of VaR, readers are referred to the literature (e.g., Dowd [15], Hull [25], and Jorion [27]).

2. While VaR is conceptually simpler to understand and operationally easier to implement than most other risk measures, it can provide a false sense of security if it is misused due to its lack of subadditivity and other limitations (see Section 4.2.7 and the literature, e.g., Artzner, Delbaen, Eber, and Heath [4]; Föllmer and Schied [20]).

$\square$

### 4.2.6 Conditional Value-at-Risk

We begin this section with a natural question:

➤ If a VaR of a portfolio is $10,000, when the portfolio loses more than $10,000, what is the expected loss (over the corresponding time horizon)?

The answer to this question is provided by the CVaR of the portfolio.

**Definition 4.5.** Given $p \in (0,1)$, the *conditional value-at-risk* of a random variable $X$ for the level of probability $p$ is denoted by $C\mathrm{VaR}_p(X)$ and defined by

$$C\mathrm{VaR}_p(X) = \mathbb{E}(X | X \geq \mathrm{VaR}_p(X)). \tag{4.21}$$

The conditional value-at-risk ($C\mathrm{VaR}$) is also called the *average value-at-risk* ($A\mathrm{VaR}$) for the following obvious reason:

$$C\,\mathrm{VaR}_p(X) = \mathbb{E}(X|X \geq \mathrm{VaR}_p(X))$$
$$= \frac{1}{1-p}\int_{F_X^{-1}(p)}^{\infty} x\,dF_X(x) = \frac{1}{1-p}\int_p^1 F_X^{-1}(y)\,dy$$
$$= \frac{1}{1-p}\int_p^1 \mathrm{VaR}_y(X)\,dy.$$

Indeed, if $X$ represents the possible loss of a portfolio, then the last expression represents the average of the VaRs on the losses in the tail, which are larger than $\mathrm{VaR}_p(X)$. This explains why notations $C\,\mathrm{VaR}_p(X)$ and $A\,\mathrm{VaR}_p(X)$ are often interchangeable and why $C\,\mathrm{VaR}$ is also called *expected tail loss, tail VaR* , or *expected shortfall*.

**Example 4.17.** The possible return information of a portfolio is given in the table below:

| $X$ | $-70$ | $-30$ | $0$ | $30$ | $70$ |
|---|---|---|---|---|---|
| $p$ | 10% | 20% | 40% | 20% | 10% |

where $X$ represents the annual return of the portfolio (in P/L form). The table indicates that, for instance, if the unit on X is one thousand dollars, the chance of the portfolio's losing \$70 thousand per year is 10%.

Determine $C\,\mathrm{Var}_p(X)$ for the cases $p = 10\%$ and $p = 20\%$.

**Solution.**

$$C\,\mathrm{Var}_{10\%}(X) = \frac{-70 \times \frac{10}{100}}{\frac{10}{100}} = -70,$$

$$C\,\mathrm{Var}_{20\%}(X) = \frac{-70 \times \frac{10}{100} - 30 \times \frac{10}{100}}{\frac{20}{100}} = -50.$$

$\square$

**Remark 4.10.** CVaR satisfies the subadditivity property, which is not valid for ordinary VaR. For further discussions of the CVaR and its applications, we refer the reader to the literature (e.g., Föllmer and Schied [20]; Goldberg, Hayes, Menchero, and Mitra [23]; Goldberg and Hayes [24]; Rockfellar and Uryasev [32, 33]).

$\square$

## 4.2.7 Coherent Risk Measures

From the standpoint of practical application (especially from an institutional perspective), risk has many different dimensions that not only should have exposure and uncertainty components but also should include sources and decision-making inputs.

Just as a coherent and integrated air quality measure needs monitoring stations to measure the presence of contaminants in the air such as carbon monoxide, ozone and particulate matter, and so on, a coherent financial risk measure needs to satisfy a set of properties that covers a number of different dimensions of risk as briefly mentioned above. A proposal of such a set of properties with mathematical clarity is postulated by Artzner *et alia* [3, 4] and described in the definition below.

**Definition 4.6.** Let random variables $X$ and $Y$ represent two portfolio returns. A *coherent risk measure* for portfolio return is a function $\varrho$ that satisfies each of the following properties with probability 1:

1. Monotonicity: If $X \leq Y$, then $\varrho(X) \geq \varrho(Y)$.
2. Subadditivity: $\varrho(X + Y) \leq \varrho(X) + \varrho(Y)$.
3. Positive homogeneity: $\varrho(cX) = c\varrho(X)$, for any $c \in (0, \infty)$.
4. Translational invariance: $\varrho(X + b) = \varrho(X) - b$, for any $b \in (-\infty, \infty)$.

**Intuition and interpretation.**

1. Monotonicity: The higher the future return is, the smaller the risk is.

2. Subadditivity: The sum of sub-portfolio risks is an upper bound of the total portfolio risk. A practical application could be for decentralized decision-making in a financial institution based on an upper bound on risk prescribed by regulation.

   Notice that subadditivity provides an incentive to diversify a portfolio.

3. Positive homogeneity: The risk of a position is proportional to the size of the position.

4. Translational invariance: Cash needed is considered a risk measure. Thus, margin account requirements and margin call concerns have been factored into a coherent risk measure.

Note that the gist of the definition of coherent risk measure is subadditivity. The rest of the properties are designed to ensure that portfolio returns under consideration are essentially well behaved.

Rockafellar and Uryasev [32] showed that *the CVaR is a coherent risk measure.* On the other hand, the VaR has been criticized for violating subadditivity by many, notably by Artzner *et alia* [4]. Thus, *the VaR is a not a coherent risk measure.* The next example elaborates this point (for more examples, see the literature Acerbi and Tasche [1], Acerbi *et alia* [2], Artzner *et alia* [4], and Dowd [15] for original ideas).

**Example 4.18.** Suppose that we bought two securities which are issued by two companies and have identical price movements. Let X and Y represent returns from each of these two securities. Assuming that each company goes bankrupt independently with probability 8%, and we lose \$10,000 if a company goes bankrupt, and we lose \$0 if no bankruptcy occurs. Therefore $\text{VaR}_{90\%}(X) = \text{VaR}_{90\%}(Y) = 0$, and we obtain

$$\text{VaR}_{90\%}(X) + \text{VaR}_{90\%}(Y) = 0.$$

On the other hand, let $A_i$ be the event that we lose $\$10,000 \times i$, $i = 0, 1, 2$, a straightforward elementary probability calculation establishes that

$$\mathbb{P}(A_0) = 0.9^2 = 0.81 = 81\%,$$
$$\mathbb{P}(A_2) = 0.1^2 = 0.01,$$
$$\mathbb{P}(A_1) = 1 - 0.81 - 0.01 = 0.18,$$

which implies $\text{Var}_{90\%}(X + Y) = 10{,}000$. The fact that

$$\text{Var}_{90\%}(X + Y) = 10{,}000 > 0 = \text{VaR}_{90\%}(X) + \text{VaR}_{90\%}(Y)$$

indicates that the VaR does not hold subadditivity for all possible random variables. Again, we note that subadditivity provides an incentive to diversify a portfolio, which the VaR clearly discourages in this case.                    □

**Remark 4.11.** For conditions under which the VaR becomes subadditive, we refer the reader to the literature (e.g., Daníelsson *et alia* [12] for subadditivity for VaR in the tail and Dhaene *et alia* [14] for risk measures and comonotonicity).

□

## 4.3 Introduction to Linear Factor Models

Pioneered by Charles Spearman [38], an English psychologist, factor analysis has more than 100 years of history and has become a principal statistical method of investigation in almost all scientific, economic, and financial fields today. In this section, we briefly introduce linear factor models in a context where only financial investment returns are under consideration.

Applying linear factor models to finance in a multifactor setting was introduced in 1976 by Stephen Ross [34] through his work on arbitrage pricing theory (APT). While two major applications of linear factor models to portfolio management are asset allocation and risk management, we briefly discuss the latter only (despite the interconnect between these two areas).

For a systematic study of factor models for their role in portfolio risk analysis, we refer the reader to the book by Connor, Goldberg, and Korajczyk [11].

### 4.3.1 Definition and Intuition

Let $P = \{S_1, S_2, \ldots, S_n\}$ be a portfolio of $n$ assets (which can be stocks, bonds, mutual funds, and so on). Let $f_1, \ldots, f_m$ represent $m$ fundamental risk factors associated with portfolio $P$.

**Definition 4.7.** A (*linear*) *factor model* relates the return[19] of an asset of the portfolio to the values of a limited number of factors, say $f_j$, $j = 1, 2, \ldots, m$, and is represented by

$$R_i = \beta_{i1} f_1 + \beta_{i2} f_2 + \cdots + \beta_{im} f_m + e_i, \tag{4.22}$$

where:

$R_i$ is the return on asset $i$, where $i = 1, 2, \ldots, n$.

$\beta_{ij}$ is the change in the return on asset $i$ per unit change in factor $j$, where $i = 1, 2, \ldots, n$ and $j = 1, 2, \ldots, m$. (In other words, $\beta_{ij}$ indicates the sensitivity of asset $i$ to factor $f_j$.)

$e_i$ is the portion of the return on asset $i$ not related to the $m$ factors. The quantity $e_i$ is called the *idiosyncratic return* of asset $i$.

Note that $f_j$ ($j = 1, 2, \ldots, m$) and $e_i$ ($i = 1, 2, \ldots, n$) are random variables and that $\beta_{ij}$ ($i = 1, 2, \ldots, n, j = 1, 2, \ldots, m$) are constants. For simplicity, we shall refer to linear factor models simply as *factor models*.

Often we write

$$e_i = \alpha_i + \varepsilon_i,$$

where $\alpha_i = \mathbb{E}(e_i)$ and $\varepsilon_i$ is the error term.[20] It follows that a factor model can be expressed by

$$R_i = \alpha_i + \sum_{k=1}^{m} \beta_{ik} f_k + \varepsilon_i, \qquad i = 1, 2, \ldots, n, \tag{4.23}$$

where $\alpha_i$ is called the *alpha value* (or simply, *alpha*) of asset $i$, $\beta_{ik}$s are called the *beta values* (or simply, *betas*) of asset $i$, and we assume that $\mathrm{Cov}(\varepsilon_i, \varepsilon_{i'}) = 0$ whenever $i \neq i'$, and that for each $i$, $\mathrm{Cov}(\varepsilon_i, f_j) = 0$ for all $j$.

**Example 4.19.** To make the concept of factor models easier to understand, let's consider a single stock portfolio. Let $f_j$, $j = 1, 2, 3, 4, 5$ represent house price-to-income ratio, new home sales, housing market index, biotech ETF performance, and FDA fast track development program, respectively. The following is a five-factor model

---

[19] Annualizing returns with compounding would make these factor models almost useless because of the linearity of the model. As we pointed out in Remark 4.1, using the logarithmic return in factor models can avoid this shortcoming and take advantage of time-additivity and statistical tools in studying properties of individual securities.

[20] Error terms are usually assumed to be normal with mean zero.

$$R = \alpha + \beta_1 f_1 + \beta_2 f_2 + \beta_3 f_3 + \beta_4 f_4 + \beta_5 f_5 + \varepsilon.$$

A least squares linear regression fit will determine whether or not the model is meaningful in practice.                                                                □

Because we imposed few conditions on the factors (for instance, no probability spaces specified or distributions required), the definition above is simple; however, in practice the "redundancy" in the factors may make our already cloudy data points cloudier and create an unnecessarily more complicated and prolonged computational process. The *Principal Component Analysis (PCA)* is a powerful factor redundancy reduction procedure that compresses the data set by keeping only important information that is represented by a likely smaller number of orthogonal and centralized new factors.

In general, a simpler model is less prone to overfitting. While the topic is beyond the scope of this book, the next example provides an intuition of it.

**Example 4.20.** Revisit our example above. Recall the model with five factors

$$R = \alpha + \beta_1 f_1 + \beta_2 f_2 + \beta_3 f_3 + \beta_4 f_4 + \beta_5 f_5 + \varepsilon,$$

where $f_j$, $j = 1,2,3,4,5$ correspond to the following factors: house price-to-income ratio, new home sales, housing market index, biotech ETF performance, and FDA fast track development program, respectively.

Now, suppose that the (sample) correlation matrix (with artificial entries) of the random vector $f^T = [f_1\ f_2\ f_3\ f_4\ f_5]$ is below:

$$\begin{bmatrix} 1.00 & 0.82 & 0.93 & 0.01 & 0.02 \\ 0.82 & 1.00 & 0.91 & 0.02 & 0.00 \\ 0.93 & 0.91 & 1.00 & 0.00 & 0.01 \\ 0.01 & 0.02 & 0.00 & 1.00 & 0.89 \\ 0.02 & 0.00 & 0.01 & 0.89 & 1.00 \end{bmatrix}.$$

The matrix suggests two underlying constructs: housing (related to $f_1, f_2$, and $f_3$, and so to their linear combinations) and the biotech sector (related to $f_4$ and $f_5$, and again, to their linear combinations). In other words, the above correlation matrix displays the redundancy of the five factors described above. There is an obvious computational advantage if $f_1, f_2$, and $f_3$ are "collapsed" into a single new factor $\tilde{f}_1$ corresponding to housing, and $f_4$ and $f_5$ are "collapsed" into another single new factor $\tilde{f}_2$ corresponding to biotech sector. We can then form a factor model with only two factors

$$R = \alpha + \beta_1 \tilde{f}_1 + \beta_2 \tilde{f}_2 + \varepsilon.$$

Intuitively, what we illustrated in this example is in essence what the PCA accomplishes.                                                                                  □

To make it more precise, PCA converts the original factors into uncorrelated linear combinations of them. These linear combinations form an orthonormal set of eigenvectors of the (sample) correlation matrix of factors. Very often in practice, a few leading eigenvalues of the correlation matrix explain, in terms of percentage, close to the total variation in the entire data set. Therefore, the original model may be reduced to a new model with fewer uncorrelated factors which are the eigenvectors belonging to those leading eigenvalues. Such reduction, in practice, not only reduces computation but also often results in more robust models with higher sustainability for the future.

This discussion leads to another statement of the definition of linear factor model:

**Definition 4.8.** A *(linear) factor model* relates the return of an asset of the portfolio to the values of a limited number of factors, say $\tilde{f}_j$, $j = 1, 2, \ldots, m$, and is represented by

$$R_i = \alpha_i + \sum_{j=1}^{m} \beta_{ij} \tilde{f}_j + \varepsilon_i, \qquad i = 1, 2, \ldots, n, \tag{4.24}$$

where:

$R_i$ is the return on asset $i$, $i = 1, 2, \ldots, n$.

$\mathbb{E}(\tilde{f}_j) = 0$, $j = 1, 2, \ldots, m$ (i.e., $\tilde{f}_j$, $j = 1, 2, \ldots, m$ are centered).

$\tilde{f}_j$, $j = 1, 2, \ldots, m$ are orthonormal factors under covariance (i.e., $\mathrm{Cov}(\tilde{f}_j, \tilde{f}_{j'}) = 0$ whenever $j \neq j'$, and $\mathrm{Var}(\tilde{f}_j) = 1$ for all $j$).

$\beta_{ij}$ indicates the sensitivity of asset $i$ to factor $\tilde{f}_j$ and is called the *factor loading* of return $R_i$.

$\mathrm{Cov}(\varepsilon_i, \varepsilon_{i'}) = 0$ whenever $i \neq i'$ and that for each $i$, $\mathrm{Cov}(\varepsilon_i, \tilde{f}_j) = 0$ for all $j$.

$\alpha_i + \varepsilon_i$ is the portion of the return on asset $i$ not related to the $m$ factors (thus it is called the idiosyncratic return of asset $i$), where $\mathbb{E}(\varepsilon_i) = 0$.

The conditions imposed on the quantities in Definition 4.8 ensure that (4.24) can be estimated by the method of least squares (see Exercises 4.30 and 4.31 on page 206). Note that the alphas and betas are both measures of risks and tools used to determine the risk-reward profile of a portfolio.

Observe that (4.23) shows that a factor model decomposes an asset's return into factors common to all assets in the portfolio and an asset-specific factor. Even though this is a considerable simplification of reality, it is computationally reasonable. Factor models are practically useful in many domains in the field of investments, particularly in analyzing historic results, because they provide a tool to allow analysts to separate components of the overall return of the asset.

Indeed, factor models provide a versatile tool for an elegant framework of trading strategy analysis and investment portfolio design. However, finding good factors is always a challenging task. Almost none of the published models remain profitable (as of the writing of this text), as the market is becoming much more efficient than 10 years ago. The idea of studying these models is to help readers to develop their own models in the future. In practice, none of these models will be applied alone, and almost certainly not in the exclusive form provided here.

### 4.3.2 Portfolio Variance Decomposition

Under the single time period framework, we rewrite (4.23)

$$R_i = \alpha_i + \sum_{k=1}^{m} \beta_{ik} f_k + \varepsilon_i, \qquad i = 1, 2, \ldots, n,$$

in a matrix form

$$X = \alpha + \beta f + \varepsilon,$$

where

$$X = [R_1, R_2, \ldots, R_n]^\top, \quad \alpha = [\alpha_1, \alpha_2, \ldots, \alpha_n]^\top, \quad \beta_i = [\beta_{i1}, \beta_{i2}, \ldots, \beta_{im}],$$
$$f = [f_1, f_2, \ldots, f_m]^\top, \quad \varepsilon = [\varepsilon_1, \varepsilon_2, \ldots, \varepsilon_n]^\top, \quad \beta = [\beta_{ik}]_{n \times m}.$$

Under the assumptions that

$$\mathbb{E}(f_k) = 0, \ k = 1, 2, \ldots, m, \qquad \mathbb{E}(\varepsilon_i) = 0, \ i = 1, 2, \ldots, n,$$
$$\mathrm{Cov}(\varepsilon_i, \varepsilon_j) = 0 \text{ whenever } i \neq j, \quad \mathrm{Cov}(\varepsilon_i, f_k) = 0 \text{ for all } i \text{ and } k,$$

we compute

$$\mathrm{Cov}(R_i, R_j) = \mathbb{E}((R_i - \mathbb{E}(R_i))(R_j - \mathbb{E}(R_j))) = \mathbb{E}((R_i - \alpha_i)(R_j - \alpha_j))$$
$$= \mathbb{E}\left(\left(\sum_{k=1}^{m} \beta_{ik} f_k + \varepsilon_i\right)\left(\sum_{l=1}^{m} \beta_{jl} f_l + \varepsilon_j\right)\right)$$
$$= \mathbb{E}\left(\left(\sum_{k=1}^{m} \beta_{ik} f_k\right)\left(\sum_{l=1}^{m} \beta_{jl} f_l\right)\right) + \sum_{k=1}^{m} \beta_{ik} \mathbb{E}(f_k \varepsilon_j) + \sum_{l=1}^{m} \beta_{jl} \mathbb{E}(\varepsilon_i f_l) + \mathbb{E}(\varepsilon_i \varepsilon_j)$$
$$= \mathbb{E}(\beta_i (f^\top f) \beta_j^\top) + \mathbb{E}(\varepsilon_i \varepsilon_j) = \beta_i \mathbb{E}(f^\top f) \beta_j^\top + \mathbb{E}(\varepsilon_i \varepsilon_j).$$

This result suggests a relation between covariance matrices. To see this, we let

$\Sigma = [\sigma_{ij}]_{n \times n}$ be the covariance matrix of a portfolio return $X$,

$F = [f_{ij}]_{m \times m}$ be the covariance matrix of factors in the model,

$\Psi = [\varepsilon_{ij}]_{n \times n}$ be the covariance matrix of the error terms,

where

$$\sigma_{ij} = \mathrm{Cov}(R_i, R_j), \quad f_{ij} = \mathrm{Cov}(f_i, f_j), \quad \varepsilon_{ij} = \mathrm{Cov}(\varepsilon_i, \varepsilon_j).$$

Then we have

$$F = \mathbb{E}(f^\top f), \quad \mathrm{Cov}(\varepsilon_i, \varepsilon_j) = \mathbb{E}(\varepsilon_i \varepsilon_j) = \begin{cases} 0 & \text{if } i \neq j \\ \mathrm{Var}(\varepsilon_i) & \text{if } i = j, \end{cases}$$

and the covariance matrix of the returns can be expressed by

$$\Sigma = \beta F \beta^\top + \Psi. \tag{4.25}$$

It is readily understood that this relation of covariance matrices implies a decomposition of portfolio variance into common factor variance and idiosyncratic variance (see Exercise 4.24 on page 205).

Note that given underlying factors, a portfolio return, and its volatility are completely determined by its factor loadings.

**Remark 4.12.**

1. Different portfolios may be exposed to different types of risk when different scenarios occur in the market. Intuitively speaking, portfolio risk management is about what to anticipate in dealing with different market scenarios. The additive decomposition in (4.25) is a significant result (or rather a key objective of factor models) for portfolio risk management. It suggests that factor models allow portfolio managers to perform risk management by approaching these very underlying factors and examining how they impact covariance matrix of returns in a direct way.

2. Achieving optimal asset allocation requires a robust understanding of portfolio risk. To pass the test for robust understanding, one needs to quantify portfolio risk. Quantitative portfolio risk management requires quantifying the overall portfolio risk (e.g., VaR and CVaR), and slicing and dicing into sources of portfolio risk (e.g., multifactor models). In this sense, factor analysis refines the risk profile of the portfolio and allows portfolio managers to perform allocation and hedging from factor risk perspective.

   For portfolio optimization, we refer the reader to the literature (e.g., Chan, Karceski, and Lakonishokk [6]; Connor and Korajczyk [8, 9]).

□

### 4.3.3 Factor Categorization

Generally speaking, financial markets will reflect the economic conditions of an economy. Empirical evidence for this can be found in the literature (e.g., Chen, Roll, and Ross [7]).

The field of economics can be divided into many subfields depending on how finely one wishes to make the division. The terms *macroeconomics* and *microeconomics* were coined in 1933. As the names suggest, the former studies how the economy as a whole works, whereas the latter studies how specific individual units function. Examples of major factors affecting macroeconomics are the rate of change of the gross domestic product (GDP), unemployment rate, and inflation (see Section 1.3), and examples of factors affecting microeconomics are companies' size, dividend yield, and price-to-earnings ratio (see Section 4.3.6).

Factors can be divided into *observable*[21] and *unobservable* (or *latent*[22]).

**Example 4.21.** A publicly traded company's market capitalization and industry classification are observable factors.

Factors that are determined by PCA may be unobservable or latent (see $\tilde{f}_1$ and $\tilde{f}_2$ in Example 4.20 on page 187). □

Depending on the different factor types and factor construction methods, multifactor models of security returns may be categorized by macroeconomic, fundamental, and statistical models:

A *macroeconomic factor model* is a factor model whose common risk factors are determined by observable macroeconomic factors.

A *fundamental factor model* is a factor model whose common risk factors are created from stock-specific fundamentals (e.g., company size and price-to-market ratio) that affect the corresponding returns.

A *statistical factor model* is a factor model whose common risk factors are extracted from historical returns by using analytical methods (e.g., PCA).

Note that such a division of factor models has blurred boundaries. Fama-French three-factor model will be introduced in Section (4.3.6) and can be viewed as a fundamental-based factor model or a combination of fundamental and macroeconomic factor model.

For an overview of the empirical procedures for the three types of factor models including inputs, outputs, and estimation technique, we refer the reader to Table 1 in [10] by Connor.

---

[21] *Observable variables*, as a statistical term, are those that can be directly measured.

[22] *Latent variables*, as opposed to observable variables in statistics, are those inferred through mathematical models and cannot be directly observed.

### 4.3.4 Alpha and Beta

In order to understand the significance of the factor model in relation to investments, let's consider its simplest form (with $n = m = 1$).

$$R = \alpha + \beta R_M + \varepsilon, \tag{4.26}$$

where $R_M$ is the stock market return, say, the return of the S&P 500 over time period $[0, T]$, and $R$ is the return of a stock over the same time period. Here $\alpha$ and $\beta$ are referred as the *the stock's $\alpha$ and $\beta$*. (If $R$ is the return of a portfolio, then $\alpha$ and $\beta$ are referred as the *the portfolio's $\alpha$ and $\beta$*.) Taking the expectation of both sides of (4.26) yields

$$\mathbb{E}(R) = \alpha + \beta \mathbb{E}(R_M).$$

Accordingly, we have

$$\bar{R} = \alpha + \beta \bar{R}_M, \tag{4.27}$$

where $\bar{R}$ and $\bar{R}_M$ are sample means of $R$ and $R_M$, respectively.

In general, $\alpha$ and $\beta$ are more stable than the equity price itself. Consequently, they provide a certain predictive value. In other words, they are not only measures of risk but also tools used to determine the risk-reward profile of a portfolio to form investment strategies.

**Example 4.22.** The mathematical expression of the single-factor model seen in (4.27) suggests the following possible investment strategies:

a) Getting stock returns from both $\alpha$ and $\beta$. Perhaps a number of investors do so without knowing the formal meanings of $\alpha$ and $\beta$.

b) Making stock investment profits based on a better judgment of the direction of major stock market (say, the S&P 500). If one can do it well, then with a certain level of leverage, one can make good returns on high $\beta$ stocks.

c) Making stock investment profits based on selecting high $\alpha$ stocks. In fact, many quantitative funds try to use index futures positions to hedge away the beta part to obtain so-called *market-neutral portfolios*.

□

Before we illustrate how to determine $\alpha$ and $\beta$ given the historical data on $\bar{R}$ and $\bar{R}_M$ in (4.27), let us understand the general idea of *the method of least squares*, which is a classical technique in finding "the approximate solution" of an overdetermined[23] system of linear equations. Such systems often occur based on raw field data and usually have no solutions.

Let $f_i$, where $i = 1, 2, \cdots, n$, be $n$ functions of $m$ variables $x_1, x_2, \cdots, x_m$, where $n > m$. Suppose that we are interested in "solving" for $m$ unknowns

---

[23] A system of linear equations is called *overdetermined* if there are more equations than unknowns.

$x_1, x_2, \cdots, x_m$ from the system of $n$ equations:

$$f_1(x_1, x_2, \cdots, x_m) = 0,$$

$$f_2(x_1, x_2, \cdots, x_m) = 0,$$

$$\cdots$$

$$f_n(x_1, x_2, \cdots, x_m) = 0.$$

Note that the system holds if and only if the equation

$$\sum_{i=1}^{n} (f_i(x_1, x_2, \cdots, x_m))^2 = 0$$

holds. Unfortunately, in general, the system does not have a solution. What we intend to do is to find points $(x_1, x_2, \cdots, x_m)$ that minimize

$$\sum_{i=1}^{n} (f_i(x_1, x_2, \cdots, x_m))^2.$$

**Example 4.23.** To simplify the notations in (4.27), we let $y = \bar{R}$ and $x = \bar{R}_M$. Rewrite (4.27) into $y = \alpha + \beta x$. Given $n$ sets of data $(x_i, y_i)$, $i = 1, 2, \cdots, n$, where $n > 2$, determine $\alpha$ and $\beta$.

**Solution.** Consider the following overdetermined system with $\alpha$ and $\beta$ as unknowns and $n > 2$:

$$y_1 = \alpha + \beta x_1,$$

$$y_2 = \alpha + \beta x_2,$$

$$\cdots$$

$$y_n = \alpha + \beta x_n.$$

By the above discussion, we need to minimize

$$L = \sum_{i=1}^{n} (y_i - \alpha - \beta x_i)^2.$$

We compute

$$\frac{\partial L}{\partial \alpha} = 2 \sum_{i=1}^{n} (y_i - \alpha - \beta x_i)(-1) = -2 \sum_{i=1}^{n} (y_i - \alpha - \beta x_i),$$

$$\frac{\partial L}{\partial \beta} = 2 \sum_{i=1}^{n} (y_i - \alpha - \beta x_i)(-x_i) = -2 \sum_{i=1}^{n} (x_i y_i - \alpha x_i - \beta x_i^2).$$

Setting[24]

---

[24] The solution to this system can only be minimum point(s) since $L$ does not have maximum point.

$$\frac{\partial L}{\partial \beta} = 0, \qquad \frac{\partial L}{\partial \alpha} = 0,$$

we obtain

$$0 = \sum_{i=1}^{n} (y_i - \alpha - \beta x_i) = \sum_{i=1}^{n} y_i - n\alpha - \beta \sum_{i=1}^{n} x_i,$$

$$0 = \sum_{i=1}^{n} (x_i y_i - \alpha x_i - \beta x_i^2) = \sum_{i=1}^{n} x_i y_i - \alpha \sum_{i=1}^{n} x_i - \beta \sum_{i=1}^{n} x_i^2.$$

Equivalently,

$$\alpha = \frac{1}{n} \sum_{i=1}^{n} y_i - \frac{1}{n} \left( \sum_{i=1}^{n} x_i \right) \beta,$$

$$\sum_{i=1}^{n} x_i y_i = \left( \sum_{i=1}^{n} x_i \right) \alpha + \left( \sum_{i=1}^{n} x_i^2 \right) \beta.$$

Straightforward algebraic manipulations yield

$$\alpha = \frac{\sum_{i=1}^{n} y_i - \left( \sum_{i=1}^{n} x_i \right) \beta}{n}, \qquad \beta = \frac{n \sum_{i=1}^{n} x_i y_i - \left( \sum_{i=1}^{n} x_i \right) \left( \sum_{i=1}^{n} y_i \right)}{n \sum_{i=1}^{n} x_i^2 - \left( \sum_{i=1}^{n} x_i \right)^2}. \qquad (4.28)$$

$\square$

As a matter of fact, what we have just accomplished is a procedure to determine a line that best fits the data $\{(x_i, y_i), \ i = 1, 2, \cdots, n\}$. This procedure is called *the method of least squares*. The best[25] line to fit the data is the line with the slope $\beta$ and $y$-intercept $\alpha$ as we derived above. More rigorously speaking, the *least squares* "approximate solution" of an overdetermined system of linear equations should be referred to as the best fit graph[26] to data.

**Remark 4.13.**

1. Although a stock's $\alpha$ and $\beta$ are more stable than the stock price in general, they change over time as the risk profile of the company changes. While the discussion of their (random) variability is beyond the scope of this book, we would like to point out that sometimes a stock's $\alpha$ and $\beta$ may become quite unstable when the volatility of the market is high.

2. The calculation of the actual values of a stock's $\alpha$ and $\beta$ depends on several factors including the length of the time period in the data sampling and the frequency of the data sampling (see Exercise 4.22 on page 204). For instance, consider a stock performance over a period of 3 years. If $\beta_{\mathrm{wk}}$ and $\beta_{\mathrm{mth}}$ denote the stock's $\beta$ calculated based on weekly returns and monthly returns,

---

[25] The *best* line to fit the data is in the sense of the least Euclidean distance.

[26] For example, if the data space is $\mathbb{R}^2$, and the unknown space is $\mathbb{R}^3$, say $y = \alpha + \beta x + \gamma x^2$, then the best fit graph is a curve in the data space. If both the data space and unknown space are $\mathbb{R}^3$, say $y = \alpha + \beta x^2 + \gamma z^2$, then the best fit graph is a surface in the data space.

respectively, then $\beta_{wk}$ may be different from $\beta_{mth}$. Also, a stock's $\beta$ calculated based on a 3-year period is likely to be different from that based on a 5-year period. Similar results apply to a stock's $\alpha$.

3. Most academics use log returns to calculate $\alpha$ and $\beta$ since log returns are additive. In practice, the difference between the log return and (total) return is small unless the return rates are large.

□

### 4.3.5 CAPM Beta Versus Linear Factor Beta

*We have the quantity beta in the CAPM:*

$$\mathbb{E}(R - r) = \beta \mathbb{E}(R_M - r), \tag{4.29}$$

where $R_M$ is the return of a market portfolio. Expression (4.29) is equivalent to

$$R - r = \beta(R_M - r) + \varepsilon_1, \tag{4.30}$$

where the error term $\varepsilon_1$ satisfies $\mathbb{E}(\varepsilon_1) = 0$.

On the other hand, *there is also a beta in the linear factor model in the form of excess returns:*

$$R - r = \alpha + \beta(R_M - r) + \varepsilon_2, \tag{4.31}$$

where the error term[27] $\varepsilon_2$ satisfies $\mathbb{E}(\varepsilon_2) = 0$. Since a portfolio or security beta plays an important role in investment strategy, one should be clear about the beta value under the consideration. We would like to point out that *these two betas are different unless $R - r$ and $R_M - r$ have zero means.*

To verify the previous claim, consider a data set $\{(R_i, R_{M,i}), \ i = 1, 2, \cdots n\}$ of $n$ observations. Plug it into (4.30) and let

$$y_i = R_i - r_{f,i}, \quad x_i = R_{M,i} - r_{f,i},$$

where $R_{M,i}$ and $r_{f,i}$ are, respectively, the market return and risk-free rate for the $i$th sample period. Also, let

$$x = [x_1 \ x_2 \ \dots \ x_n]^\top, \quad y = [y_1 \ y_2 \ \dots \ y_n]^\top, \quad e = [1 \ 1 \ \dots \ 1]^\top.$$

We then obtain the system

$$y = \beta x, \tag{4.32}$$

which has the least squares solution

---

[27] We assume that error terms are normal with mean zero.

$$\beta = \frac{x \cdot y}{x \cdot x} = \frac{\sum_{i=1}^{n} x_i y_i}{\sum_{i=1}^{n} x_i^2} \equiv \beta_{\text{CAPM}}. \tag{4.33}$$

Similarly, plug the data set into (4.31). After eliminating $\alpha$ from

$$y = \alpha e + \beta x,$$

we obtain the system

$$y - \bar{y}e = \beta(x - \bar{x}e), \tag{4.34}$$

which has the following least squares solution:

$$\beta = \frac{(x - \bar{x}e) \cdot (y - \bar{y}e)}{(x - \bar{x}e) \cdot (x - \bar{x}e)} = \frac{\sum_{i=1}^{n}(x_i - \bar{x})(y_i - \bar{y})}{\sum_{i=1}^{n}(x_i - \bar{x})^2} \equiv \beta_{\text{LF}}. \tag{4.35}$$

It follows that if $\bar{x} = 0$ (i.e., $\mathbb{E}(R_M - r) = 0$) and $\bar{y} = 0$ (i.e., $\mathbb{E}(R - r) = 0$), then by (4.33) and (4.35), we obtain $\beta_{\text{CAPM}} = \beta_{\text{LF}}$.

### 4.3.6 Fama-French Three-Factor Model

A well-known example of factor models is the Fama-French three-factor model. It requires some basic understanding of the following financial terminologies:

➤ The *book value* of a company is the total asset value of the company carried on its balance sheet.[28]

Given a publicly traded company, suppose that it has a total number of $N$ shares outstanding. Let $t_0$ be the current time, $p$ the current stock price, and $B$ its current book value. Here $p$ and $B$ carry the same units, say, in dollars:

➤ $p$ is called the *market value* or *market price* at time $t_0$.

➤ $Np$ is called the *market capitalization* of the company at time $t_0$, which is the product of the total number of its outstanding shares and its market value.

According to a company's market capitalization, stocks can be classified into small-cap, mid-cap, and big-cap. Although the definitions of these categories can vary across financial institutions, the classification ranges are generally from $300 million to $2 billion, from $2 billion to $10 billion, and more than $10 billion, respectively. In a similar way, stocks can also be further classified into mega-cap, large-cap, mid-cap, small-cap, micro-cap and nano-cap.

---

[28] A *balance sheet* gives a snapshot of the financial position of a company at a given time. Most accounting balance sheets consist of two sides: the left side indicates ASSETS (things that have value), and the right indicates LIABILITIES (things that are owed to third parties) and STOCKHOLDERS' EQUITY (which is the value of the remaining assets if the company were to go out of business immediately). For an oversimplified example, consider equity in your home = what you paid - what you owe (loan remaining). It is called balance sheet because it has to balance between both sides: Assets = Liabilities + Stockholder's Equity.

➤ $\frac{B/N}{p}$ is called the *book-to-price ratio* of the company at time $t_0$. It is also known as *book-to-market ratio*.

➤ *P/B ratio* or the *price-to-book ratio* is the reciprocal of the book-to-price ratio.

In practice, the book-to-price ratio is often calculated by dividing the latest quarter's book value per share by the current day closing price of the stock. The same time frame/units apply to the calculations of market capitalization. The book-to-price ratio is also used to identify undervalued (the ratio $> 1$) or overvalued (the ratio $< 1$) stocks (although the practical definitions of undervalued and overvalued can be in other relative terms).

**Example 4.24.** The key statistics of companies can be easily accessed online. For instance, valuation measures for IBM provided by Yahoo! Finance on November 8, 2012, indicated

$$\text{Market Cap} : 214.28B \quad \text{and} \quad \text{Price/Book (mrq)} : 10.3,$$

where mrq stands for "most recent quarter." Given that the closing price of the stock on that day was \$190.10, determine the capitalization of IBM and the book-to-price ratio of IBM on that day.

**Solution.** The market capitalization was 214.28 billion dollars and

$$\text{book-to-price ratio} = \frac{1}{10.3} = 0.097.$$

IBM's *P/B ratio* = 10.3 simply means that the stock costs 10.3 times as much as its asset could be sold for if the company were liquidated.

□

By definition, the P/B ratio provides a measure of the market's valuation of a company in relation to the value of that company indicated on its financial statements. In other words, a book-to-price ratio tells investors how the company's stock value measures up to its book value. In this sense, the higher the book-to-price ratio is, the higher the "value" investors get (assuming that the fundamentals of a company are accurately and completely reflected on its balance sheet). By this measure, a stock at \$1 per share may not be cheap, whereas a stock at \$100 per share may be a bargain.

The following terminologies are useful in describing additional returns:

➤ *Size premium* is the additional return that investors receive by investing in stocks of small-market-capitalization companies rather than in big-market-capitalization companies:

$$\text{size premium} = R_S - R_B.$$

➤ *Value premium* is the additional return that investors receive by investing in stocks of high book-to-price ratio companies to that of low book-to-price ratio companies:
$$value\ premium = R_H - R_L.$$

Both size premium and value premium are called *risk premiums* in the theory of financial investments.

Intuitively, the stock price is expected to depend on market capitalization (size), book-to-price ratio (value), and the systematic risk in stock investing that is directly associated with the general stock market (e.g., S&P 500). Thus, one natural approach to stock investments is to consider the general market, size, and value as three risk factors. As a matter of fact, the factor models we defined before are linear approximations of this dependence.

Although natural ideas do not always lead to successful investment strategies, once they are quantified, statistical testing provides a mechanism for accepting or rejecting the ideas. It is in such frameworks that factor models can come in handy.

Motivated by the desire to better explain differences in the returns of diversified equity portfolios by asset pricing models, Fama and French created a three-factor model [18] in the form of excess returns[29]:

$$R - r = \alpha + \beta_1(R_M - r) + \beta_2(R_S - R_B) + \beta_3(R_H - R_L) + \varepsilon, \qquad (4.36)$$

which implies

$$\mathbb{E}(R - r) = \alpha + \beta_1\mathbb{E}(R_M - r) + \beta_2\mathbb{E}(R_S - R_B) + \beta_3\mathbb{E}(R_H - R_L).$$

Denote $\mathbb{E}(R_S - R_B)$ and $\mathbb{E}(R_H - R_L)$ by SMB and HML, respectively. We obtain

$$\mathbb{E}(R) = r + \alpha + \beta_1\mathbb{E}(R_M - r) + \beta_2\,\text{SMB} + \beta_3\,\text{HML}, \qquad (4.37)$$

where r is the risk-free rate, $\mathbb{E}(R_M - r)$ is the expected excess return of the general market, and SMB and HML represent the expected size premium and expected value premium, respectively.

The regression test (4.36) has been widely applied to various equity markets of both developed and emerging economies. The following example is to help the reader to understand that the Fama-French three-factor model provides an effective asset pricing mechanism in the US stock markets.

**Example 4.25.** Let $S$, $M_S$, and $B$ denote the sets of stocks of companies with small, medium, and big capitalizations, respectively. Let $L$, $M$, and $H$ denote the sets of stocks of companies with low, medium, and high book-to-market

---

[29] Equation (4.36) is the form that most researchers currently use for the Fama-French three-factor model. We refer the reader to [18] and [19] for the original form of this model.

ratios, respectively. By taking the intersections among them as indicated in the table below, we obtain nine portfolios.

|  | Low b/m ratio | Medium b/m ratio | High b/m ratio |
|---|---|---|---|
| **Small size** | $S \cap L$ | $S \cap M$ | $S \cap H$ |
| **Medium size** | $M_S \cap L$ | $M_S \cap M$ | $M_S \cap H$ |
| **Big size** | $B \cap L$ | $B \cap M$ | $B \cap H$ |

In the 1990s, Davis, Fama, and French performed empirical tests on the nine portfolios above. Confirmatory factor analysis of the information in [13], which includes data source, sample size, data organization, regression results, factor loadings, and R-square values, strongly supports that value and small-cap stock portfolios outperformed markets on a regular basis in the USA from 1929 to 1996. □

As a contrast to the CAPM which explains, on average, 70% of the variability in well-diversified portfolios returns, the Fama-French three-factor model explains over 90% of the variability (see Fama and French [17, 18]).

There are many more nice discussions in the literature related to the topics presented in this chapter, e.g., [22, 26, 28, 30, 31, 35, 37, 39].

## 4.4 Exercises

### 4.4.1 Conceptual Exercises

**4.1.** State the definition of the *market portfolio*.

**4.2.** Explain how the sign of the beta of a stock indicates the direction of the movement of the stock price with respect to that of the market portfolio.

**4.3.** Let $A$ be a class of stocks with beta between 0.5 and 2. What is the main property that all stocks in $A$ have in terms of the market returns?

**4.4.** Give a financial interpretation of the mathematical expression: $\frac{\mu_M - r_f}{\sigma_M}$.

**4.5.** Can you find an example of a company in the USA with a negative beta? If so, do you think that they are as abundant as companies with a positive beta? Explain.

**4.6.** Consider random variable $X = 1,2,3$ satisfying $\mathbb{P}(X = i) = \frac{1}{3}$, $i = 1,2,3$. Find the quantile function of $X$.

**4.7.** Consider a random variable $X = 1,2,3,4,5,6,7,8,9,10$ with p.d.f. given by $\mathbb{P}(X = i) = \frac{1}{11}$ for $i \neq 7$ and $\mathbb{P}(i = 7) = \frac{2}{11}$. Find the quantile function of $X$.

**4.8.** Use the table given in Example 4.17 on page 183 to determine $\text{Var}_p(X)$ for arbitrary $p$.

**4.9.** Use the table given in Example 4.17 on page 183 to determine $C\text{Var}_p(X)$ for $p = 30\%$ and $p = 50\%$.

**4.10.** What does a negative VaR imply?

**4.11.** Returns and risks are two aspects involved in every investment. Identify each statement below as true or false or identify scenarios when it is true and when it is false. Justify your answer.

a) The quantity $\alpha$ relates to factors affecting the performance of an individual stock or the fund manager's skill in selecting the stocks.
b) The factor $\beta$ relates an individual stock-to-market risks.
c) A higher $\alpha$ stock and a lower $\beta$ stock would be preferred choices.
d) The quantity $\alpha = 0$ if a stock market is efficient.

**4.12.** Briefly justify your answer to each of the following:

a) Suppose that empirical evidence were sufficient to confirm the CAPM. What would be an investment implication?
b) Suppose that extraordinary premiums from the (Fama-French) three-factor investing during the period of 1926–1996 were to repeat today, what would be an investment implication?

**4.13.** All investments carry some form of risk. Major risks include, but are not limited to, the following:

a) Systematic risk
b) Interest rate risk
c) Liquidity risk[30]
d) Regulatory/political risk[31]
e) Leverage risk[32]
f) Credit risk
g) Currency risk
h) Counterparty risk[33]

Find an example for each type of risk listed above.

---

[30] Liquidity risk is the risk that an investor cannot execute a buy/sell order in the market due to the lack of anticipated/reasonable bid/ask spread or sufficient volume.

[31] Regulatory changes or governmental policy changes may have significant impact on asset values. Such risk can be either systematic risk or market risk.

[32] Such risk is often associated with unexpected and unfavorable volatility.

[33] Counterparty here means the other party in a financial transaction.

### 4.4.2 Application Exercises

**4.14.** Use the data in Table 4.1 below to compute the Sharpe ratio of the S&P 500 for the periods 1986 to 1999. Note that the risk-free rate is not constant.

**Table 4.1** Annual return rate data from 1986 to 1999 for the S&P 500 and 1-year treasury bills. Data Source: istockanalyst.com

| Year | S&P500 annual return | 1-year T-bill rate |
|------|---------------------|--------------------|
| 1986 | 18.82% | 7.21% |
| 1987 | 5.40% | 5.46% |
| 1988 | 15.99% | 6.52% |
| 1989 | 31.56% | 8.37% |
| 1990 | -2.97% | 7.38% |
| 1991 | 30.51% | 6.25% |
| 1992 | 7.45% | 3.95% |
| 1993 | 10.09% | 3.35% |
| 1994 | 1.33% | 3.39% |
| 1995 | 37.28% | 6.59% |
| 1996 | 22.69% | 4.82% |
| 1997 | 33.60% | 5.30% |
| 1998 | 30.73% | 4.98% |
| 1999 | 21.10% | 4.31% |

**4.15.** Assume a risk-free rate of 1.5%. Answer the questions below using the information in the following table:

| Portfolio | A | B | C | D | E | F |
|-----------|------|------|-------|------|-------|------|
| **Expected Return** | 3.2% | 8.1% | 9.8% | 5.1% | 10.7% | 4.8% |
| **Standard Deviation** | 2.7% | 9.9% | 13.7% | 6.2% | 17% | 6.1% |

a) Among the portfolios in the table, which one is closest to the market portfolio? Justify your answer.

b) Plot the capital market line (CML) based on your answer in part (a).

c) For portfolio C, what is the portfolio risk premium per unit of portfolio risk?

d) Suppose we are willing to make an investment only with $\sigma = 6.2\%$. Is a return of 6.5% a realistic expectation for us?

**4.16.**

a) Given the information in the table below,

| Stock | Beta | Expected Return |
|-------|------|-----------------|
| A | 1.25 | |
| B | 0.7 | |
| C | -0.4 | |

and assuming that $r_f$ is 3% and that the market return is 5%, find the expected returns for each stock listed in the table and plot them on an SML graph.

b) Suppose the table below provides further information about the stocks in part a):

| Stock | Current Price | Expected Price | Expected Dividend |
|-------|---------------|----------------|-------------------|
| A | 29.5 | 21.5 | 0.71 |
| B | 47 | 49.75 | 1.85 |
| C | 35.4 | 38.7 | 1.05 |

Indicate your estimated returns on each stock on the graph from part a), and decide your buy/sell/hold rating on each stock based on your graph. Justify your decisions.

**4.17.** Suppose that we borrow an amount equal to 25% of our original wealth at the risk-free rate 4.125%. Use the CML to find $\mu_P$ and $\sigma_P$.

**4.18.** Assume the risk-free rate is 1.5% and consider the information in the table below:

| Portfolio | Expected Return | Standard Deviation |
|-----------|-----------------|--------------------|
| A | 3.2% | 2.7% |
| B | 8.1% | 9.9% |
| C | 9.8% | 13.7% |
| D | 5.1% | 6.2% |
| E | 10.7% | 17% |
| F | 4.8% | 6.1% |

a) Which of these six portfolios offers investors the best combination of risk and return? Justify your answer from a capital market perspective.

b) Use the formula
$$\mu_P = w_0 r_f + (1 - w_0)\mu_M$$
to determine your investment asset allocation.

c) If you plan to invest $100,000, what is your investment strategy based on the information given in this exercise?

**4.19.** Given the following information, find the security's beta and expected return:

a) The risk-free rate is 1.72%. The market portfolio has its standard deviation as 15.92% and its expected return as 5%. The covariance of the security with the market is 0.04.

b) The risk-free rate is 2%. The market portfolio has its standard deviation as 14% and its expected return as 11%. The security is uncorrelated with the market and has a standard deviation of 39.7%.

c) The risk-free rate is 1.72%. The market portfolio has its standard deviation as 15.92% and its expected return as 5%. The covariance of the security with the market is −0.04.

**4.20.** The ticker symbol for the Goldman Sachs Group is GS. The table below provides the daily closing prices for GS and S&P 500 index on the trading days during the period between June 30 and July 14 of 2011. Let $r$ and $r_M$ be the daily log returns of GS and S&P 500 index, respectively. Using the model $r = \alpha + \beta r_M$, determine GS's $\alpha$ and $\beta$ for the period.

| Date | GS | S&P 500 |
|---|---|---|
| June 30 | $133.09 | $1320.64 |
| July 1 | $136.65 | $1339.67 |
| July 5 | $134.5 | $1337.88 |
| July 6 | $133.89 | $1339.22 |
| July 7 | $135.01 | $1353.22 |
| July 8 | $134.08 | $1343.8 |
| July 11 | $132.02 | $1319.49 |
| July 12 | $130.31 | $1313.64 |
| July 13 | $129.7 | $1317.72 |
| July 14 | $129.89 | $1308.87 |

**4.21.** The ticker symbols (stock symbols) for the Goldman Sachs Group and SPDR S&P 500 ETF are GS and SPY[34], respectively. The table below provides

[34] SPDR (Spiders) is a short form of Standard & Poor's depositary receipt, an exchange-traded fund (ETF) that tracks the Standard & Poor's 500 Index (S&P 500). Each share of SPY contains one-tenth

the daily closing prices for GS and SPY on the trading days during the period between June 30 and July 14 of 2011. Let $P$ be a portfolio consisting of longing 200 shares of SPY and shorting 100 shares of GS. Suppose that $r_f = 0$. Find the maximum drawdown and the Sharpe ratio for the portfolio in the time period indicated in the table.

| Date | GS | SPY |
|---|---|---|
| June 30 | $133.09 | $131.97 |
| July 1 | $136.65 | $133.92 |
| July 5 | $134.5 | $133.81 |
| July 6 | $133.89 | $133.97 |
| July 7 | $135.01 | $135.36 |
| July 8 | $134.08 | $134.4 |
| July 11 | $132.02 | $131.97 |
| July 12 | $130.31 | $131.4 |
| July 13 | $129.7 | $131.84 |
| July 14 | $129.89 | $130.93 |

**4.22.** Although most stocks' $\alpha$ and $\beta$ can be found online, the actual values of $\alpha$ and $\beta$ for the same stock may be different at different sites. Besides, how the actual values are calculated might be considered as proprietary information. Thus, it is critical to understand the factors that affect the calculations.

Use Yahoo Finance as a data source to complete each of the following problems:

a) The stock symbol of Apple Inc. is AAPL. Estimate $\alpha$ and $\beta$ for AAPL by using the weekly adjusted closing prices over the last 2 years and the S&P 500 index as the market portfolio.
b) Estimate $\alpha$ and $\beta$ for AAPL by using the weekly adjusted closing prices over the last 4 years and the S&P 500 index as the market portfolio.
c) Estimate $\alpha$ and $\beta$ for AAPL by using the daily adjusted closing prices over the last 2 years and the S&P 500 index as the market portfolio.
d) Estimate $\alpha$ and $\beta$ for AAPL by using the weekly adjusted closing prices over the last 2 years and the NASDAQ-100 index as the market portfolio.
e) Observe the results above and give the factors that the actual calculated value of $\beta$ depends on.

### 4.4.3 Theoretical Exercises

**4.23.** Establish (4.5) on page 158, i.e., show

---

of the S&P index and trades at approximately one-tenth of the dollar value of the S&P 500. Thus, the rate of daily returns of SPY and S&P 500 index are basically the same.

$$\mu_M = \frac{C - Br_f}{B - Ar_f}, \quad \sigma_M^2 = \frac{Ar_f^2 - 2Br_f + C}{(B - Ar_f)^2}, \quad w_M = \frac{V^{-1}(\mu_M - r_f e)}{B - Ar_f}.$$

**4.24.** Show that under linear factor model framework, portfolio variance can be decomposed into common factor variance and idiosyncratic variance. (*Hint:* Apply relation (4.25) on page 190.)

**4.25.** Use the single-factor model (4.26) on page 192, namely,

$$R = \alpha + \beta R_M + \varepsilon,$$

to express $\beta$ in terms of the variance of the total market return and the covariance of the market return with an individual security's return.

**4.26.** Use the single-factor model (4.26) on page 192, namely,

$$R = \alpha + \beta R_M + \varepsilon,$$

to obtain a corresponding asset pricing formula.

**4.27.** A self-financing or dollar-neutral portfolio is established by using the proceeds of the short sales to finance the long purchases. In other words, under the assumption of *Frictionless Trading*, a self-financing portfolio is a zero-cost portfolio. For example, the excess return $R_M - r_f$ can be viewed as the return of a portfolio that is formed by using the borrowed amount at interest rate $r_f$ to purchase shares of SPY.

A traditionally common representation of many asset pricing models is in the linear factor form:

$$E(R_i - r_f) = \alpha + \sum_{k=1}^{m} \beta_k \lambda_k,$$

where $\lambda_k$ with $k = 1, 2, \ldots, m$ are the values of the corresponding risk factors.

Given two risk factors—one is the excess return on the market portfolio and the other is an economic recession factor—use a self-financing portfolio to establish a two-factor linear model for the excess return of a security.

**4.28.** Let

$$A = \frac{\frac{1}{n} \sum_{i=1}^{n} X_i Y_i - \overline{XY}}{\sqrt{\frac{1}{n^2} \sum_{i=1}^{n} (X_i - \overline{X})^2 \sum_{i=1}^{n} (Y_i - \overline{Y})^2}},$$

$$B = \frac{n \sum_{i=1}^{n} X_i Y_i - \left(\sum_{i=1}^{n} X_i\right)\left(\sum_{i=1}^{n} Y_i\right)}{\sqrt{n \sum_{i=1}^{n} X_i^2 - \left(\sum_{i=1}^{n} X_i\right)^2} \sqrt{n \sum_{i=1}^{n} Y_i^2 - \left(\sum_{i=1}^{n} Y_i\right)^2}}.$$

Prove that $A = B$.

**4.29.** Given a portfolio $P$, show that its Sortino ratio is no less than its Sharpe ratio.

**4.30.** Let $A$ be an $n \times m$ matrix and $A^T$ be the transpose of $A$. Prove the following property:
$$\text{rank}(A^T A) = \text{rank}(A).$$

**4.31.** We continue from the last exercise. Let

$$A = \begin{bmatrix} 1 & x_1 \\ 1 & x_2 \\ \vdots & \vdots \\ 1 & x_n \end{bmatrix} \quad \text{with } x_1 \neq x_2 \quad \text{and} \quad y = \begin{bmatrix} y_1 \\ y_2 \\ \vdots \\ y_n \end{bmatrix}$$

be given. Find the best fit to the system

$$A \begin{bmatrix} \alpha \\ \beta \end{bmatrix} = y$$

such that the norm $\left| A \begin{bmatrix} \alpha \\ \beta \end{bmatrix} - y \right|$ is minimized.

*Hint*: Let $\gamma = \begin{bmatrix} \alpha \\ \beta \end{bmatrix}$ and $L = |A\gamma - y|^2$, then

$$L = (A\gamma - y)^T (A\gamma - y) = \gamma^T A^T A\gamma - \gamma^T A^T y - y^T A\gamma + y^T y.$$

## References

[1] Acerbi, C., Tasche, D.: Expected shortfall: a natural coherent alternative to value at risk. Econ. Notes **31**(2), 379–388 (2002)
[2] Acerbi, C., Nordio, C., Sirtori, C.: Expected Shortfall as a Tool for Financial Risk Management. Abaxbank Working Paper. arXiv:condmat/0102304. http://arxiv.org/pdf/cond-mat/0102304.pdf
[3] Artzner, P., Delbaen, F., Eber, J.-M., Heath, D.: Thinking coherently. Risk **10**, 68–71 (1997)
[4] Artzner, P., Delbaen, F., Eber, J.-M., Heath, D.: Coherent measures of risk. Math. Finance **9**(3), 203–228 (1999)
[5] Bernstein, P.L.: Against the Gods: The Remarkable Story of Risk. Wiley, New York (1996)
[6] Chan, L.K.C., Karceski, J., Lakonishokk, J.: On portfolio optimization: forecasting covariances and choosing the risk model. Rev. Financ. Stud. **12**(5), 937–974 (1999)
[7] Chen, N., Roll, R., Ross, S.A.: Economic forces and the stock market. J. Bus. **59**(3) 383(1986)

[8] Connor G., Korajczyk, R.A.: Risk and return in an equilibrium APT: application of a new test methodology. J. Financ. Econ. **21**(2), 255–290 (1988)

[9] Connor G., Korajczyk, R.A.: A test for the number of factors in an approximate factor model. J. Finance **48**, 1263–1291 (1993)

[10] Connor G.: The three types of factor models: a comparison of their explanatory power. Financ. Anal. J. **51**(1), 42–46 (1995)

[11] Connor G., Goldberg, L.R., Korajczyk, R.A.: Portfolio Risk Analysis. Princeton University Press, Princeton (2010)

[12] Daníelsson, J., Jorgensen, B.N., Samorodnitsky, G., Mandira, M.: Subadditivity Re-Examined: the Case for Value-at-Risk. Financial Markets Group, Discussion paper, 549. London School of Economics and Political Science, London (2005)

[13] Davis, J., Fama, E., French, K.: Characteristics, covariances, and average returns: 1929-1997. J. Finance **55**(1), 396 (2000)

[14] Dhaene, J., Vanduffel, S., Goovaerts, M.J., Kaas, R., Tang, Q., Vyncke, D.: Risk Measures and Comonotonicity: A Review. Stochastic Models. Taylor & Francis Group, LLC (2006)

[15] Dowd, K.: Measuring Market Risk. Wiley, New York (2005)

[16] Fama, E.: Efficient capital markets: a review of theory and empirical work. J. Finance **25**, 383(1970)

[17] Fama, E., French, K.: The cross-section of expected stock returns. J. Finance **47**(2), 427(1992)

[18] Fama, E., French, K.: Common risk factors in the returns on stocks and bonds. J. Financ. Econ. **33**(1), 3(1993)

[19] Fama, E., French, K.: The capital asset pricing model: theory and evidence. J. Econ. Perspect. **18**(3), 25(2004)

[20] Föllmer, H., Schied, A.: Stochastic Finance. Walter de Gruyter, Berlin (2004)

[21] Gilchrist, W.G.: Statistical Modelling with Quantile Functions. Chapman and Hall/CRC, Boca Raton (2000)

[22] Glasserman, P., Heidelberger, P., Shahabuddin, P.: Portfolio value-at-risk with heavy-tailed risk factors. Math. Finance **12**(3), 239(2002)

[23] Goldberg, L.R., Hayes, M.Y., Menchero, J., Mitra, I.: Extreme risk analysis. J. Perform. Meas. **14**, 3 (2010)

[24] Goldberg, L.R., Hayes, M.Y.: The long view of financial risk. J. Investment Manag. **8**, 39–48 (2010)

[25] Hull, J.C.: Risk Management and Financial Institutions. Prentice Hall, Upper Saddle River (2007)

[26] Ingersoll, J.: Theory of Financial Decision Making. Rowman and Littlefield, Savage (1987)

[27] Jorion, P.: Value at Risk, 3rd edn. McGraw-Hill, New York (2007)

[28] Luenberger, D.: Investment Science. Oxford University Press, New York (1998)

[29] Ma, Y., Genton, M., Parzen, E.: Asymptotic properties of sample quantiles of discrete distributions. Ann. Inst. Stat. Math. **63**, 227(2011)

[30] Markowitz, H.: Portfolio Selection. Blackwell, Cambridge (1959)

[31] Reiley, F., Brown, K.: Investment Analysis and Portfolio Management. South-Western Cengage Learning, Mason (2009)

[32] Rockafellar, R.T., Uryasev, S.: Optimization of Conditional Value-at-Risk. J. Risk **2**, 21–41 (2000)

[33] Rockafellar, R.T., Uryasev, S.: Conditional value-at-risk for general loss distributions. J. Bank. Finance **26**, 1443(2002)

[34] Ross, S.A.: The arbitrage theory of capital asset pricing. J. Econ. Theory **13**, 341(1976)

[35] Sharpe, W.: Capital asset prices: a theory of market equilibrium under conditions of risk. J. Finance **19**(3), 425(1964)

[36] Sharpe, W.: "The Sharpe Ratio" objectively determined and measured. J. Portf. Manag. **21**, 1 (1994)

[37] Sharpe, P., Alexander, G.J., Bailey, J.V.: Investments. Prentice-Hall, Upper Saddle Rive (1999)

[38] Spearman, C.: "General Intelligence" objectively determined and measured. Am. J. Psychol. **15**, 201(1904)

[39] Wilmott, P.: Paul Wilmott on Quantitative Finance. Wiley, New York (2006)

# Chapter 5

# Binomial Trees and Security Pricing Modeling

We introduce a discrete-time model of a risky security's futures price using a binomial tree. By increasing the number of time steps in the tree, the assumption is that one obtains a more and more accurate model of the random futures price of a security. The chapter starts with a general $n$-period binomial tree model and then restricts to the case of a Cox-Ross-Rubinstein tree. The latter allows us to determine a continuous-time model, i.e., one in the limit $n \to \infty$, of the random futures price of a risky security. The chapter concludes with some basic properties of the continuous-time security price model.

## 5.1 The General Binomial Tree Model of Security Prices

We shall present a general binomial tree model of the futures price of a security given its current price, which is assumed known to all market participants.

Fix a time interval $[t_0, t_f]$ over which to model a security's price. The time span of the interval $[t_0, t_f]$ is denoted by

$$\tau = t_f - t_0.$$

Let $n$ be a positive integer. Divide $[t_0, t_f]$ into $n$ subintervals $[t_0, t_1], \ldots, [t_{n-1}, t_n]$ of the same period (i.e., length) $h_n$:

$$0 \le t_0, \quad t_1 = t_0 + h_n, \quad t_2 = t_0 + 2h_n, \quad \ldots, \quad t_n = t_0 + n h_n = t_f,$$

where

$$h_n = \frac{\tau}{n}.$$

Here $t_0$ is the current time with $t_1, \ldots, t_n$ future times. The subintervals $[t_{j-1}, t_j]$, $j = 1, \ldots, n$, will be called *time steps*. Since the overall time span $\tau$ is divided into $n$ periods, the tree will also be called an *$n$-period binomial tree*. Note that as the number $n$ of time steps changes, the label $t_n$ for the fixed final time $t_f$ changes.

A.O. Petters, X. Dong, *An Introduction to Mathematical Finance with Applications*, Springer Undergraduate Texts in Mathematics and Technology, DOI 10.1007/978-1-4939-3783-7_5

Denote the price of the security at time $t_j$ by $S(t_j)$, where $j = 0, 1, \ldots, n$. As-*sume that at the current time $t_0$, the price of the security is known from the market, say, $S(t_0) = S_0$.* Then the futures prices $S(t_1), \ldots, S(t_n)$ are random.

**Remark 5.1.** To avoid our notation becoming too cumbersome, we denote the price of a security at time $t$ by $S(t)$ without additional notation to indicate the type of model used for the security. In particular, whether $S(t_j)$ represents a security's price at $t_j$ using a discrete or continuous-time model will be made clear from the context. For example, almost all of this chapter (except at the end) will employ a discrete model, while the next chapter and beyond typically will use continuous-time modeling.                                                      □

We now list some key assumptions and properties of an $n$-period binomial tree. It is helpful to reference Figure 5.1 during the presentation.

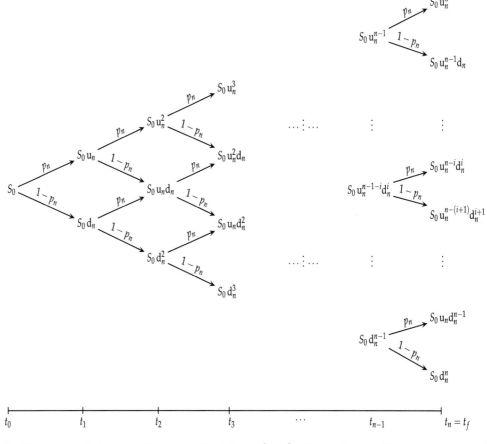

**Fig. 5.1** An $n$-period binomial tree over the interval $[t_0, t_f]$, where $n$ is a positive integer and each $[t_{j-1}, t_j]$ has the same length $\tau/n$, where $\tau = t_f - t_0$. The tree satisfies the recombining property and has independent paths. There are $n + 1$ possible values of $S(t_n)$ and $2^n$ possible price paths from $t_0$ to $t_n$. The random gross returns $S(t_j)/S(t_{j-1})$, where $j = 1, \ldots, n$ are assumed independent.

➢ **Recombining property.** If the price $S_{\text{node}}$ at any node increases over the next time step and is followed by a decrease in the subsequent time step, then we get the same value if, instead, we had a price decrease followed by an increase: $(S_{\text{node}} u_n) d_n = (S_{\text{node}} d_n) u_n$. By the recombining property, *there are then $n + 1$ possible prices at time $t_n$:*

$$S_0 d_n^n, \quad S_0 u_n d_n^{n-1}, \quad \ldots, \quad S_0 u_n^i d_n^{n-i}, \quad \ldots, \quad S_0 u_n^{n-1} d_n, \quad S_0 u_n^n.$$

*The prices increase as one moves through the list from left to right, equivalently, from the end of the bottommost branch of the tree to the topmost one.* In particular, for the $i$th price, $S_0 u_n^i d_n^{n-i}$, where $i = 0, 1, \ldots, n$, the power $i$ of $u_n$ is the number of times the price had to increase during the $n$ time steps to arrive at $S_0 u_n^i d_n^{n-i}$, while $n - i$ is the number of times the price had to decrease. In other words, we can express the price of the security at time $t_j$ as

$$S(t_j) = S(t_0) u_n^{N_{U,j}} d_n^{j - N_{U,j}}, \tag{5.1}$$

where $S(t_0) = S_0$ and $N_{U,j}$ is the random number of upticks in the security's price from time $t_0$ to $t_j$.

➢ **Gross returns, capital-gain returns, and log returns.** For each time step $[t_{j-1}, t_j]$, where $j = 1, \ldots, n$, the random up or down price movement of the security is given by its gross return $S(t_j)/S(t_{j-1})$ over the interval. *We assume that the gross returns of the $n$ time steps, namely,*

$$\frac{S(t_1)}{S(t_0)}, \quad \frac{S(t_2)}{S(t_1)}, \quad \cdots \quad \frac{S(t_n)}{S(t_{n-1})}, \tag{5.2}$$

*are independent and identically distributed (i.i.d.).* Explicitly, the gross returns are assumed to be independent Bernoulli random variables with each having the same probability distribution determined by

$$\frac{S(t_j)}{S(t_{j-1})} = \begin{cases} u_n & \text{with probability } p_n \\ \\ d_n & \text{with probability } 1 - p_n, \end{cases} \tag{5.3}$$

where $j = 1, \ldots, n$ and

$$u_n > 1, \qquad 0 < d_n < 1, \qquad 0 < p_n < 1. \tag{5.4}$$

In other words, from time $t_{j-1}$ to $t_j$, each possible value of the price $S(t_{j-1})$ at $t_{j-1}$ either goes up by the factor $u_n$ with probability $p_n > 0$ or down by the factor $d_n$ with probability $1 - p_n > 0$. *The situations $p_n = 0$ and $p_n = 1$ are excluded* since there is little interest in binomial trees with such probabilities.

The expected gross return over $[t_0, t_f]$ is

$$\mathbb{E}\left(\frac{S(t_n)}{S(t_0)}\right) = \mathbb{E}\left(\frac{S(t_1)}{S(t_0)}\frac{S(t_2)}{S(t_1)}\cdots\frac{S(t_n)}{S(t_{n-1})}\right)$$

$$= \left(\mathbb{E}\left(\frac{S(t_1)}{S(t_0)}\right)\right)^n \qquad\qquad \text{(gross returns are i.i.d.)}$$

$$= (p_n u_n + (1 - p_n) d_n)^n. \tag{5.5}$$

Since the gross returns (5.2) are i.i.d., the capital-gain returns

$$\mathfrak{R}_1 = \frac{S(t_1)}{S(t_0)} - 1, \quad \mathfrak{R}_2 = \frac{S(t_2)}{S(t_1)} - 1, \quad \ldots, \quad \mathfrak{R}_n = \frac{S(t_n)}{S(t_{n-1})} - 1,$$

are i.i.d. as well as the log returns

$$\ln\left(\frac{S(t_1)}{S(t_0)}\right), \quad \ln\left(\frac{S(t_2)}{S(t_1)}\right), \quad ,\ldots, \quad \ln\left(\frac{S(t_n)}{S(t_{n-1})}\right),$$

where

$$\ln\left(\frac{S(t_j)}{S(t_{j-1})}\right) = \begin{cases} \ln(u_n) & \text{with probability } p_n \\ \\ \ln(d_n) & \text{with probability } 1 - p_n, \end{cases} \tag{5.6}$$

for $j = 1, \ldots, n$. We are using the fact that if $X$ and $Y$ are independent random variables and if $f$ and $g$ are continuous[1] functions, then $f(X)$ and $g(Y)$ are independent.[2]

➤ **The triple $p_n, u_n, d_n$.** These quantities depend on the size $h_n$ of the time step. This is reasonable because under normal market conditions, we would expect the price of a stock to be more likely to vary less during, say, a time-step size of an hour as opposed to a day. Notationally, the subscript $n$ in $p_n, u_n, d_n$ is there as a reminder of the dependence of these quantities on $h_n$.

Additionally, *since each time step in an n-period binomial tree has the same size, the quantities $p_n, u_n, and d_n$ are assumed to have the same value over every time step in the tree.* However, though their values are the same across an $n$-period binomial tree, they are not necessarily the same for different trees. For instance, the triplet $p_n, u_n, and d_n$ does not carry over to 1-period, 2-period, ..., and $(n-1)$-period binomial trees since those trees have larger time steps and different probability spaces. In fact, each element of the

---

[1] For readers familiar with measure theory, it suffices for the functions $f$ and $g$ to be measurable, which includes the continuous functions. In fact, all the functions you will encounter in our financial applications are measurable. Some measure theory will be introduced in Chapter 6.

[2] *Proof.* $\mathbb{P}(f(X) \in A, g(Y) \in B) = \mathbb{P}(X \in f^{-1}(A), Y \in g^{-1}(B)) = \mathbb{P}(X \in f^{-1}(A))\,\mathbb{P}(Y \in g^{-1}(B)) = \mathbb{P}(f(X) \in A)\,\mathbb{P}(g(Y) \in B)$, where the independence of $X$ and $Y$ was used in the second to the last equality.

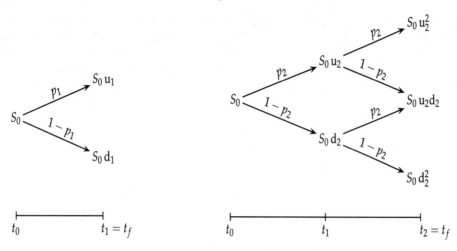

**Fig. 5.2** One- and two-period binomial trees (left to right). Since the time interval is $[t_0, t_f]$ for both trees, each time step in the tree on the right is actually half the size of the one on the left.

sample space $\Omega_n$ of an $n$-period binomial tree is a sequence of $n$ up $(U)$ and down $(D)$ movements of the security's price. Consequently, the sample spaces are different—e.g., the 1- and 2-period trees have the following sample spaces:

$$\Omega_1 = \{U, D\}, \qquad \Omega_2 = \{UU, UD, DU, DD\}.$$

➤ **Sample price paths.** Geometrically, each element $\omega_n$ of the sample space can be viewed as a discrete sample price path from $t_0$ to $t_n$, namely,

$$t \mapsto S_t(\omega_n),$$

where

$$S_t = S(t), \qquad t \in \{t_0, t_1, \ldots, t_n\},$$

and $S_{t_0}(\omega_n) = S_0$ for every $\omega_n$ in $\Omega_n$. Moreover, the realizations of the random price $S(t_j)$ at a fixed time $t_j$, where $1 \leq j \leq n$, are given by the values $S_{t_j}(\omega_n)$ as $\omega_n$ varies over the $\Omega_n$. In addition, for a fixed $t_j$ and $\omega_n$, the possible price $S_{t_j}(\omega_n)$ depends only on the portion of the sequence of $U$'s and $D$'s in $\omega_n$ up to time $t_j$, $j = 1, \ldots, n$.

**Example 5.1.** Let us illustrate the above observations using the 2-period binomial tree in Figure 5.2. The associated sample space of possible outcomes is

$$\Omega_2 = \{UU, UD, DU, DD\}.$$

Consider the following specific possible outcome:

$$\omega_2 = UD \in \Omega.$$

**Table 5.1** Possible outcomes $\omega_2$ and security price values at times $t_0$, $t_1$, and $t_2$ for a 2-period binomial tree. The sample space is $\Omega_2 = \{UU, UD, DU, DD\}$. For example, if $\omega_2 = DU$, then $S_{t_1}(\omega_2) = S_0 d_2$. The random variable $S_{t_1}$ depends only on the first slot of $\omega_2$ (e.g., $S_{t_1}(UD)$ depends only on $U$), while the random variable $S_{t_2}$ depends on both slots. Note that $S_{t_0}$ is nonrandom, i.e., it is a constant random variable.

| $\omega_2$ | $S_{t_0}(\omega_2)$ | $S_{t_1}(\omega_2)$ | $S_{t_2}(\omega_2)$ |
|------------|---------------------|---------------------|---------------------|
| $UU$ | $S_0$ | $S_0 u_2$ | $S_0 u_2^2$ |
| $UD$ | $S_0$ | $S_0 u_2$ | $S_0 u_2 d_2$ |
| $DU$ | $S_0$ | $S_0 d_2$ | $S_0 d_2 u_2 = S_0 u_2 d_2$ |
| $DD$ | $S_0$ | $S_0 d_2$ | $S_0 d_2^2$ |

Then $\omega_2$ determines the following prices at times $t_0$, $t_1$, and $t_2$:

$$S_{t_0}(\omega_2) = S_0, \qquad S_{t_1}(\omega_2) = S_0 u_2, \qquad S_{t_2}(\omega_2) = S_0 u_2 d_2.$$

In other words, we can view $\omega_2$ as a discrete curve given by

$$t \mapsto S_t(\omega_2), \qquad\qquad t = t_0, t_1, t_2.$$

We also see that the possible price $S_{t_1}(\omega_2)$ depends only on the first entry $U$ of $\omega_2 = UD$. Table 5.1 presents all the possible outcomes in $\Omega_2$ along with the prices across time that each possible outcome determines. Compare with Figure 5.2.

□

Since every path to one of the $n+1$ possible prices,

$$S_0 d_n^n, \quad S_0 u_n d_n^{n-1}, \quad \dots, \quad S_0 u_n^i d_n^{n-i}, \quad \dots, \quad S_0 u_n^{n-1} d_n, \quad S_0 u_n^n,$$

consists of a sequence of $n$ up or down price movements (i.e., gross returns) and because these movements are independent, all the paths are independent.

The number of price paths in the tree is $2^n$, which quickly becomes extremely large. For example, a 50-period binomial tree has

$$2^{50} \approx 10^{15} \quad \text{(one quadrillion!) price paths.}$$

➤ **Probability measure.** The probability of a particular price path occurring is obtained by multiplying the probabilities along the path. Explicitly, let $N_U$ be the random number of upticks in the price of the security from $t_0$ to $t_n = t_f$. Then $N_U$ is a binomial random variable. The probability measure on the sample space $\Omega_n$ of an $n$-period binomial tree is then defined at a possible outcome $\omega_n$ by

$$\mathbb{P}(\omega_n) = p_n^{N_U(\omega_n)} (1 - p_n)^{n-N_U(\omega_n)}.$$

Thinking of $\omega_n$ as a discrete price path, we see that $\mathbb{P}(\omega_n)$ is indeed the product of the probabilities along the path.

Additionally, since the price paths are independent, the probability of a specific security price occurring at time $t_f$ is the sum of the probabilities of the individual paths to the given price. For instance, the probability of the price $S_0 u_n^i d_n^{n-i}$, conditioned on the current price being $S(t_0) = S_0$, is

$$\mathbb{P}\left(S(t_n) = S_0 u_n^i d_n^{n-i}\right) = \binom{n}{i} p_n^i (1 - p_n)^{n-i}$$

$$= \text{probability of } i \text{ price increases during } [t_0, t_f],$$

$$\text{(5.7)}$$

where $i = 0, 1, 2, \ldots, n$, and

$$\mathbb{P}\left(S(t_n) \leq S_0 u_n^k d_n^{n-k}\right) = \sum_{i=0}^{k} \binom{n}{i} p_n^i (1 - p_n)^{n-i}$$

$$= \text{probability of at most } k \text{ price increases,}$$

$$\text{(5.8)}$$

where $k = 0, 1, 2, \ldots, n$.

The expected security price at time $t_f$, conditioned on the current price being $S(t_0) = S_0$, is

$$\mathbb{E}\left(S(t_f)\right) = \sum_{i=0}^{n} S_0 u_n^i d_n^{n-i} \binom{n}{i} p_n^i (1 - p_n)^{n-i}$$

$$= S_0 \sum_{i=0}^{n} \binom{n}{i} (p_n u_n)^i ((1 - p_n) d_n)^{n-i}$$

$$= S_0 (p_n u_n + (1 - p_n) d_n)^n, \qquad (5.9)$$

where the binomial formula was used for the last equality. Note that Equation (5.9) agrees with (5.5).

➤ **Cash dividends and the return rate.** *From time $t_0$ to $t_f$, assume that the security pays a constant, continuous, proportional annual dividend yield rate $q$ giving a cash dividend of*

$$D(t_{j-1}, t_j) = q S(t_{j-1}) h_n, \qquad (j = 1, 2, \ldots, n).$$

*Unless stated to the contrary, assume that in the continuous-time limit (i.e., $n \to \infty$ or $h_n \to 0$), the cash dividends are continuously reinvested in the security to* *buy more units of the security.* The (total) return rate from $t_{j-1}$ to $t_j$ is then

$$R(t_{j-1}, t_j) = \frac{S(t_j) - S(t_{j-1}) + D(t_{j-1}, t_j)}{S(t_{j-1})} = \Re_j + q h_n. \qquad (5.10)$$

Note that the return rates $R(t_0, t_1), \ldots, R(t_{n-1}, t_n)$ are i.i.d.

**Example 5.2.** Consider a nondividend-paying stock with a current price of $155. Model the futures price behavior of the stock over the next 6 months using a 100-period binomial tree with

$$u_{100} = 1.01424, \quad d_{100} = 0.985957, \quad p_{100} = 0.52299.$$

a) What is the 57th possible price 6 months out, where the 101 possible prices are counted from bottom to top in the tree?

**Solution.** The $(i+1)$st possible price is

$$S_0 \, u_{100}^i \, d_{100}^{n-i},$$

so the 57th one is $155 \, u_{100}^{56} \, d_{100}^{44} = \$183.64$.

b) What is the probability that the 57th possible price will occur?

**Solution.** By (5.7), do not even need to know the 57th possible price in order to compute the desired probability:

$$\mathbb{P}\left(S(0.5) = \$155 \, u_{100}^{56} \, d_{100}^{44}\right) = \binom{100}{56} p_{100}^{56} (1 - p_{100})^{44} = 6.1\%.$$

c) What is the probability that the stock's price increases 56 times?

**Solution.** Equation (5.7) yields that this probability is the same as that of the 57th possible price occurring, which is 6.1%.

d) What is the probability that the stock's price is at most $183.64?

**Solution.** Since

$$\$183.64 = \$155 \, u_{100}^{56} \, d_{100}^{100-56},$$

Equation (5.8) gives

$$\mathbb{P}\left(S(0.5) \leq \$155 \, u_{100}^{56} \, d_{100}^{44}\right) = \sum_{i=0}^{56} \binom{100}{i} p_{100}^i (1 - p_{100})^{100-i} = 80\%.$$

e) What is the expected price of the stock 6 months from now?

**Solution.** $\mathbb{E}(S(0.5)) = \$155 \times \left(p_{100} u_{100} + (1 - p_{100}) d_{100}\right)^{100} = \$167.05.$

□

## Log Returns and Security Price Modeling

We now conclude the section with a formal expression for the futures price of the security in terms of the log returns over the various time steps. Writing

$$S(t_n) = S_0 \frac{S(t_n)}{S_0} = S_0 e^{\ln\left(\frac{S(t_n)}{S_0}\right)}$$

and using

$$\ln\left(\frac{S(t_n)}{S_0}\right) = \ln\left(\frac{S(t_1)}{S_0} \frac{S(t_2)}{S(t_1)} \frac{S(t_3)}{S(t_2)} \cdots \frac{S(t_n)}{S(t_{n-1})}\right) = \sum_{j=1}^{n} \ln\left(\frac{S(t_j)}{S(t_{j-1})}\right), \quad (5.11)$$

we can express the price $S(t_n)$ in terms of the time-step log returns as

$$S(t_n) = S_0 \exp\left[\sum_{j=1}^{n} \ln\left(\frac{S(t_j)}{S(t_{j-1})}\right)\right], \quad (5.12)$$

where $t_n = t_f$. *Equation (5.12) highlights that, to determine the security's price, it suffices to determine the log returns. Log returns are then important.*

We can further breakdown Equation (5.12) into individual parts that can be determined. Let $X_{n,j}$ be the standardization of $\ln\left(S(t_j)/S(t_{j-1})\right)$:

$$X_{n,j} = \frac{\ln\left(\frac{S(t_j)}{S(t_{j-1})}\right) - \mathbb{E}\left(\ln\left(\frac{S(t_j)}{S(t_{j-1})}\right)\right)}{\sqrt{\mathrm{Var}\left(\ln\left(\frac{S(t_j)}{S(t_{j-1})}\right)\right)}}.$$

Equivalently,

$$\ln\left(\frac{S(t_j)}{S(t_{j-1})}\right) = \mathbb{E}\left(\ln\left(\frac{S(t_j)}{S(t_{j-1})}\right)\right) + \sqrt{\mathrm{Var}\left(\ln\left(\frac{S(t_j)}{S(t_{j-1})}\right)\right)}\, X_{n,j}.$$

Since the log returns are i.i.d., we have

$$\mathbb{E}\left(\ln\left(\frac{S(t_j)}{S(t_{j-1})}\right)\right) = \mathbb{E}\left(\ln\left(\frac{S(t_1)}{S(t_0)}\right)\right) \qquad (j = 1,\ldots,n)$$

and

$$\mathrm{Var}\left(\ln\left(\frac{S(t_j)}{S(t_{j-1})}\right)\right) = \mathrm{Var}\left(\ln\left(\frac{S(t_1)}{S(t_0)}\right)\right) \qquad (j = 1,\ldots,n).$$

It follows

$$\sum_{j=1}^{n} \ln\left(\frac{S(t_j)}{S(t_{j-1})}\right) = \sum_{j=1}^{n} \mathbb{E}\left(\ln\left(\frac{S(t_1)}{S(t_0)}\right)\right) + \sum_{j=1}^{n} \sqrt{\operatorname{Var}\left(\ln\left(\frac{S(t_1)}{S(t_0)}\right)\right)} X_{n,j}$$

$$= n\,\mathbb{E}\left(\ln\left(\frac{S(t_1)}{S(t_0)}\right)\right) + \sqrt{\operatorname{Var}\left(\ln\left(\frac{S(t_1)}{S(t_0)}\right)\right)} \sum_{j=1}^{n} X_{n,j}.$$

Let

$$\mathbb{Z}_n = \frac{1}{\sqrt{n}} \sum_{j=1}^{n} X_{n,j}.$$

Then the discrete pricing formula (5.12) can be expressed as

$$S(t_n) = S_0 \exp\left[ n\,\mathbb{E}\left(\ln\left(\frac{S(t_1)}{S(t_0)}\right)\right) + \sqrt{n\,\operatorname{Var}\left(\ln\left(\frac{S(t_1)}{S(t_0)}\right)\right)} \mathbb{Z}_n \right].$$

$$(5.13)$$

Equation (5.13) shows that, to characterize the futures price of the security in the continuous-time limit, it suffices to determine $n\mathbb{E}\left(\ln\left(\frac{S(t_1)}{S(t_0)}\right)\right)$, $n\operatorname{Var}\left(\ln\left(\frac{S(t_1)}{S(t_0)}\right)\right)$, and $\mathbb{Z}_n$ as $n \to \infty$.

The goal of Sections 5.2 and 5.3 is to characterize these quantities for $n$ sufficiently large and then in the full limit $n \to \infty$.

## 5.2 The Cox-Ross-Rubinstein Tree

The Cox-Ross-Rubinstein (CRR) tree is a special case of a binomial tree and, consequently, satisfies all the assumptions and properties of a general binomial tree. For a sufficiently large number $n$ of time steps, this tree will allow us to express $u_n$, $d_n$, and $p_n$ approximately in terms of the time-step period $h_n$ and two quantities, denoted $\mu_{\mathrm{RW}}$ and $\sigma$, which measure the respective instantaneous expectation and volatility of the security's log returns, respectively. In fact, the CRR tree will allow us to estimate $n\,\mathbb{E}\left(\ln\left(\frac{S(t_1)}{S(t_0)}\right)\right)$ and $\sqrt{n\,\operatorname{Var}\left(\ln\left(\frac{S(t_1)}{S(t_0)}\right)\right)}$ in the limit $n \to \infty$. Additionally, in Section 5.3 we shall characterize $\mathbb{Z}_n$ as $n \to \infty$ by using the CRR tree and the Lindeberg Central Limit Theorem.

We present the $n$-period CRR tree by using a $1/n$ perturbative approach. The approximations are presented informally to preserve the intuitive ideas behind the analysis. Our discussion will utilize and parallel some aspects of the development in Cox, Ross, and Rubinstein [4], Cox and Rubinstein [5], Roman [15], and Wilmott, Dewynne, and Howison [16]. See Roman [15, Chap. 9] for a very detailed mathematical treatment that keeps track of the probability spaces.

## 5.2.1 The Real-World CRR Tree

It is assumed that, as the number $n$ of time steps increases, the CRR tree becomes a more and more accurate model of a security's price. For this reason, the model is also called a *real-world* CRR tree. Of course, such terminology should not be taken literally. It is another matter whether the real-world CRR tree actually becomes an accurate fit to security prices in the marketplace as $n$ increases. These issues are beyond the scope of this introductory text, but we do note that other types of trees have been studied, e.g., trinomial trees. Interestingly, we shall see in Chapter 8 that the risk-neutral CRR tree is a natural choice when pricing derivatives.

### Assumptions and the Quantities $m$, $\mu_{RW}$, and $\sigma^2$

We now list some additional assumptions and define certain key quantities associated with the real-world CRR tree:

➤ **Computations for large $n$.** For $n$ sufficiently large, we shall compute only to first order in $1/n$, i.e., we ignore all terms of the form $(1/n)^a$ with $a > 1$.

➤ **Almost surely continuous sample price paths.** As $n \to \infty$, assume that the CRR binomial tree's probability space transforms to a continuous-time analog with sample space denoted $\Omega$. Similar to the discrete case, we identify each element $\omega$ in $\Omega$ with the sample price path $t \mapsto (S(t))(\omega)$, where $0 \leq t \leq t_f$. *The sample price paths are assumed to be continuous almost surely in the limit $n \to \infty$.* Since we shall explore probabilities of price paths, we do not need to require that literally every price path must be continuous in the limit $n \to \infty$; only that they are continuous with probability 1 in the limit.

➤ **Expected returns and the constant $m$.** *Assume that the expected time-step return rate per unit time period converges to a constant $m$ in the continuous-time limit:*

$$\frac{\mathbb{E}(R(t_0, t_1))}{h_n} \rightarrow m \quad \text{as} \quad n \to \infty. \tag{5.14}$$

In other words, since the time period shrinks to a single moment of time as $n \to \infty$, we think of $m$ as the security's instantaneous expected return rate per unit time and will call it the *instantaneous expected return rate* or, simply, the *instantaneous expected return*. The rate $m$ is quoted per annum.

We now show explicitly that $m$ applies at each instant of time in the sense that the expected price of the security is obtained by continuously compounding the current price at the rate $m - q$. First, since the (total) return rate $R(t_0, t_1)$ arises from the capital-gain return $\mathfrak{R}_1$ plus the dividend yield contribution $q h_n$, i.e.,

$$R(t_0, t_1) = \mathfrak{R}_1 + q h_n,$$

employing (5.14) gives

$$\frac{\mathbb{E}(\mathfrak{R}_1)}{h_n} = \frac{\mathbb{E}(R(t_0, t_1) - q h_n)}{h_n} \quad \rightarrow \quad m - q \quad \text{as} \quad n \rightarrow \infty. \qquad (5.15)$$

We then interpret $m - q$ as the *instantaneous capital-gain return rate* or simply the *instantaneous capital-gain return*. Note that $m - q$ is also quoted per annum. Now, by (5.15), we have

$$(m - q) h_n \approx \mathbb{E}(\mathfrak{R}_1) \qquad (n \text{ sufficiently large}) \qquad (5.16)$$

and since

$$e^{(m-q) h_n} \approx 1 + (m - q) h_n \qquad (n \text{ sufficiently large}),$$

it follows[3]

$$\mathbb{E}(S(t_1)) \approx S_0 e^{(m-q) h_n} \qquad (n \text{ sufficiently large}). \qquad (5.17)$$

Equation (5.17) shows that as $n$ increases, the gross return over the interval $[t_0, t_f]$ becomes independent of the number of time periods $n$ in a partition of the interval. In fact, recalling (5.5), we see that (5.17) yields

$$\mathbb{E}\left(\frac{S(t_f)}{S(t_0)}\right) = \left(\mathbb{E}\left(\frac{S(t_1)}{S(t_0)}\right)\right)^n \approx \left(e^{(m-q) h_n}\right)^n = e^{(m-q) \tau}$$

$$(n \text{ sufficiently large})$$

or

$$\mathbb{E}(S(t_f)) \approx S_0 e^{(m-q) \tau} \qquad (n \text{ sufficiently large}). \qquad (5.18)$$

The expectation on the left hand side of (5.18) is for discrete time.

*We assume that $m$ and $q$ are known. Additionally, Equation (5.18) shows that the instantaneous expected capital-gain return $m - q$ can be interpreted as the continuously compounded rate at which the security's expected price increases $(m - q > 0)$ or decreases $(m - q < 0)$.* In most cases, $m - q > 0$. Naturally, a security must have nontrivial promise for investors to tolerate an expected return of $m - q < 0$ for an extended period.

---

[3] Strictly speaking, we are considering the conditional expectation $\mathbb{E}(S(t_1) | S(t_0) = S_0) \approx S_0 e^{(m-q) h_n}$ for $n$ sufficiently large.

**Example 5.3.** Suppose that the time span $\tau = t_f - t_0$ is 2 years and the time period $h_n$ is a trading day. Assuming 252 trading days in a year, we then consider a 504-period CRR tree. Assume that the annual dividend yield rate is 2% and the instantaneous annual expected return is 10%. If the current security price is \$75, then (5.18) yields that the expected price of the security 1 year from now is obtained by continuously compounding \$75 at the annual rate $m - q = 8\%$. Explicitly, taking the current time to be 0, we get

$$\mathbb{E}(S(2)) \approx \$75 \, e^{0.08 \times 2} = \$88.01.$$ □

**Remark 5.2.** Estimating $m$ by using historical security prices is problematic since its value is affected by the sampling frequency (e.g., daily, weekly, monthly, etc.) and size of the data (e.g., 6 months of prices versus 1.5 years, versus 10 years, etc.) See Leunberger [11, pp. 214–216] for more. However, in the study of derivatives, this issue is avoided all together since the price of a derivative will be independent of $m$! See pages 404 and 415.

➤ **First constraint.** *Assume the following recombining condition:*

$$u_n \, d_n = 1. \tag{5.19}$$

This condition makes the CRR tree recombine along the horizontal line through the initial price $S_0$ since $S_0 \, u_n \, d_n = S_0$.

➤ **Second constraint and the constant $\mu_{RW}$.** Define $\mu_n$ to be the expected time-step log return per unit time period:

$$\mu_n = \frac{1}{h_n} \mathbb{E}\left(\ln\left(\frac{S(t_1)}{S(t_0)}\right)\right).$$

Explicitly

$$\mu_n \, h_n = p_n \ln u_n + (1 - p_n) \ln d_n. \tag{5.20}$$

Additionally, since the time-step log returns are identically distributed, Equation (5.11) yields that the expected log return over the interval $[t_0, t_f]$ is $n$ times the expected log return over the time step $[t_0, t_1]$:

$$\mathbb{E}\left(\ln\left(\frac{S(t_n)}{S(t_0)}\right)\right) = n \mathbb{E}\left(\ln\left(\frac{S(t_1)}{S(t_0)}\right)\right) = \mu_n \, \tau. \tag{5.21}$$

*A second constraint assumed for a CRR tree is that $\mu_n$ converges to a constant $\mu_{RW}$ as $n$ increases without bound:*

$$\mu_n \to \mu_{RW} \quad \text{as} \quad n \to \infty. \tag{5.22}$$

The constant $\mu_{RW}$ is called the *real-world*[4] *instantaneous drift* or, simply, the *real-world drift* of the security's price. Also, some authors refer to $\mu_{RW}$ as a *natural drift* or *physical drift*. We quote $\mu_{RW}$ per annum.

Finally, *assume that $\mu_{RW}$ is known.* As in the case of $m - q$, the quantity $\mu_{RW}$ is estimated by using historical security prices over an appropriate choice of time steps (e.g., daily) and

$$\mu_{RW} \approx \frac{1}{h_n} \mathbb{E}\left(\ln\left(\frac{S(t_1)}{S(t_0)}\right)\right) = \frac{n\,\mathbb{E}\left(\ln\left(\frac{S(t_1)}{S(t_0)}\right)\right)}{\tau} \qquad (n \text{ sufficiently large}).$$

(5.23)

**Example 5.4.** Suppose that the time interval $[t_0, t_f]$ is a year and partition the year into 252 trading days. Then

$$\tau = 1, \quad n = 252, \quad h_n = \frac{1}{252}.$$

Taking the current time to be $t_0 = 0$, one trading day later is $t_1 = \frac{1}{252}$, and 1 year away is $t_f = 1$. Historical adjusted[5] closing prices of the security can be used to estimate the expected daily log return:

$$\mathbb{E}\left(\ln\left(\frac{S(1/252)}{S(0)}\right)\right).$$

Employing (5.21), the expected log return over a year is then

$$\mathbb{E}\left(\ln\left(\frac{S(1)}{S(0)}\right)\right) = 252\,\mathbb{E}\left(\ln\left(\frac{S(1/252)}{S(0)}\right)\right). \qquad (5.24)$$

In other words, the annual expected log return on the left hand side of (5.24) is obtained by annualizing the expected daily log return, i.e., by multiplying the expected daily log return by 252. Equations (5.23) and (5.24) then show that the real-world drift $\mu_{RW}$ is estimated by the annualized expected daily log return:

$$\mu_{RW} \approx 252\,\mathbb{E}\left(\ln\left(\frac{S(1/252)}{S(0)}\right)\right).$$

Note that the daily log returns are assumed to be i.i.d. Also, we shall see later (Equation (5.30) on page 226) that $\mu_{RW}$ is a function of $m$, which as mentioned in Remark 5.2 is problematic to estimate. □

---

[4] We remind readers to be mindful of the usage of the terminology *real world*; see the introductory paragraph to this section on page 219.
[5] See page 19.

➤ **Third constraint and the constant $\sigma$.** Define $\sigma_n^2$ to be the variance of the time-step log return per unit time step:

$$\sigma_n^2 = \frac{1}{h_n} \, \mathrm{Var}\left(\ln\left(\frac{S(t_1)}{S(t_0)}\right)\right).$$

We have (Exercise 5.19)

$$\sigma_n^2 h_n = p_n(1 - p_n)\left[\ln\left(\frac{u_n}{d_n}\right)\right]^2. \tag{5.25}$$

Observe that by the i.i.d. property of the time-step log returns, it follows

$$\sigma_n^2 \tau = \mathrm{Var}\left(\ln\left(\frac{S(t_n)}{S(t_0)}\right)\right).$$

Equivalently, the variance of the log return over $[t_0, t_f]$ is $n$ times the variance of the log return over a time step:

$$\mathrm{Var}\left(\ln\left(\frac{S(t_n)}{S(t_0)}\right)\right) = n \, \mathrm{Var}\left(\ln\left(\frac{S(t_1)}{S(t_0)}\right)\right). \tag{5.26}$$

*A third assumed constraint for a CRR tree is that $\sigma_n^2$ converges to a positive constant $\sigma^2$ as $n$ increases without bound:*

$$\sigma_n^2 \to \sigma^2 > 0 \quad \text{as} \quad n \to \infty. \tag{5.27}$$

The quantity $\sigma > 0$ is called the *continuous-time volatility* or, simply, the *volatility* of the security's price. *We assume that $\sigma$ is known.*

**Example 5.5.** Employing the same inputs as in Example 5.4, Equation (5.26) shows that the annual variance of the log return is obtained by annualizing the variance of the daily log returns:

$$\mathrm{Var}\left(\ln\left(\frac{S(1)}{S(0)}\right)\right) = 252 \, \mathrm{Var}\left(\ln\left(\frac{S(1/252)}{S(0)}\right)\right).$$

Equation (5.27) then yields that if the standard deviation of the daily log returns is estimated by using historical security price data, then the annual volatility $\sigma$ of the security is estimated as follows:

$$\sigma \approx \sqrt{252}\left[\mathrm{Var}\left(\ln\left(\frac{S(1/252)}{S(0)}\right)\right)\right]^{1/2}.$$

In other words, the standard deviation of the daily log returns is annualized through multiplying by $\sqrt{252}$. As noted in Example 5.4, the daily log returns are assumed to be i.i.d. in order to apply the CRR tree. $\qquad\square$

**Remark 5.3.** The appearance of the constraints (5.19), (5.22), and (5.27) may seem mysterious. We shall see that they make it possible to determine approximate expressions of $p_n$, $u_n$, and $d_n$ for $n$ sufficiently large; see page 226. Perhaps a more practical explanation for the constraints is that they allow in the limit $n \to \infty$ for the discrete CRR pricing formula to converge to the continuous-time security pricing formula utilized in the Black-Scholes-Merton model.                                                                                                         □

➤ **Approximating the variances of the time-step log return and gross return.** A useful approximation of the variance of the log return is

$$\mathrm{Var}\left(\ln\left(\frac{S(t_1)}{S(t_0)}\right)\right) \approx \mathbb{E}\left(\mathfrak{R}_1^2\right) \qquad (n \text{ sufficiently large}). \qquad (5.28)$$

To obtain this result, first observe that

$$\mathrm{Var}\left(\ln\left(\frac{S(t_1)}{S(t_0)}\right)\right) = \mathbb{E}\left(\left[\ln\left(\frac{S(t_1)}{S(t_0)}\right)\right]^2\right) - \left[\mathbb{E}\left(\ln\left(\frac{S(t_1)}{S(t_0)}\right)\right)\right]^2$$

$$= \mathbb{E}\left(\left[\ln\left(\frac{S(t_1)}{S(t_0)}\right)\right]^2\right) - \left(\frac{\mu_n \tau}{n}\right)^2$$

$$\approx \mathbb{E}\left(\left[\ln\left(\frac{S(t_1)}{S(t_0)}\right)\right]^2\right) \qquad (n \text{ sufficiently large}).$$

Now, by the continuity assumption of the sample price paths, we have $|\mathfrak{R}_1| \to 0$ almost surely as $n \to \infty$. Using this result and the leading terms in the Taylor expansion of the natural logarithm, namely,

$$\ln x = (x - 1) - \frac{1}{2}(x - 1)^2 + \cdots \qquad (|x - 1| \ll 1),$$

with

$$x = \frac{S(t_1)}{S(t_0)}, \qquad x - 1 = \mathfrak{R}_1,$$

it follows

$$\mathbb{E}\left(\left[\ln\left(\frac{S(t_1)}{S(t_0)}\right)\right]^2\right) \approx \mathbb{E}\left(\left(\mathfrak{R}_1 - \frac{\mathfrak{R}_1^2}{2}\right)^2\right) \qquad (n \text{ sufficiently large})$$

$$\approx \mathbb{E}\left(\mathfrak{R}_1^2 - \mathfrak{R}_1^3 + \frac{\mathfrak{R}_1^4}{4}\right) \qquad (n \text{ sufficiently large}).$$

For $n$ sufficiently large, we have $|\mathfrak{R}_1| \ll 1$ almost surely and so ignore any contributions from skewness (i.e., the term $\mathbb{E}\left(\mathfrak{R}_1^3\right)$) and kurtosis (i.e., the term $\mathbb{E}\left(\mathfrak{R}_1^4\right)$). We then enforce the following approximation:

$$\mathbb{E}\left(\ln\left(\frac{S(t_1)}{S(t_0)}\right)^2\right) \approx \mathbb{E}\left(\mathfrak{R}_1^2\right) \qquad\qquad (n \text{ sufficiently large}).$$

Our heuristic arguments then imply the desired result:

$$\text{Var}\left(\ln\left(\frac{S(t_1)}{S(t_0)}\right)\right) \approx \mathbb{E}\left(\mathfrak{R}_1^2\right) \qquad (n \text{ sufficiently large}).$$

Under our assumptions, the variances of the log return and gross return are also approximately equal:

$$\text{Var}\left(\frac{S(t_1)}{S(t_0)}\right) \approx \text{Var}\left(\ln\left(\frac{S(t_1)}{S(t_0)}\right)\right) \qquad (n \text{ sufficiently large}), \quad (5.29)$$

which is a consequence of

$$\text{Var}\left(\frac{S(t_1)}{S(t_0)}\right) = \text{Var}\left(\frac{S(t_1)}{S(t_0)} - 1\right) = \mathbb{E}\left(\mathfrak{R}_1^2\right) - \left(\mathbb{E}\left(\mathfrak{R}_1\right)\right)^2$$

$$= \mathbb{E}\left(\mathfrak{R}_1^2\right) - \left(\frac{(m-q)\tau}{n}\right)^2 \qquad\qquad (\text{by } (5.16))$$

$$\approx \mathbb{E}\left(\mathfrak{R}_1^2\right) \qquad\qquad (n \text{ sufficiently large})$$

and Equation (5.28).

➤ **Expressing $\mu_{\text{RW}}$ in terms of $m - q$ and $\sigma$.** Using the Taylor expansion of the natural logarithm, we get

$$\mu_n \frac{\tau}{n} = \mathbb{E}\left(\ln\left(\frac{S(t_1)}{S(t_0)}\right)\right) \approx \mathbb{E}\left(\mathfrak{R}_1 - \frac{\mathfrak{R}_1^2}{2}\right) \qquad (n \text{ sufficiently large})$$

or

$$\mathbb{E}\left(\mathfrak{R}_1^2\right) \approx 2\mathbb{E}\left(\mathfrak{R}_1\right) - \frac{2\mu_n \tau}{n} \qquad\qquad (n \text{ sufficiently large}).$$

On the other hand, employing (5.28) yields

$$\sigma_n^2 \frac{\tau}{n} \approx \mathbb{E}\left(\mathfrak{R}_1^2\right) \qquad\qquad (n \text{ sufficiently large}).$$

Consequently

$$\sigma_n^2 \tau \approx 2n\mathbb{E}\left(\mathfrak{R}_1\right) - 2\mu_n \tau.$$

Taking the limit $n \to \infty$ and making use of the convergences (5.15), (5.22), and (5.27), namely,

$$n\mathbb{E}\left(\mathfrak{R}_1\right) \to (m-q)\tau, \qquad \mu_n \to \mu_{\text{RW}}, \qquad \sigma_n^2 \to \sigma^2,$$

we obtain the following expression for real-world drift $\mu_{\text{RW}}$ in terms of the instantaneous capital-gain return $m - q$ and volatility $\sigma$:

$$\mu_{\text{RW}} = m - q - \frac{\sigma^2}{2}. \tag{5.30}$$

## The Real-World CRR Equations

To obtain formulas for the unknowns $u_n, d_n, p_n$ of a real-world CRR tree in terms of the inputs $m - q, \sigma, h_n$, solve the three constraints (5.19), (5.22), and (5.27) for the three unknowns. For the reader's convenience, we restate these equations below:

$$u_n \, d_n \; = \; 1 \tag{5.31}$$
$$\mu_n \; \approx \; \mu_{\text{RW}} \qquad\qquad (n \text{ sufficiently large}) \tag{5.32}$$
$$\sigma_n^2 \; \approx \; \sigma^2 \qquad\qquad (n \text{ sufficiently large}), \tag{5.33}$$

where

$$\mu_n h_n \; = \; p_n \ln u_n \; + \; (1 - p_n) \ln d_n \tag{5.34}$$

$$\sigma_n^2 h_n \; = \; p_n (1 - p_n) \left[ \ln \left( \frac{u_n}{d_n} \right) \right]^2. \tag{5.35}$$

We claim that for $n$ sufficiently large, Equations (5.31)-(5.35) are solved by

$$u_n \approx e^{\sigma \sqrt{h_n}}, \qquad d_n \approx e^{-\sigma \sqrt{h_n}}, \qquad p_n \approx \frac{1}{2} \left( 1 + \frac{\mu_{\text{RW}}}{\sigma} \sqrt{h_n} \right), \tag{5.36}$$

which are called *the real-world CRR equations*. They are the governing formulas for the real-world CRR tree. Note that

$$p_n \to \frac{1}{2} \qquad\qquad \text{as } n \to \infty \tag{5.37}$$

and by (5.17),

$$p_n u_n + (1 - p_n) d_n \; \approx \; e^{(m-q) h_n} \qquad\qquad (n \text{ sufficiently large}),$$

which yields another expression for the uptick probability:

$$p_n \; \approx \; \frac{e^{(m-q) h_n} - d_n}{u_n - d_n} \qquad\qquad (n \text{ sufficiently large}). \tag{5.38}$$

Equation (5.38) and the expression for $p_n$ in (5.36) are equivalent to first order in $1/n$ (see Exercise 5.20).

To establish (5.36), first observe that by (5.31), we have $\ln u_n = -\ln d_n$, which when inserted in (5.34) yields

$$\mu_n h_n = -p_n \ln d_n + (1 - p_n) \ln d_n = -2 p_n \ln d_n + \ln d_n.$$

We then solve for the desired uptick probability:

$$p_n = \frac{\mu_n h_n - \ln d_n}{2(-\ln d_n)} = \frac{\mu_n h_n - \ln d_n}{2 \ln u_n} = \frac{1}{2}\left(1 + \frac{\mu_n h_n}{\ln u_n}\right).$$

Substituting the above expression for $p_n$ and $\ln(u_n/d_n) = 2 \ln u_n$ into (5.35) gives

$$\sigma_n^2 h_n = \frac{1}{4}\left(1 + \frac{\mu_n h_n}{\ln u_n}\right)\left(1 - \frac{\mu_n h_n}{\ln u_n}\right) 4 (\ln u_n)^2$$

$$= \left(1 - \left(\frac{\mu_n h_n}{\ln u_n}\right)^2\right)(\ln u_n)^2.$$

Employing (5.33) and computing to first order in $h_n$, it follows

$$\sigma^2 h_n \approx \sigma_n^2 h_n = (\ln u_n)^2 - (\mu_n h_n)^2 \approx (\ln u_n)^2 \qquad (n \text{ sufficiently large}).$$

Since $u_n > 1$, we have

$$\sigma \sqrt{h_n} \approx \ln u_n = -\ln d_n \qquad (n \text{ sufficiently large}). \qquad (5.39)$$

Applying (5.32) then gives

$$p_n \approx \frac{1}{2}\left(1 + \frac{\mu_{RW}}{\sigma} \sqrt{h_n}\right) \qquad (n \text{ sufficiently large}) \qquad (5.40)$$

and

$$u_n \approx e^{\sigma \sqrt{h_n}}, \quad d_n \approx e^{-\sigma \sqrt{h_n}} \qquad (n \text{ sufficiently large}).$$

**Example 5.6.** Suppose that a nondividend-paying stock with a current price of \$75 has an instantaneous annual expected return of 12% and annual volatility of 10%. Model the behavior of the stock's price using a 100-period CRR tree.

a) What is the expected price (i.e., the forecasted price) of the stock 3 months from now?

**Solution.** Given the large number of periods, we employ the CRR tree to determine $u_{100}$, $d_{100}$, and $p_{100}$ in terms of $h_{100}$, $\sigma$, and $m$:

$$u_{100} \approx e^{\sigma \sqrt{h_{100}}}, \quad d_{100} \approx e^{-\sigma \sqrt{h_{100}}}, \quad p_{100} \approx \frac{e^{(m-q) h_{100}} - d_{100}}{u_{100} - d_{100}}.$$

We have $n = 100$, $h_{100} = \frac{0.25}{100} = 0.0025$, $\sqrt{h_{100}} = 0.05$, $m = 0.12$, $q = 0$, $\sigma = 0.10$, and $S(t_0) = \$75$. Moreover

$$u_{100} = 1.00501, \quad d_{100} = 0.995012, \quad p_{100} = 0.528754.$$

Therefore, the expected price is

$$\mathbb{E}(S(0.25)) = \$75 \times \left( p_{100}\, u_{100} + (1 - p_{100})\, d_{100} \right)^{100} = \$77.28.$$

b) What is the probability that 3 months from now, the stock's price is less than or equal to its expected price?

**Solution.** First, we check whether the expected price $\$77.28$ is one of the 101 possible prices 3 months from now. To do this, we must find the number $k$ of price upticks such that

$$\$77.28 = \$75\, u_{100}^{k}\, d_{100}^{100-k}.$$

In general, the number of upticks associated with a price $S_\star$ is the nonnegative integer $k$ that solves

$$S_\star = S_0\, u_n^{k}\, d_n^{n-k}.$$

It follows

$$k = \frac{\ln\left(\frac{S_\star}{S_0\, d_n^{n}}\right)}{\ln\left(\frac{u_n}{d_n}\right)}. \tag{5.41}$$

Since $S_\star = \$77.28$, we get $k = 53$, i.e., the expected price $\$77.28$ is the 54th possible price, counting from bottom to top.

With $k$ available, we can apply (5.8) that yields

$$\mathbb{P}\left( S(0.25) \leq \$77.28 \right) = \mathbb{P}\left( S(0.25) \leq \$75\, u_{100}^{53}\, d_{100}^{47} \right)$$

$$= \sum_{i=0}^{53} \binom{100}{i} p_{100}^{i}\, (1 - p_{100})^{100-i}$$

$$= 55\%.$$

For the 100-step tree, *the probability that the price 3 months from now is less than or equal to its mean is more than 50%.* We emphasize that this property is not an accident for $n$ sufficiently large. It carries over to the continuous-time limit; see Equation (5.78) on page 248.

**Remark 5.4.** Note that if $k$ were not an integer, say, $k = 42.6785$, then the dollar amount $\$75\, u_n^{k}\, d_n^{100-k}$ would not be one of the possible price values of the stock. However, we can still compute the desired probability by omitting that dollar amount (which has probability zero) and adding the probabilities up to $k = 42$.                                                                    □

c) What is the probability that 3 months from now, the stock's price is less than or equal to its current price? What would be the probability if the volatility of the stock were much higher, say, 40%?

**Solution.** By (5.41), the number of upticks associated with the price $S_\star = \$75$ is $k = 50$, which means

$$\$75 \, u_{100}^{50} \, d_{100}^{100-50} = \$75.$$

The current price is then the 51st possible price, counting from bottom to top. The probability of the price 3 months out being less than or equal to $75 is actually less that 50%. In fact, by (5.8) the probability is

$$\mathbb{P}\left(S(0.25) \leq \$75\right) = \sum_{i=0}^{50} \binom{100}{i} p_{100}^{i} \, (1 - p_{100})^{100-i} = 32\%.$$

Now, if we increase the volatility from 10% to 40%, the CRR tree yields:

$$u_{100} = 1.0202, \quad d_{100} = 0.980199, \quad p_{100} = 0.502501.$$

In addition, the number of upticks for $S_\star = \$75$ is also $k = 50$. However, the probability that the stock price 3 months from now is less than or equal to its current price increases to 52%. In other words, *for a sufficiently large volatility, we see that there is more than a 50% probability that the price 3 months away is less than or equal to the current price.* This property is also not coincidental for $n$ and $\sigma$ sufficiently large. We shall encounter it again in our continuous-time study of security prices; see Equation (5.79) on page 248.

$\square$

**Security Price Formula for a Real-World CRR Tree**

We now express the general binomial tree security price formula (5.13) on page 218, namely,

$$S(t_n) = S_0 \exp\left[n\, \mathbb{E}\left(\ln\left(\frac{S(t_1)}{S(t_0)}\right)\right) + \sqrt{n\, \mathrm{Var}\left(\ln\left(\frac{S(t_1)}{S(t_0)}\right)\right)} \, Z_n\right],$$

in the framework of a real-world CRR tree. Recall that

$$Z_n = \frac{1}{\sqrt{n}} \sum_{j=1}^{n} X_{n,j}.$$

Here $X_{n,j}$ is the standardization of $\ln\left(\frac{S(t_j)}{S(t_{j-1})}\right)$ and, since the log returns being identically distributed, it follows

$$X_{n,j} = \frac{\ln\left(\frac{S(t_j)}{S(t_{j-1})}\right) - \mathbb{E}\left(\ln\left(\frac{S(t_j)}{S(t_{j-1})}\right)\right)}{\sqrt{\mathrm{Var}\left(\ln\left(\frac{S(t_j)}{S(t_{j-1})}\right)\right)}} = \frac{\ln\left(\frac{S(t_j)}{S(t_{j-1})}\right) - \mathbb{E}\left(\ln\left(\frac{S(t_1)}{S(t_0)}\right)\right)}{\sqrt{\mathrm{Var}\left(\ln\left(\frac{S(t_1)}{S(t_0)}\right)\right)}}.$$

Because

$$n\,\mathbb{E}\left(\ln\left(\frac{S(t_1)}{S(t_0)}\right)\right) = \mu_n\,\tau, \qquad n\,\mathrm{Var}\left(\ln\left(\frac{S(t_1)}{S(t_0)}\right)\right) = \sigma_n^2\,\tau,$$

we have

$$X_{n,j} = \frac{\ln\left(\frac{S(t_j)}{S(t_{j-1})}\right) - \mu_n\,h_n}{\sigma_n\,\sqrt{h_n}}. \tag{5.42}$$

Also, it can be shown (Exercise 5.21) using the formulas (5.31)–(5.35) that in a real-world CRR setting, the standardization $X_{n,j}$ becomes

$$X_{n,j} = \begin{cases} \dfrac{1-p_n}{\sqrt{p_n(1-p_n)}} & \text{with probability } p_n \\[3ex] \dfrac{-p_n}{\sqrt{p_n(1-p_n)}} & \text{with probability } 1-p_n. \end{cases} \tag{5.43}$$

The security pricing formula now takes the following more compact form for a real-world CRR tree:

$$S(t_n) = S_0\,e^{\mu_n\,\tau + \sigma_n\,\sqrt{\tau}\,\mathbb{Z}_n}. \tag{5.44}$$

The goal is to determine the quantity to which (5.44) converges in the limit $n \to \infty$.

### 5.2.2 The Risk-Neutral CRR Tree

The current section explores the CRR tree in the context of a risk-neutral world, that is, in a world of only risk-neutral investors (see page 137).

*In a risk-neutral world, meaning a world of risk-neutral investors (see page 137), the expected futures price of a security is assumed to be given by its current price continuously compounded at the risk-free rate r minus any cash dividend yield rate.* For such a world, there is no compensation required for the security's risk since the rate r compensates only for opportunity cost and inflation. Explicitly, if a security pays a continuous, proportional cash dividend at constant annual yield rate $q$, then in a risk-neutral world, the security's expected price at the future time $t_1$, given the current price $S(t_0) = S_0$, is

$$\mathbb{E}\left(S(t_1)\right) = S_0\,e^{(r-q)\,h_n}. \tag{5.45}$$

On the other hand, a real marketplace consists primarily of risk-averse rather than risk-neutral investors, so it may seem like a mathematical indulgence to explore risk neutrality. We shall show in Chapter 8 that the price of a derivative in the BSM model is actually independent of the instantaneous expected return $m$; see Remark 5.2 on page 221. In other words, if we assume $m = r$ for the security, which is the case in a risk-neutral world, then though our pricing model for the underlying security would be unrealistic, the resulting price of a derivative based on the security would be the same as if we had used a real-world value for $m$.[6] For this reason, we shall investigate a risk-neutral CRR tree and will do so to first order in $1/n$.

*We assume that in a risk-neutral world, the security prices and dividend yield rate are the same as those in the real marketplace.* Explicitly, the real-world prices $S(t_0),\ldots,S(t_n)$ and $q$ will also be used for the risk-neutral CRR analysis. In particular, *for a risk-neutral CRR tree, we shall employ the security's current market price $S(t_0)$ and the same up and down factors $u_n$ and $d_n$, where $u_n > 1, 0 < d_n < 1$, and also assume that the relationship between $u_n$ and $d_n$ is unchanged: $u_n = \frac{1}{d_n}$.* Because the up and down factors $u_n$ and $d_n$ and the period $h_n$ are the same for the real world and risk-neutral world, we can make use of

$$u_n \approx e^{\sigma \sqrt{h_n}}, \qquad d_n \approx e^{-\sigma \sqrt{h_n}} \qquad (n \text{ sufficiently large}), \qquad (5.46)$$

where $\sigma$ is the real-world, continuous-time volatility of the security.

### Risk-Neutral Uptick Probability $p_n^*$

*What actually changes in the switch from the real world to a risk-neutral world is the probability.* In particular, the uptick probability will no longer be $p_n$. This is because, in a risk-neutral world, Equation (5.45) is true. Can we find an uptick probability $p_n^*$ that makes the risk-neutral condition (5.45) hold? Writing out (5.45) formally using the unknown quantity $p_n^*$, we obtain

$$p_n^* S_0 u_n + (1 - p_n^*) S_0 d_n = S_0 e^{(r-q)h_n}. \qquad (5.47)$$

Equation (5.47) is readily solved:

$$p_n^* = \frac{e^{(r-q)h_n} - d_n}{u_n - d_n}. \qquad (5.48)$$

Strictly speaking, we do not yet know if the quantity $p_n^*$ given by (5.48) is actually a probability. It will be a probability if we can prove that $p_n^*$ is between zero and one. As pointed out on page 211, *we shall exclude binomial trees with $p_n^* = 0$ or $p_n^* = 1$.* By (5.47), this means that the constraints

---

[6] Recall that it is essentially impossible to determine a reliable value for $m$ in the marketplace.

$$e^{(r-q)h_n} \neq d_n, \qquad e^{(r-q)h_n} \neq u_n, \qquad (5.49)$$

are enforced.

*We claim that, for an n-period risk-neutral binomial tree, the no-arbitrage condition implies*

$$0 < p_n^* < 1.$$

To establish the claim, it suffices to prove that if there is no arbitrage, then an $n$-period risk-neutral binomial tree satisfies

➤ $d_n < e^{(r-q)h_n}$.
➤ $u_n > e^{(r-q)h_n}$.

To show these two statements, it suffices to consider the time step $[t_0, t_1]$ since the uptick probability is assumed the same across every time step in a binomial tree. Assume $d_n > e^{(r-q)h_n}$. At the current time $t_0$, short (borrow) at the risk-free rate r an amount equal to the cost $S_0 e^{-qh_n}$ of $e^{-qh_n}$ units of the security. Use these funds to long $e^{-qh_n}$ units of the security. At time $t_1$, the number of units of the security grows to 1 due to continuous cash dividend reinvesting to buy more shares. Sell the one unit of the security to receive $S(t_1)$. The amount owed on the loan at $t_1$ is $S_0 e^{(r-q)h_n}$. The net profit/loss is

$$S(t_1) - S_0 e^{(r-q)h_n}.$$

Since $u_n > 1$ and $0 < d_n < 1$, we see

$$S(t_1) = \begin{cases} S_0 u_n & \text{with probability } p_n \\ \\ S_0 d_n & \text{with probability } 1 - p_n \end{cases} \geq S_0 d_n.$$

Consequently,

$$S(t_1) - S_0 e^{(r-q)h_n} \geq S_0 d_n - S_0 e^{(r-q)h_n}.$$

If $d_n > e^{(r-q)h_n}$, then

$$S(t_1) - S_0 e^{(r-q)h_n} > 0,$$

which is an arbitrage. Hence

$$d_n \leq e^{(r-q)h_n}.$$

However, we exclude the case $d_n = e^{(r-q)h_n}$ since (5.47) shows that it leads to a binomial tree with $p_n^* = 0$. The argument for $u_n > e^{(r-q)h_n}$ is left as an exercise (Exercise 5.7).

The no-arbitrage condition (along with the constraints (5.49)) then yields

$$d_n < e^{(r-q)h_n} < u_n.$$

Therefore, by (5.48) we obtain $0 < p_n^* < 1$. The quantity $p_n^*$ is called a *risk-neutral uptick probability*.

**Notation.** For the risk-neutral CRR tree, we shall designate expectations, variances, etc. with respect to $p_n^*$ using an $*$.

### The Risk-Neutral CRR Equations

Equations (5.46) and (5.48) show that the following quantities govern the risk-neutral CRR tree for $n$ sufficiently large:

$$u_n \approx e^{\sigma \sqrt{h_n}}, \qquad d_n \approx e^{-\sigma \sqrt{h_n}}, \qquad p_n^* = \frac{e^{(r-q)h_n} - d_n}{u_n - d_n}. \qquad (5.50)$$

Let us now express $p_n^*$ in a form analogous to the expression for $p_n$ in (5.40). The quantities $m$, $\mu_n$, and $\sigma_n$ change to the following $m^*$, $\mu_n^*$, and $\sigma_n^*$ in the risk-neutral setting:

$$m^* h_n \approx \mathbb{E}_*(R(t_0, t_1)) \qquad \text{($n$ sufficiently large)}$$

$$\mu_n^* h_n = \mathbb{E}_*\left(\ln\left(\frac{S(t_1)}{S(t_0)}\right)\right)$$

$$(\sigma_n^*)^2 h_n = \text{Var}_*\left(\ln\left(\frac{S(t_1)}{S(t_0)}\right)\right).$$

We assume there are constants $\mu_*$ and $\sigma^* > 0$ such that

$$\mu_n^* \to \mu_*, \qquad \sigma_n^* \to \sigma^* \qquad \text{as } n \to \infty.$$

Since in a risk-neutral world, the expected futures price comes from compounding $S_0$ at the rate $r - q$, the instantaneous expected return rate $m^*$ is the risk-free rate. In fact, analogous to (5.17) we have

$$S_0 e^{(m^*-q)h_n} \approx \mathbb{E}_*\left(S(t_1)\right) \qquad \text{($n$ sufficiently large)},$$

which with (5.45) gives

$$S_0 e^{(m^*-q)h_n} \approx S_0 e^{(r-q)h_n} \qquad \text{($n$ sufficiently large)},$$

i.e., in the limit $n \to \infty$ we get
$$m^* = r.$$

Arguing analogously to the real-world CRR tree case (see page 225), we obtain

$$\mu_* = r - q - \frac{(\sigma^*)^2}{2},$$

which is called the *risk-neutral drift* of the security. Similarly, the analogs of (5.20) and (5.25) are

$$\mu_n^* h_n = p_n^* \ln u_n + (1 - p_n^*) \ln d_n, \qquad (\sigma_n^*)^2 h_n = p_n^*(1 - p_n^*) \left[ \ln \left( \frac{u_n}{d_n} \right) \right]^2,$$

which with $u_n = \frac{1}{d_n}$ yield the following for $n$ sufficiently large:

$$u_n \approx e^{\sigma^* \sqrt{h_n}}, \qquad d_n \approx e^{-\sigma^* \sqrt{h_n}}, \qquad p_n^* \approx \frac{1}{2} \left( 1 + \frac{\mu_*}{\sigma^*} \sqrt{h_n} \right).$$

Because the up and down factors are the same for the risk-neutral world and the real world, we have

$$e^{\sigma^* \sqrt{h_n}} \approx u_n \approx e^{\sigma \sqrt{h_n}} \qquad\qquad (n \text{ sufficiently large}).$$

In other words, *the continuous-time volatility is the same in the real world and the risk-neutral world:*

$$\sigma^* = \sigma.$$

Consequently, *the risk-neutral CRR equations become the following for n sufficiently large:*

$$u_n \approx e^{\sigma \sqrt{h_n}}, \qquad d_n \approx e^{-\sigma \sqrt{h_n}}, \qquad p_n^* \approx \frac{1}{2} \left( 1 + \frac{\mu_*}{\sigma} \sqrt{h_n} \right), \qquad (5.51)$$

where

$$\mu_* = r - q - \frac{\sigma^2}{2}. \qquad (5.52)$$

### Relationship Between $p_n^*$ and $p_n$

The risk-neutral uptick probability $p_n^*$ yields

$$\mathbb{E}_* \left( S(t_1) \right) = S_0 e^{(r-q) h_n},$$

while the real-world uptick probability $p_n$ gives

$$\mathbb{E} \left( S(t_1) \right) \approx S_0 e^{(m-q) h_n} \qquad\qquad (n \text{ sufficiently large}).$$

We can express $p_n^*$ in terms of $p_n$ through the following transformation:

$$p_n^* \approx p_n - \eta_n \sqrt{p_n(1 - p_n)} \qquad (n \text{ sufficiently large}), \qquad (5.53)$$

where

$$\eta_n = \frac{\mathbb{E}(R(t_0, t_1)) - r h_n}{\sqrt{\text{Var}(R(t_0, t_1))}}. \qquad (5.54)$$

Here $\eta_n$ is the Sharpe ratio of the security given as the ratio of the spread between the expected total return rate and the risk-free rate[7] $r h_n$ across the time step $[t_0, t_1]$, to the security risk across the same time step. The relationship (5.53) is a discrete-time example of *Girsanov theorem* (see Neftci [13, Chap. 14] and references therein).

We now give a heuristic proof of (5.53). Our approach is to compute the Sharpe ratio $\eta_n$ first. By (5.14) on page 219, the expectation in the numerator of (5.54) is

$$\mathbb{E}\left(R(t_0, t_1)\right) \approx m h_n \qquad (n \text{ sufficiently large}).$$

To determine the denominator $\sqrt{\text{Var}\left(R(t_0, t_1)\right)}$ of (5.54), employ (5.29):

$$\text{Var}\left(\frac{S(t_1)}{S(t_0)}\right) \approx \text{Var}\left(\ln\left(\frac{S(t_1)}{S(t_0)}\right)\right) \qquad (n \text{ sufficiently large}).$$

Since

$$\text{Var}\left(R(t_0, t_1)\right) = \text{Var}\left(\frac{S(t_1)}{S(t_0)} - 1 + q h_n\right) = \text{Var}\left(\frac{S(t_1)}{S(t_0)}\right),$$

we obtain

$$\text{Var}\left(R(t_0, t_1)\right) \approx \text{Var}\left(\ln\left(\frac{S(t_1)}{S(t_0)}\right)\right) = \sigma_n^2 h_n \qquad (n \text{ sufficiently large}).$$

Equations (5.25) on page 223 and (5.39) on page 227 then yield

$$\text{Var}\left(R(t_0, t_1)\right) \approx p_n(1 - p_n)\left(\ln\left(\frac{u_n}{d_n}\right)\right)^2$$
$$\approx p_n(1 - p_n)\left(4\sigma^2 h_n\right) \qquad (n \text{ sufficiently large}),$$

where

$$p_n \approx \frac{1}{2}\left(1 + \frac{\mu_{\text{RW}}}{\sigma}\sqrt{h_n}\right) \qquad (n \text{ sufficiently large})$$

with

$$\mu_{\text{RW}} = m - q - \frac{\sigma^2}{2}.$$

Consequently, the Sharpe ratio becomes

$$\eta_n \approx \frac{(m - r) h_n}{\sqrt{p_n(1 - p_n)}\,\left(2\sigma\sqrt{h_n}\right)} \qquad (n \text{ sufficiently large}).$$

Now, for $n$ sufficiently large, note that

$$\eta_n\sqrt{p_n(1 - p_n)} \approx \frac{((m - q) - (r - q)) h_n}{2\sigma\sqrt{h_n}} = \frac{((m - q) - (r - q))\sqrt{h_n}}{2\sigma}$$

---

[7] Recall that r is the risk-free rate per annum (by default).

and

$$p_n \approx \frac{1}{2}\left(1 + \frac{(m-q)}{\sigma}\sqrt{h_n} - \frac{\sigma^2}{2\sigma}\sqrt{h_n}\right).$$

Hence, for $n$ sufficiently large, it follows

$$
\begin{aligned}
p_n - \eta_n\sqrt{p_n(1-p_n)} &\approx \frac{1}{2}\left(1 + \frac{(r-q)}{\sigma}\sqrt{h_n} - \frac{\sigma^2}{2\sigma}\sqrt{h_n}\right) \\
&= \frac{1}{2}\left(1 + \frac{\mu_*}{\sigma}\sqrt{h_n}\right) \\
&\approx p_n^*,
\end{aligned}
$$

where

$$\mu_* = r - q - \frac{\sigma^2}{2}.$$

### Security Price Formula for a Risk-Neutral CRR Tree

Though the price of a security has the same value in the real world and the risk-neutral world, its mathematical expression can also be given in terms of quantities in a risk-neutral CRR tree. In parallel to the discussion for the security price formula of a real-world CRR tree (see page 229), we obtain

$$S(t_n) \approx S_0\, e^{\mu_n^* \tau + \sigma_n\sqrt{\tau}\, Z_n^*} \qquad (n \text{ sufficiently large}), \qquad (5.55)$$

where we used $\sigma_n^* \approx \sigma_n$ for sufficiently large $n$ and employed the quantity

$$Z_n^* = \frac{1}{\sqrt{n}} \sum_{j=1}^{n} X_{n,j}^*$$

with

$$
X_{n,j}^* =
\begin{cases}
\dfrac{1-p_n^*}{\sqrt{p_n^*(1-p_n^*)}} & \text{with probability } p_n^* \\[2ex]
\dfrac{-p_n^*}{\sqrt{p_n^*(1-p_n^*)}} & \text{with probability } 1-p_n^*.
\end{cases}
$$

We shall obtain the function to which (5.55) converges in the limit $n \to \infty$. Observe that since the risk-neutral security price expression (5.55) equals the real-world one in (5.44), we have a transformation between $Z_n^*$ and $Z_n$:

$$Z_n^* \approx Z_n + \frac{(\mu_n - \mu_n^*)\,\tau}{\sigma_n\sqrt{\tau}} \qquad (n \text{ sufficiently large}). \qquad (5.56)$$

We shall also show this transformation explicitly in the continuous-time limit (see (5.70) on page 243).

## 5.3 Continuous-Time Limit of the CRR Pricing Formula

The goal is to determine an explicit expression for the random price of a security in the continuous-time limit $n \to \infty$ of the discrete CRR pricing formulas for the real-world and risk-neutral world CRR trees. We shall carry out the analysis in detail for the real-world CRR tree. The risk-neutral case is the same, except for a minor relabeling of the notation.

### 5.3.1 The Lindeberg Central Limit Theorem

For a real-world CRR tree, Equation (5.44) on page 230 expresses the price of a security as follows:

$$S(t_n) = S_0 e^{\mu_n \tau + \sigma_n \sqrt{\tau} \, \mathbb{Z}_n}.$$

Here

$$\mathbb{Z}_n = \frac{1}{\sqrt{n}} \sum_{j=1}^{n} X_{n,j},$$

with $X_{n,j}$ the standardization of the log return $\ln \left( \frac{S(t_j)}{S(t_{j-1})} \right)$ (see (5.42)):

$$X_{n,j} = \frac{\ln \left( \frac{S(t_j)}{S(t_{j-1})} \right) - \mu_n h_n}{\sigma_n \sqrt{h_n}},$$

where $j = 1, \ldots, n$. Note that

$$\mathrm{Var}(\mathbb{Z}_n) = \frac{1}{n} \sum_{j=1}^{n} \mathbb{E} \left( X_{n,j}^2 \right) = 1. \tag{5.57}$$

In fact, each $\mathbb{Z}_n$ is the standardization of $\sum_{j=1}^{n} X_{n,j}$. In terms of the uptick probability $p_n$, we have explicitly (see (5.43))

$$X_{n,j} = \begin{cases} \dfrac{1-p_n}{\sqrt{p_n(1-p_n)}} & \text{with probability } p_n \\[4mm] \dfrac{-p_n}{\sqrt{p_n(1-p_n)}} & \text{with probability } 1 - p_n. \end{cases} \tag{5.58}$$

Table 5.2 lists some of the random variables $X_{n,j}$ and $\mathbb{Z}_n$. Note that the row sequences of $X_{n,j}$'s form a triangular-array pattern. *For each $n \geq 1$, the random variables $X_{n,1}, X_{n,2}, \ldots, X_{n,n}$ in the nth row are independent (since the time-step log returns are independent) and identically distributed,* i.e., the probability measures $\mathbb{P}_{n,1}, \mathbb{P}_{n,2}, \ldots, \mathbb{P}_{n,n}$ of $X_{n,1}, X_{n,2}, \ldots, X_{n,n}$, respectively, are the same:

**Table 5.2** A triangular array of the random variables $X_{n,j}$, where $n = 1,2,3,\ldots$ and $j = 1,\ldots,n$. Each row sequence $X_{n,1}, X_{n,2}, \ldots, X_{n,n}$, where $n = 1,2,\ldots$, consists of i.i.d. standardized random variables, whereas the sequence $\mathbb{Z}_1, \mathbb{Z}_2, \ldots$, is not i.i.d.

| $n$ | Sequence | $\mathbb{Z}_n$ |
|---|---|---|
| 1 | $X_{1,1}$ | $X_{1,1}$ |
| 2 | $X_{2,1}, \; X_{2,2}$ | $\dfrac{1}{\sqrt{2}}(X_{2,1} + X_{2,2})$ |
| 3 | $X_{3,1}, \; X_{3,2}, \; X_{3,3}$ | $\dfrac{1}{\sqrt{3}}(X_{3,1} + X_{3,2} + X_{3,3})$ |
| $\vdots$ | $\vdots$ | $\vdots$ |
| $n$ | $X_{n,1}, \; X_{n,2}, \; X_{n,3}, \; \ldots, \; X_{n,n}$ | $\dfrac{1}{\sqrt{n}}(X_{n,1} + X_{n,2} + X_{n,3} + \cdots + X_{n,n})$ |
| $\vdots$ | $\vdots$ | $\vdots$ |

$$\mathbb{P}_{n,j}\left(X_{n,j} = \frac{1 - p_n}{\sqrt{p_n(1-p_n)}}\right) = p_n, \qquad \mathbb{P}_{n,j}\left(X_{n,j} = \frac{-p_n}{\sqrt{p_n(1-p_n)}}\right) = 1 - p_n$$

for $j = 1,\ldots,n$. *It may also be tempting to apply the classical Central Limit Theorem* (CLT)[8] *to conclude that* $\mathbb{Z}_n = \frac{1}{\sqrt{n}}\sum_{j=1}^{n} X_{n,j}$ *converges in distribution to a standard normal random variable as* $n \to \infty$. *However, this is not allowed* because $X_{1,1}, X_{2,1}, X_{2,2}, X_{3,1}, X_{3,2}, X_{3,3}, X_{4,1}, \ldots$ are not identically distributed. In fact, if $n \neq n'$, then $X_{n,j}$ and $X_{n',j}$ have different probability distributions since $p_n \neq p_{n'}$.

To determine the convergence of $\mathbb{Z}_n$ as $n \to \infty$, we employ a generalization of the classical CLT due to Lindeberg. Consider a triangular array of random variables, namely, the following collection of rows of random variables:

$$
\begin{array}{ccccc}
\mathfrak{X}_{1,1} & & & & \\
\mathfrak{X}_{2,1}, & \mathfrak{X}_{2,2} & & & \\
\mathfrak{X}_{3,1}, & \mathfrak{X}_{3,2}, & \mathfrak{X}_{3,3} & & \\
& & \vdots & & \\
\mathfrak{X}_{n,1}, & \mathfrak{X}_{n,2}, & \mathfrak{X}_{n,3}, & \ldots, & \mathfrak{X}_{n,n} \\
& & \vdots & &
\end{array}
\tag{5.59}
$$

---

[8] **Theorem. (Classical CLT)** Assume that $X_1, X_2, \ldots$ are i.i.d. random variables with each having a finite mean $\mathbb{E}(X_i) = \mu_0$ and finite variance $\text{Var}(X_i) = \sigma_0^2 > 0$. Let $Z_n$ be the standardization of the sample mean $\bar{X}_n = \frac{1}{n}(X_1 + \cdots + X_n)$, namely,

$$Z_n = \frac{\bar{X}_n - \mu_0}{(\sigma_0/\sqrt{n})} = \frac{1}{\sqrt{n}}\sum_{j=1}^{n}\left(\frac{X_j - \mu_0}{\sigma_0}\right).$$

Then $Z_n$ converges in distribution to a standard normal $Z$ as $n \to \infty$.

*Assume that the random variables $\mathfrak{X}_{n,1}, \mathfrak{X}_{n,2}, \ldots, \mathfrak{X}_{n,n}$ in the nth row are independent for every n (i.e., every row) and satisfy*

$$\mathbb{E}\left(\mathfrak{X}_{n,j}\right) = 0 \qquad \text{for } j = 1, \ldots, n, \qquad \text{and} \qquad \sum_{j=1}^{n} \mathbb{E}\left(\mathfrak{X}_{n,j}^2\right) = 1. \tag{5.60}$$

*For the array (5.59), we do not require that the random variables in any row be identically distributed and nor do we require that those in different rows are independent.*

### The Lindeberg Condition

Before we can get the desired generalization of the classical CLT, we need a constraint, called the *Lindeberg condition*, on the random variables in the rows of (5.59). Our discussion will be in parallel with the treatment in Pitman's lecture [14]; see Section 27 and Theorem 27.2 of Billingsley [1] for more.

Our assumptions about the array in Equation (5.59) yield that the summation in (5.60) is actually the total variance of the $n$th row:

$$\text{Var}\left(\sum_{j=1}^{n} \mathfrak{X}_{n,j}\right) = \sum_{j=1}^{n} \mathbb{E}\left(\mathfrak{X}_{n,j}^2\right) = 1, \tag{5.61}$$

where $\text{Var}\left(\mathfrak{X}_{n,j}\right) = \mathbb{E}\left(\mathfrak{X}_{n,j}^2\right)$. Note that the largest value of the variance terms in the summation (5.61) is less than or equal to 1, and, if the largest value equals 1, then it is attained by only one variance term and the other terms must vanish. Intuitively speaking, the Lindeberg condition is the requirement that as one moves further and further down the array in (5.59) (i.e., as $n \to \infty$), the variances of the random variables in each row become smaller and smaller.

To state the Lindeberg condition precisely, recall the definition of the indicator function on $\{X > c\}$, where $X$ is a random variable and $c$ a real number:

$$\mathbf{1}_{\{X > c\}} = \begin{cases} 1 & \text{if } X > c \\ \\ 0 & \text{if } X \leq c. \end{cases}$$

Then the *Lindeberg condition* is defined as follows:

$$\text{For every fixed } \epsilon > 0, \qquad \lim_{n \to \infty} \sum_{j=1}^{n} \mathbb{E}\left(\mathfrak{X}_{n,j}^2 \, \mathbf{1}_{\{|\mathfrak{X}_{n,j}| > \epsilon\}}\right) = 0. \tag{5.62}$$

Note that as $n \to \infty$ in (5.62), the value of $\epsilon$ remains unchanged since it is fixed during the limit.

To see how the Lindeberg condition implies that the variance of each element in a row of the array gets smaller as $n \to \infty$, first observe that for every random variable $\mathcal{X}_{n,i}$ in the $n$th row, we have

$$\mathcal{X}_{n,i}^2 \leq \epsilon^2 + \mathcal{X}_{n,i}^2 \, \mathbf{1}_{\{|\mathcal{X}_{n,i}| > \epsilon\}} \qquad\qquad (i = 1, \ldots, n).$$

Since the inequality is between nonnegative quantities, taking the expectation yields an upper bound on the variance of $\mathcal{X}_{n,i}$:

$$\operatorname{Var}\left(\mathcal{X}_{n,i}\right) \leq \epsilon^2 + \mathbb{E}\left(\mathcal{X}_{n,i}^2 \, \mathbf{1}_{\{|\mathcal{X}_{n,i}| > \epsilon\}}\right) \qquad\qquad (i = 1, \ldots, n). \quad (5.63)$$

Adding to the upper bound in Equation (5.63) the contributions (each of which is also nonnegative) from the remaining terms in the $n$th row, namely, the sum $\sum_{1 \leq j \neq i \leq n} \mathbb{E}\left(\mathcal{X}_{n,j}^2 \, \mathbf{1}_{\{|\mathcal{X}_{n,j}| > \epsilon\}}\right)$, we find

$$\operatorname{Var}\left(\mathcal{X}_{n,i}\right) \leq \epsilon^2 + \sum_{j=1}^{n} \mathbb{E}\left(\mathcal{X}_{n,j}^2 \, \mathbf{1}_{\{|\mathcal{X}_{n,j}| > \epsilon\}}\right) \qquad\qquad (i = 1, \ldots, n).$$

Because this upper bound holds for the variance of each random variable in the $n$th row, it holds in particular for the largest value of those variances:

$$\max_{1 \leq i \leq n} \operatorname{Var}\left(\mathcal{X}_{n,i}\right) \leq \epsilon^2 + \sum_{j=1}^{n} \mathbb{E}\left(\mathcal{X}_{n,j}^2 \, \mathbf{1}_{\{|\mathcal{X}_{n,j}| > \epsilon\}}\right) \qquad (i = 1, \ldots, n). \quad (5.64)$$

By Equation (5.64), if the array (5.59) satisfies the Lindeberg condition, then as one moves down the array, the maximum of the variances of the random variables in each row starts to approach zero:

$$\lim_{n \to \infty} \max_{1 \leq i \leq n} \operatorname{Var}\left(\mathcal{X}_{n,i}\right) \leq \epsilon^2 \qquad \text{for } \underline{\text{all}} \quad \epsilon > 0 \quad \text{(no matter how small)}, \quad (5.65)$$

i.e.,

$$\lim_{n \to \infty} \max_{1 \leq i \leq n} \operatorname{Var}\left(\mathcal{X}_{n,i}\right) = 0.$$

Note: since we always have $\operatorname{Var}\left(\mathcal{X}_{n,j}\right) \leq 1$, Equation (5.65) gives no new constraint on the maximum variance for $\epsilon \geq 1$, but it does for $0 < \epsilon < 1$.

## The Lindeberg CLT

With the Lindeberg condition available, we can now state the following generalization of the classical CLT:

**Theorem 5.1. (The Lindeberg Central Limit Theorem)** *Assume that the triangular array of random variables,*

$$\mathfrak{X}_{1,1}$$

$$\mathfrak{X}_{2,1}, \quad \mathfrak{X}_{2,2}$$

$$\mathfrak{X}_{3,1}, \quad \mathfrak{X}_{3,2}, \quad \mathfrak{X}_{3,3}$$

$$\vdots$$

$$\mathfrak{X}_{n,1}, \quad \mathfrak{X}_{n,2}, \quad \mathfrak{X}_{n,3}, \quad \ldots, \quad \mathfrak{X}_{n,n}$$

$$\vdots$$

*is such that the random variables $\mathfrak{X}_{n,1}, \mathfrak{X}_{n,2}, \ldots, \mathfrak{X}_{n,n}$ in the nth row are independent for every $n \geq 1$ (i.e., every row) and satisfy*

$$\mathbb{E}\left(\mathfrak{X}_{n,j}\right) = 0 \quad \text{for } j = 1,\ldots,n, \quad \text{and} \quad \sum_{j=1}^{n} \mathbb{E}\left(\mathfrak{X}_{n,j}^2\right) = 1.$$

*If the triangular array obeys the Lindeberg condition, then*

$$\mathcal{Z}_n = \sum_{j=1}^{n} \mathfrak{X}_{n,j} \xrightarrow{\ d\ } Z \qquad\qquad \text{as } n \to \infty,$$

*where Z is the standard normal random variable.*

Readers are referred to Billingsley [1, p. 360], Durrett [6, p. 129], and Pitman [14] for proofs of Theorem 5.1.[9]

The Lindeberg CLT tells us that, for $n$ sufficiently large, all probabilities about $\mathcal{Z}_n$ can be approximated by those of a standard normal random variable $Z$. Explicitly, for $n$ sufficiently large and for every real number $x$, we have

$$\mathbb{P}_n\left(\mathcal{Z}_n \leq x\right) \approx \mathbb{P}\left(Z \leq x\right) = N(x) \qquad (-\infty < x < \infty, \quad n \text{ sufficiently large}),$$

where $N(x)$ is the c.d.f. of $Z$ at $x$, i.e., $N(x) = \frac{1}{\sqrt{2\pi}} \int_{-\infty}^{x} e^{-\frac{y^2}{2}} \, dy$.

### 5.3.2 The Continuous-Time Security Price Formula

We now show that the Lindeberg CLT applies to both the real-world and risk-neutral CRR security price formulas.

**Continuous-Time Real-World Security Price Formula**

Starting with the real-world CRR tree, consider the triangular array of random variables $X_{n,j}$ in Table 5.2 associated with the security price formula (5.44) on page 230. Let

---

[9] The classical CLT follows from the Lindeberg CLT through the Dominated Convergence Theorem (expectation form) from measure theory; see Durrett [6, pp. 29,129].

$$\mathfrak{X}_{n,j} = \frac{X_{n,j}}{\sqrt{n}}.$$

To check that these random variables obey the hypotheses of the Lindeberg CLT, recall that for each $n \geq 1$, the random variables $X_{n,1}, X_{n,2}, \ldots, X_{n,n}$ are independent and standardized, and satisfy (5.57) on page 237. Then for every $n \geq 1$, the random variables $\mathfrak{X}_{n,1}, \mathfrak{X}_{n,2}, \ldots, \mathfrak{X}_{n,n}$ are also independent and satisfy (5.60). Next, to see that the Lindeberg condition holds, Equation (5.58) yields

$$\mathfrak{X}_{n,j} = \frac{1}{\sqrt{n}} \begin{cases} \dfrac{1-p_n}{\sqrt{p_n(1-p_n)}} & \text{with probability } p_n \\[3mm] \dfrac{-p_n}{\sqrt{p_n(1-p_n)}} & \text{with probability } 1-p_n. \end{cases}$$

Since $p_n \to 1/2$ as $n \to \infty$ (see (5.37) on page 226), we get

$$\mathfrak{X}_{n,j} \longrightarrow 0 \quad \text{as} \quad n \to \infty.$$

This implies that, for any fixed $\epsilon > 0$, the following condition does *not* hold for $n$ sufficiently large:

$$\left| \mathfrak{X}_{n,j} \right| > \epsilon > 0.$$

Consequently, $\mathbf{1}_{\{|\mathfrak{X}_{n,j}| > \epsilon\}} = 0$ for $n$ sufficiently large and so

$$\mathbb{E}\left( \mathfrak{X}_{n,j}^2 \, \mathbf{1}_{\{|\mathfrak{X}_{n,j}| > \epsilon\}} \right) = 0 \qquad\qquad (n \text{ sufficiently large}).$$

Hence, as $n \to \infty$, the Lindeberg condition (5.62) holds. By the Lindeberg CLT, we obtain

$$\mathcal{Z}_n = \sum_{j=1}^{n} \mathfrak{X}_{n,j} = \mathbb{Z}_n \xrightarrow{\text{d}} Z \qquad\qquad \text{as } n \to \infty.$$

Thus, *the discrete-time real-world CRR security price at the future time $t_f = t_0 + \tau$, which is labeled by $t_n$ in an $n$-period real-world CRR tree, converges in distribution to the following continuous-time security price formula at $t_f$:*

$$S_0 \, e^{\mu_n \tau + \sigma_n \sqrt{\tau}\, \mathbb{Z}_n} \xrightarrow{\text{d}} S_0 \, e^{\mu_{\text{RW}} \tau + \sigma \sqrt{\tau}\, Z} \qquad \text{as } n \to \infty. \qquad (5.66)$$

Here $\tau = t_f - t_0$, $\mu_{\text{RW}} = m - q - \dfrac{\sigma^2}{2}$ (i.e., the real-world drift), and the standard normal $Z$ is randomly drawn at a time span $\tau$ from the present.

**Notation.** *In security price modeling, we typically choose the current time to be $t_0 = 0$ and allow the final time $t_f$ to be a variable $t$. By (5.66), the continuous-time security price at time $t$ is then written as*

$$S(t) \stackrel{\text{d}}{=} S_0 \, e^{\mu_{\text{RW}} t + \sigma \sqrt{t}\, Z_t} \qquad\qquad (t \geq 0), \qquad (5.67)$$

where $S_0 = S(0)$ and $Z_t \sim \mathcal{N}(0,1)$. Interpret the standard normal $Z_t$ to mean that its realizations are randomly drawn at time $t$. We insert the subscript "$t$" rather than use the notation $Z$ to avoid some potential notational and conceptual confusion later; see page 246. Notice the dependency of $S(t)$ on the length $t$ of the interval $[0,t]$. We shall also switch freely between $S(t)$ and $S_t$ as notation for the security's price at time $t$ and, more generally, between $X(t)$ and $X_t$ for a random variable dependent on $t$:

$$S(t) = S_t, \qquad X(t) = X_t.$$

## Continuous-Time Risk-Neutral Security Price Formula

The security price is the same in the real world and the risk-neutral world, but has different expressions and probabilities. Explicitly, we have

$$\mathfrak{X}^*_{n,j} = \frac{X^*_{n,j}}{\sqrt{n}}, \qquad \mathcal{Z}^*_n = \sum_{j=1}^n \mathfrak{X}^*_{n,j},$$

where

$$X^*_{n,j} = \begin{cases} \dfrac{1-p^*_n}{\sqrt{p^*_n(1-p^*_n)}} & \text{with probability } p^*_n \\[2mm] \dfrac{-p^*_n}{\sqrt{p^*_n(1-p^*_n)}} & \text{with probability } 1 - p^*_n. \end{cases}$$

Essentially the same arguments as in the real-world case carry over to the risk-neutral setting to allow application of the Lindeberg CLT. For example, Equation (5.51) on page 234 shows that $p^*_n \to 1/2$ as $n \to \infty$, which in turn yields the following condition used to verify the Lindeberg condition: $\mathfrak{X}^*_{n,j} \longrightarrow 0$ as $n \to \infty$. Since $\sigma^*_n \to \sigma^* = \sigma$ as $n \to \infty$, the Lindeberg CLT yields

$$S_0\, e^{\mu^*_n \tau + \sigma^*_n \sqrt{\tau}\, \mathcal{Z}^*_n} \quad \xrightarrow{\ \mathrm{d}\ } \quad S_0\, e^{\mu_* \tau + \sigma \sqrt{\tau}\, Z^*_\tau} \qquad \text{as } n \to \infty, \qquad (5.68)$$

where $\mu_* = r - q - \frac{\sigma^2}{2}$ (i.e., the real-neutral drift) and $Z^*_\tau$ is a standard normal random variable drawn at a time span, $\tau = t_f - t_0$, from the present relative to a risk-neutral probability. By our continuous-time convention that $t_0 = 0$ and $t_f = t$, where $t$ is a variable, the resulting continuous-time risk-neutral security price is

$$S(t) \overset{\mathrm{d}}{=} S_0\, e^{\mu_* t + \sigma \sqrt{t}\, Z^*_t} \qquad (t \geq 0). \qquad (5.69)$$

Equations (5.67) and (5.69) yield the relationship between the real world and risk-neutral standard normal random variables:

$$Z^*_t \overset{\mathrm{d}}{=} Z_t + \frac{(\mu_{\mathrm{RW}} - \mu_*)\, t}{\sigma \sqrt{t}} = Z_t + \left(\frac{m - r}{\sigma}\right) \sqrt{t}, \qquad (5.70)$$

which is the continuous-time analog of Equation (5.56) on page 236. Note that $\frac{m-r}{\sigma}$ is the security's Sharpe ratio.

**A Remark About the Notation for Security Prices**

As mentioned in Remark 5.1 on page 210, to avoid notation that is too cumbersome, we do not introduce any new notation to distinguish the security price formula in discrete time versus continuous time. For example, given a partition, $0 = t_0 < t_1 < \cdots < t_n = t$, of the interval $[0, t]$, it will be made clear from the context whether the real-world security price $S(t_j)$ at time $t_j$ is given by

$$S_0\, e^{\mu_n t_j\, +\, \sigma_n \sqrt{t_j}\, \mathbb{Z}_n} \text{ (discrete time)} \quad \text{or} \quad S_0\, e^{\mu_{\mathrm{RW}} t_j\, +\, \sigma \sqrt{t_j}\, Z_{t_j}} \text{ (continuous time)}.$$

Similarly, for a risk-neutral world, we shall clarify whether we intend

$$S_0\, e^{\mu_* t_j\, +\, \sigma_n^* \sqrt{t_j}\, \mathbb{Z}_n^*} \text{ (discrete time)} \quad \text{or} \quad S_0\, e^{\mu_* t_j\, +\, \sigma \sqrt{t_j}\, Z_{t_j}^*} \text{ (continuous time)}.$$

In fact, for most of our study going forward, we shall use the continuous-time formula and abide by the following:

> *Unless stated to the contrary, assume that the price of a security is expressed by the continuous-time formula (5.67) for the real-world setting or formula (5.69) for a risk-neutral world.*

**Some Assumed Properties of the Continuous-Time Security Price**

We now motivate and present several properties we shall assume about security prices for continuous time.

➢ **Continuous sample paths**. For an $n$-period CRR tree, a possible outcome $\omega_0$ is a sequence of $n$ up and down movements in a security's price. There are $2^n$ such possible outcomes giving rise to $2^n$ possible security price paths. *In the continuous-time limit $n \to \infty$, we have assumed that the resulting uncountably many sample paths of the security, whether in a real world or risk-neutral world, are continuous functions of time t with probability 1.* Later, we shall explore the discontinuous case, where the price paths can have jumps; see the Merton jump-diffusion model in Section 8.9 (page 448).

➢ **Markov property and the Weak Efficient Market Hypothesis**. *The futures price of a security depends on the current price $S(0)$ and does not depend explicitly on the security's past prices*; see Equations (5.67) and (5.69). That is, probabilities about futures prices of a security are independent of the security's past price path (*Markov property*). Nonetheless, information about the past is not totally excluded in the sense that such information is included in the current price. Indeed, one of our key assumptions is *the Weak Efficient Market Hypothesis*, which states that the current price of a security reflects all market information concerning the security. In particular, the security's past price path and present market expectations about the security's future behavior are already taken into account in the current price of the security.

➤ **Stationarity of log returns.** By Equations (5.67) and (5.69), the security's continuous-time log return from time 0 to $t$ does not depend on the price of the security at the start of the interval $[0,t]$. It depends on the length of the interval:

$$\ln\left(\frac{S(t)}{S(0)}\right) \overset{\mathrm{d}}{=} \mu_{\mathrm{RW}} t + \sigma\sqrt{t}Z_t,$$

where $t \geq 0$. Similarly, in a risk-neutral world $\ln\left(\frac{S(t)}{S(0)}\right) \overset{\mathrm{d}}{=} \mu_* t + \sigma\sqrt{t}Z_t^*$. Motivated by this, *assume that in the continuous-time setting, the log return of a security is stationary,* which means

$$\ln\left(\frac{S(t)}{S(x)}\right) \overset{\mathrm{d}}{=} \ln\left(\frac{S(t+u)}{S(x+u)}\right)$$

for all $0 \leq x < t$ and every u such that $x + u \geq 0$ and $t + u \geq 0$. In other words, *rigidly shifting the interval does not change the log return probabilities.*

➤ **Independence of log returns over nonoverlapping[10] intervals.** For the $n$-period CRR tree, the discrete-time log returns over different time steps are independent. Motivated by the latter, *we assume in the continuous-time setting that for any finite sequence of times,* $0 \leq t_1 < t_2 < \cdots < t_{k-1} < t_k$, *which is not required to coincide with the time steps of the CRR tree we studied earlier, the log returns* $\ln\left(\frac{S(t_j)}{S(t_{j-1})}\right)$ *are independent for* $j = 2,\ldots,k$.

➤ **Geometric Brownian motion model of security prices.** The price of a security is an example of a stochastic process, which, intuitively speaking, is a random variable that is a function of time. In Chapter 6, we shall model the continuous-time price of a security more precisely as a stochastic process called *geometric Brownian motion,* namely,

$$S(t) = S(0)\,e^{\mu_{\mathrm{RW}} t + \sigma\mathfrak{B}(t)} \qquad\qquad (t \geq 0).$$

Here $\mathfrak{B}(t)$, where $t \geq 0$, is also a stochastic process, called *standard Brownian motion,* and equals in distribution the following:

$$\mathfrak{B}(t) \overset{\mathrm{d}}{=} \sqrt{t}\,Z_t.$$

An important property (stationarity) of standard Brownian motion is

---

[10] Recall that two intervals are *nonoverlapping* if their interiors are disjoint.

$$\mathfrak{B}(t) - \mathfrak{B}(v) \stackrel{\mathrm{d}}{=} \mathfrak{B}(t - v) \sim \mathcal{N}(0, t - v) \qquad (0 \le v \le t).$$

Note that if we used the notation $\mathfrak{B}(t) \stackrel{\mathrm{d}}{=} \sqrt{t}\, Z$, then it is tempting to write

$$\mathfrak{B}(t) - \mathfrak{B}(v) \stackrel{\mathrm{d}}{=} \sqrt{t}\, Z - \sqrt{v}\, Z = (\sqrt{t} - \sqrt{v}) Z,$$

which is *incorrect* since it states $\mathfrak{B}(t) - \mathfrak{B}(v) \sim \mathcal{N}\left(0, t + v - 2\sqrt{tv}\right)$.

The risk-neutral expression of the security's price is

$$S(t) = S(0)\, e^{\mu_* t + \sigma \mathfrak{B}^*(t)} \qquad (t \ge 0),$$

where $\mathfrak{B}^*(t)$ is also standard Brownian motion, but in a risk-neutral world. We have $\mathfrak{B}^*(t) \stackrel{\mathrm{d}}{=} \sqrt{t}\, Z_t^*$, where the asterisk indicates that the probability distribution is risk neutral. Chapter 6 and Section 8.4 will have more on risk-neutral probabilities. Note that Equation (5.70) gives the link between $Z_t$ and $Z_t^*$.

## 5.4  Basic Properties of Continuous-Time Security Prices

For each future moment of time $t > 0$, the continuous-time security price,

$$S(t) \stackrel{\mathrm{d}}{=} S(0)\, e^{\mu_{\mathrm{RW}} t + \sigma \sqrt{t}\, Z_t},$$

is a lognormal random variable with p.d.f.

$$f_{S(t)}(x) = \frac{1}{\sigma \sqrt{2\pi t}\, x} \exp\left[ -\frac{\left( \ln\left(\frac{x}{S(0)}\right) - \mu_{\mathrm{RW}} t \right)^2}{2\sigma^2 t} \right] \qquad (t > 0).$$

### 5.4.1  Some Statistical Formulas for Continuous-Time Security Prices

The statistics of a lognormal random variable are well known and readily yield formulas for the mean, median,[11] variance, and covariance of continuous-time security prices:

---

[11] The *median* of an absolutely continuous random variable $X$ is a number, denoted $\mathrm{Med}(X)$, such that $\mathbb{P}\left(X > \mathrm{Med}(X)\right) = \frac{1}{2} = \mathbb{P}\left(X < \mathrm{Med}(X)\right)$. Additionally, if $X \sim \mathcal{N}(a, b^2)$, where $a$ and $b > 0$ are constants, then the mean and median of $X$ satisfy $\mathbb{E}\left(e^X\right) = e^{a + b^2/2}$ and $\mathrm{Med}\left(A\, e^{aX}\right) = A\, e^{a\,\mathrm{Med}(X)}$.

$$\mathbb{E}(S(t)) = S(0)e^{(m-q)t} \qquad\qquad (t \geq 0) \qquad (5.71)$$

$$\text{Med}(S(t)) = e^{-\frac{\sigma^2 t}{2}} \mathbb{E}(S(t)) \qquad\qquad (t \geq 0) \qquad (5.72)$$

$$\text{Var}(S(t)) = \left(e^{\sigma^2 t} - 1\right)\left(\mathbb{E}(S(t))\right)^2 \qquad\qquad (t \geq 0) \qquad (5.73)$$

$$\text{Cov}(S(t),S(u)) = \left(e^{\sigma^2 u} - 1\right)\mathbb{E}(S(u))\mathbb{E}(S(t)) \qquad (t > u \geq 0). \quad (5.74)$$

Equation (5.71) shows that *the expected price of the continuous-time security at a future time t is obtained by continuously compounding the current price S(0) at the instantaneous capital-gain rate m − q over the time span t.* Also, since $\sigma > 0$, Equation (5.72) readily shows that *the median of a continuous-time security price at t > 0 is always below its mean:*

$$0 < \text{Med}(S(t)) < \mathbb{E}(S(t)) \qquad\qquad (t > 0). \qquad (5.75)$$

This is due to the skewness of the lognormal p.d.f. Equations (5.73) and (5.74) yield, since $\sigma$ and the mean price $\mathbb{E}(S(t))$ are positive, that the variance and covariance of the security's prices are positive for all future times $t > u > 0$.

We can also obtain formulas for the conditional expectations of a continuous-time security price (Exercise 5.23):[12]

$$\mathbb{E}\left(S(t) \mid S(t) > K\right) = \mathbb{E}(S(t))\frac{N(d_{+,\text{RW}}(t))}{N(d_{-,\text{RW}}(t))} \qquad\qquad (5.76)$$

$$\mathbb{E}\left(S(t) \mid S(t) < K\right) = \mathbb{E}(S(t))\frac{N(-d_{+,\text{RW}}(t))}{N(-d_{-,\text{RW}}(t))}, \qquad\qquad (5.77)$$

where $t > 0, K > 0$, and $d_{-,\text{RW}}(t) = \frac{\ln\left(\frac{S(0)}{K}\right) + \mu_{\text{RW}}t}{\sigma\sqrt{t}}$ with $d_{+,\text{RW}}(t) = d_{-,\text{RW}}(t) + \sigma\sqrt{t}$.

### 5.4.2 Some Probability Formulas for Continuous-Time Security Prices

We present some probability results for certain basic continuous-time security price behavior.

➤ *Since the median of a security's price is below the mean, there is more than a 50% probability that the futures price of a security will be below its mean price.* To see this, observe that

$$\mathbb{P}\left(S(t) < \mathbb{E}(S(t))\right) = \mathbb{P}\left(S(t) < \text{Med}(S(t))\right)$$
$$+ \quad \mathbb{P}\left(\text{Med}(S(t)) \leq S(t) \leq \mathbb{E}(S(t))\right).$$

---

[12] Recall that the conditional expectation of an absolutely continuous random variable $V$ given an event $A$ with probability $\mathbb{P}(A) > 0$ is defined by $\mathbb{E}(V|A) = \frac{\mathbb{E}(V\mathbf{1}_A)}{\mathbb{P}(A)} = \frac{1}{\mathbb{P}(A)}\int_A v f_V(v)dv$, where $f_V$ is the p.d.f. of $V$.

By definition of the median, we have $\mathbb{P}(S(t) < \text{Med}(S(t))) = \frac{1}{2}$ and by (5.75) we see that $\mathbb{P}(\text{Med}(S(t)) \leq S(t) \leq \mathbb{E}(S(t))) > 0$. Hence

$$\mathbb{P}(S(t) < \mathbb{E}(S(t))) > \frac{1}{2}. \tag{5.78}$$

We made a similar observation in Example 5.6 (page 227) using a 100-period CRR tree.

*If the security's volatility $\sigma$ is sufficiently large, then there is also more than a 50% probability that the futures price of the security will be below its current price.* In fact, note that (5.72) implies

$$\text{Med}(S(t)) = S(0)\, e^{\left(-\frac{\sigma^2}{2} + m - q\right) t} \approx S(0)\, e^{-\frac{\sigma^2 t}{2}} < S(0)$$

for $\sigma$ sufficiently large. Then an argument similar to the one above yields

$$\mathbb{P}(S(t) < S(0)) > \frac{1}{2} \qquad (\sigma \text{ sufficiently large}). \tag{5.79}$$

Example 5.6 (page 227) also pointed out this property using a CRR tree.

➤ *For any $K > 0$, the probability that $S(t)$ is less than $K$ is* (Exercise 5.24)

$$\mathbb{P}(S(t) < K) = \mathsf{N}(-\mathsf{d}_{-,\text{RW}}(t)) \qquad (t > 0). \tag{5.80}$$

Here $\mathsf{N}(x)$ is the standard normal c.d.f. at $x$. Since $\mathsf{N}(-x) = 1 - \mathsf{N}(x)$, *the probability that the security's price at $t > 0$ exceeds $K$ is*

$$\mathbb{P}(S(t) > K) = 1 - \mathbb{P}(S(t) < K) = 1 - \mathsf{N}(-\mathsf{d}_{-,\text{RW}}(t)) = \mathsf{N}(\mathsf{d}_{-,\text{RW}}(t)). \tag{5.81}$$

➤ *The probability of the security's price lying between two values is* (Exercise 5.24)

$$\mathbb{P}(K_1 < S(t) < K_2) = \mathsf{N}(\mathsf{d}^{K_1}_{-,\text{RW}}(t)) - \mathsf{N}(\mathsf{d}^{K_2}_{-,\text{RW}}(t)), \tag{5.82}$$

where $0 < t$, $0 < K_1 < K_2$, and $\mathsf{d}^x_{-,\text{RW}}(t) = \dfrac{\ln\left(\frac{S(0)}{x}\right) + \mu_{\text{RW}}\, t}{\sigma \sqrt{t}}$.

We add that the risk-neutral analogs[13] of the conditional expectation in Equation (5.76) and probability in Equation (5.81) appear in the pricing formula for a European call. Similarly, risk-neutral analogs of Equations (5.77) and (5.80) are part of the pricing formula for a European put. Readers are referred to McDonald [12, Sec. 18.4] for a detailed treatment.

---

[13] Replace $m$ by r and (hence) $\mu_{\text{RW}}$ by $\mu_*$ in the formulas.

## 5.5 Exercises

### 5.5.1 Conceptual Exercises

**5.1.** Suppose that a binomial tree has an initial price of \$80. If the tree has 21 periods, then is \$80 one of the possible prices at time $t_{21}$? If the tree has 300 periods, then is \$80 one of the possible prices at time $t_{300}$? Justify your answers.

**5.2.** For an $n$-period binomial tree, give a financial interpretation of each of the following: $u_n d_n = 1$, $u_n - 1$, and $d_n - 1$.

**5.3.** Explain why the condition, $d_n < e^{(m-q)h_n} < u_n$, holds for $n$-period CRR trees with $n$ sufficiently large.

**5.4.** How many 1-period subtrees are in an $n$-period binomial tree?

**5.5.** For an $n$-period binomial tree, let $N_U$ be the number of security price upticks from time $t_0$ to $t_n$. Explain why $N_U$ is a binomial random variable. What are its expected value and variance if the tree has 40 steps and the uptick probability is 60%?

**5.6.** An $n$-period CRR tree has reflection symmetry about the horizontal line $S(t) = S_0$ since the tree recombines. Agree or disagree? Justify your answer.

**5.7.** For an $n$-period risk-neutral binomial tree, show that if $u_n < e^{(r-q)h_n}$, then there is an arbitrage.

**5.8.** The risk-neutral uptick probability $p_n^*$ is related to the real-world probability $p_n$ by $p_n^* = p_n - \eta_n \sqrt{p_n(1 - p_n)}$, where $\eta_n = \dfrac{\mathbb{E}(R(t_0,t_1)) - r h_n}{\sqrt{\mathrm{Var}(R(t_0,t_1))}}$. Interpret $\eta_n$.

**5.9.** If in an $n$-period real-world CRR tree, the real-world probability $p_n$ is replaced by the risk-neutral uptick probability $p_n^*$, then the expected annualized return rate $m$ is unchanged, but the annualized variance $\sigma^2$ changes. Agree or disagree? Justify your answer.

**5.10.** For an $n$-period binomial tree, the collection of all paths has $2^{2^n}$ subcollections of paths, where the empty subcollection is included. Agree or disagree? Justify your answer.

### 5.5.2 Application Exercises

**5.11.** A trader believes that a certain stock currently at \$51.25 per share has by the end of the trading day a 70% chance of increasing by 50¢ and a 30% chance of decreasing by 25¢. Using a 1-step binomial tree with this information, what is the expected price of the stock at the end of the day?

**5.12.** Assume that the current share price of a stock is $100 with a volatility of 10%. Using a CRR tree model over a year with each being one trading day, predict the maximum spread in the stock's possible prices a trading day from now. Is your prediction impacted if you employ a Taylor approximation to $e^{\pm\sigma\sqrt{h_{252}}}$ using the CRR assumptions?

**5.13.** Suppose that a nondividend-paying stock with current price of $45 has an instantaneous annual expected return of 8% and annual volatility of 15%.

a) Using a 100-period CRR tree, forecast the price of the stock 3 months from now, i.e., find the expected price of the stock 3 months from now.
b) What is your forecast if you use an 80-period CRR tree?
c) Using an 80-period CRR tree, determine the probability of your forecasted price occurring.

**5.14.** Consider a nondividend-paying stock with a current price of $45, an instantaneous annual expected return of 8%, and annual volatility of 15%. Assume an 80-period CRR tree.

a) Can the stock price be at the same value 3 months from now? If so, how many times would the price have to increase and decrease for this to happen?
b) What is the probability that the stock price will be at the same value 3 months?
c) What is the probability that in 3 months the stock price is greater than its current price?

**5.15.** Suppose that the current date is January 21, 2016. Estimate the volatility $\sigma$ and instantaneous drift $\mu_{RW}$ for Google (ticker symbol GOOGL) using its adjusted closing prices from Yahoo! Finance for the period from January 20, 2016 to September 15, 2015. The data will consist of 90 daily log returns over the past 91 trading days. Carry out similar estimates with the past 60 daily log returns and then the past 30 daily log returns. Annualize your results using 252 trading days in a year. Discuss your findings.

**5.16.** Assume that a nondividend-paying security has a current price of $50, instantaneous expected return of 8%, and volatility of 15%. Estimate the probability, as a fraction (not percentage) to two decimal places, that 3 months from now, the price of the real-world, continuous-time price of the security is greater than $50. What fractional probability to two decimal places does a 100-period real-world CRR tree predict? How about a 1,000-period real-world CRR tree? Find a value of $n$ for which an $n$-period real-world CRR tree gives the same fractional probability to two, three, and four decimal places as that obtained from the continuous-time security price model. Use a software for this problem.

### 5.5.3 Theoretical Exercises

**5.17.** For an $n$-period binomial tree, show that

$$\mathbb{P}\left(S(t_n) = S_0 u_n^i d_n^{n-i} \mid S(t_0) = S_0\right) = \binom{n}{i} p_n^i (1 - p_n)^{n-i}, \qquad i = 0, 1, 2, \ldots, n.$$

**5.18.** Determine the number of elements in the sample space $\Omega_n$ of price paths of an $n$-period binomial tree.

**5.19.** Show that $\sigma_n^2 h_n = p_n(1 - p_n)\left[\ln\left(\frac{u_n}{d_n}\right)\right]^2$.

**5.20.** Show that for a CRR tree, the uptick probability $p_n \approx \frac{e^{(m-q)h_n} - d_n}{u_n - d_n}$ satisfies $p_n \approx \frac{1}{2}\left(1 + \frac{\mu_{\mathrm{RW}}}{\sigma}\sqrt{h_n}\right)$ for $n$ sufficiently large.

**5.21.** For a CRR tree, show $X_{n,j} = \begin{cases} \dfrac{1 - p_n}{\sqrt{p_n(1-p_n)}} & \text{with probability } p_n \\[2ex] \dfrac{-p_n}{\sqrt{p_n(1-p_n)}} & \text{with probability } 1 - p_n. \end{cases}$

**5.22.** A *Jarrow-Rudd (JR) tree* for the price of a security paying a continuous dividend yield rate $q$ is a binomial tree where $p_n = \frac{1}{2}$ and the per-period expectation $\mu_n^*$ and variance $\sigma_n^2$ defined by

$$\mu_n^* = \frac{1}{h_n}\mathbb{E}\left(\ln\left(\frac{S(t_j)}{S(t_{j-1})}\right)\right), \qquad (\sigma_n^*)^2 = \frac{1}{h_n}\mathrm{Var}\left(\ln\left(\frac{S(t_j)}{S(t_{j-1})}\right)\right),$$

satisfy $\mu_n^* \to \mu_*$ and $\sigma_n^* \to \sigma$ as $n \to \infty$, where $\mu_* = r - q - \frac{\sigma^2}{2}$. Then for $n$ sufficiently large, the constraint equations for $u_n, d_n, p_n$ in a JR tree are

$$p_n = \frac{1}{2}, \quad \mu_* h_n \approx p_n \ln u_n + (1 - p_n)\ln d_n, \quad \sigma^2 h_n \approx p_n(1 - p_n)\left[\ln\left(\frac{u_n}{d_n}\right)\right]^2.$$

a) Show that $u_n \approx e^{\mu_* h_n + \sigma\sqrt{h_n}}$ and $d_n \approx e^{\mu_* h_n - \sigma\sqrt{h_n}}$ for $n$ sufficiently large.

b) Show that the security price formula for a JR tree has the following form for $n$ sufficiently large: $S^{\mathrm{JR}}(t_n) \approx S_0 e^{\mu_* \tau + \sigma\sqrt{\tau} Z_n^{\mathrm{JR}}}$, where $Z_n^{\mathrm{JR}} = \frac{1}{\sqrt{n}}\sum_{j=1}^{n} X_{n,j}^{\mathrm{JR}}$ and $X_{n,j}^{\mathrm{JR}} \approx \pm 1$ with probability $1/2$ for each possibility.

c) Verify that the triangular array $\mathfrak{X}_{n,j}^{\mathrm{JR}} = \frac{X_{n,j}^{\mathrm{JR}}}{\sqrt{n}}$, where $n \geq 1$ and $j = 1, \ldots, n$, satisfies the hypotheses of the Lindeberg CLT.

d) Using the Lindeberg CLT, determine the continuous-time security price formula to which the JR tree security price converges in distribution. Is this the same formula obtained by using a risk-neutral CRR tree?

e) **(Risk-Neutral JR Tree)** Is the JR tree risk neutral? If not, then how would you make it risk neutral? For a sufficiently large $n$, what would be the approximate governing equations of a risk-neutral JR tree, i.e., the equations expressing $u_n, d_n, p_n$ in terms of the inputs $r - q, \sigma, h_n$?

**5.23.** For $t > 0$ and $K > 0$, show that $\mathbb{E}\left(S(t) \mid S(t) > K\right) = \mathbb{E}(S(t)) \frac{N(d_{+,\text{RW}}(t))}{N(d_{-,\text{RW}}(t))}$ and $\mathbb{E}\left(S(t) \mid S(t) < K\right) = \mathbb{E}(S(t)) \frac{N(-d_{+,\text{RW}}(t))}{N(-d_{-,\text{RW}}(t))}$.

**5.24.** For $t > 0$, $K > 0$, and $K_2 > K_1 > 0$, show that $\mathbb{P}\left(S(t) < K\right) = N(-d_{-,\text{RW}}(t))$ and $\mathbb{P}\left(K_1 < S(t) < K_2\right) = N(d^{K_1}_{-,\text{RW}}(t)) - N(d^{K_2}_{-,\text{RW}}(t))$.

# References

[1] Billingsley, P.: Probability and Measure, 3rd edn. Wiley, New York (1995)
[2] Chance, D.: Proofs and derivations of binomial models. Technical Finance Notes, Louisiana State University (2007)
[3] Chance, D.: Risk neutral pricing in discrete time. Teaching Note 96-02, Louisiana State University (2008)
[4] Cox, J., Ross, S., Rubinstein, M.: Option pricing: a simplified approach. J. Financ. Econ. **7**, 229 (1979)
[5] Cox, J., Rubinstein, M.: Options Markets. Prentice Hall, Upper Saddle River (1985)
[6] Durrett, R.: Probability: Theory and Examples, 4th edn. University Press, Cambridge (2010)
[7] Epps, T.: Pricing Derivative Securities. World Scientific, Singapore (2007)
[8] Ghahramani, S.: Fundamentals of Probability with Stochastic Processes. Pearson Prentice Hall, Upper Saddle River (2005)
[9] Jarrow, R., Rudd, A.: Option Pricing. Richard Irwin, Homewood (1983)
[10] Jarrow, R., Turnbull, S.: Derivative Securities. South-Western College Publishing, Cincinnati (2000)
[11] Leunberger, D.: Investment Science. Oxford University Press, Oxford (1998)
[12] McDonald, R.: Derivative Markets. Addison-Wesley, Boston (2006)
[13] Neftci, S.: An Introduction to the Mathematics of Financial Derivatives. Academic, San Diego (2000)
[14] Pitman, J.: Setup for the Central Limit Theorem. STAT 205 Lecture Note 10, scribe D. Rosenberg. University of California, Berkeley (2003)
[15] Roman, S.: Introduction to the Mathematics of Finance. Springer, New York (2004)
[16] Wilmott, P., Dewynne, N., Howison, S.: Mathematics of Financial Derivatives: A Student Introduction. Cambridge University Press, Cambridge (1995)

# Chapter 6
# Stochastic Calculus and Geometric Brownian Motion Model

Stochastic calculus is the methodology of choice in modern finance: it provides intuitive constructions of financial objects and is the most identified mathematical tool with financial engineers.

The materials in stochastic calculus covered in this chapter mainly serve as a preparation for pricing derivatives in a later chapter.

## 6.1 Stochastic Processes: The Evolution of Randomness

### 6.1.1 Notation for Probability Spaces

A *probability space* is denoted by the triple $(\Omega, \mathfrak{F}, \mathbb{P})$, where:

1. $\Omega$ denotes the *sample space*, that is, a nonempty set of all the possible *outcomes* of a random experiment;

2. $\mathfrak{F}$ denotes a *$\sigma$-algebra* on $\Omega$. For our purpose, it is sufficient to know that $\mathfrak{F}$ represents a collection of subsets of $\Omega$, called *events*;

3. $\mathbb{P}$ denotes a *probability measure* on $\mathfrak{F}$ that assigns *probabilities* to events. (Note that $\mathbb{P} : \mathfrak{F} \to [0,1]$ is a function whose domain is $\mathfrak{F}$.)

The next remark provides intuitive explanations of 2 and 3 and motivates the concept of $\sigma$-algebra.

**Remark 6.1.** Ideally, one would like to assign a probability for every subset of $\Omega$. Unfortunately, because of some obstacles in measure theory, this cannot always be done. It is often the case that one can only assign a probability $\mathbb{P}$ on a collection $\mathfrak{F}$ of subsets of $\Omega$. An element in $\mathfrak{F}$ is called an event in probability theory. For further development of the probability theory, we need additional algebraic structures for $\mathfrak{F}$ as intuitively explained below:

© Arlie O. Petters and Xiaoying Dong 2016
A.O. Petters, X. Dong, *An Introduction to Mathematical Finance with Applications*, Springer Undergraduate Texts in Mathematics and Technology, DOI 10.1007/978-1-4939-3783-7_6

A probability $\mathbb{P}$ should be additive in the sense that if $A$ and $B$ are disjoint events, then

$$\mathbb{P}(A \cup B) = \mathbb{P}(A) + \mathbb{P}(B),$$

and therefore $\mathfrak{F}$ should be closed under the union operation. For the application of modern integration theory, one actually wants $\mathbb{P}$ to have a slightly deeper nature that for a countable pairwise disjoint collection of events $A_i$, $i = 1, 2, \ldots,$

$$\mathbb{P}(A_1 \cup A_2 \cup \cdots) = \mathbb{P}(A_1) + \mathbb{P}(A_2) + \cdots,$$

and therefore $\mathfrak{F}$ should be closed under countable union. Furthermore, the empty set $\emptyset$ should be an event with probability zero, and if $A$ is an event, then the complement of $A^c$ should also be an event such that

$$\mathbb{P}(A) + \mathbb{P}(A^c) = 1.$$

By Morgan's law, finite or countable intersections of events should also be events.                                                                                                $\square$

Although intuitive explanations of the $\sigma$-algebra will be provided in great detail shortly, the definition's scope best captures the intuitive ideas.

Let $\Omega$ be a nonempty set. The *power set* of $\Omega$, denote by $2^{\Omega}$, is a set consisting of all subsets of $\Omega$.

**Definition 6.1.** Let $\Omega$ be a nonempty set. $\mathcal{F} \subseteq 2^{\Omega}$ is called an *algebra* over $\Omega$ if it satisfies the following conditions:

1. $\Omega \in \mathcal{F}$;

2. $A \in \mathcal{F} \Rightarrow A^c \in \mathcal{F}$;

3. $A, B \in \mathcal{F} \Rightarrow A \cup B \in \mathcal{F}$.

Note that the last condition is equivalent to closure under finite union, i.e.,

$$A_i \in \mathcal{F}, \ i = 1, 2, \ldots, n \ \Rightarrow \ \cup_{i=1}^{n} A_i \in \mathcal{F}.$$

If the last condition is replaced with closure under countable union, i.e.,

$$A_i \in \mathcal{F}, \ i = 1, 2, \ldots, \ \Rightarrow \ \cup_{i=1}^{\infty} A_i \in \mathcal{F},$$

then $\mathcal{F}$ is called a *$\sigma$-algebra* over $\Omega$.

Note that the definition of $\sigma$-algebra does not rely on the existence of a probability measure.

The next definition is a natural extension.

**Definition 6.2.** Let $\mathcal{F}_i$, $i = 1, 2$, be two $\sigma$-algebras on a nonempty set $\Omega$. $\mathcal{F}_2$ is said to be a *sub-$\sigma$-algebra* of $\mathcal{F}_1$ if $\mathcal{F}_2 \subseteq \mathcal{F}_1$.

**Definition 6.3.** For $C \subseteq 2^{\Omega}$, the smallest $\sigma$-algebra that includes $C$ is called the *$\sigma$-algebra generated by $C$* and denoted by $\sigma(C)$.

In fact, $\sigma(C)$ is the intersection of all $\sigma$-algebras including $C$.

**Example 6.1.** Let $A \subset \Omega$. The collection $\{\emptyset, A, A^c, \Omega\}$ is the smallest $\sigma$-algebra on $\Omega$ containing $A$, i.e., $\sigma(A)$, the $\sigma$-algebra generated by $A$. □

**Example 6.2.** $\{\emptyset, \Omega\}$ is the smallest $\sigma$-algebra on $\Omega$, whereas $2^{\Omega}$, the power set of $\Omega$, is the largest $\sigma$-algebra on $\Omega$.

Clearly, for any $C \subseteq 2^{\Omega}$, $\{\emptyset, \Omega\}$ is a sub-$\sigma$-algebra of $\sigma(C)$, and $\sigma(C)$ is a sub-$\sigma$-algebra of $2^{\Omega}$. □

In modern probability theory, a probability is defined on a $\sigma$-algebra as a *measure* with whole space measure equal to one. Precisely, we define the probability measure as follows:

**Definition 6.4.** Let $\Omega$ be a probability space and $\mathfrak{F}$ a $\sigma$-algebra over $\Omega$. A real-valued function $\mathbb{P}$ from $\mathfrak{F}$ to $[0,1]$ is said to be a *probability measure* on $\mathfrak{F}$ if it satisfies the following conditions:

1. $\mathbb{P}(\Omega) = 1$;

2. For each countable collection $\{A_i \in \mathfrak{F}, i \in I\}$ of pairwise disjoint sets,

$$\mathbb{P}(\cup_{i \in I} A_i) = \sum_{i \in I} \mathbb{P}(A_i).$$

This property is referred to as the *countable additivity*.

Note that items 1 and 2 in the definition imply $\mathbb{P}(\emptyset) = 0$.

A proper *sub-$\sigma$-algebra* $\mathcal{F}$ may be considered as a coarsification of $\mathfrak{F}$, in the sense that fewer events are observable and this in turn, as we will see in the next section, implies that fewer "random variables" are "measurable." In this regard, $\mathcal{F}$ gives rise to lower resolution when one tries to observe a "random variable."

**Example 6.3. (One-Period Binomial Tree)** Consider a random experiment of tossing a biased coin. Assume that the chance of landing heads up is 30% and that of landing tails up is 70%. Then, the corresponding probability space is represented by $(\Omega, \mathfrak{F}, \mathbb{P})$, where

$$\Omega = \{H, T\},$$
$$\mathfrak{F} = \{\emptyset, H, T, \Omega\},$$
$$\mathbb{P} : \mathcal{F} \to [0,1] \text{ is defined by}$$
$$\mathbb{P}(\emptyset) = 0, \ \mathbb{P}(H) = 0.3, \ \mathbb{P}(T) = 0.7, \ \mathbb{P}(\Omega) = 1.$$

In a parallel fashion, if we consider an oversimplified stock price behavior in which the stock price either goes up by a factor u with probability p or goes down by a factor d with probability $1 - p$ over the time period $[t_0, t_1]$, then the corresponding probability space is represented by $(\Omega, \mathfrak{F}, \mathbb{P})$, where

$$\Omega = \{U, D\},$$
$$\mathfrak{F} = \{\emptyset, U, D, \Omega\},$$
$$\mathbb{P} : \mathfrak{F} \to [0, 1] \text{ is defined by}$$
$$\mathbb{P}(\emptyset) = 0, \ \mathbb{P}(U) = p, \ \mathbb{P}(D) = 1 - p, \ \mathbb{P}(\Omega) = 1.$$

□

The random experiment associated with the next example is parallel to repeatedly flipping a biased coin.

**Example 6.4. (Two-Period Binomial Tree)** Consider an oversimplified stock price behavior as described by a two-period ($[t_0, t_1]$ and $[t_1, t_2]$ with $t_0 < t_1 < t_2$) binomial tree. In each period the stock price either goes up by a factor u with probability p or goes down by a factor d with probability $1 - p$. Then, the corresponding probability space is represented by $(\Omega, \mathfrak{F}, \mathbb{P})$, where

$$\Omega = \{\omega_1, \omega_2, \omega_3, \omega_4\}, \text{ where}$$
$$\omega_1 = UU, \ \omega_2 = UD, \ \omega_3 = DU, \ \omega_4 = DD,$$

$$\mathfrak{F} = \{\emptyset, \{\omega_1\}, \{\omega_2\}, \{\omega_3\}, \{\omega_4\},$$
$$\{\omega_1, \omega_2\}, \{\omega_1, \omega_3\}, \{\omega_1, \omega_4\}, \{\omega_2, \omega_3\}, \{\omega_2, \omega_4\}, \{\omega_3, \omega_4\},$$
$$\{\omega_1, \omega_2, \omega_3\}, \{\omega_1, \omega_2, \omega_4\}, \{\omega_1, \omega_3, \omega_4\}, \{\omega_2, \omega_3, \omega_4\}, \Omega\},$$

$$\mathbb{P} : \mathfrak{F} \to [0, 1] \text{ is defined by}$$
$$\mathbb{P}(\{\omega_1\}) = p^2, \mathbb{P}(\{\omega_2\}) = \mathbb{P}(\{\omega_3\}) = p(1 - p), \mathbb{P}(\{\omega_4\}) = (1 - p)^2,$$
which along with additive property of $\mathbb{P}$ (see Remark 6.1) imply
$$\mathbb{P}(\{\omega_1, \omega_2, \omega_3\}) = 1 - (1 - p)^2, \ \mathbb{P}(\{\omega_2, \omega_3, \omega_4\}) = 1 - p^2,$$
$$\mathbb{P}(\{\omega_1, \omega_3, \omega_4\}) = \mathbb{P}(\{\omega_1, \omega_2, \omega_4\}) = 1 - p(1 - p),$$
$$\mathbb{P}(\{\omega_1, \omega_2\}) = \mathbb{P}(\{\omega_1, \omega_3\}) = p, \ \mathbb{P}(\{\omega_2, \omega_3\}) = 2p(1 - p),$$
$$\mathbb{P}(\{\omega_2, \omega_4\}) = \mathbb{P}(\{\omega_3, \omega_4\}) = 1 - p, \ \mathbb{P}(\{\omega_1, \omega_4\}) = p^2 + (1 - p)^2,$$
$$\mathbb{P}(\emptyset) = 0, \ \mathbb{P}(\Omega) = 1.$$

Note that each simple event in $\Omega$ is a representation of (or corresponds to) a *path* (i.e., $(S(t_0), S(t_1), S(t_2))$, where $S(t_i)$, $i = 0, 1, 2$ denote the stock price at time $t_i$).

Also, note that $\mathfrak{F}$ is closed under countable union and intersection, and

$$\mathcal{F} = \{\emptyset, \{\omega_1\}, \{\omega_2, \omega_3, \omega_4\}, \Omega\}$$

is a sub-$\sigma$-algebra.

□

## 6.1.2 Basic Concepts of Random Variables

A *random variable* on a probability space $(\Omega, \mathfrak{F}, \mathbb{P})$ is a *measurable* real-valued function from $\Omega$ to $\mathbb{R}$. That is

$$X : \Omega \to \mathbb{R},$$
$$\omega \mapsto X(\omega),$$

where measurability refers to the $\sigma$-algebra $\mathfrak{F}$ and means that for each real number $a$, the set $\{X \le a\}$ is an event in the probability space. That is,

$$X^{-1}((-\infty, a]) = \{X \le a\} = \{\omega \in \Omega \mid X(\omega) \le a\} \in \mathfrak{F}, \tag{6.1}$$

where $X^{-1}((-\infty, a])$ is the pre-image of $(-\infty, a]$ under $X$.

In words, (6.1) says that $X$ is $\mathfrak{F}$-*measurable*. Thus,

➤ a *random variable* $X$ on $(\Omega, \mathfrak{F}, \mathbb{P})$ is a $\mathfrak{F}$-*measurable* function from $\Omega$ to $\mathbb{R}$;

➤ the notion of $X$ being $\mathfrak{F}$-measurable assures that the distribution function of $X$: $F_X(a) \equiv \mathbb{P}(X \le a) = \mathbb{P}(X^{-1}((-\infty, a]))$ is well defined. In fact, a significant attribute of random variables is that they allow us to work on distributions without recognizing sample spaces.

For our purpose, if a random variable $X$ is "known" given a sub-$\sigma$-algebra $\mathcal{F} \subseteq \mathfrak{F}$, then we say that $X$ is *measurable with respect to* $\mathcal{F}$ and write $X \in \mathcal{F}$.

**Example 6.5. (One-Period Binomial Tree Again)** Let us revisit Example 6.3, the one-period binomial tree model of an oversimplified representation of stock price behavior in which the stock price either goes up by a factor u with probability p or goes down by a factor d with probability $1 - p$ over the time period $[t_0, t_1]$. Then the corresponding probability space is represented by $(\Omega, \mathfrak{F}, \mathbb{P})$, where

$$\Omega = \{U, D\},$$
$$\mathfrak{F} = \{\emptyset, U, D, \Omega\},$$
$$\mathbb{P} : \mathfrak{F} \to [0, 1] \text{ is defined by}$$
$$\mathbb{P}(\emptyset) = 0, \ \ \mathbb{P}(U) = p, \ \ \mathbb{P}(D) = 1 - p, \ \ \mathbb{P}(\Omega) = 1.$$

Given $S(t_0) = S_0$ and factors $u > 0$ and $d > 0$, we define $X : \Omega \to \mathbb{R}$ by

$$X(\omega) = \begin{cases} S_0 u & \text{if } \omega = U \\ S_0 d & \text{if } \omega = D. \end{cases} \tag{6.2}$$

Suppose that $S_0 = 100$, $u = \frac{6}{5}$ and $d = \frac{5}{6}$. Then

$$X(\omega) = \begin{cases} 120 & \text{if } \omega = U \\ 83.33 & \text{if } \omega = D. \end{cases}$$

To show that $X$ is a random variable on the probability space $(\Omega, \mathfrak{F}, \mathbb{P})$, we need to verify (6.1):

$$X^{-1}((-\infty, a]) = \{X \le a\} = \{\omega \in \Omega \mid X(\omega) \le a\}$$
$$= \begin{cases} \varnothing & \text{if} & a < 83.33 \\ D & \text{if } 83.33 \le a < 120 \\ \Omega & \text{if} & 120 \le a. \end{cases}$$

Indeed, $X^{-1}((-\infty, a]) \in \mathfrak{F}$ for each real number $a$:

➢ Using mathematical language, we say that $X$ is $\mathfrak{F}$-measurable. Translating it into ordinary English, we say that $\mathfrak{F}$ contains *sufficient information about* $X$. This intuition of measurability is important for gaining a correct intuition of the concept of conditional expectation with respect to $\sigma$-algebra in a later section.

It follows from similar arguments that $X$ defined by (6.2) is a random variable on $(\Omega, \mathfrak{F}, \mathbb{P})$.

Note that the smallest $\sigma$-algebra containing $\{X \le a, \ a \in \mathbb{R}\} = \{\varnothing, D, \Omega\}$ is $\mathfrak{F}$.

□

➢ In general, $\sigma(X)$ is the notation for the smallest $\sigma$-algebra containing the events $\{X \le a, \ a \in \mathbb{R}\}$ and called *the $\sigma$-algebra generated by X*. In the last example, $\sigma(X)$ happens to be $\mathfrak{F}$, which, in fact, is also the largest $\sigma$-algebra containing $\{X \le a, \ a \in \mathbb{R}\}$.

➢ Let $X$ be an arbitrary random variable. By definition, $X$ is $\sigma(X)$-measurable.

**Example 6.6.** Let $X$ be a random variable. Without using mathematical jargon for the notation $\sigma(X)$, how one can *intuitively describe $\sigma(X)$*?

**Intuitive Explanation.**

The power set $2^X$ is the largest $\sigma$-algebra. On this $\sigma$-algebra, every random variable, no matter how complex, is measurable. A particular random variable may be *simpler* and therefore measurable on a smaller $\sigma$-algebra. By *simpler* we mean that the random variable $X$ is constant on more events. For example, the simplest random variable $X \equiv c$, where $c$ is a real constant and is measurable even on the smallest $\sigma$-algebra $\{\varnothing, \Omega\}$. Most random variables require $\sigma$-algebras in between so that they are measurable. For an economic reason, we sometimes may want to have a smallest $\sigma$-algebra which is sufficient for a random variable $X$ to be measurable. Such a smallest $\sigma$-algebra, denoted by $\sigma(X)$, always exists, as the intersection of all $\sigma$-algebras sufficient for $X$ to be measurable is itself a $\sigma$-algebra.

In the language of information, this is tantamount to saying that we use the least amount of information to determine $X$. □

Sometimes random variable $X$ and a $\sigma$-algebra $\mathcal{F}$ are both given, but $X$ is not measurable on $\mathcal{F}$. Intuitively, we say that $\mathcal{F}$ does not have enough resolution to read off all information contained in $X$. The meaning of $\mathbb{E}(X|\mathcal{F})$ is the least coarsification of $X$ so that $\mathcal{F}$ has enough resolution for $\mathbb{E}(X|\mathcal{F})$ (i.e., $\mathbb{E}(X|\mathcal{F})$ is measurable on $\mathcal{F}$). More discussion on conditional expectation with respect to a $\sigma$-algebra will be given shortly.

**Remark 6.2.** Now we summarize the above intuition of $X$ being $\mathcal{F}$-measurable ($X \in \mathcal{F}$) in the following:

$$X \text{ is } \mathcal{F}\text{-measurable} \iff \mathcal{F} \text{ contains sufficient information of } X$$
$$\iff X \text{ is } determined \text{ or } known \text{ given } \mathcal{F}.$$

1. The intuition of the measurability of a random variable as we have provided in the equivalent statements above has fundamental importance in studying modern finance theory. For instance, it is important for understanding adapted processes, which in turn is important for understanding martingales. Readers should always have such an intuition in their mental model.

2. If $X$ is not $\mathcal{F}$-measurable, then probabilities of events described by $X$ may not be computed based on the information contained in $\mathcal{F}$. In this sense, $X$ is *unknown*.

□

Recall that two events A and B are independent if and only if their joint probability equals the product of their probabilities: $\mathbb{P}(A \cap B) = \mathbb{P}(A)\mathbb{P}(B)$.

Since $\sigma$-algebras are collections of events, the next definition is natural.

**Definition 6.5.** Two $\sigma$-algebras $\mathcal{F}$ and $\mathcal{G}$ are said to be *independent* if

$$\mathbb{P}(A \cap B) = \mathbb{P}(A)\mathbb{P}(B), \quad \forall A \in \mathcal{F}, B \in \mathcal{G}.$$

In words, two $\sigma$-algebras are independent, if any two events, one from each $\sigma$-algebra, are independent.

Using the language of independent $\sigma$-algebras, we have an equivalent statement for the definition of independent random variables:

**Definition 6.6.** Two random variables $X$ and $Y$ are *independent* if and only if two corresponding $\sigma$-algebras $\sigma(X)$ and $\sigma(Y)$ are independent.

**Definition 6.7.** A random variable $X$ and a $\sigma$-algebra $\mathcal{F}$ are said to be *independent*, denoted by $X \perp \mathcal{F}$, if two $\sigma$-algebras $\sigma(X)$ and $\mathcal{F}$ are independent.

### 6.1.3 Basic Concepts of Stochastic Processes

We begin with a continuation of Example 6.5.

**Example 6.7.** Recall from Example 6.5 that the probability space is given by $(\Omega, \mathcal{F}, \mathbb{P})$, where

$$\Omega = \{U, D\},$$
$$\mathcal{F} = \{\emptyset, U, D, \Omega\},$$
$$\mathbb{P} : \mathcal{F} \to [0,1] \text{ is defined by}$$
$$\mathbb{P}(\emptyset) = 0, \ \mathbb{P}(U) = p, \ \mathbb{P}(D) = 1 - p, \ \mathbb{P}(\Omega) = 1,$$

and a random variable $X$ is defined by

$$X(\omega) = \begin{cases} S_0 u & \text{if } \omega = U \\ S_0 d & \text{if } \omega = D, \end{cases}$$

where $S(t_0) = S_0$ and factors $u > 0$ and $d > 0$.

To extend a one-period tree to a multi-period tree, we rewrite random variable $X$ into

$$X(\omega) = S_0 Y(\omega) \quad \text{where} \quad Y(\omega) = \begin{cases} u & \text{if } \omega = U \\ d & \text{if } \omega = D. \end{cases}$$

If we use a three-period binomial tree (with $[t_i, t_{i+1}]$, $t_i < t_{i+1}$ and $i = 0,1,2,3$, as time periods), to model an oversimplified stock price behavior (with $S(t_i)$ as the price at time $t_i$) so that in each period the stock price either goes up by a factor $u$ with probability $p$ or goes down by a factor $d$ with probability $1 - p$, then we can introduce three random variables on $(\Omega, \mathcal{F}, \mathbb{P})$, $Y_i : \Omega \to \mathbb{R}$ defined by

$$Y_i = Y_i(\omega) = \begin{cases} u & \text{if } \omega = U \\ d & \text{if } \omega = D, \end{cases} \quad \text{where} \quad i = 1,2,3.$$

Then

$$X_1 = S_0 Y_1 = S(t_1), \quad X_2 = S_0 Y_1 Y_2 = S(t_2), \quad X_3 = S_0 Y_1 Y_2 Y_3 = S(t_3).$$

In a similar fashion, we can extend the binomial tree over infinitely many time periods: $[t_0, t_1], [t_1, t_2], [t_2, t_3], \ldots, [t_{i-1}, t_i], \ldots$ by introducing infinitely many random variables on $(\Omega, \mathcal{F}, \mathbb{P})$, $Y_i : \Omega \to \mathbb{R}$ defined by

$$Y_i = Y_i(\omega) = \begin{cases} u & \text{if } \omega = U \\ d & \text{if } \omega = D, \end{cases} \quad \text{where} \quad i = 1,2,3,\ldots, \tag{6.3}$$

which form a sequence of random variables denoted by

$$Y = \{Y_1, Y_2, Y_3,\ldots\}, \quad \text{or} \quad Y = \{Y_1(\omega), Y_2(\omega), Y_3(\omega),\ldots\}, \ \omega \in \Omega.$$

A deterministic counterpart of $Y$ is a sequence of functions denoted by $y$:

$$y = \{f_1, f_2, f_3,\ldots\}, \quad \text{or} \quad y = \{f_1(x), f_2(x), f_3(x),\ldots\}, \ x \in \mathbb{R},$$

where we assume that $f_i : \mathbb{R} \to \mathbb{R}$ are deterministic functions. In vector calculus, $y$ can be viewed as a function of two variables with domain $I \times \mathbb{R}$ and range $S \subseteq \mathbb{R}$, where $I = \{1,2,\ldots\}$, and expressed as follows:

$$y : I \times \mathbb{R} \to S, \quad \text{defined by} \quad (i,x) \mapsto f_i(x) \ \text{(i.e., } y(i,x) = f_i(x)),$$

where $I$ is naturally called an *index set*, and range $S = \cup_{i=1}^{\infty} S_i$ if we denote the range of function $f_i$ by $S_i$. □

Mimicking the familiar notation from vector calculus, we give the following definition.

**Definition 6.8.** A *stochastic process* (or a *random process*, or simply a *process*) on the probability space $(\Omega, \mathfrak{F}, \mathbb{P})$, denoted by $X = \{X_t : t \in \mathcal{J}\}$, is a function of two variables with domain $\mathcal{J} \times \Omega$ and range $S \subseteq \mathbb{R}$ and is expressed by

$X : \mathcal{J} \times \Omega \to S,$

$\quad (t,\omega) \mapsto X_t \quad$ (i.e., $X(t,\omega) = X_t(\omega) = X_t$ ($\omega$ is dropped to ease notation)),

where $\mathcal{J} \subseteq \mathbb{R}$ is nonempty and called the *index set*[1] of the process $X$, and the range $S$ is called the *state space*[2] of the process $X$.

In short, a *stochastic process* is a collection of random variables on the same probability space, representing the evolution of randomness over time.

**Example 6.8.** Let $\mathcal{J} = \{1,2,\ldots\}$. The process $X = \{X_t : t \in \mathcal{J}\}$ is called a *stochastic sequence* and sometimes written in the form $\{X_n\}_{n=1}^{\infty}$ or

$$X = \{X_1, X_2, X_3,\ldots\}.$$

If $X$ is defined by $X_i = S_0 Y_1 \cdots Y_i$, where $S_0$ and $Y_i$ are defined in Example 6.7 (see (6.3)), then clearly, the random variable $X_i$ will depend on earlier value $X_{i-1}$. □

---

[1] For our purpose, the index set is always a time index set although it is not necessarily so by definition.

[2] In mathematics, the set of objects that we are considering is often referred to as a space.

**Example 6.9.** Let $\mathcal{J} = [0, \infty)$. We simply write the process $X = \{X_t : t \in \mathcal{J}\}$ into $X = \{X(t) : t \geq 0\}$.

Suppose that $X(0) \overset{\text{a.s.}}{=} 0$, i.e., $\mathbb{P}(\omega \in \Omega \,|\, X(0, \omega) = 0) = 1$. We say that process $X$ *starts at 0 almost surely*.

By Definition 6.8, the process $X + 7 = \{X(t) + 7 : t \geq 0\}$ *starts at 7 almost surely.*                                                                              □

Since the index set $\mathcal{J}$ usually represents time indeed, and the random variable describes the state of the process at time $t$, the index $t$ admits a natural interpretation: if $X_t = s$, we say that *the process is in state s at time t*.

## Classifications of Stochastic Processes

Stochastic processes can be classified according to the index set $\mathcal{J}$ into *discrete-time processes* and *continuous-time processes*.

**Example 6.10.** $\{X_t : t \in \mathcal{J}\}$ is a discrete-time process if $X_t$ represents the closing price of a stock on the $t$-th trading day, and $\mathcal{J} = \{1, 2, \dots\}$.                       □

**Example 6.11.** $\{X_t : t \in \mathcal{J}\}$ is a continuous-time process if $X_t$ represents the intraday price of a stock at time $t$, and $\mathcal{J} = [9 : 30am, \ 4 : 00pm]$ on July 7, 2015.
                                                                                                          □

Stochastic processes can also be classified according to the state space $S$ into *discrete-state processes* and *continuous-state processes*. The state space is discrete if it consists of a finite number of points or a countably infinite number of points; otherwise, it is continuous.

**Example 6.12.** $\{X_t : t \in \mathcal{J}\}$ is a discrete-state process if $X_t$ represents the total number of heads in the first $t$ flips of a coin and $\mathcal{J} = \{1, 2, \dots\}$.                    □

**Example 6.13.** $\{X_t : t \in \mathcal{J}\}$ is a continuous-state process if $X_t$ represents the log return from investing in a stock and is normally distributed with parameters $(t, 2t^2)$ and $\mathcal{J} = \{t \geq 0\}$:                                                                          □

➤ In this chapter, we focus on continuous-time and continuous-state processes $\{X_t : t \geq 0\}$, although we may use discrete-time processes as examples to ease unnecessary technicalities.

## Sample Paths of Stochastic Processes

Since sample paths of a process may have a direction almost nowhere (noise), one can "visualize" them only through imagination; however, an intuition of the notion of sample paths may come from some idea in analytic geometry.

**Example 6.14.** Recall from analytic geometry and vector calculus that the graph of the function of two variables is a surface in $\mathbb{R}^3$. For example, the graph of

$$y = f(t,x) = t^2 + x^2, \quad (t,x) \in \mathbb{R}^2$$

is an elliptic paraboloid.

However, either the $t$-cross section or the $x$-cross section of $y = f(t,x)$ is a continuous path (curve) in $\mathbb{R}^3$. For example, let $x = 1$. The graph of

$$y = f(t,1) = t^2 + 1, \quad t \in \mathbb{R},$$

which is a function of $t$ alone, is a parabola. In fact, for each fixed $x$, the graph of $y = f(t,x) = t^2 + x^2$, $t \in \mathbb{R}$, a function of $t$ alone, is a parabola, a continuous path in $\mathbb{R}^3$. □

Let $X = \{X_t : t \geq 0\}$ be a process on $(\Omega, \mathfrak{F}, \mathbb{P})$. Observe that for each fixed $t \geq 0$, $X$ defines a random variable

$$X_t : \Omega \to \mathbb{R},$$
$$\omega \mapsto X(t,\omega) \quad (\text{i.e., } \omega \mapsto X_t(\omega)),$$

and for each fixed $\omega \in \Omega$, $X$ defines a deterministic function of time

$$X(\omega) : \mathcal{J} \to \mathbb{R},$$
$$t \mapsto X(t,\omega) \quad (\text{i.e., } t \mapsto X_t(\omega)). \tag{6.4}$$

The latter part of the observation leads to the next definition.

**Definition 6.9.** For each fixed $\omega \in \Omega$, the *sample path* (or *realization*, or *trajectory*, or *sample function*) of a stochastic process $X = \{X_t : t \in \mathcal{J}\}$ on $(\Omega, \mathfrak{F}, \mathbb{P})$ is the (graph of) function defined in (6.4) and denoted by either $X_t(\omega)$ or $X(t,\omega)$ (where $\mathcal{J}$ may be discrete or continuous).

Note that for a fixed $\omega$, a *continuous sample path* of the process $\{X_t : t \geq 0\}$ is defined in the ordinary calculus sense. That is,

$$\lim_{s \to t} X(s,\omega) = X(t,\omega) \quad \text{for each } t > 0,$$
$$\lim_{s \to 0, s>0} X(s,\omega) = X(0,\omega) \quad \text{for } t = 0.$$

It is important for the reader to keep the following notational remarks in mind throughout the rest of this book.

**Notational Remark**

There may be several notations for the same mathematical concept for good reasons. Consider a familiar environment as in one variable calculus, where $f'(x)$, $\frac{dy}{dx}$, $\frac{df}{dx}$, and $Df(x)$ may all represent the derivative of function $y = f(x)$ with respect to $x$.

In our discussion, for the sake of convenience or clarity relative to different contexts, different notation for the same mathematical object may be used. The reader should keep the following in mind in the rest of this chapter:

1. *Notation for sample paths.* $X_t(\omega)$, $X(t,\omega)$, $X(\omega)$ (where $t$ is dropped to ease the notation and to emphasize the (relation) rule of function rather than the value of the function) and $X(t)$ (where $\omega$ is dropped to ease the notation and to emphasize the fact that a path is defined by a function of $t$ alone) may all represent a sample path of the stochastic process $X$.

2. *Notation for random variables in a process.* $X_t$ and $X(t)$ may be used for a random variable in process $X$.

3. *Background probability space.* Unless stated otherwise $(\Omega, \mathfrak{F}, \mathbb{P})$ represents the background probability space in the rest of our discussions involving stochastic processes.

**Continuous stochastic process**

**Definition 6.10.** A stochastic process $X = \{X_t : t \geq 0\}$ is said to be *sample-continuous* (or *almost surely continuous*, or simply *continuous*) if almost surely all sample paths are continuous. That is,

$$X(\omega) : [0, \infty) \to \mathbb{R},$$
$$t \quad \mapsto X(t, \omega)$$

is a continuous sample path for *almost surely (a.s.)* all $\omega \in \Omega$, which means that

$$\mathbb{P}(\omega \in \Omega | X(\omega) \text{ is not a continuous sample path}) = 0.$$

The sample continuity is a nice property for a stochastic process to possess as it implies that sample paths of the process are well behaved in some sense and therefore easier to analyze than general processes.

**Example 6.15.** Recall Example 6.13:

$$X = \{X_t : t \geq 0\} \quad \text{where} \quad X_t \sim \mathcal{N}(t, 2t^2),$$

is sample-continuous. Explanations will be given in the section on Brownian motion.                                                                               □

Note that continuity is a convergence property.[3]

## 6.1.4 Convergence of Random Variables

We will make use of the following definitions and relationships among different notions of convergence of random variables.

**Definition 6.11.** A sequence $\{X_n\}$ of random variables is said to *converge almost surely* (or *converge almost everywhere* or *converge with probability* 1) to a random variable $X$, written $X_n \xrightarrow{a.s.} X$, if

$$\mathbb{P}(\lim_{n\to\infty} X_n = X) = 1. \tag{6.5}$$

**Definition 6.12.** A sequence $\{X_n\}$ of random variables is said to *converge in probability* to a random variable $X$, written $X_n \xrightarrow{P} X$, if for each $\varepsilon > 0$,

$$\lim_{n\to\infty} \mathbb{P}(|X_n - X| \geq \varepsilon) = 0. \tag{6.6}$$

Equivalently, $X_n \xrightarrow{P} X$ if and only if for each $\varepsilon > 0$,

$$\lim_{n\to\infty} \mathbb{P}(|X_n - X| < \varepsilon) = 1. \tag{6.7}$$

**Definition 6.13.** A sequence $\{X_n\}$ of random variables is said to *converge in mean square* (or *converge in the $L^2$-norm*, or simply *converge in $L^2$*) to a random variable $X$, written $X_n \xrightarrow{m.s.} X$, if

$$\lim_{n\to\infty} \mathbb{E}(|X_n - X|^2) = 0. \tag{6.8}$$

**Definition 6.14.** A sequence $\{X_n\}$ of random variables is said to *converge in distribution* (or *converge in law* or *converge weakly*) to a random variable $X$, written $X_n \xrightarrow{d} X$, if

$$\lim_{n\to\infty} F_n(x) = F(x), \tag{6.9}$$

for each $x$ at which $F$ is continuous, where $F_n$ and $F$ are the cumulative distribution functions of random variables $X_n$ and $X$, respectively.

The next property is often applied. We provide the statement without proofs.

*Property 6.1.* Let $X_n$, $n = 1, 2, \ldots$ and $X$ be random variables. The following relationships hold:

---

[3] There are different kinds of continuity for stochastic processes although we have only introduced one (because we will only use one). Other kinds such as continuity in mean, continuity in probability (or stochastic continuity), and cadlag continuity are also important concepts in the study of mathematical finance. We recommend that the reader who has a serious interest in mathematical finance begin with studying measure theory.

1. $X_n \xrightarrow{\text{a.s.}} X \implies X_n \xrightarrow{\mathbb{P}} X \implies X_n \xrightarrow{d} X$.

2. $X_n \xrightarrow{\text{m.s.}} X \implies X_n \xrightarrow{\mathbb{P}} X$.

Consequently, $X_n \xrightarrow{\text{m.s.}} X \implies X_n \xrightarrow{d} X$.

### 6.1.5 Skewness and Kurtosis

Although both the Markowitz portfolio mean-variance analysis (see Chapter3) and the Black-Scholes option pricing model (see Chapter 8), which are among the most important models in quantitative finance, use only the first and second central moments of a distribution, the empirical evidence shows that higher moments are also important in financial modeling.

Given a random variable $X$, the first moment is a measure of the center of the distribution; the second moment is a measure of spread. In the following we'll see that the third moment measures the amount and direction of the skewness, and the fourth moment measures the height and sharpness of the central peak. Thus, the third and fourth moments are measures of shape.

**Skewness of Random Variables**

Let $X$ be a random variable with expectation $\mu$ and variance $\sigma^2$. The concept of skewness of a distribution is defined by a normalized form of the third central moment of a distribution as follows:

**Definition 6.15.** The *skewness* of $X$, denoted by skew$(X)$, is defined by

$$\text{skew}(X) = \mathbb{E}\left(\frac{(X-\mu)^3}{\sigma^3}\right).$$

$X$ is called *positively skewed* or *skewed right* if skew$(X) > 0$. $X$ is called *negatively skewed* or *skewed left* if skew$(X) < 0$.

The skewness of a random variable is a measure of the symmetry of its distribution. Clearly, $X$ is symmetric if skew$(X) = 0$.

The following intuition provides an easy visualization of the direction of skewness: let $f_X$ be the p.d.f. of $X$. Suppose that a distribution of a data set $X$ is unimodal. Skew$(X) > 0$ implies that more data are located on the right side of the center (i.e., $\mathbb{E}[X]$, see Figure 6.4 on page 320 and the rightmost graph in Figure 8.5 on page 443). Since the area under the graph of $f_X$ is 1, $f_X$ peaks on the left side of the center. This is to say that a skewed right distribution has the right tail longer than the left tail as the p.d.f. is tilted to the left.

Similarly, one can see that a skewed left distribution has the left tail longer than the right tail, and the p.d.f. is tilted to the right (see the leftmost graph in Figure 8.5 on page 443).

## Kurtosis of Random Variables

The concept of kurtosis is a measure of the degree of peakedness of a distribution and is defined by a normalized form of the fourth central moment of a distribution as follows:

**Definition 6.16.** The *kurtosis* of $X$, denoted by kurt$(X)$, is defined by

$$\text{kurt}(X) = \mathbb{E}\left(\frac{(X-\mu)^4}{\sigma^4}\right).$$

The following intuition provides an easy visualization of a distribution with large kurtosis: again, suppose that a distribution of a data set $X$ is unimodal. Let $a$ be a real number. Note that $a^4$ is much larger than $|a|$ if $|a| > 1$, and much smaller than $|a|$ if $|a| < 1$, and that $\frac{X-\mu}{\sigma}$ normalizes $X$. Thus, a large kurt$(X)$ means that $|X|$ can take large values with relatively high probability. This is to say that the peak of $f_X$ is tall and sharp as the area under the graph of $f_X$ is 1 (see Figure 8.7 on page 444).

**Remark 6.3.** Although both first and second moments of a random variable $X$ have units, skew$(X)$ and kurt$(X)$ do not. □

**Example 6.16.** If $X$ is a normal random variable with parameters $\mu$ and $\sigma^2$, then skew$(X) = 0$ and kurt$(X) = 3$. The detailed computation is left as an exercise for the reader. □

The following terminologies appear frequently in the literature of financial asset modeling:

**Definition 6.17.** The *kurtosis excess* of a random variable $X$ is defined by

$$\text{KurtExcess}(X) = \text{kurt}(X) - 3.$$

**Definition 6.18.** A random variable $X$ is called *leptokurtic* if kurt$(X) > 3$.

Equivalently, a random variable $X$ is leptokurtic if KurtExcess$(X) > 0$.

Note that relative to normal distributions, a unimodal distribution with kurtosis larger than 3 has a taller and sharper peak and longer and fatter tails.

Financial data are often collected from observations of the same variable (e.g., a stock price) in a form of discrete-time series $\{X_t : t = 0,1,2,\ldots n\}$. Although these data are always discretely sampled, we often establish continuous-time series to model them as mathematical tools can be more easily applied.

## 6.2 Filtrations and Adapted Processes

### 6.2.1 Filtrations: The Evolution of Information

Stochastic processes are often viewed as useful tools to model randomness evolving over a period of time. For example, we let $X = \{X_t : t \geq 0\}$ represent a process of a stock price at a future time. In normal circumstances, the market price of a stock is the price at which a buyer and a seller agree to trade (without any official arbiter of stock prices). Thus, the change of the market price of a stock is caused by the supply and demand, which reflect expectations of the company's profitability. To figure out such expectations, one has to figure out how any news about the company will be interpreted by traders and investors. Therefore, we need a time-evolving information structure to study a random process of a stock price. Another part of the motivation of establishing such a structure comes from the mathematical object itself: if the increment $X_{t+h} - X_t$, where $h > 0$, is assumed to be independent of $t$ (to make a full description[4] of the process $X$ mathematically manageable), then what we really say is that $X_{t+h} - X_t$ is independent of information up to time $t$. This observation suggests we describe the evolution of information as information propagation over time. The mathematical concept that serves this purpose is called *filtrations*, which give a rigorous definition for the past at a given time and captures the desired intuition in the above discussion.

**Definition 6.19.** A *filtered probability space* is a quadruple $(\Omega, \mathfrak{F}, \{\mathcal{F}_t\}, \mathbb{P})$, where $(\Omega, \mathfrak{F}, \mathbb{P})$ is a probability space and $\{\mathcal{F}_t\}$ is a filtration. A *filtration* is a nondecreasing collection of $\sigma$-algebras $\{\mathcal{F}_t \subseteq \mathfrak{F} : t \geq 0\}$ such that

$$\mathcal{F}_s \subseteq \mathcal{F}_t, \quad \text{for } \forall\, s, t \geq 0 \text{ with } s \leq t. \tag{6.10}$$

One intuitive way to describe a filtration is that it functions like a filter of information flow to control information propagation. For our purpose, it is sufficient to know the following:

1. $\mathcal{F}_t$ represents the information available at time $t$.

2. The information structure designated by (6.10) assures that the amount of information grows as time evolves and that no information is lost with increasing time (e.g., no computer crashes) in the sense that whatever information available at time $s$ is still available at time $t$ as long as $t \geq s$.

> In short, we translate the mathematical language "the $\sigma$-algebra $\mathcal{F}_t$" to the ordinary English as "*information set $\mathcal{F}_t$*" under the structure (6.10).

---

[4] A legitimate full description of the stochastic process is based on the notion of *finite-dimensional distributions*.

Note that filtrations can also be defined on a discrete-time index set by the same idea. An example of such filtrations is provided below.

**Example 6.17.** Recall Example 6.4, where we considered an oversimplified stock price behavior described by a binomial tree with two periods $[t_0, t_1]$ and $[t_1, t_2]$ such that in each period, the stock price either goes up by a factor u with probability p or goes down by a factor d with probability $1 - p$. The corresponding probability space is represented by $(\Omega, \mathfrak{F}, \mathbb{P})$, where

$$\Omega = \{\omega_1, \omega_2, \omega_3, \omega_4\}, \text{ where}$$
$$\omega_1 = UU, \ \omega_2 = UD, \ \omega_3 = DU, \ \omega_4 = DD,$$

$$\mathfrak{F} = \{\emptyset, \{\omega_1\}, \{\omega_2\}, \{\omega_3\}, \{\omega_4\},$$
$$\{\omega_1, \omega_2\}, \{\omega_1, \omega_3\}, \{\omega_1, \omega_4\}, \{\omega_2, \omega_3\}, \{\omega_2, \omega_4\}, \{\omega_3, \omega_4\},$$
$$\{\omega_1, \omega_2, \omega_3\}, \{\omega_1, \omega_2, \omega_4\}, \{\omega_1, \omega_3, \omega_4\}, \{\omega_2, \omega_3, \omega_4\}, \Omega\}.$$

Now, if we let

$$\mathcal{F}_{t_0} = \{\emptyset, \Omega\},$$
$$\mathcal{F}_{t_1} = \{\emptyset, \{\omega_1, \omega_2\}, \{\omega_3, \omega_4\}, \Omega\},$$
$$\mathcal{F}_{t_2} = \mathfrak{F},$$

then $\{\mathcal{F}_t : t = t_0, t_1, t_2\}$ is a filtration. Note that $\mathcal{F}_{t_0} \subseteq \mathcal{F}_{t_1} \subseteq \mathcal{F}_{t_2} \subseteq \mathfrak{F}$ indeed.

Also, note that as $t$ increases, information set $\mathcal{F}_t$ becomes finer and reveals more information about the evolution of stock price in the following sense:

| $S(t_0)$ | $S(t_1)$ | $S(t_2)$, outcomes |
|---|---|---|
| | | $S_0 u^2, \quad \omega_1$ |
| | $S_0 u, \ \{\omega_1, \omega_2\}$ | |
| | | $S_0 ud, \quad \omega_2$ |
| $S_0$ | | |
| | | $S_0 du, \quad \omega_3$ |
| | $S_0 d, \ \{\omega_3, \omega_4\}$ | |
| | | $S_0 dd, \quad \omega_4$ |

at $t = t_0$ we have no information available about outcomes,

at $t = t_1$ we know whether we will have $\{\omega_1, \omega_2\}$ or $\{\omega_3, \omega_4\}$,

at $t = t_2$ we know which $\omega_i$ we have.

We say that $\mathcal{F}_{t_2}$ is finer than $\mathcal{F}_{t_1}$, and $\mathcal{F}_{t_1}$ is finer than $\mathcal{F}_{t_0}$, or equivalently, $\mathcal{F}_{t_0}$ is coarser than $\mathcal{F}_{t_1}$, and $\mathcal{F}_{t_1}$ is coarser than $\mathcal{F}_{t_2}$. □

**Definition 6.20.** The filtration $\{\mathcal{F}_t \subseteq \mathfrak{F} : t \geq 0\}$ is called a *natural (or standard) filtration of process $X$* if

$$\mathcal{F}_t = \sigma(X_s, \, 0 \leq s \leq t), \quad t \geq 0, \tag{6.11}$$

where $\sigma(X_s, \, 0 \leq s \leq t)$ is called *the $\sigma$-algebra generated by random variables $X_s, \, 0 \leq s \leq t$.*

We also say that the filtration $\{\mathcal{F}_t : t \geq 0\}$ defined by $\mathcal{F}_t = \sigma(X_s : s \leq t)$ is the *filtration induced by $\{X_t : t \geq 0\}$.* Note that

$$\sigma(X_u, \, 0 \leq u \leq s) \subseteq \sigma(X_u, \, 0 \leq u \leq t), \quad 0 \leq s \leq t.$$

**Example 6.18.** The *$\sigma$-algebra generated by two random variables $X_1$ and $X_2$,* written $\sigma(X_1, X_2)$, is the smallest $\sigma$-algebra which is sufficient for both random variables $X_1$ and $X_2$ to be measurable (see page 257 and Example 6.6).

In the language of information, this is tantamount to saying that we use the least amount of information to determine both $X_1$ and $X_2$. For this reason, we say that:

➢ $\sigma(X_1, X_2)$ represents the information set that contains the least amount of information to determine both $X_1$ and $X_2$.

Clearly, $\sigma(X_1) \subseteq \sigma(X_1, X_2)$ and we see that given a discrete-time stochastic process $\{X_t, t = 1, 2, 3 \ldots\}$, we have $\sigma(X_1) \subseteq \sigma(X_1, X_2) \subseteq \sigma(X_1, X_2, X_3) \subseteq \cdots$.

□

### 6.2.2 Conditional Expectations: Properties and Intuition

The significance of conditional expectation with respect to a $\sigma$-algebra will be discussed at an intuitive level while suppressing the technicalities.

The notion of information set described in Section 6.2.1 is used in the concept of conditional expectation of random variable $X$ on $(\Omega, \mathfrak{F}, \mathbb{P})$ with respect to a $\sigma$-algebra (an information set) $\mathcal{F} \subseteq \mathfrak{F}$, denoted by $\mathbb{E}(X|\mathcal{F})$.

The conditional expectation $\mathbb{E}(X|\mathcal{F})$ is itself a random variable.[5]

We will make use of the following properties of conditional expectation:

*Property 6.2.* **(Measurability)** If $X$ is $\mathcal{F}$-measurable, then

$$\mathbb{E}(X|\mathcal{F}) \overset{\text{a.s.}}{=} X. \tag{6.12}$$

---

[5] A rigorous definition of conditional expectation can be found in standard graduate-level textbook of probability theory.

**Intuition and interpretation.** As we explained in Example 6.6 and on page 257, $X$ being $\mathcal{F}$-measurable simply means that $\mathcal{F}$ contains sufficient information of $X$ (sufficient to determine $X$). That is to say that $X$ is known given $\mathcal{F}$. Since the best guess (or estimate or predication) of the known is itself, (6.12) holds. In mathematical language the best guess (or estimate or predication) of a random variable $Y$ is expressed by $\mathbb{E}(Y)$ (in the context of our discussion).

**Example 6.19.** Given two $\sigma$-algebras $\mathcal{F}$ and $\mathcal{G}$ with $\mathcal{F} \subseteq \mathcal{G}$ and a $\mathcal{F}$-measurable random variable $X$, compute $\mathbb{E}(X|\mathcal{G})$.

**Solution.** Since $X$ being $\mathcal{F}$-measurable implies that $X$ is $\mathcal{G}$-measurable for $\mathcal{F} \subseteq \mathcal{G}$ (stated in common language: if the smaller information set contains sufficient information of $X$, so does the larger one), it follows by the measurability property of conditional expectation that $\mathbb{E}(X|\mathcal{G}) \stackrel{\text{a.s.}}{=} X$.

□

**Example 6.20.** Given two $\sigma$-algebras $\mathcal{F}$ and $\mathcal{G}$ with $\mathcal{F} \subsetneq \mathcal{G}$ and a $\mathcal{G}$-measurable random variable $X$, what can you say about $\mathbb{E}(X|\mathcal{F})$ or the prediction of $X$?

**Solution.** Unknown (because a smaller information set may not contain sufficient information of $X$ even the larger one does).

□

**Example 6.21.** Since $\sigma(X)$ contains sufficient information of $X$, (6.12) implies

$$\mathbb{E}(X|\sigma(X)) \stackrel{\text{a.s.}}{=} X.$$

Similarly,

$$\mathbb{E}(X_1|\sigma(X_1,X_2)) \stackrel{\text{a.s.}}{=} X_1, \quad \text{and} \quad \mathbb{E}(X_2|\sigma(X_1,X_2)) \stackrel{\text{a.s.}}{=} X_2.$$

More generally, let $X = \{X_t, t \geq 0\}$ be a stochastic process, then the natural filtration of $X$ is defined by $\mathcal{F}_t = \sigma(X_s, 0 \leq s \leq t), t \geq 0$. We have

$$\mathbb{E}(X_s|\mathcal{F}_t) \stackrel{\text{a.s.}}{=} X_s, \text{ for each } s \in [0,t].$$

But $\mathbb{E}(X_u|\mathcal{F}_t)$ is unknown for each $u > t$.

□

*Property 6.3.* **(Taking Out What Is Known)** If $X$ and $Y$ are random variables and $X$ is $\mathcal{F}$-measurable, then

$$\mathbb{E}(XY|\mathcal{F}) \stackrel{\text{a.s.}}{=} X\mathbb{E}(Y|\mathcal{F}). \tag{6.13}$$

**Intuition and interpretation.** Again, $X$ being $\mathcal{F}$-measurable means that $X$ is known given $\mathcal{F}$. That is to say that $X$ is being treated as a constant given information set $\mathcal{F}$. It follows that (6.13) holds.

**Example 6.22.** We continue from Example 6.21 and have

$$\mathbb{E}(X_t X_{t+0.2}|\mathcal{F}_t) \overset{\text{a.s.}}{=} X_t \mathbb{E}(X_{t+0.2}|\mathcal{F}_t).$$

$\square$

*Property 6.4.* **(Computing Expectations by Conditioning)** If $X$ is a random variable and $\mathcal{F}$ is a $\sigma$-algebra, then

$$\mathbb{E}(\mathbb{E}(X|\mathcal{F})) = \mathbb{E}(X). \tag{6.14}$$

*Property 6.5.* **(Tower Property)** If $\mathcal{F} \subset \mathcal{G}$, i.e., $\mathcal{F}$ is a sub-$\sigma$-algebra of $\mathcal{G}$, then

$$\mathbb{E}(\mathbb{E}(X|\mathcal{G})|\mathcal{F}) \overset{\text{a.s.}}{=} \mathbb{E}(X|\mathcal{F}). \tag{6.15}$$

**Intuition and interpretation.** Calculation of conditional expectation on an information set may be taken by two steps. First, calculate conditional expectation on a less coarse (or more detailed) information set than the original one. Then, finish the calculation by conditioning on the original information set, which is coarser (or in less detail).

*Property 6.6.* **(Linearity)** Let $a_i$ be constants and $X_i$ be random variables, then

$$\mathbb{E}(a_1 X_1 + a_2 X_2|\mathcal{F}) \overset{\text{a.s.}}{=} a_1\mathbb{E}(X_1|\mathcal{F}) + a_2\mathbb{E}(X_2|\mathcal{F}). \tag{6.16}$$

**More intuition of $\mathbb{E}(X|\mathcal{F})$.**

Sometimes random variable $X$ and a $\sigma$-algebra $\mathcal{F}$ are both given, but $X$ is not measurable on $\mathcal{F}$. Intuitively, we say that $\mathcal{F}$ does not have enough (contrast) resolution to read off all information contained in $X$. In this sense, $X$ being $\mathcal{F}$-measurable is at one end of a yardstick, where $\mathcal{F}$ has highest resolution to read off all information contained in $X$, whereas $X$ being independent of $\mathcal{F}$ (see Definition 6.7 on page 260) is at the other end of the yardstick, where $\mathcal{F}$ has no resolution (thus read off no information of $X$). In other words, we say that $X$ being independent of $\mathcal{F}$ means that $\mathcal{F}$ contains no information of $X$.

The meaning of $\mathbb{E}(X|\mathcal{F})$ is the least coarsification of $X$ so that $\mathcal{F}$ has enough resolution for $\mathbb{E}(X|\mathcal{F})$ (i.e., $\mathbb{E}(X|\mathcal{F})$ is measurable on $\mathcal{F}$).

Now, we are ready for the next property.

*Property 6.7.* **(Independence)** If $X$ is independent of $\mathcal{F}$, then

$$\mathbb{E}(X|\mathcal{F}) = \mathbb{E}(X). \tag{6.17}$$

Finally, to avoid notational confusion, we have the following definition:

**Definition 6.21.** Let $X$ and $Y$ be two random variables with finite expectations. we define the *conditional expectation of X given Y* by

$$\mathbb{E}(X|Y) = \mathbb{E}(X|\sigma(Y)). \tag{6.18}$$

**Interpretation.** In a special case of $\mathbb{E}(X|\mathcal{F})$, when $\mathcal{F} = \sigma(Y)$, we have a more concise notation: $\mathbb{E}(X|Y)$. Thus, $\mathbb{E}(X|Y) \neq \mathbb{E}(X|Y = y)$ since (6.18) says that the random variable $\mathbb{E}(X|Y)$ depends only on the events that $Y$ defines (e.g., $\{0.2 < \sin(Y^2 + e^Y) \leq 0.7\}$) instead of just on specific value of $Y$.

### 6.2.3 Adapted Processes: Definition and Intuition

**Definition 6.22.** A stochastic process $\{X_t : t \geq 0\}$ defined on a filtered probability space $(\Omega, \mathcal{F}, \{\mathcal{F}_t\}, \mathbb{P})$ is said to be *adapted* (or *non-anticipating*) if $X_t$ is $\mathcal{F}_t$-measurable for each $t \geq 0$.

**Intuition and interpretation.**

1. On the one hand, by this definition, Property 6.2 and Example 6.19,

$$X_s \overset{\text{a.s.}}{=} \mathbb{E}(X_s|\mathcal{F}_t), \quad \text{for each } t \geq s,$$

particularly,

$$X_s \overset{\text{a.s.}}{=} \mathbb{E}(X_s|\mathcal{F}_s). \tag{6.19}$$

Using the language in Remark 6.2 on page 259, we say that $X_s$ is considered "known" at any time $t$, $t \geq s$ if $\{X_t : t \geq 0\}$ is adapted to the filtration $\{\mathcal{F}_t\}$. In other words:

➤ A stochastic process $\{X_t : t \geq 0\}$ being adapted to the filtration implies that the value of $X_t$ is (almost surely) completely determined by the filtration $\mathcal{F}_t$ in the sense of (6.19).

2. On the other hand, since for each $t > s$, $X_t$ may not be $\mathcal{F}_s$-measurable, at time $s$ (see Example 6.20 or 6.21 on page 271), $X_t$ is considered "unknown" because probabilities of events described by $X_t$ may not be computed based on the information available at any time earlier than the moment $t$. In this sense:

➤ The notion of adaptedness can be interpreted as inability to have knowledge about future events. For this reason an adapted process is also called *non-anticipating* because the propagation or progressive revelation of information under adaptedness allows no anticipation of future information. An illustration of this concept is provided in the next example.

3. By definition of natural filtrations (see (6.11) on page 270):

   ➤ *A stochastic process is always adapted to its natural filtration.*

**Example 6.23.** Let us revisit Example 6.17 (see page 269) where we have

$$\Omega = \{\omega_1, \omega_2, \omega_3, \omega_4\},$$

and $\{\mathcal{F}_t \subseteq \mathfrak{F} : t = t_0, t_1, t_2\}$ with $t_0 < t_1 < t_2$, a filtration defined by

$$\mathcal{F}_t = \begin{cases} \{\emptyset, \Omega\} & \text{if } t = t_0 \\ \sigma(\{\omega_1, \omega_2\}) & \text{if } t = t_1 \\ \mathfrak{F} & \text{if } t = t_2, \end{cases}$$

where

$$\sigma(\{\omega_1, \omega_2\}) = \{\emptyset, \{\omega_1, \omega_2\}, \{\omega_3, \omega_4\}, \Omega\}.$$

If we define

$$X_{t_0}(\omega_i) = 1 \quad \text{if } i = 1, 2, 3, 4,$$

$$X_{t_1}(\omega_i) = \begin{cases} 2 & \text{if } i = 1, 2 \\ 3 & \text{if } i = 3, 4, \end{cases}$$

$$X_{t_2}(\omega_i) = \begin{cases} 4 & \text{if } i = 1 \\ 5 & \text{if } i = 2 \\ 6 & \text{if } i = 3 \\ 7 & \text{if } i = 4, \end{cases}$$

then it can be easily verified that $X_t$ is $\mathcal{F}_t$-measurable where $t = t_0, t_1, t_2$ respectively. Thus $X = \{X_t : t = t_0, t_1, t_2\}$ is adapted.

Observe that events represented by $\{X_{t_1} \leq a\}$, $a < 2$ or $a \geq 3$ (which are either $\emptyset$ or $\Omega$) are information that was propagated or revealed at time $t = t_0$, and none of them is $\{\omega_1, \omega_2\}$ or $\{\omega_3, \omega_4\}$, both of which are information revealed at time $t = t_1$. In a similar fashion, the reader may observe that events represented by $\{X_{t_1} \leq a\}$, $2 \leq a < 3$ (which are $\{\omega_1, \omega_2\}$) are information that was revealed at time $t = t_1$, and none of them is $\{\omega_i\}$, $i = 1, 2, 3, 4$, which are information revealed at time $t = t_2$.

In this sense, we say that the propagation or progressive revelation of information under adaptedness ensures no anticipation of future information:   □

➤ It is also worth noting that the concept of information in financial theory is not the same as that in the ordinary sense. Under the circumstances surrounding our discussion, by propagation of information, we mean the progressive revelation of the events represented in the form of filtrations. In a similar context, by observable realities (by time $t$), we mean those events that are observable (precisely, $\mathcal{F}_t$-measurable or in $\mathcal{F}_t$) by time $t$

(e.g., events represented by $\{X_t \leq a\}$ assuming that $\{X_t\}$ is adapted to the filtration).[6]

**Example 6.24.** In financial theory, the mathematical concept of a filtered probability space $(\Omega, \mathfrak{F}, \{\mathcal{F}_t\}, \mathbb{P})$ can be used to represent an economy, where each simple event $\omega \in \Omega$ represents an economic state, and the mathematical concept of adapted stochastic processes $\{X_t : t \geq 0\}$ to the filtration can be used to represent the evolution of a security price and to ensure that events such as $\{X_t \leq a\} = \{\omega : X_t(\omega) \leq a\}$ are not anticipated at any time $s < t$. Also, note that the concept of filtration suggests that the economic states $\omega \in \Omega$ are not instantaneous states but represent entire possible histories of the economy. □

**Remark 6.4.** The concept of an adapted stochastic process is essential in studying modern finance theory. For instance, it is essential in the definition of martingales. □

## 6.3 Martingales: A Brief Introduction

### 6.3.1 Basic Concepts

Note that in our next definition, $t$ may be an integer for a discrete-time process or a real number for a continuous-time process.

**Definition 6.23.** A stochastic process $\{X_t : t \geq 0\}$ on a filtered probability space $(\Omega, \mathfrak{F}, \{\mathcal{F}_t\}, \mathbb{P})$ is said to be a *martingale with respect to the filtration* $\{\mathcal{F}_t\}$ if $\{X_t\}$ is adapted to $\{\mathcal{F}_t\}$ and satisfies the condition $\mathbb{E}(|X_t|) < \infty$ for $\forall t \geq 0$ and the property

$$\mathbb{E}(X_t|\mathcal{F}_s) = X_s \quad \text{for } \forall s < t, \ t \geq 0. \tag{6.20}$$

Stating the defining characteristic of the martingale property given by (6.20) in common language:

➤ The best prediction for a future realization is the current value of the process.

Taking expectation on both sides of (6.20) yields

$$\mathbb{E}(X_t) = \mathbb{E}(X_s),$$

which is equivalent to $\mathbb{E}(X_t - X_s) = 0$. That is, the expected future gain (or loss) is zero. In this sense, *martingales model a fair game*.[7] For this reason the

---

[6] We also say that $\mathcal{F}_t$ consists of all events that are observable (i.e., $\mathcal{F}_t$-measurable) by time $t$.

[7] For this reason, a risk-neutral probability measure is referred to as an *equivalent martingale measure*, which is a key concept in derivative pricing (see Remark 7.3, 2 on page 336).

importance of martingales in modern finance is self-explanatory since security valuation is the determination of the fair price of a security.

**Intuition.** If $X_t$ represents the log return from investing in a stock at time $t$, and $\{\mathcal{F}_t \subseteq \mathfrak{F} : t \geq 0\}$ is a natural filtration of $\{X_t : t \geq 0\}$, where $\mathcal{F}_t = \sigma(X_s : s \leq t)$ for every $t \geq 0$ is the filtration induced by $\{X_t\}$ (see (6.11) on page 270), expression (6.20) implies that different styles of investing will not produce different expected returns regardless of the amount of financial research that investors have access to. In other words, active management or passive management, buy-and-hold investing, or market-timing trading based on technical analysis will all have the same expected returns.

**Remark 6.5.** Since all Itô integrals with respect to Brownian motion are martingales, martingales form an important class of stochastic processes.     □

   Martingales are defined for discrete-time stochastic processes by Definition 6.23. However, for the special case when $t$ takes nonnegative integer values, we have a more intuitive statement:

**Definition 6.24.** A stochastic sequence $\{X_n : n = 0,1,2,\dots\}$ on a filtered probability space $(\Omega, \mathfrak{F}, \{\mathcal{F}_n\}, \mathbb{P})$, where $\mathcal{F}_n = \sigma(X_k, 0 \leq k \leq n)$, is said to be a *martingale with respect to its natural filtration* if $\{X_n\}$ satisfies the condition $\mathbb{E}(|X_n|) < \infty$ for all $n \geq 0$ and the property

$$\mathbb{E}(X_{n+1}|\mathcal{F}_n) = X_n \quad \text{for all } n \geq 0. \tag{6.21}$$

Using a more concise notation of conditional expectation (in the spirit of Definition 6.21), (6.21) can be written into

$$\mathbb{E}(X_{n+1}|X_0, X_1, \dots, X_n) = X_n \quad \text{for all } n \geq 0. \tag{6.22}$$

Expression (6.22) in ordinary English says that:

➢ Knowing the past history of the process, the best prediction for one step ahead is the current observation.

**Example 6.25. (Random Walk)** Let $\{R_n : n \geq 1\}$ be a sequence of independent identically distributed random variables with mean $\mu = 0$ and finite variance $\sigma^2$. Let

$$S_n = R_1 + R_2 + \cdots + R_n. \tag{6.23}$$

The sequence $\{S_n : n \geq 1\}$ is a *random walk*. For example, if

$$R_1 = \begin{cases} 1 & \text{with } p = \frac{1}{2} \\ -1 & \text{with } p = \frac{1}{2} \end{cases} \quad \text{and} \quad \text{set } S_0 = 0,$$

the stochastic sequence $\{S_n : n \geq 0\}$ is called the *simple random walk* on $\mathbb{Z}$. Clearly, the walk starts at 0 and at each step moves either $+1$ (to the right) or $-1$ (to the left) with equal probability.

We are interested in showing that $\{S_n : n \geq 1\}$ defined by (6.23) is a martingale.

*Proof.* We verify two conditions required by Definition 6.24:

**Step 1.**

$$\mathbb{E}(S_n) = n\mathbb{E}(R_1) = 0.$$

$$\mathbb{E}(S_n^2) = \sum_{j=1}^{n}\sum_{i=1}^{n} \mathbb{E}(R_i R_j) = \sum_{j=1}^{n} \sigma^2 = n\sigma^2$$

implies[8]

$$\mathbb{E}(|S_n|) < \infty.$$

**Step 2.**

$$\mathbb{E}(S_{n+1}|S_1, S_2, \ldots, S_n) = \mathbb{E}(R_1 + R_2 + \cdots + R_n + R_{n+1}|S_1, S_2, \ldots, S_n)$$
$$= R_1 + R_2 + \cdots + R_n + \mathbb{E}(R_{n+1}) = S_n + 0 = S_n.$$

$\square$

### 6.3.2 Martingale as a Necessary Condition of an Efficient Market

According to Fama (1970), a market in which prices always fully reflect available information is called *efficient* (see Fama [11]).

Recall (6.20)

$$\mathbb{E}(X_t|\mathcal{F}_s) = X_s \quad \text{for } \forall s < t.$$

Let $s$ be the current time and $\mathcal{F}_s = \sigma(X_u, u \leq s)$. If $X_t$ represents a security price at time t, (6.20) indicates that the information contained in the past prices is instantly and fully reflected in the security current price. For this reason, the martingale is considered to be a necessary condition for efficient security market.

---

[8] A proof can be done either by applying Cauchy-Schwarz inequality or by using the result at the link
http://mathworld.wolfram.com/RandomWalk1-Dimensional.html to show that

$$\lim_{n\to\infty} \frac{\mathbb{E}(|S_n|)}{\sqrt{n}} = \sqrt{\frac{2}{\pi}}.$$

Since the efficient market hypothesis (EMH) is viewed as a fundamental assumption in theoretical finance, showing that a security price process is a martingale becomes highly important as otherwise establishing any model based on the study of the process behavior could be in a framework that violates the EMH. A further in-depth discussion on this subject will lead to the fundamental theorem of finance.

## 6.4 Modeling Security Price Behavior

### 6.4.1 From Deterministic Model to Stochastic Model

We will illustrate deterministic versus stochastic models by an example.

**Deterministic model**

The continuously compounding interest model can be described by an ordinary differential equation (o.d.e.)

$$\frac{dS}{dt} = rS \quad \text{with} \quad S(0) = S_0,$$

which is equivalent to

$$dS = rS\,dt \quad \text{with} \quad S(0) = S_0, \tag{6.24}$$

where $S = S(t)$ is the amount at time $t$ and $r$ is a constant interest rate.

Letting $X(t) = \ln\frac{S(t)}{S_0}$ and substituting in (6.24), we obtain a simplified model:

$$dX = r\,dt \quad \text{with} \quad X(0) = 0, \tag{6.25}$$

where $X = X(t)$ is the log return over the time period $[0, t]$.

**Stochastic model**

We consider the scenario of having a perturbation to the constant interest rate. In other words, we decompose $r$ into a sum of a nominal value $\mu$ and its perturbation $\varepsilon$ (which is unpredictable and causes $r$ to change infinitely fast):

$$r = \mu + \varepsilon \quad \text{or} \quad r_t = \mu + \varepsilon_t, \tag{6.26}$$

where the subscript $t$ is to emphasize the change with respect to time.

Corresponding to the perturbation term, the (deterministic) o.d.e. in (6.25) becomes

$$dX = \mu\,dt + \text{noise},$$

where the noise term is often considered as a Gaussian white noise stochastic process (see Definition 6.25 on page 280), which is related to Brownian motion (see Section 6.5) since it has been shown that, although Brownian motion paths are nowhere differentiable, their (formal) "time derivatives" form the white noise process (see Definition 6.25). Thus, we can write

$$dX_t = \mu dt + \sigma d\mathfrak{B}_t, \tag{6.27}$$

where $\sigma$ is a positive constant and $\mathfrak{B} = \{\mathfrak{B}_t\}$ is the standard Brownian motion which will be introduced shortly.[9]

Notice that $X = X(t)$ in (6.25) is a deterministic function, whereas $X = X_t$ in (6.27) becomes a stochastic process.

Using a mathematical jargon, we say that the process $X$ in (6.27) is *driven by* the standard Brownian motion $\mathfrak{B}$.

### 6.4.2 Innovation Processes: An Intuition

An *innovation (process)* is obtained by taking the difference between the observed value of a random variable at time $t$ (e.g., $x_t$ where $t$ is in discrete-time) and the optimal forecast of that value based on information available prior to time $t$ (e.g., $\mathbb{E}(x_t|\mathcal{F}_{t-1})$).

In other words, the quantity modeled by a process can be decomposed into two components; one is predictable and another is unpredictable (e.g., (6.26)). *The unpredictable component is called an innovation (process).* If a stochastic model successfully captures the predictable component structure in the data, then the innovation process constitutes a white noise process (see Definition 6.25 below) or an martingale difference sequence (not to be discussed in this book).

Before the next example, we need the definition of autocovariance of a process.

Let $X = \{X_t : t \in \mathcal{J}\}$ be a stochastic process with discrete or continuous time. The *autocovariance (function) of process* $X$ is defined by

$$\gamma(t,s) = \text{Cov}(X_t, X_s) \quad \text{for } \forall s,t \in \mathcal{J}, \tag{6.28}$$

which is a function of two variables and satisfies

$$\gamma(s,t) = \gamma(t,s).$$

For each $t$, we let $\tau$ be a time lag such that $t + \tau \in \mathcal{J}$ and denote by $\gamma_\tau(t)$ the autocovariance of $X$ (where $\gamma(t,s) = \gamma(t,t+\tau)$ with $\tau = s - t$). That is,

$$\gamma_\tau(t) = \gamma(t,t+\tau) = \text{Cov}(X_t, X_{t+\tau}), \quad t, t+\tau \in \mathcal{J}.$$

---

[9] *White noise* thought of as the derivative of Brownian motion, $\frac{d\mathfrak{B}_t}{dt}$, does not exist in the ordinary sense. It is related to the notion of *generalized stochastic process*, since $\frac{d\mathfrak{B}}{dt}$ is well defined as a *generalized function* on an infinite dimensional space, which is a topic beyond the scope of this book.

Note that a process has no linear forecasting value if its autocovariance function $\gamma_\tau(t)$ is identically equal to zero.

Since a "noise" is presumably unpredictable, we expect the covariance function of a white noise process at all nontrivial time lags to show a value of zero.

We define the *mean value function* of a process $\{X_t\}$ by $m(t) = \mathbb{E}(X_t)$ and give the following definitions.

**Definition 6.25.**

1. The process $\{\varepsilon_t\}$ is said to be a *white noise (process)* if its mean value function and autocovariance function respectively are

$$m(t) = \mathbb{E}(\varepsilon_t) = 0, \quad \text{and} \quad \gamma_\tau(t) = \begin{cases} \sigma_\varepsilon^2 & \text{if } \tau = 0 \\ 0 & \text{if } \tau \neq 0, \end{cases} \quad \text{for } \forall\, t,$$

   where $\sigma_\varepsilon^2 > 0$ is a constant. We write

$$\varepsilon_t \sim WN(0, \sigma_\varepsilon^2).$$

2. An *independent white noise (process)* $\{\varepsilon_t\}$ is a white noise process consisting of mutually independent random variables. We write

$$\varepsilon_t \sim i.WN(0, \sigma_\varepsilon^2).$$

3. A *strict white noise (process)* $\{\varepsilon_t\}$ is a white noise process consisting of independent and identically distributed (i.i.d.) random variables. We denote it by

$$\varepsilon_t \sim i.i.d.WN(0, \sigma_\varepsilon^2).$$

4. A white noise process $\{\varepsilon_t\}$ is said to be a *Gaussian white noise* process if

$$\varepsilon_t \sim N(0, \sigma_\varepsilon^2).$$

A white noise process has no linear prediction value (it is unpredictable) because it is serially uncorrelated.

**Example 6.26.** $\{\varepsilon_i\}$ in (4.23) on page 186, is a white noise.                                  □

**Example 6.27.** A stochastic process $\{W_i\}$ defined by

$$W_i = W_0 + \sum_{k=1}^{i} R_k = W_{i-1} + R_i, \quad i = 1, 2, \ldots$$

is said to be a *random walk* on a probability space $(\Omega, \mathfrak{F}, \mathbb{P})$ if $\{R_i, i = 0, 1, 2, \ldots\}$ is a sequence of i.i.d. random variables and $W_0$ a random variable that is independent of each $R_i$:

1. A random walk is not a white noise. (*Hint*: compute covariance.)

2. A random walk is a martingale with respect to its natural filtration if $\mathbb{E}(R_i) = 0$ for all $i$ (see Example 6.25 on page 276).

The proof is left as an exercise for the reader.                              $\square$

### 6.4.3 Securities Paying a Continuous Cash Dividend

Let us first define the dividend yield with continuously compounding. To do so intuitively, recall that the (annualized constant) interest rate $r$ under continuously compounding is a constant satisfying the equation

$$I(t) = A(t) r \, dt,$$

where $I(t)$ is the simple interest on the time interval $[t, t + dt]$, and $A(t)$ is the initial principal on the same time period. Note that the infinitesimal change on the principal is

$$d A(t) = I(t) = A(t) r \, dt.$$

Similarly, if $S(t)$ is the price of the underlier, without loss of generality, e.g., a stock at time t, and $\mathcal{D}(t)$ is the dividend paid by the stock over the time interval $[t, t + dt]$, we define the (annualized) *dividend yield*, denoted by $q$, to be the constant satisfying

$$\mathcal{D}(t) = S(t) q \, dt. \tag{6.29}$$

If we reinvest the dividends immediately in the stock, (6.29) says that the dividend paid on $[t, t + dt]$ buys us $q\,dt$ shares of the stock. It follows that over the time period $[t, t + dt]$, if we own $N(t)$ shares of the stock initially (i.e., at time t), the dividends from these $N(t)$ shares of the stock on the small interval buy us $N(t) q\,dt$ more shares than that we initially owned. Restated in mathematical language,

$$d N(t) = q N(t) \, dt, \quad t \in [0, T]. \tag{6.30}$$

In words, the infinitesimal change in the shares held on $[t, t + dt]$ is $q N(t)\,dt$.

Solving the initial value problem of o.d.e.

$$d N(t) = q N(t) \, dt, \quad N(0) = N_0, \quad t \in [0, T], \tag{6.31}$$

we obtain $N(T) = N_0 e^{qT}$. In particular,

$$N_0 = 1 \;\Rightarrow\; N(T) = e^{qT}, \quad \text{and} \quad N_0 = e^{-qT} \;\Rightarrow\; N(T) = 1. \tag{6.32}$$

In words:

➤ *The dividend reinvestment yields $e^{qT} - 1$ more shares at time T from one share of the stock at time 0;*

➤ *The dividend reinvestment yields* $1 - e^{-qT}$ *more shares at time T from* $e^{-qT}$ *shares of the stock at time 0.*

This result will be used again and again, particularly in Chapter 7.

**Example 6.28.** There are two ways to profit from a dividend-paying security: capital gains and dividends. Let $\{S(t)\}$ be the price process of such a security with dividend yield $q$. Assuming that the risk-neutrality hypothesis holds, the expected return from the security is the risk-free interest rate r. Therefore, the expected return of the capital gains (i.e., based solely on the appreciation of the stock price) must be $r - q$. This is to say that if $S(t_0)$ is the security price at the beginning of time interval $[t_0, t_0 + \Delta t]$, then the expected value of the security price at the end of the interval becomes $S(t_0)e^{(r-q)\Delta t}$. If $\{S(t)\}$ is modeled by a binomial tree with parameters p, u, and d, then the expectation of capital gains over this time period can be expressed by

$$S(t_0)(e^{(r-q)\Delta t} - 1) = S(t_0)((pu + (1-p)d) - 1).$$

□

## 6.5 Brownian Motion

### 6.5.1 Definition of Brownian Motion

The following familiar properties of normal random variables and notation will be frequently used:

*Property 6.8.* **(Normal Distributions)** We denote a normally distributed random variable $X$ with mean $\mu$ and variance $\sigma^2$ by $X \sim \mathcal{N}(\mu, \sigma^2)$. Let $a$ and $b$ be constants. The following properties hold:

1. $X \sim \mathcal{N}(0, 1) \Rightarrow a + bX \sim \mathcal{N}(a, b^2)$.
2. $X \sim \mathcal{N}(\mu, \sigma^2) \Rightarrow a + bX \sim \mathcal{N}(a + b\mu, b^2\sigma^2)$.
3. $X \sim \mathcal{N}(\mu, \sigma^2) \Rightarrow M_X(t) = e^{\mu t + \frac{1}{2}\sigma^2 t^2}$,
   where $M_X(t) \equiv \mathbb{E}(e^{tX})$ is the moment generating function of $X$.

**Definition 6.26.** A *standard* (one-dimensional) *Brownian motion*[10] or *Wiener process*,[11] denoted by $\mathfrak{B} = \{\mathfrak{B}(t) : t \geq 0\}$, is a stochastic process on $(\Omega, \mathfrak{F}, \mathbb{P})$[12] satisfying the following properties:

1. $\mathfrak{B}(0) \overset{\text{a.s.}}{=} 0$ (i.e., $\mathbb{P}(\omega \in \Omega \mid \mathfrak{B}(0) \neq 0) = 0$).

2. With probability 1, sample paths of $\mathfrak{B}$ are continuous, i.e.,

$$\mathbb{P}(\omega \in \Omega \mid \mathfrak{B}(t, \omega) \text{ is not continuous at each } t) = 0.$$

3. For every choice of nonnegative real numbers

$$0 \leq t_1 < t_2 < t_3 < \cdots t_{n-1} < t_n < \infty,$$

the increments

$$\mathfrak{B}(t_2) - \mathfrak{B}(t_1), \ \mathfrak{B}(t_3) - \mathfrak{B}(t_2), \ \ldots, \ \mathfrak{B}(t_n) - \mathfrak{B}(t_{n-1})$$

are mutually independent random variables.

4. For each $0 \leq s < t < \infty$, the increment $\mathfrak{B}(t) - \mathfrak{B}(s)$ is a normal random variable with mean 0 and variance $t - s$, written $\mathfrak{B}(t) - \mathfrak{B}(s) \sim \mathcal{N}(0, t - s)$.

It is easier to remember the properties in Definition 6.26 if we can concisely describe them in words:

1. $\mathfrak{B}$ starts at 0 almost surely (i.e., the initial state of the process is 0 a.s.).

2. $\mathfrak{B}$ is a sample-continuous process (see Definition 6.10).

3. $\mathfrak{B}$ has *independent increments*.

4. $\mathfrak{B}$ has *stationary increments* (i.e., *time homogeneity*), which are normally distributed.
   The reason for the names (i.e., stationary increments or time homogeneity) is due to the fact that Definition 6.26, 4 implies that for all $s, t, h \geq 0$,

$$\mathfrak{B}(t) - \mathfrak{B}(s) \overset{\text{d}}{=} \mathfrak{B}(t + h) - \mathfrak{B}(s + h).$$

That is, the increments over the time intervals with same length have the same probability distributions.

---

[10] Robert Brown (1773 - 1858).

[11] Nobert Wiener (1894 - 1964). *Wiener process* is a more popular name among mathematicians than among physicists whereas *Brownian motion* is vice versa.

[12] $\mathfrak{B} = \{\mathfrak{B}(t) : t \geq 0\}$ on $(\Omega, \mathfrak{F}, \mathbb{P})$ is often referred to as a $\mathbb{P}$-*Brownian motion* where $\mathbb{P}$ represents the probability measure in the real world (or physical world) in contrast to $\mathbb{Q}$-*Brownian motion*. $\mathbb{Q}$-Brownian motion means that $\mathfrak{B} = \{\mathfrak{B}(t) : t \geq 0\}$ is a process on $(\Omega, \mathfrak{F}, \mathbb{Q})$ where $\mathbb{Q}$ represents the probability measure in the risk-neutral world. More detailed explanation is given in Section 6.8.3.

**Definition 6.27.** A *Brownian motion with starting point b* is a stochastic process that can be expressed by $b + \mathfrak{B}$ where $b \in \mathbb{R}$ is a constant and $\mathfrak{B}$ is a standard Brownian motion.

In words, a Brownian motion $X = \{X(t)\}$ with initial value $X(0) \stackrel{\text{a.s.}}{=} b$ is obtained by adding $b$ to a standard Brownian motion $\mathfrak{B} = \{\mathfrak{B}(t)\}$, i.e., $X = \mathfrak{B} + b$ or equivalently $X(t) = b + \mathfrak{B}(t)$, $t \geq 0$.

**Definition 6.28.** A *Brownian motion with drift and scaling* is a stochastic process that can be expressed by $b + \mu t + \sigma \mathfrak{B}$ where $b, \mu \in \mathbb{R}$ and $\sigma > 0$ are constant and $\mathfrak{B}$ is a standard Brownian motion.

**Remark 6.6.** Although Section 6.10 provides a brief discussion on a Brownian motion as a limit of a random walk, it is worth noting that the existence of Brownian motion is not a trivial fact.[13] For a constructive proof of this non-trivial fact, we refer the reader to the literature (e.g., Durrett [9]; Mörters, and Peres [22]; and Paley, Wiener, and Zygmund [25]).

### 6.5.2 Some Properties of Brownian Motion Paths

For the sake of convenience of conversation, we refer to a sample path of a Brownian motion as a *Brownian path*.

Three of the most important Brownian path properties are given below, which will help us to visualize Brownian motion.

*Property 6.9.*

1. With probability 1, Brownian paths are continuous.

2. With probability 1, Brownian paths are nowhere differentiable.

3. With probability 1, Brownian paths do not have bounded total variation[14] on $[0, t]$ (nor on any interval by the time-homogenous property of Brownian motion).

*Proof.* The first property follows from the fact that a Brownian motion $\mathfrak{B} + b$ is sample-continuous if and only if the corresponding standard Brownian motion $\mathfrak{B}$ is sample-continuous (and this is straightforward from Definition 6.26).

---

[13] It is not obvious that all four properties in Definition 6.26 are compatible with each other. For instance, it is not obvious that stationary independent increments and sample continuity are compatible properties.

[14] The total variation is a way to measure the "variation" of a (deterministic) real-valued function (see Section (6.6)).

Since a proof of the second property requires deeper understanding of analysis that is beyond the scope of this book, we refer the reader to the literature (e.g., Billingsley [3]).

For a proof of the third property, see page 295. ☐

Scaling properties of Brownian motion are invariant properties under transformations.

*Property 6.10.* **(Scaling Invariance)** For $\forall\, a > 0$, process $\{X(t) : t \geq 0\}$ with

$$X(t) = \frac{1}{a}\mathcal{B}(a^2 t)$$

is a standard Brownian motion.

*Proof.* It is sufficient to show that increments are normally distributed. In fact,

$$X(t) - X(s) = \frac{1}{a}\mathcal{B}(a^2 t) - \frac{1}{a}\mathcal{B}(a^2 s) = \frac{1}{a}(\mathcal{B}(a^2 t) - \mathcal{B}(a^2 s))$$

$$\sim \frac{1}{a}\mathcal{N}(0, a^2(t-s)) \overset{d}{=} \frac{a}{a}\mathcal{N}(0, t-s) = \mathcal{N}(0, t-s),$$

where the third equal sign holds due to Property 6.8, 2. ☐

**Intuition.** Scaling invariance simply means that the geometric structure of Brownian paths has a *fractal*[15] nature or *self-similarity* property. In loose terms, if we "zoom in or zoom out" on a Brownian path, we always see a Brownian path.

**Example 6.29.** Show that if $\mathcal{B}$ is a standard Brownian motion, so is $-\mathcal{B}$.
What is your visualization of $-\mathcal{B}$? Is it different from that of $\mathcal{B}$?

Straightforward verifications of continuity of sample paths, independence, normal distribution, and proper means and variances of all the increments are left as an exercise to the reader.

*Hint:* Notice that $F_{-X}(x) = \mathbb{P}(-X \leq x) = 1 - \mathbb{P}(X < -x) = 1 - F_X(-x)$ implies that $f_{-X}(x) = f_X(-x)$ and that $f_X(-x) = f_X(x)$ if $f_X$ is an even function (e.g., $X \sim \mathcal{N}(0, \sigma^2)$). ☐

### 6.5.3 *Visualization of Brownian Paths*

Although any graphs can only depict a Brownian motion traveling in a manner far from desirable due to a host of microscopic random effects, a mental visualization of them may be achieved. The following explanation may be helpful.

---

[15] In short, "a *fractal* is a natural phenomenon or a mathematical set that exhibits a repeating pattern that displays at every scale." For more explanation, we refer the reader to http://en.wikipedia.org/wiki/Fractal.

**Three Keys to Reading Brownian Motion Paths**

Mimicking the notion of graphs of cross-sections of a function, say

$$x = f(t, w)$$

in analytic geometry, we try to visualize a stochastic process

$$X = \{X(t, \omega)\}$$

in terms of its sample paths ("cross-sections" when $\omega$ is fixed) and the p.d.f. of random variables (cross-sections when $t$ is fixed):

*1. Visualization of Brownian paths (cross-sections when $\omega$ is fixed)*

➤ Each Brownian path is continuous but at almost nowhere directional, i.e., makes "sharp turns" (or forms "sharp corners") everywhere due to Property 6.9, 1 and 2. In fact, it can be shown that with probability 1, a Brownian path is not monotonic on any interval. This is a manifestation of Property 6.9, 1 and 2 combined.

   To elaborate this, if a path is drawn by a physical pen, then at most points the path has directions as driven by the force exerted on the pen. This explains why a Brownian path cannot be drawn by a pen and therefore quite nonintuitive.

➤ Each Brownian path has an infinitely large variation over any time interval due to Property 6.9, 3.

   To elaborate this, we might imagine a piece of Brownian path over a time interval merely the diameter of a thread of hair yet with infinite length.

➤ All sample paths of the same Brownian motion diffuse (or radiate) from the (same) initial state (at $t = 0$) due to Definition 6.26, 1.

➤ If you zoom in or zoom out on a Brownian path, you always see a Brownian path.

*2. Visualization of distributions of random variables (cross-sections when $t$ is fixed)*

➤ The graph of probability density function of each random variable $\mathfrak{B}(t)$ is a bell-shaped curve due to Definition 6.26, 4. That is, the distributions of the "dots," which are $t$-cross-sections of Brownian paths, are governed by bell-shaped curves.

   In other words, since $\mathfrak{B}(t) \sim \mathcal{N}(0, t)$[16], for each fixed $t$, we know that
68% of Brownian paths are within $\pm\sqrt{t}$ units from the time-axis;
95% of Brownian paths are within $\pm 2\sqrt{t}$ units from the time-axis;
99.7% of Brownian paths are within $\pm 3\sqrt{t}$ units from the time-axis.

---

[16] Note that 1 standard deviation from 0 is $\sqrt{t}$.

*3. Combining sample paths and graphs of p.d.f. of (state) random variables*

Just as we can achieve 3-D visualization of the graph of $z = f(x,y)$ by mentally combining the graphs of cross-sections in analytic geometry, we can obtain some 2-D intuition of a Brownian motion by mentally combining its sample paths and those "bell-shaped curves" (see Figure 6.1).

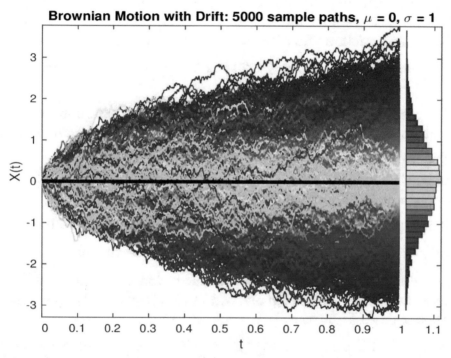

**Fig. 6.1** Standard Brownian motion is shown using 5,000 randomly selected sample paths over the interval $[0,1]$, where $X(t) = \mathfrak{B}(t)$ is plotted on the vertical axis. The current time is 0 and for each future $t > 0$, the random variable $\mathfrak{B}(t)$ is normal with mean 0 and variance $t$, which can be seen to increase with time in the figure. At time $t = 1$, the frequency distribution of the sample paths is shown as a histogram, which indeed has the shape of a normal distribution with mean 0 and variance 1. The horizontal line shows the mean value of standard Brownian motion, namely, $\mathbb{E}(\mathfrak{B}(t)) = 0$ for all $t \geq 0$

**Example 6.30.** Because of a host of microscopic random effects (e.g., see scaling invariance Property 6.10), graphs can depict a Brownian motion traveling only in a manner far from desirable; however, to visualize the Brownian motion $\mathfrak{B} + b$, one may vertically translate the graph in Figure 6.1 by $b$ units, and imagine that Brownian paths are diffusing from its initial state $\mathfrak{B}(0) \stackrel{\text{a.s.}}{=} b$, and travel in unpredictable directions all the time (therefore impossible to draw).  □

### 6.5.4 Markov Property for Brownian Motion

One may view the Markov property as a somewhat extended independent increment requirement in the definition of Brownian motion.

*Observation.*
Let $\{\mathcal{F}_t\}$ be the natural filtration of the standard Brownian motion $\{\mathfrak{B}_t\}$. That is, $\mathcal{F}_t = \sigma(\mathfrak{B}_s : 0 \leq s \leq t)$, for $\forall\, t \geq 0$.

Given $s \geq 0$, it follows from the definition of Brownian motion that:

➤ The pre-s process $\{\mathfrak{B}_t : 0 \leq t \leq s\}$ is independent of the post-s process $\{\mathfrak{B}_{s+t} - \mathfrak{B}_s : t \geq 0\}$.

➤ The post-s process $\{\mathfrak{B}_{s+t} - \mathfrak{B}_s : t \geq 0\}$ has the same distribution as the original process $\{\mathfrak{B}_t : t \geq 0\}$ (i.e., the post-s process is also a standard Brownian motion).

Thus, by Definition 6.6, $\sigma(\mathfrak{B}_{s+t} - \mathfrak{B}_s : t \geq 0)$ is independent of $\mathcal{F}_t$, $0 \leq t \leq s$. Consequently, it follows from Definition 6.7 that

➤ the post-s process $\{\mathfrak{B}_{s+t} - \mathfrak{B}_s : t \geq 0\}$ is independent of the filtration $\mathcal{F}_s$.

Now, we formally state the last result from our observation:

*Property 6.11.* **(Markov Property for Brownian Motion)** Let $\mathfrak{B} = \{\mathfrak{B}_t : t \geq 0\}$ be a standard Brownian motion on $(\Omega, \mathfrak{F}, \mathbb{P}, \{\mathcal{F}_t\})$, where $\{\mathcal{F}_t\}$ is the natural filtration of process $\mathfrak{B}$. Then:

1. For all $s \geq 0$, the post-s process $\mathfrak{B}_{\text{post-s}} = \{\mathfrak{B}_{s+t} - \mathfrak{B}_s : t \geq 0\}$ is independent of $\mathcal{F}_s$.

2. The post-s process $\mathfrak{B}_{\text{post-s}}$ and the original process $\mathfrak{B}$ have equivalent distributions.

**Remark 6.7.** We emphasize that the last property is crucial in the definition of the Itô integral with respect to Brownian motion.                    □

Intuitively, a process $\{X_t : t \geq 0\}$ is said to be a *Markov process* if, given t, all information in the truncated process $\{X_s : s \leq t\}$ relevant to the probability distribution of a future $X_u$, $u > t$ is all contained in $X_t$. In mathematical terms,

$$\mathbb{P}(X_u \leq a | \sigma(X_s, s \leq t)) = \mathbb{P}(X_u \leq a | X_t) \quad \text{for } a \in (-\infty, \infty).$$

As an example, a Brownian motion is a Markov process.

**Example 6.31.** Show that the Brownian motion $\{\mathfrak{B}_t\}$ is a martingale with respect to its natural filtration (i.e., $\mathcal{F}_t = \sigma(\mathfrak{B}_s : s \leq t)$,[17] the filtration induced by $\{\mathfrak{B}_t\}$).

---

[17] It is worth noting that both filtrations below are used frequently in the literature:

**Proof.**

$$\mathbb{E}(\mathfrak{B}_t|\mathcal{F}_s) = \mathbb{E}(\mathfrak{B}_s + \mathfrak{B}_t - \mathfrak{B}_s|\mathcal{F}_s)$$
$$= \mathbb{E}(\mathfrak{B}_s|\mathcal{F}_s) + \mathbb{E}(\mathfrak{B}_t - \mathfrak{B}_s|\mathcal{F}_s) = \mathbb{E}(\mathfrak{B}_s|\mathcal{F}_s) = \mathfrak{B}_s,$$

where the third equal sign holds due to the fact that $(\mathfrak{B}_t - \mathfrak{B}_s)\perp\mathcal{F}_s$ by Properties 6.11, 1 and 6.7, and $\mathbb{E}(\mathfrak{B}_t - \mathfrak{B}_s) = 0$. $\qquad\square$

**Remark 6.8.** Since all Itô integrals with respect to Brownian motion are martingales, martingales form an important class of stochastic processes. $\qquad\square$

Finally, we note that Brownian motion is a basic building block for the construction of a *diffusion process* (or a *diffusion* for short), which is a continuous-time Markov process with (almost surely) continuous sample paths.

The simplest and most fundamental diffusion process is Brownian motion. A more general example of diffusion processes is a *Brownian motion with drift* (e.g., $X_t = \mu t + \sigma\mathfrak{B}_t$, also see Figure 6.2).

## 6.6 Quadratic Variation and Covariation

### 6.6.1 Motivation, Definition, and Notation

There are different ways to measure the "variation" of a function. *Total variation* is a tool for us to measure the total, therefore the absolute value (as you will see in the definition below), up-and-down movement of a function. The total variation is used for deterministic real-valued functions (not as sample paths of random processes), whereas quadratic variation is used for stochastic processes (we will show that processes such as Brownian motions do not have finite total variations, but can be dealt with if one uses quadratic variation).

The *total variation* of a real-valued function on an interval is denoted by $V_a^b(f)$ and defined as

$$V_a^b(f) = \lim_{|\mathbf{P}|\to 0} \sum_{k=0}^{n-1} |f(x_{k+1}) - f(x_k)|, \tag{6.33}$$

---

(1) $\mathcal{F}_t^o = \sigma(\mathfrak{B}_s : 0 \le s \le t)$, (2) $\mathcal{F}_t^+ = \cap_{s \ge t}\mathcal{F}_t^o$, for $\forall\, t \ge 0$.

The first is the smallest filtration that makes $\{\mathfrak{B}_t\}$ adapted. The second is an extension of the first by including some zero-probability subsets and has advantages of being *complete* and *right-continuous*, which are convenient and important properties to have. The second filtration is referred to as the *Brownian filtration*. We use the first filtration in this example to suppress some conceptual and technical details involved in the second.

In loose terms and for our purpose, both filtrations are denoted by $\{\mathcal{F}_t\}$, where $\mathcal{F}_t$ is interpreted as the set of information generated by the standard Brownian motion on the time interval $[0,t]$.

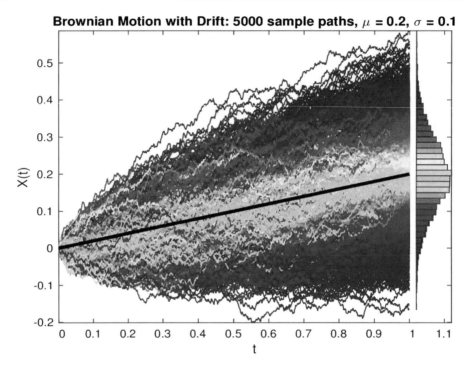

**Fig. 6.2** Brownian motion with drift parameter $\mu = 0.2$ and volatility parameter $\sigma = 0.1$ is illustrated using 5,000 randomly selected sample paths over the interval $[0,1]$. Note how the over structure drifts upward about the mean line $\mu t$. At time $t = 1$, the frequency distribution of $X(1)$ is shown as a histogram, which approximates a normal distribution with mean 0.2 and variance 0.1. The current time is 0 and the solid line is a plot of the expected value $\mathbb{E}(X(t)) = 0.2t$, where $t \geq 0$

where $\mathbf{P} : a = x_0 < x_1 < x_2 < \cdots < x_n = b$ is a partition on $[a,b]$.

The quadratic variation process is one of the central concepts in classical continuous-time martingale theory. Martingale processes will be introduced in a later section.

**Definition 6.29.** Let $X = \{X_t, \ t \geq 0\}$ and $Y = \{Y_t, \ t \geq 0\}$ be two real-valued stochastic processes defined on a probability space $(\Omega, \mathfrak{F}, \mathbb{P})$.

The *quadratic variation* of $X$ is the process denoted by $\{[X]_t\}$ and defined by

$$[X]_t = \lim_{|\mathbf{P}| \to 0} \sum_{k=1}^{n} (X_{t_k} - X_{t_{k-1}})^2, \quad t \geq 0, \tag{6.34}$$

where the limit is taken in probability[18] (see Definition 6.12), and $\mathbf{P}$ represents partitions over the interval $[0,t]$ and $|\mathbf{P}|$ is the length of the longest subintervals associated to $\mathbf{P}$.

---

[18] Notice that, among different concepts of convergence of a sequence of random variables (i.e., convergence in probability, almost sure convergence, and convergence in mean square), convergence in probability is the weakest.

More generally, the *quadratic covariation* or *cross-variation* (or just *covariation*) of two processes $X$ and $Y$ is the process denoted by $\{[X,Y]_t\}$ and defined by

$$[X,Y]_t = \lim_{|\mathbf{P}|\to 0} \sum_{k=1}^{n} (X_{t_k} - X_{t_{k-1}})(Y_{t_k} - Y_{t_{k-1}}), \quad t \geq 0, \tag{6.35}$$

where the limit is taken in probability, and $\mathbf{P}$ and $|\mathbf{P}|$ are the same as stated above.

Clearly, $[X]_t = [X,X]_t$. If there is no ambiguity, to ease notation, $[X]_t$ and $[X,Y]_t$ may be used to represent *quadratic variation process* (i.e., $\{[X]_t\}$) and *covariation process* (i.e., $\{[X,Y]_t\}$), respectively.

If $X$ and $Y$ have finite quadratic variations, we rewrite (6.34) and (6.35) with substitutions

$$\Delta_k X = X_{t_k} - X_{t_{k-1}} \quad \text{and} \quad \Delta_k Y = Y_{t_k} - Y_{t_{k-1}}$$

and obtain

$$[X]_t = \lim_{\Delta t \to 0} \sum_{k=1}^{n} (\Delta_k X)^2,$$

$$[X,Y]_t = \lim_{\Delta t \to 0} \sum_{k=1}^{n} (\Delta_k X)(\Delta_k Y),$$

which motivate the following *notation*, respectively:

$$d[X]_t = (dX_t)^2, \tag{6.36}$$
$$d[X,Y]_t = (dX_t)(dY_t). \tag{6.37}$$

### 6.6.2 Basic Properties

The next two properties of the covariation process can be easily verified.

*Property 6.12.* **(Symmetry and Bilinearity)** Let $X^{(i)} = \{X_t^{(i)}, t \geq 0\}, i = 1,2,3$ be three processes with finite quadratic variations. The following properties hold:

1. The covariation process is *symmetric*. That is,

$$[X^{(1)}, X^{(2)}]_t = [X^{(2)}, X^{(1)}]_t, \quad t \geq 0.$$

2. The covariation process is *bilinear*. That is, for any constants $a,b \in (-\infty,\infty)$,

$$[X^{(1)}, aX^{(2)} + bX^{(3)}]_t = a[X^{(1)}, X^{(2)}]_t + b[X^{(1)}, X^{(3)}]_t, \quad t \geq 0,$$
$$[aX^{(1)} + bX^{(2)}, X^{(3)}]_t = a[X^{(1)}, X^{(3)}]_t + b[X^{(2)}, X^{(3)}]_t, \quad t \geq 0.$$

Property 6.13 below is motivated by the following:

$$\lim_{\Delta t \to 0} \sum_{k=1}^{n} (\Delta_k X + \Delta_k Y)^2$$

$$= \lim_{\Delta t \to 0} \sum_{k=1}^{n} (\Delta_k X)^2 + 2 \lim_{\Delta t \to 0} \sum_{k=1}^{n} (\Delta_k X)(\Delta_k Y) + \lim_{\Delta t \to 0} \sum_{k=1}^{n} (\Delta_k Y)^2,$$

which is obtained by applying the binomial formula and shows that the covariation can be defined by quadratic variations.

*Property 6.13.* **(Covariation Expressed in Terms of Quadratic Variations)** Let $X = \{X_t\}$ and $Y = \{Y_t\}$ be two processes with finite quadratic variations. Then

$$[X,Y]_t = \frac{1}{2}([X + Y]_t - [X]_t - [Y]_t).$$

Again, let $X = \{X_t\}$ and $Y = \{Y_t\}$ be two processes. As a motivation for the next property, let us verify the following identity:

$$\Delta(X_t Y_t) = X_t \Delta Y_t + Y_t \Delta X_t + \Delta X_t \Delta Y_t, \tag{6.38}$$

where $\Delta X_t, \Delta Y_t$, and $\Delta(X_t Y_t)$ are defined as the corresponding increments over time interval $[t, t + \Delta t]$. In fact,

$$\text{LHS} = X_{t+\Delta t} Y_{t+\Delta t} - X_t Y_t,$$
$$\text{RHS} = X_t (Y_{t+\Delta t} - Y_t) + Y_t (X_{t+\Delta t} - X_t) + (X_{t+\Delta t} - X_t)(Y_{t+\Delta t} - Y_t).$$

A verification of LHS = RHS is straightforward.

The infinitesimal version of (6.38) is a useful *product rule*:

$$d(X_t Y_t) = X_t dY_t + Y_t dX_t + dX_t dY_t \tag{6.39}$$

holds under certain condition (e.g., both $X$ and $Y$ are Itô processes, which will be introduced later) and has an equivalent form:

$$d(X_t Y_t) = X_t dY_t + Y_t dX_t + d[X,Y]_t. \tag{6.40}$$

A proof of the product rule can be done by applying the two-dimensional Itô's lemma (see Exercises 6.32 and 6.33 on page 326). For this the product rule is also referred to as the *Itô product rule*.

If $f$ is a deterministic function, we denote $[f]_a^b$ the quadratic variation of $f$ over interval $[a,b]$. That is, $[f]_a^b \equiv [f]_t$, $t \in [a,b]$.

We will make use of the next two properties.

*Property 6.14.* Let $f$ be a continuous (deterministic) function on $[0,T]$. If $f$ has finite total variation, then $[f]_a^b$, the quadratic variation of $f$, is identically equal to zero on $[a,b]$. That is,

$$[f]_a^b \equiv 0. \tag{6.41}$$

In words, the quadratic variation of a continuous $f$ is identically equal to zero.

*Proof.* A proof is provided in Remark 6.9 on page 297. □

*Property 6.15.* Let $f$ be a continuous (deterministic) function and $X = \{X_t\}$ be a sample-continuous process on $[0, T]$ (see Definition 6.10 on page 264). Then

$$[X, f]_t \equiv 0, \qquad t \in [0, T].$$

*Proof.* A proof is provided in the Remark 6.10 on page 297. □

**Example 6.32.** Let $X = \mathfrak{B}$, a standard Brownian motion, and $f(t) = t$. It follows from Property 6.15 that

$$[\mathfrak{B}, t]_t \equiv 0, \quad t \geq 0.$$

Using the notation $d[X, Y]_t = (dX_t)(dY_t)$ defined by (6.37), we write

$$d[\mathfrak{B}, t]_t = d\mathfrak{B}(t)\, dt$$

(keep in mind: $\mathfrak{B}(t) \equiv \mathfrak{B}_t$ by the notational remark on page 264) and obtain

$$d\mathfrak{B}(t)\, dt = 0. \tag{6.42}$$

□

As a good exercise, the reader is encouraged to derive identity (6.42) by directly using the definition of covariation.

**Example 6.33.** Compute $d(e^{-t^2+t}S(t))$ given $dS(t) = 0.2\, dt + 0.095\, d\mathfrak{B}(t)$.

**Solution.** Let $f(t) = e^{-t^2+t}$.

Applying the product rule provided by (6.40), we have

$$
\begin{aligned}
d(e^{-t^2+t}S(t)) &= e^{-t^2+t}dS(t) + d(e^{-t^2+t})S(t) + d[S, f]_t \\
&= e^{-t^2+t}dS(t) + (-2t+1)e^{-t^2+t}S(t)dt + 0 \\
&= e^{-t^2+t}(0.2\, dt + 0.095\, d\mathfrak{B}(t)) + (-2t+1)e^{-t^2+t}S(t)dt \\
&= (0.2 + (-2t+1)S(t))e^{-t^2+t}dt + 0.095e^{-t^2+t}d\mathfrak{B}(t).
\end{aligned}
$$

□

### 6.6.3 Quadratic Variation and Covariation Properties of BM

*Property 6.16.* (**Quadratic Variation Property of Brownian Motion**) Let $\mathfrak{B}$ be a standard Brownian motion on a probability space $(\Omega, \mathfrak{F}, \mathbb{P})$ (i.e., $\mathfrak{B} = \{\mathfrak{B}_t, t \geq 0\}$ is a $\mathbb{P}$-standard Brownian motion). Then

$$[\mathfrak{B}]_t = t. \tag{6.43}$$

Note that the equality (6.43) can be expressed by

$$(d\mathfrak{B}_t)^2 = dt \quad \text{or} \quad d\mathfrak{B}_t^2 = dt. \tag{6.44}$$

*Proof.* It is sufficient to show that for partitions $\mathbf{P}_n : 0 = t_0 < t_1 < t_2 < \cdots < t_n = t$ on the interval $[0,t]$ with $t_k = \frac{kt}{n}$, $k = 0,1,\ldots,n-1$, $n > 0$,

$$\lim_{|\mathbf{P}_n| \to 0} \sum_{k=1}^{n} (\mathfrak{B}_{t_k} - \mathfrak{B}_{t_{k-1}})^2 = \lim_{n \to \infty} \sum_{k=0}^{n-1} (\mathfrak{B}_{t_k} - \mathfrak{B}_{t_{k-1}})^2 = t,$$

where the limit is taken in the mean square sense (see Definition 6.13 and Property 6.1, 2 on page 265).

We let $\Delta\mathfrak{B}_k = \mathfrak{B}_{t_{k+1}} - \mathfrak{B}_{t_k}$, and $\Delta t = t_{k+1} - t_k$, $k = 0,1,\ldots,n-1$, and

$$S_n = \sum_{k=0}^{n-1} (\mathfrak{B}_{t_k} - \mathfrak{B}_{t_{k-1}})^2 = \sum_{k=0}^{n-1} \Delta\mathfrak{B}_k^2.$$

Since

$$\mathbb{E}(S_n) = \mathbb{E}\left(\sum_{k=0}^{n-1} \Delta\mathfrak{B}_k^2\right) = \sum_{k=0}^{n-1} \mathbb{E}(\Delta\mathfrak{B}_k^2) = \sum_{k=0}^{n-1} \mathrm{Var}(\Delta\mathfrak{B}_k) = \sum_{k=0}^{n-1} \Delta t = t,$$

we have

$$\mathbb{E}((S_n - t)^2) = \mathrm{Var}(S_n) = \sum_{k=0}^{n-1} \mathrm{Var}(\Delta\mathfrak{B}_k^2) = \sum_{k=0}^{n-1} [\mathbb{E}(\Delta\mathfrak{B}_k^4) - (\mathbb{E}(\Delta\mathfrak{B}_k^2))^2]$$

$$= \sum_{k=0}^{n-1} [(\mathrm{Var}(\Delta\mathfrak{B}_k))^2 \mathrm{kurt}(\Delta\mathfrak{B}_k) - (\mathrm{Var}(\Delta\mathfrak{B}_k))^2]$$

$$= \sum_{k=0}^{n-1} [(\mathrm{Var}(\Delta\mathfrak{B}_k))^2 (3-1)] = 2\sum_{k=0}^{n-1} \Delta t^2 = 2\frac{t^2}{n} \to 0 \quad \text{as } n \to \infty.$$

Since the convergence in mean square implies the convergence in probability, we have proved $[\mathfrak{B}]_t = t$ indeed. $\qquad\square$

➢ It follows from $[\mathfrak{B}]_t = t$ and Property 6.14 that with probability 1, sample paths of Brownian motion do not have bounded variation on $[0,t]$ (nor on any interval by the time-homogenous property of Brownian motion).

In fact, we have just proved Property 6.9, 3 on page 284.

Now, consider two standard Brownian motions

$$\mathfrak{B}^{(i)} = \{\mathfrak{B}_t^{(i)}, t \geq 0\}, \quad i = 1,2,$$

which are defined on the same filtered probability space $(\Omega, \mathfrak{F}, \{\mathcal{F}_t\}, \mathbb{P})$ and adapted to the filtration $\{\mathcal{F}_t\}$. Let

$$\Delta \mathfrak{B}^{(i)} = \mathfrak{B}_t^{(i)} - \mathfrak{B}_s^{(i)}, \quad \text{where } t > s \text{ and } i = 1, 2. \tag{6.45}$$

We say that $\mathfrak{B}^{(1)}$ and $\mathfrak{B}^{(2)}$ have correlation $\rho$ if, for $\forall\, t > s \geq 0$,

1. $\operatorname{Cov}\left(\Delta \mathfrak{B}^{(1)}, \Delta \mathfrak{B}^{(2)}\right) = \rho \Delta t$,
2. $\Delta \mathfrak{B}^{(1)}$ and $\Delta \mathfrak{B}^{(2)}$ are independent of $\mathcal{F}_s$, and
3. $\Delta \mathfrak{B}^{(1)}$ and $\Delta \mathfrak{B}^{(2)}$ have a *bivariate normal distribution*.[19]

*Property 6.17.* **(Quadratic Covariation Property of Brownian Motion)** If $\mathfrak{B}^{(1)}$ and $\mathfrak{B}^{(2)}$ have correlation $\rho$, then

$$[\mathfrak{B}^{(1)}, \mathfrak{B}^{(2)}]_t = \rho t. \tag{6.46}$$

Note that (6.46) can be expressed by

$$(d\mathfrak{B}^{(1)})(d\mathfrak{B}^{(2)}) = \rho\, dt \quad \text{or} \quad d\mathfrak{B}^{(1)}\, d\mathfrak{B}^{(2)} = \rho\, dt. \tag{6.47}$$

*Proof.* Let

$$X = a(\mathfrak{B}^{(1)} + \mathfrak{B}^{(2)}) \quad \text{with} \quad a^2 = \frac{1}{2 + 2\rho}.$$

Thus, $\frac{1}{2a^2} = 1 + \rho$.

**Step 1.** We verify that $X$ is a standard Brownian motion as follows:

Let $\Delta X = a(\Delta \mathfrak{B}^{(1)} + \Delta \mathfrak{B}^{(2)})$, where $\Delta \mathfrak{B}^{(i)}$ are defined in (6.45). We establish

$$\begin{aligned}
\operatorname{Var}(\Delta X) &= \mathbb{E}(\Delta X^2) = \mathbb{E}(a^2(\Delta \mathfrak{B}^{(1)} + \Delta \mathfrak{B}^{(2)})^2) \\
&= a^2 \mathbb{E}((\Delta \mathfrak{B}^{(1)})^2 + 2(\Delta \mathfrak{B}^{(1)})(\Delta \mathfrak{B}^{(2)}) + (\Delta \mathfrak{B}^{(2)})^2) \\
&= a^2 [\mathbb{E}((\Delta \mathfrak{B}^{(1)})^2) + 2\mathbb{E}((\Delta \mathfrak{B}^{(1)})(\Delta \mathfrak{B}^{(2)})) + \mathbb{E}((\Delta \mathfrak{B}^{(2)})^2)] \\
&= a^2 [\operatorname{Var}(\Delta \mathfrak{B}^{(1)}) + 2 \operatorname{Cov}(\Delta \mathfrak{B}^{(1)}, \Delta \mathfrak{B}^{(2)}) + \operatorname{Var}(\Delta \mathfrak{B}^{(2)})] \\
&= a^2 [t - s + 2\rho(t - s) + t - s] \\
&= \frac{1}{2 + 2\rho}(2 + 2\rho)(t - s) = t - s,
\end{aligned}$$

$$\begin{aligned}
\mathbb{E}(\Delta X) &= a(\mathbb{E}(\Delta \mathfrak{B}^{(1)}) + \mathbb{E}(\Delta \mathfrak{B}^{(2)})) = 0, \\
X_0 &= a(\mathfrak{B}_0^{(1)} + \mathfrak{B}_0^{(2)}) \stackrel{\text{a.s.}}{=} 0.
\end{aligned}$$

By Property 6.16, $[X]_t = t$.

**Step 2.** Applying Property 6.12, the symmetry and bilinearity of covariation process yield

---

[19] The random $n$-vector $\mathbf{X}^{\mathsf{T}} = [X_1\ X_2 \dots X_n]$ is (or the random variables $X_1, X_2, \dots, X_n$ are) said to have a *multivariate normal distribution* if and only if all linear combinations of $X_1, X_2, \dots, X_n$ are normally distributed. When $n = 2$, $\mathbf{X}^{\mathsf{T}}$ is said to have a *bivariate normal distribution*.

$$t = [X]_t = [X, X]_t = [a(\mathfrak{B}^{(1)} + \mathfrak{B}^{(2)}), a(\mathfrak{B}^{(1)} + \mathfrak{B}^{(2)})]_t$$
$$= a[\mathfrak{B}^{(1)}, a(\mathfrak{B}^{(1)} + \mathfrak{B}^{(2)})]_t + a[\mathfrak{B}^{(2)}, a(\mathfrak{B}^{(1)} + \mathfrak{B}^{(2)})]_t$$
$$= a^2([\mathfrak{B}^{(1)}, \mathfrak{B}^{(1)}]_t + 2[\mathfrak{B}^{(1)}, \mathfrak{B}^{(2)}]_t + [\mathfrak{B}^{(2)}, \mathfrak{B}^{(2)}]_t)$$
$$= a^2([\mathfrak{B}^{(1)}]_t + 2[\mathfrak{B}^{(1)}, \mathfrak{B}^{(2)}]_t + [\mathfrak{B}^{(2)}]_t)$$
$$= a^2(t + 2[\mathfrak{B}^{(1)}, \mathfrak{B}^{(2)}]_t + t) = 2a^2(t + [\mathfrak{B}^{(1)}, \mathfrak{B}^{(2)}]_t).$$

Solving for $[\mathfrak{B}^{(1)}, \mathfrak{B}^{(2)}]_t$, we obtain

$$[\mathfrak{B}^{(1)}, \mathfrak{B}^{(2)}]_t = \frac{t}{2a^2} - t = t\left(\frac{1}{2a^2} - 1\right) = t(1 + \rho - 1) = \rho t.$$

$\square$

Clearly, if $\mathfrak{B}^{(1)}$ and $\mathfrak{B}^{(2)}$ are uncorrelated, then $[\mathfrak{B}^{(1)}, \mathfrak{B}^{(2)}]_t = 0$.

For future convenience, we summarize the results represented by the identities (6.41), (6.42), (6.44), and (6.47) in the following multiplication table for Brownian motion variation and covariation:

|                    | $dt$ | $d\mathfrak{B}^{(1)}$ | $d\mathfrak{B}^{(2)}$ |
|--------------------|------|-----------------------|-----------------------|
| $dt$               | 0    | 0                     | 0                     |
| $d\mathfrak{B}^{(1)}$ | 0    | $dt$                  | $\rho\, dt$           |
| $d\mathfrak{B}^{(2)}$ | 0    | $\rho\, dt$           | $dt$                  |

In fact, the multiplication table can be enlarged to any size we wish. More specifically, given that the correlation of Brownian motions $\mathfrak{B}^{(i)}$ and $\mathfrak{B}^{(j)}$ is $\rho_{i,j}$, $i, j = 1, 2, 3$, we can establish a table of size $5 \times 5$ as below:

|                    | $dt$ | $d\mathfrak{B}^{(1)}$ | $d\mathfrak{B}^{(2)}$ | $d\mathfrak{B}^{(3)}$ |
|--------------------|------|-----------------------|-----------------------|-----------------------|
| $dt$               | 0    | 0                     | 0                     | 0                     |
| $d\mathfrak{B}^{(1)}$ | 0    | $dt$                  | $\rho_{1,2}\, dt$     | $\rho_{1,3}\, dt$     |
| $d\mathfrak{B}^{(2)}$ | 0    | $\rho_{1,2}\, dt$     | $dt$                  | $\rho_{2,3}\, dt$     |
| $d\mathfrak{B}^{(3)}$ | 0    | $\rho_{1,3}\, dt$     | $\rho_{2,3}\, dt$     | $dt$                  |

Such tables function like a handy computational tool in studying stochastic models involving more than one random source.

**Example 6.34. (Interpretation of the Multiplication Product Table)** In the $(2,2)$ entry of the table,

$$(dt)^2 = d[\, t\, ]_t \equiv 0$$

by Property 6.14, which states that $[f]_t \equiv 0$ if $f$ is a continuous (deterministic) function (in our case, $f(t) = t$).                    $\square$

## 6.6.4 *Significance of Quadratic Variation*

The following three remarks provide some insights into the concept of quadratic variation and are for the interested reader.

**Remark 6.9.** Recall from (6.33) on page 289 that *the total variation* of a real-valued function on interval is denoted $V_a^b(f)$ and defined as

$$V_a^b(f) = \lim_{|\mathbf{P}| \to 0} \sum_{k=0}^{n-1} |f(x_{k+1}) - f(x_k)|,$$

where $\mathbf{P} : a = x_0 < x_1 < x_2 < \cdots < x_n = b$ is a partition on $[a,b]$.

Let $f$ be a continuous function on $[a,b]$. We claim that if $f$ has finite total variation, then $[f]_a^b$, the quadratic variation of $f$, is identically equal to zero on $[a,b]$ (this is why quadratic variation is not defined for deterministic functions).

In fact, $V_a^b(f)$ being finite implies that $\exists\, M > 0$ such that

$$\lim_{|\mathbf{P}| \to 0} \sum_{k=0}^{n-1} |f(x_{k+1}) - f(x_k)| \leq M.$$

Since $f$ being continuous on a closed interval $[a,b]$ implies the uniform continuity of $f$ on $[a,b]$: $\forall\, \varepsilon > 0$, $\exists\, N > 0$ such that $|f(x_{k+1}) - f(x_k)| < \varepsilon$ whenever $|x_{k+1} - x_k| < \frac{1}{N}$. Consequently,

$$\sum_{k=0}^{n-1} (f(x_{k+1}) - f(x_k))^2 < \varepsilon \sum_{k=0}^{n-1} |f(x_{k+1}) - f(x_k)| \leq \varepsilon M,$$

which implies that

$$0 < [f]_a^x \leq [f]_a^b \leq \lim\sup \sum_{k=0}^{n-1} (f(x_{k+1}) - f(x_k))^2 \leq \varepsilon M \to 0.$$

Thus, $[f]_a^x \equiv 0$ where $a \leq x \leq b$. $\qquad\qquad\square$

**Remark 6.10.** A proof of Property 6.15 on page 293 is provided below for the interested reader.

Since for each fixed $\omega$, sample path $X(t,\omega) = X(t)$ of $X$ being continuous on a closed interval $[0,T]$ implies $X(t)$ being absolutely continuous on $[0,T]$, $\forall \varepsilon > 0$, $\exists\, N > 0$, such that

$$|X(t_{k+1}) - X(t_k)| < \varepsilon \quad \text{whenever} \quad |t_{k+1} - t_k| < \frac{1}{N}.$$

Consequently,

$$\left| \sum_{k=0}^{n-1} (X(t_{k+1}) - X(t_k))(f(t_{k+1}) - f(t_k)) \right| < \varepsilon \sum_{k=0}^{n-1} |f(t_{k+1}) - f(t_k)| \leq \varepsilon V_0^T(f),$$

where $V_0^T(f)$ is the total variation of $f$ on $[0, T]$, which is finite due to the continuity of $f$. It follows that

$$0 \leq |[X, f]_t| \leq \varepsilon \lim \sum_{k=0}^{n-1} |f(x_{k+1}) - f(x_k)| \leq \varepsilon V_0^T(f) \to 0.$$

Thus, $[X, f]_t = 0$.                                                                       □

**Remark 6.11.** Since a stochastic differential equation (s.d.e.) is defined by a stochastic integration equation (s.i.e.) which we have not yet defined, our discussion here can only be formal. However, it may help to immediately satisfy the curiosity of those who are wondering about the significance of quadratic variation.

Loosely speaking, the local dynamics of a stochastic process driven by Brownian motion over time interval $[t, t + \Delta t]$ can be represented by approximations in the form below

$$\Delta(X(t)) = \mu(t) \Delta t + \sigma(t) \Delta \mathfrak{B}(t), \tag{6.48}$$

where $\{\mu(t)\}$ is an adapted drift process, $\{\sigma(t)\}$ is an adapted volatility process, and $\Delta \mathfrak{B}(t)$ causes fluctuations. The adaptedness refers to Brownian filtration (the set of information of the past history of the Brownian motion $\mathfrak{B}$).

The infinitesimal version of (6.48) is

$$dX(t) = \mu(t) dt + \sigma(t) d\mathfrak{B}(t). \tag{6.49}$$

When both $\mu(t)$ and $\sigma(t)$ are functions of $X(t)$ only, we replace $\mu(t)$ and $\sigma(t)$ by $\mu(X(t))$ and $\sigma(X(t))$ respectively and have

$$dX(t) = \mu(X(t)) dt + \sigma(X(t)) d\mathfrak{B}(t), \tag{6.50}$$

which defines, in loose terms, a *diffusion process*.

Note that (6.27) on page 279 is a special case of (6.50) when both $\mu(X(t))$ and $\sigma(X(t))$ are constant, and (6.50) is a special case of (6.49).

The corresponding s.i.e. to s.d.e. (6.49) can be expressed by

$$X(t) = X(t, \omega) = \int_0^t \mu(s, \omega) ds + \int_0^t \sigma(s, \omega) d\mathfrak{B}(s)$$
$$= \text{f.t.v. term} + \text{i.t.v. term},$$

where f.t.v. and i.t.v. stand for finite total variation and infinite total variation respectively. The second term on the right labeled by the infinite total variation term is because, with probability 1, the total variation of a Brownian path on any interval no matter how small is infinite. It is the infinite total variation term that complicates our interpretation of the limiting procedure in the ordinary sense.

A straightforward computation shows that for a diffusion process $X$ (i.e., for $X$ satisfying diffusion equation (6.50))

$$[X]_t = \int_0^t \sigma^2(s)\,ds. \tag{6.51}$$

This is why we impose the condition of $\sigma^2(t) = \sigma^2(X(t))$ in the diffusion equation being square integrable.                                                   □

## 6.7 Itô Integral: A Brief Introduction

### 6.7.1 Importance of Itô Integral with Respect to BM

Stochastic integrations may be taken with respect to different stochastic processes. For both theoretical and practical considerations, we will introduce only the one with respect to the standard Brownian motion:

➤ Brownian motion is theoretically important because its sample paths are (almost surely) continuous but nowhere differentiable (see Billingsley [3]; Durrett [9]; and Paley, Wiener, and Zygmund [25]).

➤ The integration with respect to Brownian motion is practically important because $\{d\mathcal{B}(t)\}$ is a white noise process (see Definition 6.25 on page 280).[20]

### 6.7.2 Basic Concepts

To avoid involving too many technicalities, we will approach the concept of the Itô integral with respect to standard Brownian motion by emphasizing the key idea of the definition under which certain computations become easier and the integral of a stochastic process will produce a martingale.

The mathematical rigor in the definition of a stochastic integral can be perfected when we display a similar drill to the familiar one in Riemann integrals over interval $[a,b]$.[21]

---

[20] *Brownian noise* (also known as *brown noise* or *red noise*) is the kind of signal noise produced by Brownian motion. Naturally, it is also called *random walk noise* as a Brownian motion can be viewed as a limit of random walks. Note that a random walk noise is not a white noise (see Example 6.27).

[21] That is, following an approximation procedure: *Step 1.* Divide the interval into finitely many subintervals (the partition). *Step 2.* Construct a simple function (use step functions for intuition) that has a constant value on each of the subintervals of the partition (the upper and lower sums). *Step 3.* Define integrals of simple functions (simple processes are random step functions). *Step 4.* Take the limit of these simple functions as more and more dividing points are added to the partition. If the limit exists,

Just as the indefinite integral defines a function in the deterministic calculus, the Itô integral defines a stochastic process.

Let $\mathfrak{B} = \{\mathfrak{B}_t, \, t \geq 0\}$ be the standard Brownian motion adapted to its natural filtration $\{\mathcal{F}_t\}$ and keep in mind that "independent increments" mean that for $\forall \, s, t$ with $s < t$, the random variable $\mathfrak{B}_t - \mathfrak{B}_s$ is independent of $\mathcal{F}_s$.

Let $f = \{f_t, \, t \geq 0\}$ be a stochastic process to integrate and adapted to the same filtration $\{\mathcal{F}_t\}$, where the idea of the adaptedness requirement is that $f$ is allowed to depend on the history of $\mathfrak{B}$ but not on the future development of $\mathfrak{B}$. That is, $f_s$, $s < t$ do not contain any information of future increments $\mathfrak{B}_t - \mathfrak{B}_s$, $t > s$. Consequently,

$$\mathbb{E}(f_s(\mathfrak{B}_t - \mathfrak{B}_s)|\mathcal{F}_s) = f_s \mathbb{E}(\mathfrak{B}_t - \mathfrak{B}_s|\mathcal{F}_s) = f_s \mathbb{E}(\mathfrak{B}_t - \mathfrak{B}_s) = 0,$$

where the first equal sign holds due to a property of conditional expectation (taking out $f_s$, which is known), the second equal sign holds due to the Markov property of Brownian motion, and the last equal sign holds due to the definition of Brownian motion. Thus, if we let $d\mathfrak{B}_s = \mathfrak{B}_{s+ds} - \mathfrak{B}_s$ (where $ds > 0$ is considered to be infinitesimal), then

$$\mathbb{E}(f_s \, d\mathfrak{B}_s|\mathcal{F}_s) = 0.$$

To introduce the Itô integral with respect to Brownian motion $\mathfrak{B}$, we begin with dividing interval $[0, t]$ into $n$ subintervals $0 = t_0 < t_1 < t_2 \cdots < t_n = t$ and forming a sum associated with the partition

$$S_n = \sum_{i=1}^{n} f_{t_{i-1}}(\mathfrak{B}_{t_i} - \mathfrak{B}_{t_{i-1}}).$$

The *Itô integral of the process* $\{f_t\}$ *with respect to Brownian process* $\{\mathfrak{B}_t\}$ is denoted by the process $\{Y_t\}$ and defined by

$$Y_t = \int_0^t f_s \, d\mathfrak{B}_s = \lim_{n \to \infty} S_n, \tag{6.52}$$

where the limit is taken in terms of the mean squared errors:

$$\lim_{n \to \infty} \mathbb{E}((Y_t - S_n)^2) = 0.$$

$\{Y_t\}$ is called a process *driven by Brownian motion* $\{\mathfrak{B}_t\}$ *and transformed by the integrand process* $\{f_t\}$.

Before considering computational details, let us understand (6.52) conceptually:

---

it is called the Riemann integral and the function is called the Riemann integrable (Itô integrability requires convergence in mean square).

Note that a *simple function* is a finite linear combination of indicator functions of measurable sets. Thus all step functions are simple functions.

1. $S_n$ by definition is a random variable for each $n$, so is $Y_t$ for each $t$. Just as an indefinite integral transforms an integrand function to another function in the deterministic calculus, an Itô integral transforms an integrand process to another process in the stochastic calculus. One application of Itô integrals is to construct a new process.

2. An important property of Itô integrals with respect to Brownian motion is that

   ➤ $\{Y_t\}$ is a martingale.

   In fact, for $\forall\, a,b \in [0,t]$ with $b > a$, if $a$ and $b$ are not among the dividing points $t_i$ in the partition, then we can insert them into the sequence of dividing points to obtain that

$$\mathbb{E}(Y_b - Y_a | \mathcal{F}_a) = \mathbb{E}\left( \int_a^b f_s \, d\mathcal{B}_s | \mathcal{F}_a \right)$$

$$= \int_a^b \mathbb{E}(f_s \, d\mathcal{B}_s | \mathcal{F}_a) = \int_a^b \mathbb{E}(\mathbb{E}(f_s \, d\mathcal{B}_s | \mathcal{F}_s) | \mathcal{F}_a)$$

$$= \int_a^b \mathbb{E}(0 | \mathcal{F}_a) = 0.$$

3. In a more general version of stochastic integrals, the driven process does not have to be Brownian motion. In that case, from the argument in item 2, we see that $\{Y_t\}$ is always a martingale as long as the driven process is a martingale.

### 6.7.3 A Famous Example

The example below is well known.

**Example 6.35.** To calculate $\int_0^t \mathcal{B}_s \, d\mathcal{B}_s$, we divide interval $[0,t]$ into $n$ subintervals with equal length $\Delta t = \frac{t}{n}$,

$$\frac{1}{2}\mathcal{B}_t^2 = \frac{1}{2}(\mathcal{B}_t^2 - \mathcal{B}_0^2) = \frac{1}{2}\sum_{i=1}^n (\mathcal{B}_{i\Delta t}^2 - \mathcal{B}_{(i-1)\Delta t}^2)$$

$$= \frac{1}{2}\sum_{i=1}^n (\mathcal{B}_{i\Delta t} - \mathcal{B}_{(i-1)\Delta t})(\mathcal{B}_{i\Delta t} + \mathcal{B}_{(i-1)\Delta t})$$

$$= \frac{1}{2}\sum_{i=1}^n (\mathcal{B}_{i\Delta t} - \mathcal{B}_{(i-1)\Delta t})(\mathcal{B}_{i\Delta t} - \mathcal{B}_{(i-1)\Delta t} + 2\mathcal{B}_{(i-1)\Delta t})$$

$$= \frac{1}{2}\sum_{i=1}^n (\mathcal{B}_{i\Delta t} - \mathcal{B}_{(i-1)\Delta t})^2 + \sum_{i=1}^n (\mathcal{B}_{i\Delta t} - \mathcal{B}_{(i-1)\Delta t})\mathcal{B}_{(i-1)\Delta t}$$

$$= I_1 + I_2$$

$$\rightarrow \frac{1}{2}[\mathcal{B}]_t + \int_0^t \mathcal{B}_s \, d\mathcal{B}_s$$

$$= \frac{t}{2} + \int_0^t \mathcal{B}_s \, d\mathcal{B}_s,$$

where the convergence as $n \to \infty$ is in probability,

$$I_1 = \frac{1}{2} \sum_{i=1}^{n} (\mathfrak{B}_{i\Delta t} - \mathfrak{B}_{(i-1)\Delta t})^2,$$

$$I_2 = \sum_{i=1}^{n} (\mathfrak{B}_{i\Delta t} - \mathfrak{B}_{(i-1)\Delta t}) \mathfrak{B}_{(i-1)\Delta t}.$$

We obtain

$$\int_0^t \mathfrak{B}_s \, d\mathfrak{B}_s = \frac{1}{2} \mathfrak{B}_t^2 - \frac{t}{2}.$$

Equivalently,

$$d(\mathfrak{B}_t^2) = 2\mathfrak{B}_t \, d\mathfrak{B}_t + dt.$$

$\square$

## 6.8 Itô's Formula for Brownian Motion

### 6.8.1 Itô Processes

**Definition 6.30.** A one-dimensional *Itô process* is a stochastic process $X = \{X(t)\}$ defined on a probability space $(\Omega, \mathfrak{F}, \mathbb{P})$ and has an expression

$$X(t) = X(0) + \int_0^t \mu(X(s),s) \, ds + \int_0^t \sigma(X(s),s) \, d\mathfrak{B}(s), \quad 0 \le t \le T, \quad (6.53)$$

where $\{\mu(X(t),t)\}$ is an adapted *drift process*, and $\{\sigma(X(t),t)\}$ is an adapted *volatility process* and square integrable. The adaptedness refers to the Brownian filtration (the set of information of the past history of the Brownian motion $\mathfrak{B}$).

An Itô process $X = \{X(t)\}$ is said to be an *Itô diffusion* if both drift process and volatility process are functions of $X(t)$ only. That is, $X$ has the expression

$$X(t) = X(0) + \int_0^t \mu(X(s)) \, ds + \int_0^t \sigma(X(s)) \, d\mathfrak{B}(s), \quad 0 \le t \le T, \quad (6.54)$$

where $\mu(X(t))$ and $\sigma(X(t))$ are also known as the *drift coefficient* and *diffusion coefficient* of $X$, respectively.

➤ As a *shorthand notation*, we write stochastic integral equation (6.53) in the form of the stochastic differential equation:

$$dX(t) = \mu(X(t),t) \, dt + \sigma(X(t),t) \, d\mathfrak{B}(t), \quad 0 \le t \le T. \quad (6.55)$$

## Remark 6.12.

1. An Itô process is a process of a sum of the initial state of the process and two integrals:

$$X(t,\omega) = X(0,\omega) + I_1(t,\omega) + I_2(t,\omega), \quad 0 \le t \le T, \quad \text{where}$$

$$I_1(t,\omega) = \int_0^t \mu(X(s,\omega),s)\,ds, \qquad 0 \le t \le T,$$

$$I_2(t,\omega) = \int_0^t \sigma(X(s,\omega),s)\,d\mathfrak{B}(s), \quad 0 \le t \le T.$$

For each fixed $\omega$, $I_1$ is an ordinary integral and $I_2$ an Itô integral. For (6.53) to be well defined, $\mu(t,X(t))$ must be integrable in the ordinary sense, and $\sigma(X(t),t)$ must be integrable in the stochastic sense as we defined in Section 6.7.2. The square integrability of $\sigma^2(X(t),t)$ ensures that the quadratic variation of the process $X$ is finite. In fact, a straightforward calculation shows that

$$[X]_t = \int_0^t \sigma^2(X(u),u)\,du.$$

The detailed verification is left as an exercise for the reader.

2. Although the form of stochastic differential equations has the advantage of being intuitive in modern financial theory, stochastic differential equations acquire mathematical meanings only through their corresponding stochastic integral equations. Equation (6.55) alone is not well defined, and dividing both sides of the equation by $dt$ is forbidden because the Brownian path is non-differentiable.[22]

□

**Example 6.36.** Both familiar processes $X$ and $S$ with constants $\mu$ and $\sigma$ on page 278 defined by

$$dX(t) = \mu\,dt + \sigma\,d\mathfrak{B}(t),$$

$$dS(t) = \mu S(t)\,dt + \sigma S(t)\,d\mathfrak{B}(t),$$

respectively, are Itô diffusions.

➤ We emphasize that both s.d.e.'s should be understood as defined by their corresponding s.i.e.'s:

$$X(t) = X(0) + \int_0^t \mu\,dt + \int_0^t \sigma\,d\mathfrak{B}(s),$$

$$S(t) = S(0) + \int_0^t \mu S(u)\,du + \int_0^t \sigma S(u)\,d\mathfrak{B}(u).$$

□

---

[22] With probability 1 the Brownian path is non-differentiable in the ordinary sense, but in the context of *generalized stochastic process*, $\frac{d\mathfrak{B}}{dt}$ is well defined as a *generalized function* on an infinite dimensional space, which is a topic beyond the scope of this book.

### 6.8.2 Itô's Lemma for Brownian Motion

Ito's lemma is the tool of the trade in continuous-time stochastic process modeling.

**Theorem 6.1. (Itô's Lemma)** *Let $f(x,t)$ be a function that is continuously differentiable in $t$ and twice continuously differentiable in $x$.*

*Let $X = \{X(t)\}$ be an Itô process represented by the s.d.e.*

$$dX(t) = \mu(X(t),t)\, dt \; + \; \sigma(X(t),t)\, d\mathcal{B}(t), \quad 0 \le t \le T. \tag{6.56}$$

*Define a process $Y = \{Y(t)\}$ by $Y(t) = f(X(t),t)$, $0 \le t \le T$.*

*Then $Y = \{Y(t)\}$ is an Itô process that has a s.d.e. representation with*

$$dY(t) = \left( \frac{\partial f}{\partial t}(X(t),t) + \mu(X(t),t)\frac{\partial f}{\partial x}(X(t),t) + \frac{1}{2}\sigma^2(X(t),t)\frac{\partial^2 f}{\partial x^2}(X(t),t) \right) dt$$

$$+ \sigma(t)\frac{\partial f}{\partial x}(X(t),t)\, d\mathcal{B}(t). \tag{6.57}$$

In less detail, formula (6.57), which is referred to as *Itô's formula*, can be written into

$$dY = \left( f_t + \mu f_x + \frac{1}{2}\sigma^2 f_{xx} \right) dt + \sigma f_x\, d\mathcal{B}. \tag{6.58}$$

An informal proof of the case when $X = \mathcal{B}$ (i.e., $\mu = 0$ and $\sigma = 1$) will be given shortly (see Example 6.39 on page 307). The idea of the proof for the general case is similar but with many more tedious computational details.

**Remark 6.13.**

1. Note that $f(x,t)$ in Theorem 6.1 is a smooth function, and that

$$\mu_Y = f_t + \mu f_x + \frac{1}{2}\sigma^2 f_{xx}$$
$$\sigma_Y = \sigma f_x$$

   are the drift and volatility processes, respectively, for the new process $Y$.

2. In words, Itô's lemma says that Itô processes are stable under smooth maps in the sense that any smooth function maps (sends) an Itô process $X$ in terms of its driven process $\mathcal{B}$, drift $\mu_X$, and volatility coefficient $\sigma_X$ to another Itô process in terms of its driven process $\mathcal{B}$, drift $\mu_Y$, and volatility coefficient $\sigma_Y$. In short, a smooth function of an Itô process is an Itô process.

3. There are different versions of Itô's lemma (e.g., Itô's lemma for jump-diffusion processes, whereas Theorem 6.1 for Brownian motions), which are widely employed in modern financial theory. The best known application of Itô's lemma is in the derivation of the Black-Scholes-Merton equation for option values, which will be introduced in a later chapter.

□

**Example 6.37.** Suppose that a stock price is modeled by a process $S = \{S(t)\}$ where

$$S(t) = e^{\mu t + \sigma \mathcal{B}(t)}$$

with $\mu \neq 0$ and $\sigma > 0$ being constant. What is the expected growth rate of the stock at any given time $t \leq T$?

**Solution 1.**

*Step 1.* Identify the given process $X$ and smooth function $f$ in Itô's formula:

$$X(t) = \mathcal{B}(t) \text{ (thus, } \mu_X = 0, \quad \sigma_X = 1) \quad \text{and} \quad f(x,t) = e^{\mu t + \sigma x},$$

where $\mu_X = \mu(X(t),t)$ and $\sigma_X = \sigma(X(t),t)$ as defined in (6.56). Clearly, the conditions of Itô's lemma are satisfied.

*Step 2.* Compute partial derivatives of $f$ in (6.58):

$$f_t = \mu f(\mathcal{B}(t),t) = \mu S(t),$$
$$f_x = \sigma f(\mathcal{B}(t),t) = \sigma S(t),$$
$$f_{xx} = \sigma^2 f(\mathcal{B}(t),t) = \sigma^2 S(t).$$

*Step 3.* Apply Itô's lemma to obtain $dY = dS(t)$:

$$dS(t) = dS = \left( f_t + \mu_X f_x + \frac{1}{2}\sigma_X^2 f_{xx} \right) dt + \sigma_X f_x \, d\mathcal{B}$$

$$= \left( \mu S(t) + 0 + \frac{1}{2}\sigma^2 S(t) \right) dt + \sigma S(t) \, d\mathcal{B}(t)$$

$$= \left( \mu + \frac{1}{2}\sigma^2 \right) S(t) \, dt + \sigma S(t) \, d\mathcal{B}(t).$$

*Step 4.* The answer to the question is $\mu + \frac{1}{2}\sigma^2$.

As a postscript, we point out that in a risk-neutral world, if the stock pays a known dividend yield $q$, then

$$\frac{dS(t)}{S(t)} = \left( \mu + \frac{1}{2}\sigma^2 \right) dt + \sigma \, d\mathcal{B}(t)$$

implies $\mu + \frac{1}{2}\sigma^2 = r - q$. We obtain $\mu = r - q - \frac{1}{2}\sigma^2$.

**Solution 2.**

*Step 1.* Identify the given process $X$ and smooth function $f$ in Itô's lemma:
   Keep s.d.e. (6.56) in mind.

$$dX(t) = \mu_X(X(t),t) \, dt + \sigma_X(X(t),t) \, d\mathcal{B}(t), \quad 0 \leq t \leq T$$

is a shorthand of s.i.e. (6.53):

$$X(t) = X(0) + \int_0^t \mu_X(X(s),s)\,ds + \int_0^t \sigma_X(X(s),s)\,d\mathcal{B}(s), \quad 0 \le t \le T.$$

It is natural to attempt $X(t) = \mu t + \sigma \mathcal{B}(t)$, which is equivalent to

$$X(t) = 0 + \int_0^t \mu\,ds + \int_0^t \sigma\,d\mathcal{B}(s), \quad 0 \le t \le T.$$

Thus, we identify $\mu_X = \mu$ and $\sigma_X = \sigma$ and summarize all we need for applying Itô's lemma below:

$$X(t) = \mu t + \sigma \mathcal{B}(t), \quad \mu_X = \mu, \quad \sigma_X = \sigma, \quad \text{and} \quad f(x,t) = f(x) = e^x.$$

Notice that the conditions of Itô's lemma are satisfied.

*Step 2.* Compute partial derivatives of $f$ in (6.58):

$$f_t = 0, \quad f_x = f_{xx} = f(X(t)) = S(t).$$

*Step 3.* Apply Itô's lemma to obtain $dY = dS(t)$:

$$\begin{aligned}
dS(t) = dS &= \left( f_t + \mu_X f_x + \frac{1}{2}\sigma_X^2 f_{xx} \right) dt + \sigma_X f_x\,d\mathcal{B} \\
&= \left( 0 + \mu S(t) + \frac{1}{2}\sigma^2 S(t) \right) dt + \sigma S(t)\,d\mathcal{B}(t) \\
&= \left( \mu + \frac{1}{2}\sigma^2 \right) S(t)\,dt + \sigma S(t)\,d\mathcal{B}(t).
\end{aligned}$$

*Step 4.* The answer to the question is $\mu + \frac{1}{2}\sigma^2$, which is the same as the one we obtained earlier.                                                                                     □

Using (6.58)

$$dY = \underbrace{\left( f_t + \mu f_x + \frac{1}{2}\sigma^2 f_{xx} \right) dt}_{\mu_Y} \quad + \quad \underbrace{(\sigma f_x)\,d\mathcal{B},}_{\sigma_Y}$$

we have a corollary of Itô's lemma:

**Corollary 6.1.** *Let $\{\mathcal{F}_t\}$ be the Brownian filtration.*
*Let $Y$ be an Itô process expressed by (6.58). Then*

$$\mathbb{E}(dY|\mathcal{F}_t) = \left( f_t + \mu f_x + \frac{1}{2}\sigma^2 f_{xx} \right) dt,$$

$$\text{Var}(dY|\mathcal{F}_t) = (\sigma f_x)^2\,dt.$$

*Proof.* It follows from:

1. Itô's lemma[23] (conditions are satisfied indeed),
2. Properties 6.2, 6.3, 6.6, and 6.11,
3. $d\mathfrak{B} \sim \mathcal{N}(0, dt)$,

that both formulas in Corollary 6.1 hold. □

**Example 6.38.** Suppose that a stock price is modeled by a process $S = \{S(t)\}$ where

$$S(t) = e^{\frac{1}{2}t^2 + 2\mathfrak{B}(t)}.$$

What is the (conditional) expected growth rate of the stock at any given time $t \leq T$?

**Solution 1.**
Since $X(t) = \frac{1}{2}t^2 + 2\mathfrak{B}(t)$ is an Itô process with $\mu_X = t$ and $\sigma_X = 2$,

$$Y = f(X(t)) = e^{\frac{1}{2}t^2 + 2\mathfrak{B}(t)} = S(t), \quad \text{where} \quad f(x) = e^x$$

is also an Itô process. Note that the conditions of Itô's lemma must be verified before the lemma can be applied.

Applying Corollary 6.1, we obtain

$$\mathbb{E}(dS(t)|\mathcal{F}_t) = \mathbb{E}(dY|\mathcal{F}_t) = \left(f_t + \mu f_x + \frac{1}{2}\sigma^2 f_{xx}\right) dt$$
$$= (0 + tS(t) + 2S(t)) dt$$
$$= (2 + t)S(t) dt.$$

The answer to the question is $2 + t$. □

**Solution 2.**
Take the same steps in Example 6.37. The detailed work is left as an exercise to the reader. □

**Example 6.39. (Informal Proof of Itô's Lemma)** Consider the case $\mu = 0, \sigma = 1$ and $X = \mathfrak{B}$.

Let $f(x)$ be a function that is twice continuously differentiable. Then

$$f(x + \Delta x) - f(x) = f'(x)\Delta x + \frac{1}{2}f''(x)(\Delta x)^2 + \sum_{i=1}^{n} o((\Delta x)^2),$$

where the little $o$ is a popular notation in analysis: $o(\xi)$ refers to a quantity such that $\lim_{\xi \to 0} o(\xi)/\xi = 0$. This notation is very convenient as in many cases such a quantity needs not be specified.

We recall the multiplication table:

---

[23] Itô's lemma ensures that $Y$ in (6.58) is an Itô process. Thus, both drift process $\mu_Y$ and volatility process $\sigma_Y$ are adapted to the Brownian filtration by definition.

|         | $dt$ | $d\mathcal{B}_t$ |
|---------|------|------|
| $dt$    | 0    | 0    |
| $d\mathcal{B}_t$ | 0    | $dt$ |

which implies that $(d\mathcal{B}_t)^m \to 0$ as $dt \to 0$ for $m \geq 2$.

For an arbitrary partition on interval $[0,t]$: $0 = t_0 < t_1 < t_2 \cdots < t_n = t$,

$$f(\mathcal{B}_t) - f(\mathcal{B}_s) = \sum_{i=1}^{n} (f(\mathcal{B}_{t_i}) - f(\mathcal{B}_{t_{i-1}}))$$

$$= \sum_{i=1}^{n} f'(\mathcal{B}_{t_i})(\mathcal{B}_{t_i} - \mathcal{B}_{t_{i-1}}) + \frac{1}{2} \sum_{i=1}^{n} f''(\mathcal{B}_{t_{i-1}})(\mathcal{B}_{t_i} - \mathcal{B}_{t_{i-1}})^2$$

$$+ \sum_{i=1}^{n} o((\mathcal{B}_{t_i} - \mathcal{B}_{t_{i-1}})^2)$$

$$= I_1 + I_2 + I_3,$$

where

$$I_1 = \sum_{i=1}^{n} f'(\mathcal{B}_{t_i})(\mathcal{B}_{t_i} - \mathcal{B}_{t_{i-1}}) \to \int_0^t f'(\mathcal{B}_u)\, d\mathcal{B}_u,$$

$$I_2 = \frac{1}{2} \sum_{i=1}^{n} f''(\mathcal{B}_{t_{i-1}})(\mathcal{B}_{t_i} - \mathcal{B}_{t_{i-1}})^2 \to \frac{1}{2} \int_0^t f''(\mathcal{B}_u)\, du,$$

$$I_3 = \sum_{i=1}^{n} o((\mathcal{B}_{t_i} - \mathcal{B}_{t_{i-1}})^2) \to 0,$$

where, by applying the multiplication table, $I_3 \to 0$ as the partition becomes finer and finer so that $\max(t_{i+1} - t_i) \to 0$.

Consequently, for twice differentiable function $f$, the basic form of Itô's lemma can be understood as a Taylor expansion to second order

$$df(X) = f'(X)dX + \frac{1}{2}f''(X)(dX)^2,$$

where $X = \mathcal{B} = \{\mathcal{B}_t\}$, the standard Brownian motion, and the quadratic term $(dX)^2$ is the quadratic variation of the process $X$, i.e., $(d\mathcal{B})^2 = dt$. That is,

$$df(\mathcal{B}) = f'(\mathcal{B})d\mathcal{B} + \frac{1}{2}f''(\mathcal{B})dt. \tag{6.59}$$

Assume that all partial derivatives of $g(x,t)$ exist and are continuous at $(x,t)$. Define $Y_t = g(\mathcal{B}(t), t)$, then

$$dY_t = g(\mathcal{B}(t+dt), t+dt) - g(\mathcal{B}(t), t)$$

$$= (g(\mathcal{B}(t+dt), t+dt) - g(\mathcal{B}(t+dt), t)) + [g(\mathcal{B}(t+dt), t) - g(\mathcal{B}(t), t)]$$

$$= \frac{\partial g}{\partial t}dt + \frac{1}{2}\frac{\partial^2 g}{\partial t^2}(dt)^2 + \text{higher order terms} + \left[\frac{\partial g}{\partial x}d\mathcal{B} + \frac{1}{2}\frac{\partial^2 g}{\partial x^2}dt\right],$$

where we apply (6.59) to the inside of brackets with

$$df(\mathfrak{B}) = g(\mathfrak{B}(t+dt), t) - g(\mathfrak{B}(t), t),$$

$$f'(\mathfrak{B}) = \frac{\partial g(\mathfrak{B}(t), t)}{\partial x},$$

$$f''(\mathfrak{B}) = \frac{\partial^2 g(\mathfrak{B}(t), t)}{\partial x^2}.$$

We discard all terms involving $dt$ to a power higher then 1 and obtain

$$
\begin{aligned}
dY_t &= \frac{\partial g(\mathfrak{B}(t+dt), t)}{\partial t} dt + \frac{1}{2} \frac{\partial^2 g(\mathfrak{B}(t), t)}{\partial x^2} dt + \frac{\partial g(\mathfrak{B}(t), t)}{\partial x} d\mathfrak{B} \\
&= \left[ \frac{\partial g(\mathfrak{B}(t+dt), t)}{\partial t} - \frac{\partial g(\mathfrak{B}(t), t)}{\partial t} \right] dt + \frac{\partial g(\mathfrak{B}(t), t)}{\partial t} dt \\
&\quad + \frac{1}{2} \frac{\partial^2 g(\mathfrak{B}(t), t)}{\partial x^2} dt + \frac{\partial g(\mathfrak{B}(t), t)}{\partial x} d\mathfrak{B} \\[2mm]
&= \left[ \frac{\partial^2 g(\mathfrak{B}(t), t)}{\partial x \partial t} d\mathfrak{B} + \frac{1}{2} \frac{\partial^3 g(\mathfrak{B}(t), t)}{\partial x^2 \partial t} dt \right] dt + \frac{\partial g(\mathfrak{B}(t), t)}{\partial t} dt \\
&\quad + \frac{1}{2} \frac{\partial^2 g(\mathfrak{B}(t), t)}{\partial x^2} dt + \frac{\partial g(\mathfrak{B}(t), t)}{\partial x} d\mathfrak{B},
\end{aligned}
$$

where we apply (6.59) to the inside of the last brackets with

$$df(\mathfrak{B}) = \frac{\partial g(\mathfrak{B}(t+dt), t)}{\partial t} - \frac{\partial g(\mathfrak{B}(t), t)}{\partial t},$$

$$f'(\mathfrak{B}) = \frac{\partial g^2(\mathfrak{B}(t), t)}{\partial x \partial t},$$

$$f''(\mathfrak{B}) = \frac{\partial^3 g(\mathfrak{B}(t), t)}{\partial x^2 \partial t}.$$

Once again, we discard all terms involving $dt$ to a power higher than 1 and obtain the form of Itô's lemma in the case where $X = \mathfrak{B}$ (i.e., $\mu = 0$ and $\sigma = 1$) given in (6.58):

$$dY_t = \left( \frac{\partial g}{\partial t} + \frac{1}{2} \frac{\partial^2 g}{\partial x^2} \right) dt + \frac{\partial g}{\partial x} d\mathfrak{B}.$$

$\square$

### 6.8.3 Risk-Neutral Probability Measure

Those who on the sell side of the security industry (e.g., market makers) and policy makers (e.g., the Federal Reserve) usually work with the risk-neutral probability measure.

In the risk-neutral world, if we ignore dividends, the (conditional) expectation of the stock returns must be equal to the risk-free rate. To interpret this statement in mathematical language, let us recall (6.55) with $\mu(t, X(t)) = \mu(t)$ and $\sigma(t, X(t)) = \sigma(t)$:

$$dX(t) = \mu(t)\,dt + \sigma(t)\,d\mathfrak{B}(t), \tag{6.60}$$

where $\{\mathfrak{B}(t)\}$ is a $\mathbb{P}$-Brownian motion. To emphasize this fact, we rewrite $\mathfrak{B}(t)$ into $\mathfrak{B}_{\mathbb{P}}(t)$. Accordingly, (6.60) becomes

$$dX(t) = \mu(t)\,dt + \sigma(t)\,d\mathfrak{B}_{\mathbb{P}}(t). \tag{6.61}$$

Thus, in the risk-neutral world (i.e., under the risk-neutral probability measure $\mathbb{Q}$),

$$dX(t) = r(t)\,dt + \sigma(t)\,d\mathfrak{B}_{\mathbb{Q}}(t), \tag{6.62}$$

where $\mathfrak{B}_{\mathbb{Q}}$ is a $\mathbb{Q}$-Brownian motion. This $\mathbb{Q}$ is called the *risk-neutral measure* because the expected appreciation rate of the log return on the stock is identical to the risk-free rate despite the presence of risk in the form of $\sigma(t)\,d\mathfrak{B}_{\mathbb{Q}}(t)$.

Note that under our consideration, drift parameter $\mu(t)$, scale parameter $\sigma(t)$, and the risk-free rate $r(t)$ are all constants. Thus, we write

$$\mu(t) = \mu, \quad \sigma(t) = \sigma, \quad r(t) = r.$$

Recall the definition of the Sharpe ratio (i.e., the *market price of risk* from Section 4.2.1 on page 166), written $S$, and we have

$$S = \frac{\mu - r}{\sigma} \quad \Leftrightarrow \quad r = \mu - \sigma S.$$

Taking the difference of (6.61) and (6.62) yields

$$d\mathfrak{B}_{\mathbb{Q}}(t) = S\,dt + d\mathfrak{B}_{\mathbb{P}}(t) = \frac{\mu - r}{\sigma}dt + d\mathfrak{B}_{\mathbb{P}}(t), \tag{6.63}$$

where $\mathfrak{B}_{\mathbb{Q}}$ and $\mathfrak{B}_{\mathbb{P}}$ are Brownian motions under probability measure $\mathbb{Q}$ and $\mathbb{P}$, respectively. Notice that (6.63) connects $\mathbb{P}$ and $\mathbb{Q}$ only implicitly (an explicit connection will be given shortly). Nevertheless, since (6.63) is equivalent to

$$\mathfrak{B}_{\mathbb{Q}}(t) = \frac{\mu - r}{\sigma}t + \mathfrak{B}_{\mathbb{P}}(t), \tag{6.64}$$

$\mathfrak{B}_{\mathbb{P}}$ is adapted if and only if $\mathfrak{B}_{\mathbb{Q}}$ is adapted to the same filtration.

It follows from the martingale property that $\mathbb{E}(d\mathfrak{B}(t)|\mathcal{F}_t) = 0$ if $\mathfrak{B}$ is adapted to the filtration $\{\mathcal{F}_t\}$ that

$$\mathbb{E}_{\mathbb{P}}(dX(t)|\mathcal{F}_t) - \mathbb{E}_{\mathbb{Q}}(dX(t)|\mathcal{F}_t) = \mathbb{E}_{\mathbb{P}}(dX(t)|\mathcal{F}_t) - r\,dt,$$

which is equivalent to

$$\mathbb{E}_{\mathbb{P}}\left(\frac{dS(t)}{S(t)}\Big|\mathcal{F}_t\right) - \mathbb{E}_{\mathbb{Q}}\left(\frac{dS(t)}{S(t)}\Big|\mathcal{F}_t\right) = \mathbb{E}_{\mathbb{P}}\left(\frac{dS(t)}{S(t)}\Big|\mathcal{F}_t\right) - rdt. \qquad (6.65)$$

In words, (6.65) says that the difference between the (conditional) expected log returns over time period $[t, t + dt]$ (of a stock) under real-world and risk-neutral probability measures is the risk premium.

**Example 6.40.** Equality (6.65) assumes that the stock pays no dividend. How should it be modified if a dividend-paying stock is under consideration?

**Solution.** For a dividend-paying stock, (6.62) needs to be modified to

$$dX(t) = (r - q)\,dt + \sigma(t)\,d\mathfrak{B}_{\mathbb{Q}}(t),$$

where $q$ is the annual dividend yield of the stock. Consequently, (6.65) is changed accordingly to

$$\mathbb{E}_{\mathbb{P}}\left(\frac{dS(t)}{S(t)}\Big|\mathcal{F}_t\right) - \mathbb{E}_{\mathbb{Q}}\left(\frac{dS(t)}{S(t)}\Big|\mathcal{F}_t\right) = \mathbb{E}_{\mathbb{P}}\left(\frac{dS(t)}{S(t)}\Big|\mathcal{F}_t\right) - (r - q)dt.$$

$\square$

### 6.8.4 Girsanov Theorem for a Single Brownian Motion

Relation (6.64)

$$\mathfrak{B}_{\mathbb{Q}}(t) = \frac{\mu - r}{\sigma}t + \mathfrak{B}_{\mathbb{P}}(t)$$

changes Brownian motion with no drift to Brownian motion with drift, which can be described by the simplest version of *Girsanov theorem* for a single Brownian motion.

The *Girsanov theorem*, also referred to as the *Cameron-Martin-Girsanov theorem*, is a tool of changing probability measures. Changes of probability measures can be used for changes of the expectation of a random variable, which in turn can be used for security pricing in finance, particularly for derivative pricing. After all, establishing a probability measure in practice may not always be done in an objective way. It is desirable to look at the effect of different probability measures on expectations.

**Theorem 6.2. (Girsanov Theorem for a Single Brownian Motion with Drift)**
*Let $\mathfrak{B} = \{\mathfrak{B}(t)\}$ be a standard Brownian motion on $(\Omega, \mathfrak{F}, \{\mathcal{F}_t\}, \mathbb{P})$, where $\{\mathcal{F}_t\}$ is the natural filtration of $\mathfrak{B}$. Let $\{\theta(t)\}$ be an adapted process to $\{\mathcal{F}_t\}$ satisfying $e^{\frac{1}{2}\int_0^T (\theta(s))^2\,ds} < \infty$ (Novikov's condition). For each $t \in [0, T]$, define*

*1. $D(t) = e^{-\int_0^t \theta(s)\,d\mathfrak{B}(s) - \frac{1}{2}\int_0^t (\theta(s))^2\,ds}$,*

2. $W(t) = \mathcal{B}(t) + \int_0^t \theta(s)\,ds$,

3. a measure $Q$ with $\frac{dQ}{d\mathbb{P}} = D(T)$.

Then $\{D(t)\}$ is a martingale under $\mathbb{P}$ and $\{W(t)\}$ is a standard Brownian motion under $Q$.[24]

Novikov's condition is a sufficient condition for $\{D(t)\}$ to be a martingale.
   For our purpose, it is sufficient to know a special case of Theorem 6.2:

➤ If $\{\mathcal{B}(t)\}$ is a standard Brownian motion under $\mathbb{P}$ and $\theta(t) = \theta$ is constant, then $\{\theta t + \mathcal{B}(t)\}$ is a standard Brownian motion under $Q$.

The Girsanov theorem (the version in Theorem 6.2, change of measure) describes how the dynamics of stochastic processes change when the original probability measure $\mathbb{P}$ is changed to an equivalent probability measure $Q$. In other words, it describes the distribution of the process $\{W(t)\}$ under the new probability measure $Q$. As a by-product, Girsanov theorem also provides a tool for the martingale approach to pricing derivatives.
   The following note is for the interested reader.

**Interpretation and explanation.**

1. To ease the notation, we drop $T$ and let $D = D(T)$. Note that

$$\frac{dQ}{d\mathbb{P}} = D \;\Leftrightarrow\; \frac{dQ}{d\mathbb{P}}(\omega) = D(\omega), \quad \omega \in \Omega$$

$$\Leftrightarrow\; Q(A) = \int_A D(\omega)\,d\mathbb{P}(\omega) \quad \text{for each } A \in \mathfrak{F}.$$

   At this point of our formal observation, $D$ functions like a probability density (in fact, it is, as we will explain shortly). For this $\{D(t)\}$ is also called the *density process* of $Q$ relative to $\mathbb{P}$.

2. There are two basic concepts involved in the hypotheses of more rigorous versions of Girsanov theorem. One is called the *absolute continuity of measures*.[25] This relation between $\mathbb{P}$ and $Q$ assures the existence of a nonnegative random variable $D$ such that

$$Q(A) = \int_A D(\omega)\,d\mathbb{P}(\omega) \quad \text{for all } A \in \mathfrak{F}.$$

---

[24] (a) Let $X_t = -\int_0^t \theta(s)\,d\mathcal{B}(s)$. A straightforward computation leads to $D(t) = e^{X_t - \frac{1}{2}[X]_t}$ and $W(t) = \mathcal{B}(t) - [\mathcal{B}, X]_t$. (b) If $\theta(t) = \theta$ is constant, then $D(t) = e^{-\theta \mathcal{B}(t) - \frac{1}{2}\theta^2 t}$ and $W(t) = \theta t + \mathcal{B}(t)$.
[25] Given a filtered probability space $(\Omega, \mathfrak{F}, \{\mathcal{F}_t\}, \mathbb{P})$ and a probability measure $Q$ defined on measurable space $(\Omega, \mathfrak{F})$, $Q$ is said to be *absolutely continuous* with respect to probability measure $\mathbb{P}$ if any $A \in \mathfrak{F}$ with $\mathbb{P}(A) = 0$ implies $Q(A) = 0$ (i.e., every $\mathbb{P}$-null event is a $Q$-null event). This is one of the reasons we prefer another filtration (infinitesimally larger than the natural filtration) for the Brownian motion (see the footnote on page 289) and work with complete probability space.

$D$ is called the *Radon-Nikodym derivative* (or *density*) of $\mathbb{Q}$ with respect to $\mathbb{P}$, denoted by

$$\frac{d\mathbb{Q}}{d\mathbb{P}}(\omega) = D(\omega).$$

The Radon-Nikodym derivative connects probabilities in one measure $\mathbb{P}$ to probabilities in an *equivalent measure*[26] $\mathbb{Q}$, which is another basic concept involved in changing measures. This relation between $\mathbb{P}$ and $\mathbb{Q}$ assures that $\mathbb{P}(\omega|D(\omega) > 0) = 1$. Indeed, Radon-Nikodym derivative gives a probability density.

The Radon-Nikodym derivative is a topic beyond the scope of this book. We refer the reader to any standard graduate-level textbook on measure theory so that it can be learned properly.

3. Keep in mind that our model assumes that $X(t) = \ln \frac{S(t)}{S_0}$, the log return on a stock over time period $[0,t]$, is an Itô process satisfying s.d.e. (6.27) on page 279:

$$dX(t) = \mu dt + \sigma d\mathcal{B}(t).$$

Equivalently, the stock process $S(t)$ is a geometric Brownian motion (to be defined shortly) governed by the s.d.e.

$$dS(t) = \mu S(t) dt + \sigma S(t) d\mathcal{B}(t).$$

The conditions of Itô's lemma are satisfied if we let $f(x,t) = e^{-rt}x$ and $S(t)$ satisfy the last equation. We apply Itô's lemma to the deflated (discounted) stock price, $Y(t) = e^{-rt}S(t)$, and obtain

$$\begin{aligned}
dY(t) &= (-re^{-rt}S(t) + \mu S(t)e^{-rt} + 0) dt + \sigma S(t)e^{-rt} d\mathcal{B}_\mathbb{P}(t) \\
&= e^{-rt}S(t)[(-r + \mu)dt + \sigma d\mathcal{B}_\mathbb{P}(t)] \\
&= \sigma Y(t) \left( \frac{\mu - r}{\sigma} dt + d\mathcal{B}_\mathbb{P}(t) \right) \\
&= \sigma Y(t) d\mathcal{B}_\mathbb{Q}(t),
\end{aligned}$$

where the last equal sign holds due to Girsanov theorem with $\theta = \frac{\mu - r}{\sigma}$, which assures that $\{\mathcal{B}_\mathbb{Q}(t)\}$ defined by

$$\mathcal{B}_\mathbb{Q}(t) = \frac{\mu - r}{\sigma} t + \mathcal{B}_\mathbb{P}(t)$$

---

[26] Two measures $\mathbb{P}$ *and* $\mathbb{Q}$ *are equivalent if for any* $A \in \mathfrak{F}$, $\mathbb{P}(A) = 0 \Leftrightarrow \mathbb{Q}(A) = 0$.

is a standard Brownian motion under $Q$, therefore a martingale process with respect to its natural filtration under $Q$. Consequently, so is the deflated price process $\{Y(t)\}$ under $Q$ for[27]

$$dY(t) = \sigma Y(t)\, d\mathfrak{B}_Q(t). \tag{6.66}$$

For these reasons we note that $Q$ is also referred to as an *equivalent martingale measure*.

It is worth noting that:

➤ s.d.e. (6.66) says that the risk-neutral probability measure $Q$ renders the price process $\{S(t)\}$ to be a martingale $\{Y(t)\}$, and when probability measures are changed in the way described by Girsanov theorem, the volatility is not altered.

**Remark 6.14.** In practice, an equivalent martingale measure $Q$ is derived based on the market data of various financial instruments rather than assumed by the derivative pricing model. From the stand point of the construction of $Q$, it is an empirical measure. For the technique of estimating risk-neutral distributions, we refer the reader to the literature (e.g., Hunt and Kennedy [15], and Malz [20]). □

A formal statement of Girsanov theorem must operate at a higher level of mathematical abstraction. We refer the reader to standard graduate-level stochastic calculus textbooks. For the proof of Theorem 6.2, we refer the reader to the literature (e.g., Karatzas and Shreve [17]).

## 6.9 Geometric Brownian Motion

### 6.9.1 GBM: Definition

In terms of stochastic differential equations, the definition of Brownian motion with drift and scaling can be restated in the following:

➤ A stochastic process $\{X(t)\}$ is said to be a *Brownian motion with drift and scaling* if it is a solution to the stochastic differential equation

$$dX(t) = \mu\, dt + \sigma\, d\mathfrak{B}(t), \tag{6.67}$$

where $\mu$ and $\sigma > 0$ are constant.

Since taking the integral on both sides of (6.67) with initial condition $X(0) \overset{\text{a.s.}}{=} x_0$:

---

[27] This is significant because martingales model a fair game, and security valuation is the determination of the fair price of a security.

$$\int_0^t dX(s) = \int_0^t \mu \, ds + \int_0^t \sigma \, d\mathfrak{B}(s)$$

yields

$$X(t) = x_0 + \mu t + \sigma \mathfrak{B}(t),$$

$\{X(t)\}$ is a Brownian motion with drift $\mu$ and scaling $\sigma$ if and only if

$$X(t) = x_0 + \mu t + \sigma \mathfrak{B}(t) \quad \text{for some } x_0 \in \mathbb{R}. \tag{6.68}$$

Note that $\sigma$ is also referred to as a *scale parameter* or *volatility parameter* or *diffusion coefficient* for obvious reasons. Also, note that (6.68) is equivalent to the expression in Definition 6.28 on page 284.

A class of stochastic processes that are closely related to Brownian motion with drift and scaling are called the geometric Brownian motion.

**Definition 6.31.** A stochastic process $\{X(t)\}$ is said to be a *geometric Brownian motion (GBM)* if it is a solution to the stochastic differential equation

$$dX(t) = \mu X(t) \, dt + \sigma X(t) \, d\mathfrak{B}(t), \tag{6.69}$$

where $\mu$ and $\sigma > 0$ are constant.

Since solving (6.69) with initial condition $X(0) = x_0$, where $x_0 > 0$, by applying Itô's lemma with function $f(x) = \ln(x)$ (see Exercise 6.20 on page 324) yields

$$X(t) = x_0 e^{(\mu - \frac{1}{2}\sigma^2)t + \sigma \mathfrak{B}(t)}, \tag{6.70}$$

➤ $\{X(t)\}$ with $X(t) = e^{Y(t)}$ is a geometric Brownian motion if and only if $\{Y(t)\}$ is a Brownian motion with drift and scaling.

### 6.9.2  GBM: Basic Properties

Let us recall (6.27) on page 279

$$dX(t) = \mu \, dt + \sigma \, d\mathfrak{B}(t), \quad \text{where } X(t) = \ln\left(\frac{S(t)}{S(0)}\right),$$

which is equivalent to

$$X(t) = \mu t + \sigma \mathfrak{B}(t), \tag{6.71}$$

for $X(0) = \ln 1 = 0$. Clearly, $\{X(t) : t \geq 0\}$ is a Brownian motion with drift parameter $\mu$ and volatility parameter $\sigma$.

For each fixed $t$, since $\mathfrak{B}(t) \sim \mathcal{N}(0, t)$, applying Property 6.8 on page 282, we have

$$X(t) \sim \mathcal{N}(\mu t, \sigma^2 t).$$

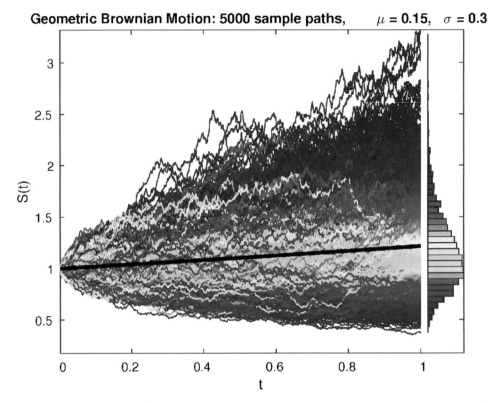

**Fig. 6.3** Geometric Brownian motion with drift parameter $\mu = 0.15$ and volatility parameter $\sigma = 0.3$ is illustrated using 5,000 randomly selected sample paths over the interval $[0,1]$. The current time is 0 and $S(0) = 1$. At time $t = 1$, the frequency distribution of $S(1)$ is shown as a histogram, which approximates a lognormal distribution. The solid curve shows the plot of the expected value $\mathbb{E}(S(t)) = e^{\left(0.15 + \frac{1}{2}(0.3)^2\right)t}$, where $t \geq 0$

That is, $X(t) = \ln\left(\frac{S(t)}{S(0)}\right)$ is a normal random variable. Let $S(0) = S_0$ be a positive number. Note that the nonnegative random variables

$$S(t) = S_0 e^{X(t)} = S_0 e^{\mu t + \sigma \mathcal{B}(t)}, \quad t \geq 0 \tag{6.72}$$

define a stochastic process $\{S(t) : t \geq 0\}$, which is a geometric Brownian motion with parameters $\mu$ and $\sigma$ (see Figure 6.3). We say that random variable $S(t)$ has a *lognormal distribution*, written

$$S(t) \sim \text{log-normal}\,(\mu t,\ \sigma^2 t). \tag{6.73}$$

➤ Notice that $\mathbb{E}(S(t))$ and $\text{Var}(S(t))$ are determined by (6.74) and (6.75), respectively (which are not $\mu t$ and $\sigma^2 t$).

Property 6.8, 3 on page 282 indicates

$$X(t) \sim \mathcal{N}(\mu t, \sigma^2 t) \Rightarrow M_{X(t)}(s) \equiv \mathbb{E}(e^{sX(t)}) = e^{\mu t s + \frac{1}{2}\sigma^2 t s^2},$$

and therefore with $s = 1$,

$$\mathbb{E}\left(\frac{S(t)}{S_0}\right) = \mathbb{E}(e^{X(t)}) = M_{X(t)}(1) = e^{\mu t + \frac{1}{2}\sigma^2 t}, \tag{6.74}$$

and with $s = 2$,

$$\mathbb{E}\left(\left(\frac{S(t)}{S_0}\right)^2\right) = \mathbb{E}(e^{2X(t)}) = M_{X(t)}(2) = e^{2\mu t + 2\sigma^2 t}. \tag{6.75}$$

That is,

$$\mathbb{E}\left(\frac{S(t)}{S_0}\right) = e^{(\mu + \frac{1}{2}\sigma^2)t} \quad \text{and} \quad \mathbb{E}\left(\left(\frac{S(t)}{S_0}\right)^2\right) = e^{2(\mu + \sigma^2)t}. \tag{6.76}$$

**Remark 6.15.** Since stock prices are never negative, it is more reasonable to use geometric Brownian motions to model stock price dynamics than to use Brownian motions as the latter may take on negative values.

Similar arguments apply to a comparison between the binomial tree model and random walk model (the latter, in fact, converges to a Brownian motion; see Theorem 6.3 on page 321). □

### 6.9.3 Relation Between Binomial Tree Model and GBM Model

Recall the Brownian motion with drift and scaling represented by (6.71)

$$X(t) = \mu t + \sigma \mathcal{B}(t) \quad (\text{note that } X(0) = 0),$$

which implies that increments have normal distributions:

$$X(t) - X(s) = \mu(t - s) + \sigma(\mathcal{B}(t) - \mathcal{B}(s)) \sim \mathcal{N}(\mu(t - s), \sigma^2(t - s)).$$

Recall (6.72) and consider geometric Brownian motion $S(t) = S_0 e^{X(t)}$. We obtain

$$\frac{S(t)}{S(s)} = \frac{e^{X(t)}}{e^{X(s)}} = e^{X(t) - X(s)} \sim \text{log-normal}(\mu(t - s), \sigma^2(t - s)). \tag{6.77}$$

Given $t > 0$ and $S(0) = S_0$, let **P** be a partition on $[0, t]$ with equal length:

$$\mathbf{P} : 0 = t_0 < t_1 < t_2 < \cdots < t_n = t \quad \text{where} \quad t_i = \frac{it}{n}, i = 0, 1, \ldots n.$$

To establish a binomial tree model, we let

$$S_i = \begin{cases} S_{i-1}u & \text{with probability } p \\ S_{i-1}d & \text{with probability } 1 - p, \end{cases} \quad \text{where} \quad i = 1, 2, \ldots, n.$$

Given the stock price at current time, $S(0) = S_0$, the evolution of a stock price process from time 0 to time t governed by each of the two different models is illustrated in the table below.

|  | $t_1$ | $t_2$ | $t_3$ | $\cdots$ | $t_{n-1}$ | $t$ |
|---|---|---|---|---|---|---|
| Binomial tree model forecast | $S_1$ | $S_2$ | $S_3$ | $\cdots$ | $S_{n-1}$ | $S_n$ |
| GBM model forecast | $S(t_1)$ | $S(t_2)$ | $S(t_3)$ | $\cdots$ | $S(t_{n-1})$ | $S(t)$ |

Again, $\{S(t),\ t \geq 0\}$, is a geometric Brownian motion with parameters $(\mu, \sigma)$ given in (6.72). Our goal is to find a relation between the binomial tree model and the geometric Brownian motion model when the partition $\mathbf{P}$ becomes finer and finer.

Recall that the parameters u, d, and p in the binomial tree model satisfy the relations

$$0 < d < 1 + r < u, \qquad 0 < p < 1, \qquad ud = 1. \tag{6.78}$$

To achieve our goal, we are going to look for specific values of these parameters u, d, and p such that they can be expressed in terms of $\mu$ and $\sigma$ as well as satisfy relations in (6.78). To do so, we take four steps:

**Step 1.** Let us recall (6.3) on page 261 and write

$$Y_i = Y_i(\omega) = \begin{cases} u & \text{if } \omega = U \text{ with probability p} \\ d & \text{if } \omega = D \text{ with probability } 1 - p, \end{cases} \quad \text{where} \quad i = 1, 2, \ldots, n. \tag{6.79}$$

Thus

$$S_i = S_0 Y_1 \cdots Y_i, \quad i = 1, 2, \ldots, n. \tag{6.80}$$

Note that $Y_i$s are independent and identically distributed, and straightforward computation yields

$$Y_i = \frac{S_{t_i}}{S_{t_{i-1}}}, \quad i = 1, 2, \ldots, n, \tag{6.81}$$

$$\mathbb{E}(Y_i) = pu + (1 - p)d, \tag{6.82}$$

$$\mathbb{E}((Y_i)^2) = pu^2 + (1 - p)d^2. \tag{6.83}$$

**Step 2.** Notice that (6.77) implies

$$\frac{S(t_i)}{S(t_{i-1})} \sim \text{log-normal}\left(\mu \frac{t}{n}, \sigma^2 \frac{t}{n}\right), \quad i = 1, 2, \ldots, n,$$

and that the formulas in (6.76) imply

$$\mathbb{E}\left(\frac{S(t_i)}{S(t_{i-1})}\right) = e^{(\mu + \frac{1}{2}\sigma^2)\frac{t}{n}}, \quad i = 1, 2, \ldots, n, \tag{6.84}$$

$$\mathbb{E}\left(\left(\frac{S(t_i)}{S(t_{i-1})}\right)^2\right) = e^{2(\mu + \sigma^2)\frac{t}{n}}, \quad i = 1, 2, \ldots, n. \tag{6.85}$$

**Step 3.** Setting the expectations in (6.82) and (6.84) equal, with corresponding i, yields[28]

$$pu + (1-p)d = e^{(\mu + \frac{1}{2}\sigma^2)\frac{t}{n}},$$

which is equivalent to

$$p = \frac{e^{(\mu + \frac{1}{2}\sigma^2)\frac{t}{n}} - d}{u - d}.$$

Setting the expectations in (6.83) and (6.85) equal, with corresponding i, yields

$$pu^2 + (1-p)d^2 = e^{2(\mu + \sigma^2)\frac{t}{n}}.$$

**Step 4.** By solving for parameters p, u, and d from the system of equations

$$
\begin{cases}
pu + (1-p)d = e^{(\mu + \frac{1}{2}\sigma^2)\frac{t}{n}} \\
pu^2 + (1-p)d^2 = e^{2(\mu + \sigma^2)\frac{t}{n}} \\
ud = 1
\end{cases}
\tag{6.86}
$$

we can express p, u, and d in terms of $\mu$, $\sigma$, and $n$.

It can be shown that for large n,

$$p = \frac{e^{(\mu + \frac{1}{2}\sigma^2)\frac{t}{n}} - d}{u - d}, \quad u = e^{\sigma\sqrt{\frac{t}{n}}}, \quad d = e^{-\sigma\sqrt{\frac{t}{n}}} \tag{6.87}$$

provide an approximation to the solution of the system.

Recall the comment in a solution provided in Example 6.37 on page 305 that in a risk-neutral world, if the stock pays a known dividend yield $q$, then

$$\mu + \frac{1}{2}\sigma^2 = r - q.$$

This explains why we choose the binomial model parameters as

$$p = \frac{e^{(r-q)\Delta t} - d}{u - d}, \quad u = e^{\sigma\sqrt{\Delta t}}, \quad d = e^{-\sigma\sqrt{\Delta t}}. \tag{6.88}$$

By making such a choice of parameters, the geometric Brownian motion can be simulated by the binomial tree model (see Figure 6.4, which depicts at time $t = 1$ a security price's lognormal density, mean, and median along with a histogram approximation to the density and illustrates that $S(1)$ is skewed right (see Definition 6.15 on page 266), thus mode < median < mean).

Verification of p, u, and d in (6.87) is left as an exercise for the reader.

Finally, we point out that it is readily understood from (6.80) that the binomial tree approximates a lognormal distribution.

---

[28] The subscript $n$ is dropped from $p_n$, $u_n$, and $d_n$ to ease the notation.

**Fig. 6.4** Lognormal density of $S(1)$ with initial price $S(0) = \$1$, drift parameter $\mu_{RW} = 0.1$, and volatility parameter $\sigma = 0.4$. This right-skewed density has a median (dashed vertical line) and mean (solid vertical line) at 1.1 and 1.2, respectively. The histogram approximation is for 100,000 values randomly drawn from a lognormal distribution with the same drift and volatility parameters

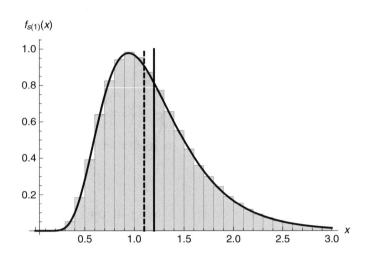

## 6.10 BM as a Limit of Simple Symmetric RW

A *random walk* consists of a succession of random steps, which means that either the direction or size of each step (or both) is chosen at random. To make a mathematical formalization of this notion, we associate each step with a random variable, say $R_i$. Thus, the dynamics of the walk are governed by the distribution of these random variables.

*Construction of Brownian motion from simple symmetric random walk.*

Recall Example 6.27 on page 280. In mathematical language, a stochastic process $\{W_i\}$ on $(\Omega, \mathfrak{F}, \mathbb{P})$ is said to be a one-dimensional *random walk* (or random walk on $\mathbb{Z}$) if

$$W_i = W_0 + \sum_{k=1}^{i} R_k = W_{i-1} + R_i, \quad i = 1, 2, \ldots$$

where $\{R_i, i = 1, 2, \ldots\}$ is a sequence of independent, identically distributed random variables and $W_0$ a random variable that is independent of each $R_i$.

A (one-dimensional) random walk is *simple* if

$$W_0 = 0 \quad \text{and} \quad R_i = \begin{cases} 1 & \text{with probability } p \\ -1 & \text{with probability } 1 - p, \end{cases} \quad i = 1, 2, \ldots,$$

i.e., a walk starting from its mean 0 with unit step size only. A (one-dimensional) simple random walk is *symmetric* if $p = \frac{1}{2}$, i.e., a walk taking each step equally likely to be in all possible directions (for a one-dimensional simple symmetric walk, it means that moving to the left is as likely to occur as moving to the right at each step).

**Example 6.41.** For each positive integer $n$, let

$$S_m^{(n)} = \frac{1}{\sqrt{n}} \sum_{i=1}^{nm} R_i \quad \text{where} \quad R_i = \begin{cases} 1 & \text{with probability } \frac{1}{2} \\ -1 & \text{with probability } \frac{1}{2}, \end{cases} \quad i = 1, 2, \ldots.$$

with $\mu_{RW} \equiv \mathbb{E}(R_i) = 0$ and $\sigma_{RW}^2 \equiv \text{Var}(R_i) = 1$.

Apply the central limit theorem to obtain

$$\frac{1}{\sqrt{m}} S_m^{(n)} = \frac{1}{\sigma_{RW}\sqrt{nm}} \left( \sum_{i=1}^{nm} R_i - nm\mu_{RW} \right) \xrightarrow{d} \mathcal{N}(0,1) \quad \text{as } n \to \infty,$$

and observe

$$S_m^{(n)} = \sqrt{m} \frac{1}{\sigma_{RW}\sqrt{nm}} \left( \sum_{i=1}^{nm} R_i - nm\mu_{RW} \right) \xrightarrow{d} \mathcal{N}(0,m) \quad \text{as } n \to \infty.$$

Notice that different scalings lead to different limits and $\mathcal{B}(m) \sim \mathcal{N}(0,m)$. This observation makes the next theorem plausible. □

For each positive integer $n$, we define a continuous-time stochastic process $\mathbb{W}^{(n)} = \left\{ W_t^{(n)} : t \geq 0 \right\}$ by

$$W_t^{(n)} = \frac{1}{\sqrt{n}} \sum_{i=1}^{\lfloor nt \rfloor} R_i, \tag{6.89}$$

whereby convention $\lfloor nt \rfloor$ is the greatest integer less than or equal to $nt$, and random variables $R_i$ are defined in Example 6.41.

In words, (6.89) defines the rescaled random walk with steps of size $\frac{1}{\sqrt{n}}$ taken every $\frac{1}{n}$ time units (i.e., steps are taken at time $i/n$, $i = 1, 2, \ldots, \lfloor nt \rfloor$).

**Theorem 6.3.** *Standard Brownian motion can be approximated by the rescaled random walk:*

$$\mathcal{B}(t) \stackrel{d}{=} \lim_{n \to \infty} \frac{1}{\sqrt{n}} \sum_{i=1}^{\lfloor nt \rfloor} R_i, \quad t \geq 0.$$

Intuitively, the theorem implies that Brownian motion has the microscopic structure that may emerge from a random walk. Although Example 6.41 might seem intuitively reasonable in suggesting a proof of this theorem by applying the (ordinary version of) central limit theorem, a proper proof of the theorem involves sophisticated technicality at the mathematical level beyond the scope of this book (it requires functional central limit theorem). We refer the reader to the literature (e.g., Billingsley [2]).

There are many more nice discussions in the literature related to the topics presented in this chapter, e.g., [1, 5, 6, 7, 8, 10, 12, 13, 14, 16, 18, 19, 21, 23, 24, 26, 27, 28].

## 6.11 Exercises

### 6.11.1 Conceptual Exercises

**6.1.** Consider an oversimplified stock price behavior as described by a two-period binomial tree. In each period the stock price either goes up by a factor u with probability p or goes down by a factor d with probability $1 - p$. Identify the corresponding probability space $(\Omega, \mathfrak{F}, \mathbb{P})$.

**6.2.** Continue from Exercise 6.1.
   Let $\omega_1 = UU$ (i.e., the event that the stock price goes up in both periods). Construct a sub-$\sigma$-algebra $\mathcal{F}$ such that $\{\omega_1\} \in \mathcal{F} \subsetneq 2^\Omega$.

**6.3.** Continue from Exercise 6.1.
   Construct a filtration $\{\mathcal{F}_t \subseteq \mathfrak{F}, \ t = t_0, t_1, t_2\}$ such that as time $t$ increases, $\mathcal{F}_t$ reveals more information about the evolution of the stock price.

**6.4.** Consider an oversimplified stock price behavior as described by a three-period binomial tree. In each period the stock price either goes up by a factor u with probability p or goes down by a factor d with probability $1 - p$. Identify the corresponding sample space $\Omega$ (in the notation for the probability space $(\Omega, \mathfrak{F}, \mathbb{P})$). How many elements are there in the $\sigma$-algebra $\mathfrak{F} = 2^\Omega$? (*Hint:* Each simple event corresponds to a path.)

**6.5.** Continue from Exercise 6.4 on page 322.
   Given the corresponding sample space

$$\Omega = \{\omega_1, \omega_2, \omega_3, \omega_4, \omega_5, \omega_6, \omega_7, \omega_8\}, \text{ where}$$
$$\omega_1 = UUU, \ \omega_2 = UUD, \ \omega_3 = UDU, \ \omega_4 = UDD,$$
$$\omega_5 = DUU, \ \omega_6 = DUD, \ \omega_7 = DDU, \ \omega_8 = DDD,$$

Construct a sub-$\sigma$-algebra $\mathcal{F}$ such that

$$\{\omega_1, \omega_2, \omega_3, \omega_4\} \in \mathcal{F} \subsetneq 2^\Omega.$$

**6.6.** Continue from Exercise 6.4.
   Construct a filtration $\{\mathcal{F}_t \subseteq \mathfrak{F}, \ t = t_0, t_1, t_2, t_3\}$ such that as time $t$ increases, $\mathcal{F}_t$ reveals more information about the evolution of the stock price.

**6.7.** Let $\mathfrak{B} = \{\mathfrak{B}(t)\}$ be standard Brownian motion. What is the probability that $\mathfrak{B}(1)$ lies between $-1$ and $1$?

**6.8.** Let $X = \{X_t\}$ and $Y = \{Y_t\}$ be two processes. Verify the following identity:

$$\Delta(X_t Y_t) = X_t \Delta Y_t + Y_t \Delta X_t + \Delta X_t \Delta Y_t$$

where $\Delta X_t$, $\Delta Y_t$, and $\Delta(X_t Y_t)$ are defined as the corresponding increments over time interval $[t, t + \Delta t]$. (Hint: see (6.38) on page 292.)

**6.9.** Let $\mathfrak{B} = \{\mathfrak{B}(t)\}$ be standard Brownian motion. Describe your visualization of $-\mathfrak{B}$. Is it different from that of $\mathfrak{B}$?

**6.10.** Use the definition of covariation to derive

$$d\mathfrak{B}(t)\,dt = 0.$$

**6.11.** Suppose that a stock price is modeled by a process $S = \{S(t)\}$ where

$$S(t) = e^{0.7t^2 + 2.3\mathfrak{B}(t)}.$$

What is the expected growth rate of the stock at any given time $t$?

**6.12.** Let $\{S(t)\}$ be governed by the s.d.e.

$$dS(t) = \mu S(t)dt + \sigma S(t)d\mathfrak{B}(t),$$

where $\mu$ and $\sigma > 0$ are constants. Is $\{S(t)\}$ an Itô diffusion process?

**6.13.** Recall s.d.e. (6.58) on page 304:

$$dY = \left( f_t + \mu f_x + \frac{1}{2}\sigma^2 f_{xx} \right) dt + \sigma f_x\, d\mathfrak{B}.$$

Compute $\mathbb{E}(dY|\mathcal{F}_t)$ and $\mathrm{Var}(dY|\mathcal{F}_t)$, where $\{\mathcal{F}_t\}$ is the Brownian filtration described in Definition 6.30. Indicate properties of conditional expectation that you applied.

**6.14.** Let $\{S(t)\}$ be governed by the s.d.e.

$$dS(t) = \mu S(t)dt + \sigma S(t)d\mathfrak{B}(t),$$

where $\mu$ and $\sigma > 0$ are constants. Let $Y(t) = e^{-rt}S(t)$. Use the Itô product rule (see (6.40) on page 292) to compute $dY(t)$.

**6.15.** Continue from the last exercise. Is $\{Y(t)\}$ defined in Exercise 6.14 a martingale under a risk-neutral probability measure?

**6.16.** Compute $\int_0^t \mathfrak{B}(s)\,d\mathfrak{B}(s)$ first, then express the s.i.e. in the form of s.d.e.

**6.17.** Is it true that $X = \{X(t), t \geq 0\}$ is a Brownian motion process if and only if $Y = \{Y(t), t \geq 0\}$, where $Y(t) = e^{X(t)}$, is a geometric Brownian motion process?

**6.18.** In continuous-time financial mathematics, in what situations is geometric Brownian motion useful?

## 6.11.2 Application Exercises

**6.19.** Rewrite s.d.e. (6.27) on page 279 into

$$dS(t) = \mu S(t)\,dt + \sigma S(t)\,d\mathcal{B}_t.$$

Consider a time period of length $\Delta t$. Compute the ratio of the per-period standard deviation to the per-period drift, i.e., $\frac{\sqrt{\mathrm{Var}(\Delta S(t))}}{\mathbb{E}(\Delta S(t))}$, and interpret your result.

**6.20.** Find a solution to the s.d.e. $dX(t) = \mu X(t)dt + \sigma X(t)d\mathcal{B}(t)$ by applying Itô's lemma with $f(x) = \ln x$.

**6.21.** Verify that p, u, and d given by (6.88) on page 319

$$p = \frac{e^{(r-q)\frac{t}{n}} - d}{u - d}, \quad u = e^{\sigma\sqrt{\frac{t}{n}}} \quad d = e^{-\sigma\sqrt{\frac{t}{n}}}$$

are an approximation to the solution of system (6.86):

$$\begin{cases} pu + (1-p)d = e^{(\mu+\frac{1}{2}\sigma^2)\frac{t}{n}} \\ pu^2 + (1-p)d^2 = e^{2(\mu+\sigma^2)\frac{t}{n}} \\ ud = 1 \end{cases}$$

## 6.11.3 Theoretical Exercises

**6.22.** Given a standard Brownian motion $\mathcal{B}(t)$, show that each of following stochastic processes is also a standard Brownian motion:

a) $X(t) = \frac{1}{\sqrt{c}}\mathcal{B}(ct)$ for all constants $c > 0$.
b) $Y(t) = \mathcal{B}(t+c) - \mathcal{B}(c)$ for all constants $c > 0$.

**6.23.** Let $\{\mathcal{B}(t)\}$ be a standard Brownian motion. Let $t_1, t_2, \ldots, t_n \in (0, \infty)$ with $0 < t_1 < t_2 < \cdots < t_n$. Show that the random vector (or multivariate random variable) $(\mathcal{B}(t_1), \mathcal{B}(t_2), \ldots, \mathcal{B}(t_n))$ has a multivariate normal distribution for any fixed choice of $n$ time points $0 < t_1 < t_2 < \cdots < t_n$, $n \geq 1$.

**6.24.** Continue from the last exercise. Show that the joint probability density function of $(\mathcal{B}(t_1), \mathcal{B}(t_2), \ldots, \mathcal{B}(t_n))$ is

$$f(x_1, \ldots, x_n) = \frac{\exp\left(-\frac{1}{2}\left(\frac{x_1^2}{t_1} + \sum_{j=1}^{n-1}\frac{(x_{j+1}-x_j)^2}{t_{j+1}-t_j}\right)\right)}{\sqrt{(2\pi)^k\, t_1\,(t_2-t_1)\cdots(t_n-t_{n-1})}},$$

where $-\infty < x_j < \infty$ for $j = 1, \ldots, n$.

**6.25.** Let $\{\mathfrak{B}(t)\}$ be a standard Brownian motion. Compute $\mathbb{E}(\mathfrak{B}(t))$ for $t \geq 0$ and $\mathrm{Cov}(\mathfrak{B}(s), \mathfrak{B}(t))$ for $s, t \geq 0$. Is Brownian motion a white noise?

**6.26.** Let $\{Z_n : n \geq 1\}$ be a sequence of independent identically distributed random variables with mean $\mu = 0$ and finite variance $\sigma^2$. Let

$$S_n = Z_1 + Z_2 + \cdots + Z_n \quad \text{and} \quad X_n = S_n^2 - n\sigma^2.$$

Show that $\{X_n\}$ is a martingale with respect to the natural filtration of the sequence $\sigma(S_n : n \geq 1)$.

**6.27.** Let $W = \{W_i\}$ be a random walk defined in Example 6.27. Show that $W$ is not a white noise.

**6.28.** Prove (6.51) on page 299:

$$[X]_t = \int_0^t \sigma^2(s)\, ds,$$

where $X$ satisfies (6.50) on page 298 and $\mu(t)$ is continuous on $[0, \infty)$.

**6.29.** Show that $\mathfrak{B}_t^2 - [\mathfrak{B}]_t$ is a martingale with respect to the filtration generated by the Brownian motion itself. That is, $\mathfrak{B}$ is adapted to its natural filtration $\{\mathcal{F}_t\}$, where $\mathcal{F}_t = \sigma(\{\mathfrak{B}_s, s \leq t\})$.

**6.30.** Show that as positive integer $n \to \infty$, the sequence of random variables $\{W_t^{(n)}\}$, defined by (6.89) on page 321, converges in distribution to a normal random variable in $\mathcal{N}(0, t)$, where $t > 0$ is an integer.

**6.31.** Let $X = \{X(t)\}$ be an Itô process represented by s.d.e. (6.56)

$$dX(t) = \mu(X(t), t)\, dt + \sigma(X(t), t)\, d\mathfrak{B}(t), \quad 0 \leq t \leq T.$$

Show that if we define a process $Y = \{Y(t)\}$ by $Y(t) = f(X(t))$, $0 \leq t \leq T$, then Itô's formula (see (6.57) on page 304) has a convenient form:

$$dY = f'(X)dX + \frac{1}{2}f''(X)(dX)^2,$$

where $X = X(t)$. Clearly, this form is easier to remember than the form in (6.57) on page 304 because it bears greater similarity to the Taylor expansion.

**6.32.** Let $X = (X_1, X_2)$, where $X_1$ and $X_2$ are Itô processes governed respectively by the s.d.e.'s

$$dX_i = \mu_i dt + \sigma_i d\mathfrak{B}_i, \quad i = 1, 2.$$

Two-dimensional Itô's lemma states that if $\mathfrak{B}_1$ and $\mathfrak{B}_2$ are standard Brownian motion processes with correlation $\rho_{12}$, then $Y = f(X)$, where $f : \mathbb{R}^2 \to \mathbb{R}$ is twice continuously differentiable, is an Itô process satisfying the s.d.e.

$$df(X) = f_1(X)dX_1 + f_2(X)dX_2$$
$$+ \frac{1}{2}\left(f_{11}(X)dX_1^2 + 2f_{12}(X)dX_1dX_2 + f_{22}(X)dX_2^2\right), \qquad (6.90)$$

where $f_i = \frac{\partial f}{\partial X_i}$ and $f_{ij} = \frac{\partial^2 f}{\partial X_i \partial X_j}$, $i,j = 1,2$.[29]

Show that (6.90) is equivalent to (6.57) if $X_2 = t$ (i.e., $\mu_2 = 1$ and $\sigma_2 = 0$).

**6.33.** Prove the Itô product rule. (*Hint:* prove (6.39) on page 292 by applying two-dimensional Itô's lemma given in Exercise 6.32).

**6.34.** An investment in a foreign asset carries exchange risk. The model under our consideration is introduced by Briys and Solnik [4] in study of hedging such risk.

Let $V(t)$ be the local (domestic) currency value of a foreign asset at time t. Let $S(t)$ be the exchange rate at time t expressed as the local currency value of one unit of foreign currency (e.g., 1.11 USD/Euro). The model assumes that both $\{V(t)\}$ and $\{S(t)\}$ are geometric Brownian motion processes:

$$\frac{dV}{V} = \mu_V dt + \sigma_V d\mathfrak{B}_V$$
$$\frac{dS}{S} = \mu_S dt + \sigma_S d\mathfrak{B}_S,$$

where two standard Brownian motion processes $\mathfrak{B}_V$ and $\mathfrak{B}_S$ have correlation $\rho_{VS}$.

Let $V^* = VS$, the value of the foreign investment expressed in domestic currency. Compute $\frac{dV^*}{V^*}$ and interpret your answer. (*Hint*: apply Itô product rule.)

## References

[1] Baxter, M., Rennie, A.: Financial Calculus: AnIntroduction to Derivative Pricing. Cambridge University Press, Cambridge (1996)

[2] Billingsley, P.: Weak convergence of measures. In: CBMS-NSF Regional Conference Series in Applied Mathematics (1971)

[3] Billingsley, P.: Probability and Measures. Wiley, New York (1995) or (2012)

[4] Briys, E., Solnik, B.: Optimal currency hedge ratios and interest rate risk. J. Int. Money Finance **11**, 431–446 (1992)

[5] Chung, K.L.: A Course in Probability. Academic, New York (1974)

[6] Clark, J.M.C.: The representation of functionals of Brownian motion by stochastic integrals. Ann. Math. Stat. **41**(4), 1282(1970)

[7] Dellacherie, C., Meyer, P.-A.: Probabilities and Potential C. North-Holland, Amsterdam (1988)

---

[29] More precisely speaking, we assume that $\mathfrak{B}_1$ and $\mathfrak{B}_2$ are defined on the same filtered probability space $\{\Omega, \mathfrak{F}, \{\mathcal{F}_t\}, \mathbb{P}\}$ and adapted to the filtration $\{\mathcal{F}_t\}$.

[8] Durrett, R.: Probability: Theory and Examples, 4th edn. Cambridge University Press, Cambridge (2010)

[9] Durrett, R.: Brownian Motion and Martingales in Analysis. Wadsworth Advanced Books and Software, Belmont (1984)

[10] Epps, T.: Pricing Derivative Securities. World Scientific, Singapore (2007)

[11] Fama, E.: Efficient capital markets: a review of theory and empirical work. J. Finance **25**(2), 383 (1969)

[12] Girsanov, I.V.: On transforming a certain class of stochastic processes by absolutely continuous substitution of measures. Theory Probab. Appl. **5**, 285(1960)

[13] Harrison, J.M., Pliska, S.R.: Martingales and stochastic integrals in the theory of continuous trading. Stoch. Process. Appl. **11**(3), 215(1981)

[14] Hull, J.: Options, Futures, and Other Derivatives, 7th edn. Pearson Prentice Hall, Upper Saddle River (2009)

[15] Hunt, P., Kennedy, J.: Financial Derivatives in Theory and Practice. Wiley Series in Probability and Statistics. Wiley, New York (2004)

[16] Itô, K.: Multiple wiener integral. J. Math. Soc. Japan, **3**, 157–169 (1951)

[17] Karatzas, I., Shreve, S.: Brownian Motion and Stochastic Calculus. Springer, New York (1991)

[18] Korn, R., Korn E.: Option Pricing and Portfolio Optimization. American Mathematical Society, Providence (2001)

[19] Mackean, H.: Stochastic Integrals. Academic, New York/London (1969)

[20] Malz, A.: A Simple and Reliable Way to Compute Option-Based Risk-Neutral Distributions. Federal Reserve Board of New York Staff Reports (2014)

[21] Mikosch, T.: Elementary Stochastic Calculus with Finance in View. World-Scientific, Singapore (1998)

[22] Mörters, P., Peres, Y.: Brownian Motion. Cambridge University Press, Cambridge (2010)

[23] Musiela, M., Rutkowsk, M.: Martingale Methods in Financial Modelling. Springer, New York (2004)

[24] Neftci, S.: An Introduction to the Mathematics of Financial Derivatives. Academic, San Diego (2000)

[25] Paley, R., Wiener, N., Zygmund, A. Note on some random functions. Math. Z. **37**, 647–668 (1993)

[26] Roman, S.: Introduction to the Mathematics of Finance, 1st edn. Springer, New York (2004)

[27] Shreve, S.: Stochastic Calculus for Finance II: Continuous-Time Models. Springer, New York (2004)

[28] Wilmott, P., Dewynne, N., Howison, S.: Mathematics of Financial Derivatives: a Student Introduction. Cambridge University Press, Cambridge (1995)

# Chapter 7
# Derivatives: Forwards, Futures, Swaps, and Options

To introduce major concepts and ideas about derivatives in a simple and concise fashion, we make some assumptions[1] throughout this chapter so that we can concentrate on the task at hand and make the main ideas easier to understand.

Unless otherwise stated, we impose the following assumptions:

- There are no transaction costs.
- There are no taxes.
- The risk-free interest rate is the same for borrowing and lending.
- There are no restrictions (e.g., margin requirements) to the short seller.
- There is sufficient liquidity in all markets (e.g., stocks, bonds, derivatives, foreign exchange, and so on).

## 7.1 Derivative Securities: An Overview

### 7.1.1 Basic Concepts

In financial accounting, an *asset* represents value of ownership that can be converted into cash.

A *financial asset* is an (intangible) asset whose value is derived from a contractual claim, such as bank deposits, bonds, and stocks. Unlike real-estate properties and commodities (which are tangible, physical assets), financial assets do not necessarily have physical worth.

A *security* is a tradable financial asset. Thus, the set of all securities is a subset of the set of all assets as any security is an asset by definition.

---

[1] These assumptions are often either unreasonable or unrealistic from a practical point of view (e.g., a time t is assumed to be continuous, but in reality it is discrete).

© Arlie O. Petters and Xiaoying Dong 2016

A.O. Petters, X. Dong, *An Introduction to Mathematical Finance with Applications*, Springer Undergraduate Texts in Mathematics and Technology, DOI 10.1007/978-1-4939-3783-7_7

Securities are represented by certificates (this explains the name "securities") which may be either in paper form, in electronic form, or in book-entry form.

**Example 7.1.** Cash is an asset but not a security.                                    □

Securities may be classified into three categories:

1. *debt* securities or *debts* (e.g., bonds)
2. *equity* securities or *equities* (e.g., common stocks)
3. *derivative* securities or *derivatives* (e.g., forwards, futures, options, and swaps).

*Derivatives* are securities in the form of contracts between two parties, the *buyer* and the *seller*. Since these contracts are either for *contingent claims* or for *forward commitments*, derivatives can be classified into *contingent claims* and *forward commitments* (or *noncontingent claims*) according to the type of contracts.

A *contingent claim* is a contract which gives the buyer the *right*, but not the obligation, to buy or sell a security, called *underlying security* or *underlier*, at a specified price, called *strike price*, on or before a specified date, called *expiration date*. Examples of contingent claims are options.[2]

A *forward commitment* is a contract by which the buyer and seller have the *obligation* to buy or to sell (to deliver) an underlying security at a predetermined price in the future, called the *delivery date*. Examples of forward commitments are forwards and futures.

Derivative contracts can be created on and traded in some exchanges such as the Chicago Board of Trade (CBOT), which is the world's oldest futures and options exchange, or on OTC markets, which are less transparent than exchanges. For example, a futures contract is considered to be a standardized version of a forward commitment and therefore traded on CBOT, whereas a forward contract can be customized to any commodity, amount, and delivery date and, therefore, is nonstandardized and traded on OTC markets.

**Remark 7.1.** One might attempt to classify securities in terms of underlying securities and derivative securities. If you do so, think again, for underlying securities are often real assets such as gold, oil, and metals or financial assets such as bonds, stocks, and currencies. But underlying securities can also be derivatives on other underliers; for example, options on futures which are on metals are derivatives on derivatives.

---

[2] Later we will show that the definition of options coincides with the definition of contingent claim in the financial dictionary: "a claim that can be made only if one or more specified outcomes occur."

Keep in mind that derivative contracts can be written on real assets such as *commodities*[3] as well as on financial assets such as bonds, stocks, currencies, and other derivatives. Derivative contracts written on precious metals or agricultural products are called *commodity derivatives*, whereas those written on securities are called *financial derivatives*.                                                     □

Although we will primarily focus on stock options and have not yet provided all the details in the financial jargon used in this section, the next example will help the reader to understand Remark 7.1.

**Example 7.2.** The premium of an option on futures tracks the price of its underlying futures contract which, in turn, tracks the price of the underlying cash. In fact, regardless of the underlying security, it is a fairly safe bet that the future's price will generally converge to the spot price of the underlying security as the delivery month of a futures contract approaches because otherwise there would be arbitrage opportunities.

For instance, the July copper option tracks the July copper futures contract. The March S&P 500 index option follows the March S&P 500 index futures. Furthermore, the prices of these futures converge to the spot prices of copper and S&P 500 index with high probability at least in their delivery months.   □

## 7.1.2 Basic Functions of Derivatives

Derivatives may serve three basic functions, which are price discovery, speculative activity, and hedging activity:

*Price discovery* is a process involving buyers and sellers arriving at a transaction price for a given product with given quality and quantity at a given time in a given location. Although they are interrelated, price discovery and price valuation are different concepts. The former is a mechanism, whereas the latter is a determination. Price discovery plays an important role in economic decision-making that involves either entrepreneurs or policy makers.

*Speculators* are those who take calculated risks in the hope of making large short-term profits. By definition, speculators are usually not interested in holding possession of the underlying securities. They are typically sophisticated investors with expertise in the markets in which they are trading who usually use highly leveraged investments such as futures and options.

*Hedgers* are those who take steps to reduce the risk of an investment by making an offsetting investment. By definition, hedgers do not usually seek a profit but rather seek to stabilize the performance of their portfolios or the revenues

---

[3] A *commodity* is a raw material used in commerce or primary agricultural product that can be bought and sold such as copper, silver, crude oil, natural gas, wheat, beef cattle, and coffee.

or costs of their business operations. Their gains or losses are usually offset to some degree by a corresponding loss or gain in the market for the underlying securities.

**Example 7.3.** Price discovery begins with market price information. Imagine how difficult business decision-making would be in the extreme scenario in which no one knows at what price competitors have sold a given product. When more varieties of a product are sold through more venues, more market price information becomes available. Derivatives serve this purpose by providing a wider variety of assets through many venues, including cash settlements. □

**Example 7.4.** Speculators provide very important market information. Arbitrage makes prices converge to the fair price as a riskless way of making a profit would only be a transient opportunity. In other words, free-lunch opportunities will not last long as other people may seek and grab them quickly. □

**Example 7.5.** If you are in a manufacturing business and sell products to foreign countries, you may have specific business ideas in mind. But foreign exchange markets may be in turmoil and highly volatile. With the help from derivatives, you can minimize or even eliminate the unwanted volatility. □

### 7.1.3 Characteristics of Derivative Valuation

Security valuation is the determination of the fair price of a security.

If we focus only on a straight bond (a bond with no embedded options[4]), the valuation is based on discounting its expected cash flows at the appropriate discount rate.[5]

If we focus only on a single (dividend-paying) stock, valuation models include the dividend discount and discounted cash flow model.[6]

To summarize in loose terms, the valuation methods above are based on forecasting future cash flows, growth rates, and risks.

A derivative valuation, by contrast, involves no forecasting and is directly linked to the price of the underlier in the sense that, in relative terms, the higher the volatility of the price movement of the underlier, the higher the value of the derivative.

Derivative valuation models are derived mostly by using replication or hedging arguments under the Law of One Price (see Section 7.1.4).

---

[4] A callable bond is an example of a bond with an embedded call option.

[5] Section 2.10

[6] http://www.investopedia.com/terms/d/ddm.asp

**Example 7.6. (Price by Hedging)** Consider a derivative[7] with expiration $T$ on a stock whose price is modeled by a one-step binomial tree (where the probability the stock price will go up is p and the probability it will go down is $1 - p$). Let $C(t)$ and $S(t)$ denote the prices of the derivative and stock at time $t \in [0, T]$, respectively. What is the fair value of $C(0)$ if it is priced by hedging based on the following assumptions:

(a) $S(0) = \$50$,
(b) Only two possible states at time $T$: either $S(T) = \$60$ or $S(T) = \$40$,
(c) $C(T) \equiv \max(S(T) - 50, 0)$,
(d) The risk-free interest rate $r = 0$.

**Solution.** Note that an equivalence between the no-arbitrage condition and the existence of the risk-neutral probability measure implies that $p = \frac{1}{2}$ (which is the solution to the equation $50 = 60p + 40(1 - p)$).

Suppose that we sold one derivative contract and need to buy $\Delta$ shares of the underlying stock to hedge away any portfolio risk, where the portfolio consists of a long position in $\Delta$ shares of the stock and a short position in one derivative contract. Let $\Pi(T)$ denote the value of the portfolio at time $T$. Since

$$C(T) \equiv \max\{S(T) - 50, 0\} = \begin{cases} \$10 & \text{if } S(T) = \$60 \\ 0 & \text{if } S(T) = \$40, \end{cases}$$

we obtain

$$\Pi(T) = \begin{cases} 60\Delta - 10 & \text{if } S(T) = 60 \\ 40\Delta & \text{if } S(T) = 40, \end{cases}$$

which should be a constant if the risk is completely hedged away. It follows from the equation $60\Delta - 10 = 40\Delta$ that $\Delta = \frac{1}{2}$.

Thus, $50\Delta - C(0) = \Pi(0) = \mathbb{E}(\Pi(T)|\Pi(0)) = \Pi(T) = 20$ with $\Delta = \frac{1}{2}$ implies that $C(0) = \$5$.

$\square$

---

[7] One can think of a call option (see Definition 7.8). It is defined in a later section, but this example should be readily understandable.

### 7.1.4 No-Arbitrage Principle and Law of One Price

The key principle of mathematical finance is the principle of no arbitrage.

We consider a portfolio[8] $\Pi$ on time interval $[0, T]$ and denote by $\Pi(t)$ the value of portfolio $\Pi$ at time $t$. There are different versions[9] of the notion of arbitrage. We will use the one below.

**Definition 7.1.** A portfolio $\Pi$ is said to be an *arbitrage portfolio*, or simply an *arbitrage*, if it satisfies either of the two sets of conditions:

1. $\Pi(0) = 0,$ $\quad$ $\mathbb{P}(\Pi(T) \geq 0) = 1,$ and $\mathbb{P}(\Pi(T) > 0) > 0;$

2. $\Pi(0) < 0$ and $\mathbb{P}(\Pi(T) \geq 0) = 1.$

In words, the first set of conditions describes a portfolio that is at no cost, never loses money, and sometimes makes money. The second set of conditions describes a portfolio that starts from a debt and guarantees to finish with a non-negative amount of money.

**Example 7.7.** A no-arbitrage portfolio $\Pi$ must satisfy the property that

$$\Pi(T) \geq 0 \quad \text{implies} \quad \Pi(t) \geq 0 \quad \text{for } \forall t < T,$$

for otherwise, $\exists t_0 < T$ such that $\Pi(t_0) < 0$ and $\Pi(T) \geq 0$ with probability 1, contradicting to the no-arbitrage assumption! $\quad\square$

**Definition 7.2.** An *arbitrage opportunity* in a market is an opportunity to construct an arbitrage portfolio consisting of securities in the market.

An *arbitrage-free market* is a market providing no arbitrage opportunities.

Let $\Pi_A$ and $\Pi_B$ be two portfolios on the same time interval.

The *law of one price* is said to hold if $\Pi_A(T) = \Pi_B(T)$ implies $\Pi_A(t) = \Pi_B(t)$ for $\forall t < T$.

Equivalently, the *law of one price (LOP)* is said to hold if there do not exist two portfolios, say $\Pi_A$ and $\Pi_B$, such that $\Pi_A(T) = \Pi_B(T)$ but $\Pi_A(t) \neq \Pi_B(t)$ for some $t < T$.

In other words, *the law of one price simply says that two portfolios that will produce exactly the same cash flows in the future must have the same value to begin with.*

---

[8] It is understood that by a portfolio we mean a self-financing portfolio (i.e., no money is added to or withdrawn from the portfolio after time 0).

[9] One of those versions is given as follows:

A portfolio $\Pi$ is said to be a *statistical arbitrage portfolio*, or simply a *statistical arbitrage* if it satisfies either of the two sets of conditions:

$\quad$ (a) $\Pi(0) = 0$ $\quad$ and $\quad$ $\mathbb{E}(\Pi(T) > 0) > 0,$ $\quad$ (b) $\Pi(0) < 0$ $\quad$ and $\quad$ $\mathbb{E}(\Pi(T) \geq 0) > 0.$

The next example shows that no arbitrage is a sufficient condition of the law of one price. Equivalently, the law of one price is a necessary condition of no arbitrage.

**Example 7.8.** Show that the law of one price holds if there are no arbitrage portfolios.

**Proof.** Suppose otherwise, i.e., there exist two portfolios $\Pi_A$ and $\Pi_B$ and some $t_0 < T$ such that $\Pi_A(T) = \Pi_B(T)$ and $\Pi_A(t_0) \neq \Pi_B(t_0)$.

Without loss of generality, say $\Pi_A(t_0) < \Pi_B(t_0)$. Then we construct a portfolio $\Pi$ by long portfolio A and short portfolio B, written $\Pi = \Pi_A - \Pi_B$, which leads to

$$\Pi(t_0) = \Pi_A(t_0) - \Pi_B(t_0) < 0, \quad \text{and}$$
$$\Pi(T) = \Pi_A(T) - \Pi_B(T) = 0 \quad \text{with probability 1.}$$

Thus, by Definition 7.1, 2, $\Pi$ is an arbitrage! This contradicts the no-arbitrage assumption. □

**Example 7.9.** Let $\Pi_A$ and $\Pi_B$ be two portfolios on $[0, T]$ with

$$\Pi_A(0) = \Pi_B(0) \quad \text{and} \quad \Pi_A(T) < \Pi_B(T).$$

Construct an arbitrage portfolio.

**Solution.** Consider the portfolio $\Pi$ defined by $\Pi_B - \Pi_A$. We have

$$\Pi(0) = \Pi_B(0) - \Pi_A(0) = 0,$$
$$\Pi(T) = \Pi_B(T) - \Pi_A(T) > 0 \quad \text{with probability 1.}$$

Thus, $\Pi$ is an arbitrage by Definition 7.1, 1.

If the given information becomes

$$\Pi_A(0) = \Pi_B(0) \quad \text{and} \quad \Pi_A(T) > \Pi_B(T),$$

is $\Pi$ defined by $\Pi_A - \Pi_B$ an arbitrage?

□

**Remark 7.2. (Law of One Price)** The approaches to constructing an arbitrage demonstrated in Examples 7.8 and 7.9 provide a general guideline which will be applied throughout this chapter. Combining these two examples, we conclude the following:

If each of two investment portfolios produces a deterministic stream of cash flows as indicated below,

|  | $t = 0$ | $t = T$ |
|---|---|---|
| cash flows from $\Pi_A$ | $\Pi_A(0)$ | $\Pi_A(T)$ |
| cash flows from $\Pi_B$ | $\Pi_B(0)$ | $\Pi_B(T)$ |

then

$$\Pi_A(T) = \Pi_B(T) \;\Rightarrow\; \Pi_A(0) = \Pi_B(0) \text{ by the law of one price,}$$
$$\Pi_A(T) > \Pi_B(T) \text{ and } \Pi_A(0) = \Pi_B(0) \;\Rightarrow\; \Pi = \Pi_A - \Pi_B \text{ is an arbitrage,}$$
$$\Pi_A(T) < \Pi_B(T) \text{ and } \Pi_A(0) = \Pi_B(0) \;\Rightarrow\; \Pi = \Pi_B - \Pi_A \text{ is an arbitrage.}$$

□

**Remark 7.3.**

1. Recall that under the continuous-time framework, the investors are allowed to trade up to time $T < \infty$. Let $\mathbf{S}(t) = (\,S_0(t),\ S_1(t),\ S_2(t),\ldots,S_N(t)\,)$ be a market, where price process $\{S_i(t)\}$ is defined on filtered probability space $(\Omega, \mathfrak{F}, P, \{\mathcal{F}_t \subset \mathfrak{F} : 0 \le t \le T\})$ with $\mathcal{F}_T = \mathfrak{F}$ for each $i = 0,1,2,\ldots,N$. Suppose that positions of a portfolio P respectively on securities

$$S_0(t),\ S_1(t),\ S_2(t),\ldots,S_N(t)$$

at time t are expressed by $n_0(t), n_1(t), n_2(t), \ldots, n_N(t)$.
In order to understand the *law of one price* in a more precise fashion, let us define a *self-financing portfolio trading strategy* (or simply a *trading strategy*) to be a vector-valued function of $t$ denoted by $\mathbf{n}(t)$ and expressed by

$$\mathbf{n}(t) = (n_0(t), n_1(t), n_2(t), \ldots, n_N(t)).$$

Also, let us denote by $\Pi_P^{\mathbf{n}}(t)$ the process of the value of a portfolio P constructed by trading strategy $\mathbf{n}$. Then

$$\Pi_P^{\mathbf{n}}(t) = \mathbf{n}(t) \cdot \mathbf{S}(t) = \sum_{i=0}^{N} n_i(t) S_i(t).$$

The *law of one price* is said to hold if there do not exist two trading strategies, say $\mathbf{n}$ and $\mathbf{n}'$ such that $\Pi_P^{\mathbf{n}}(T) = \Pi_P^{\mathbf{n}'}(T)$ but $\Pi_P^{\mathbf{n}}(t) \ne \Pi_P^{\mathbf{n}'}(t)$ for some $t < T$. An interpretation of this more precise version of the *law of one price* along with Example 7.8 is that *arbitrage opportunities exist when the prices of similar assets are set at different levels.*

2. A relationship between no arbitrage and the equivalent martingale measure is given by the *First Fundamental Theorem of Asset Pricing*, which states: *The market is arbitrage free if and only if there exists an equivalent martingale measure.* (See page 314.)

3. The non-arbitrage assumption is a reasonable one for financial theory: In the real world, arbitrage opportunities do exist, but they are only transient because, once more investors jump in to share the free lunch, soon the free lunch will be over. The market price will be adjusted and move from an old equilibrium to a new one.

4. Although the no arbitrage assumption is reasonable theoretically, there are many different arbitrage strategies which are practically efficient and are often applied by quantitative models.

$\square$

## 7.2 Forwards

If today, denoted by time 0, you want to secure a transaction on a future date, denoted by time $T$, rather than waiting with no certainty of getting the market price, you may consider locking in the terms of the transaction by entering into a forward contract today.

Forward contracts offer users the ability to lock in a purchase or sale price without incurring any direct cost. In fact, forward contracts are the foundation of all derivatives as futures contracts are standardized forward contracts, swaps are series of forward contracts, and options are a variation of forward contracts.

### 7.2.1 Basic Concepts

**Definition 7.3.** A *forward contract* is an OTC-traded agreement between two parties to buy or sell a specified quantity of a specified asset at a specified future time at a price agreed today.

The *forward price* is the agreed unit price of an asset in a forward contract.
The *long position* is the position that the buyer enters in a forward contract.
The *short position* is the position that the seller enters in a forward contract.
The *delivery date* or *expiry* is the specified future time in a forward contract.
The *underlier* is the asset to be delivered on the expiry in a forward contract.
The *contract size* is the specified quantity in a forward contract.

There is no payment by either party when the contract is first entered into.[10] Thus the value of a forward contract at the time the contract is entered into is zero. The delivery date or expiry is also called the *exercise date* or *maturity* or *expiration date*, the time at which the asset changes hands. The seller is also called the *writer* of the contract. The forward price is also called the *exercise price*.

The *spot market* or *cash market* or *physical market* is a financial market where assets are traded for cash and immediately delivered on spot (e.g., the stock

---

[10] In old times, people would most often shake hands when agreeing on deals.

market, commodity market, and foreign exchange market), whereas a *derivative market* (e.g., forward market) is a market in which the delivery of the underlier asset is due at a later date(s). The spot market price is also called the *spot price*.

Since forward contracts are not standardized, the payments can be customized to fit a given situation. The *delivery price* is a price negotiated at the time the contract is entered into, which may or may not coincide with the forward price (see Example 7.13).

Forward contracts offer users the ability to lock in a purchase or sale price without incurring any direct cost.

**Example 7.10.** It is a well-known fact that grain prices may swing substantially between highs and lows.

In order to secure a smooth wheat supply, a flour mill A enters into a forward contract with a farmer B on June 1 to buy 100 (metric) tons of wheat at $222 per ton on September 30.

To become familiar with the terminologies in Definition 7.3, we identify that in this contract, the buyer is A, the seller is B, the underlying asset is wheat, *the forward price is $222*, the expiration date is September 30, the contract size is 100, and the delivery price is $222 as well (which happens to be the same as the forward price).

On June 1, A and B sign the contract and shake hands. No money changes hands. However, on September 30, A will pay $22,200, irrespective of the price of wheat in the spot market, and B will deliver 100 tons of wheat to the flour mill.

Both parties A and B are bound by the contract and have to honor their commitments.                                                                            □

Let

$t = 0$ be the time the forward contract is entered into (e.g., June 1),
$t = T$ be the expiration of the forward contract (e.g., September 30),
$S(t)$ be the spot price of the underlier in the forward contract at time $t$.

➤ We denote by $F_T(0)$ the forward price specified in a forward contract initiated at time $0$ and expiring on time $T$ (e.g., $F_T(0) = \$222$).

**Example 7.11.** We continue from the last example.

The wheat spot price fluctuations either naturally or artificially become an interplay between supply and demand. Keep in mind that the spot wheat price is a stochastic process. Thus the forward price specified in the forward contract on wheat is a stochastic process as well.

Suppose that starting from July 10, massive droughts (or floods) negatively affect the supply of wheat, which in turn triggers the spot price of wheat to surge (say, to $322). Imagine that another flour mill X enters into a forward contract on July 10 to buy 100 tons of wheat on September 30. Would X be able to negotiate with any counterparty to get the same forward price, \$222? Of course not!

If we denote July 10 by time $t_1$, in this new contract, denoted by $F_T(t_1)$, the forward price $F_T(t_1)$ would be much higher than $F_T(0)$. □

Notice that two forward contracts from the last two examples have the same terms except the forward price and the initial time of the contract.

➤ To emphasize the fact that forward prices are functions of time, we denote by $F_T(t)$ the forward price.

**Example 7.12.** We continue from Example 7.10.

In order to understand the profit and loss (P&L) from the deal *in terms of a market value*, we argue that if, on September 30, the spot price of wheat were \$232, A could sell 100 tons of wheat immediately and profit \$1000 (i.e., (\$232 - \$222)×100). However, if on September 30, the spot price of wheat were \$202, A could sell 100 tons of wheat immediately and lose \$2000. That is,

$$\text{the payoff for the buyer} = S(T) - F_T(0),$$
$$\text{the payoff for the seller} = F_T(0) - S(T).$$

□

For the convenience of conversation,

➤ we define the *terminal payoff of a forward contract* to be the payoff from a long forward contract, that is,

$$\text{forward payoff} = \text{long forward payoff} = S(T) - F_T(0). \quad (7.1)$$

The payoff from a short forward contract is the negative value given by (7.1):

$$\text{short forward payoff} = F_T(0) - S(T) = -\text{forward payoff}.$$

**Definition 7.4.** A *forward payoff diagram* is a graph of the terminal payoff from long position of a forward contract as a function of the underlier price at $T$.

In short, a forward payoff diagram is a graph of long forward payoff against $S(T)$.

If we simplify notation by letting $K = F_T(0)$, $x = S(T)$, and $y = x - K$, then a forward payoff diagram is a graph on xy-plane, which is a straight line that can be obtained by translating the graph of $y = x$ to the right by $K$ units.

In a parallel way, one can define and visualize a payoff diagram from the short position of a forward contract. It is the graph of the straight line $y = -x + K$, where $x = S(T)$ and $K = F_T(0)$ on xy-plane and can be obtained by translating the line $y = -x$ to the right by $K$ units (see diagrams below).

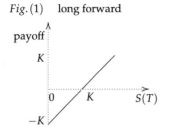

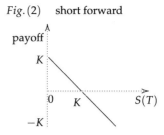

**Example 7.13. (Delivery Price of a Forward)** Once again, we continue from Example 7.10.

Now, keep in mind that forward contracts are not standardized. This means that payments can be made in many different ways. Suppose that there were an upfront payment of $2000 (e.g., a down payment) from party A to party B at the time when the contract is entered into; the delivery price would become $202 (different from the forward price).                                               □

### 7.2.2 Forwards on Assets Paying a Continuous Cash Dividend

The spot-forward parity (see Section 7.2.3) for underliers with continuously paid cash dividends is useful when the underliers are stock indexes containing many stocks,[11] since such an index can be modeled as the dividend being paid continuously at a rate that is proportional to the level of the index.

To obtain such a spot-forward parity in the next section, let us now recall (6.29) on page 281 from Section 6.4.3: $\mathcal{D}(t) = S(t)\,q\,dt$, where $S(t)$ is the spot price of the underlier of a forward contract and $q$ is the (annualized) *dividend yield* of the underlier. In the rest of this chapter, we will use the familiar results on page 281:

➤ The dividend reinvestment yields $e^{qT} - 1$ more units at time T from 1 unit of the underlier at time 0,
➤ The dividend reinvestment yields $1 - e^{-qT}$ more units at time T from $e^{-qT}$ units of the underlier at time 0,

and the following assumptions: *Unless stated otherwise,*

1. *By stating that "an asset with annual dividend yield q," we mean that "an asset pays a constant, continuous proportional annual dividend yield rate q that is continuously reinvested to buy more units of the asset and that holding the asset neither incurs cost of carry nor provides any other convenience yield."*

2. *r is the risk-free interest rate compounded continuously.*

---

[11] We leave out of the theoretical arguments such as whether a no-arbitrage condition can be verified when the underlier is the S&P 500 Index.

### 7.2.3 Forward Price Formula and the Spot-Forward Parity

We note that the assumption of cost of carry[12] and convenience yield[13] other than the dividend yield in the next theorem is not a reasonable one.

**Theorem 7.1.** *Suppose that a forward contract entered into at time 0 with expiration T is on "an asset with annual dividend yield q" (see page 341) and that r is the risk-free interest rate, the fair forward price is given by*

$$F_T(0) = S(0) e^{(r-q)T}, \tag{7.2}$$

*where $S(0)$ is the price of the asset at the time 0.*

In words, (7.2) says that under an arbitrage-free assumption, buying the forward contract and taking delivery is equivalent to buying the underlying asset from its spot market today and holding in the sense that the cost of both strategies must have the same present value.[14]

*Proof.*
**Case 1.** If $F_T(0) < S(0) e^{(r-q)T}$, we construct two portfolios A and B, denoted by $\Pi_A$ and $\Pi_B$, respectively, with positions established at time 0 indicated below:

$\Pi_A$ : short 1 forward.
$\Pi_B$ : short $e^{-qT}$ units of the asset to long bond[15] with $S(0)e^{-qT}$ at rate r.

We denote by $\Pi_A(t)$ the value of portfolio $A$ and $\Pi_B(t)$ the value of portfolio $B$ at time $t$. Then the initial and terminal values of two portfolios are

$\Pi_A(0) = 0$ (no initial cost for a forward contract),
$\Pi_B(0) = 0$ (the proceed from shorting is used to buy the bond immediately),
$\Pi_A(T) = F_T(0)$ ($F_T(0)$ is the forward price, the cash flow of $\Pi_A$ at T),
$\Pi_B(T) = S(0) e^{(r-q)T}$ (the principle $S(0)e^{-qT}$ earns interest at r),

---

[12] The *cost of carry* is the cost incurred by holding the underlying asset such as storage costs or insurance as well as other incidental costs.

[13] For example, being able to take advantage of shortages of the underlying asset.

[14] Note that in the equality $F_T(0)e^{-rT} = S(0) e^{-qT}$, the LHS is the present value of $F_T(0)$, and the RHS is the present value of $S(T)$ because $e^{-qT}$ shares at time 0 grow to 1 share at time $T$ and, during the same time period, the stock value changes from S(0) to $S(T)$.

[15] The bond here is a zero-coupon bond. For our purpose, the position of long 1 par zero maturing at time $T$ can be interpreted as "lend \$$S(0)e^{-qT}$ at interest rate r." Similarly, a position of short zero here would be interpreted as "borrow \$$S(0)e^{-qT}$ at interest rate r." Using the bond terminologies will provide us convenience (e.g., for letter expression of long or short position) later.

which are summarized in the table below.

|                        | At time $t$ | At time $T$ |
|------------------------|-------------|-------------|
| Cash flows from $\Pi_A$ | 0           | $F_T(0)$    |
| Cash flows from $\Pi_B$ | 0           | $S(0)e^{(r-q)T}$ |

An arbitrage portfolio $\Pi = \Pi_B - \Pi_A$ (note: $-\Pi_A$ has position: long 1 forward) has been established (see Remark 7.2). This contradicts the no-arbitrage assumption.

**Case 2.** If $F_T(0) > S(0)e^{(r-q)T}$, using the table in case 1, let $\Pi = \Pi_A - \Pi_B$, then $\Pi$ is an arbitrage portfolio by Definition 7.1, 1 (for $\Pi$, at time 0, consisting of the positions: short 1 forward, short bond with $S(0)e^{-qT}$ at rate $r$ to long $e^{-qT}$ units of the asset).

Thus, under the assumption of no arbitrage, (7.2) holds.                    □

For a given $t \in [0,T]$, if we replace 0 by $t$ in (7.2), then we establish a relation between the underlier spot price $S(t)$ and the forward price $F_T(t)$. This relation is referred to as the spot-forward parity and shows how a forward can be replicated.

**Corollary 7.1.** *The spot-forward parity is given by*

$$F_T(t) = S(t)e^{(r-q)(T-t)}, \quad t \in [0,T]. \tag{7.3}$$

**Example 7.14.** Suppose that the current spot price of a continually paying dividend asset is \$222, the interest rate is $r = 3\%$, and the dividend yield is $q = 2\%$.

1. What are the one-month and seven-month forward prices for the asset in an arbitrage-free market?
2. Let $\Pi$ be a portfolio on time interval $[0,T]$ consisting of three positions starting from time 0: borrow \$222 at the rate 3%, long 1 unit of the asset, and short the three-month forward. Is $\Pi$ an arbitrage portfolio?

**Solution.** Denote by $K_1$ and $K_7$ the one-month and seven-month forward prices respectively. Applying (7.2), we obtain

$$K_1 = 222\,e^{(0.03-0.02)\times\frac{1}{12}} = \$222.19,$$
$$K_7 = 222\,e^{(0.03-0.02)\times\frac{7}{12}} = \$223.30.$$

In order to answer the second part of questions, one needs to fill out the table below first.

|                      | At time 0 | At time $T$ |
|----------------------|-----------|-------------|
| Cash flows from $\Pi$ |           |             |

The detailed work is left as an exercise for the reader.

□

**Example 7.15.** Let $F = F(S,t) = F_T(t) = S(t)e^{(r-q)(T-t)}$. Suppose that the underlier's price follows a geometric Brownian motion:

$$dS = \mu S\,dt + \sigma S\,d\mathfrak{B}.$$

We claim that so does the forward price. In fact,

$$dF = (\mu - r + q)F\,dt + \sigma F\,d\mathfrak{B}.$$

The detailed computation is straightforward and is left for the reader as an exercise (see Exercise 7.26 on page 381).                                                    □

**Remark 7.4.**

1. Using the language of stochastic calculus that we introduced in the last chapter and only considering the forward contracts that can be replicated by self-financing trading strategies, the forward price can be expressed by

$$F_T(t) = S(t)/d_{t,T}, \tag{7.4}$$

where $d_{t,T} = \mathbb{E}_{\mathbb{Q}}(B(t)/B(T)\,|\,\mathcal{F}_t)$ with $B(0) = 1$ and $B(t) = e^{\int_0^t r(u)du}$, in which $\{r(t)\}$ represents a stochastic interest rate process adapted to the filtration $\{\mathcal{F}_t\}$, and $\mathbb{Q}$ is an equivalent martingale measure. It is worth noting that formula (7.4) can be verified, or proved, by establishing the equality $\mathbb{E}_{\mathbb{Q}}((S(t') - F_T(t))B(t)/B(t')\,|\,\mathcal{F}_t) = 0$, where $t < t' \le T$. (*Hint*: The time $t'$ value of the forward contract entered into at time $t$ is $S(t') - F_T(t)$ and $F_T(t)$ is $\mathcal{F}_t$-measurable.)

2. The treatment of forwards and futures in the current and next sections emphasizes the intuitive nature of these financial products. For a systematically derivational approach to them, we refer the reader to the literature (e.g., Anderson and Kercheval [1]).

                                                                                        □

### 7.2.4 Forward Value Formula

We begin this section with a natural question:

➤ How much is a particular forward contract worth today that was entered into in the past but has not yet expired?

Let $t \in [0,T]$ be the current time; a natural way to evaluate a forward contract initiated at 0 and expiring at $T$ is to consider the difference between two forward terminal payoffs: one is entered into at time 0 with forward price $F_T(0)$ and the other is entered into currently, at time $t$, with forward price $F_T(t)$.

Both forwards must be written on the same asset and have the same expiration and contract size. Recalling the definition of a (long) forward payoff from (7.1) on page 339, we obtain

$$\text{Difference between two forward payoffs} = S(T) - F_T(0) - (S(T) - F_T(t))$$
$$= F_T(t) - F_T(0).$$

To discount this difference to get its value at the current time, we obtain the current (market) value of the (long) forward contract that was entered into at time 0 in an arbitrage-free market:

**Theorem 7.2.** *The value of a (long) forward contract at time t entered into at time 0 with expiration at time T, denoted by $f_T(t)$, is given by*

$$f_T(t) = (F_T(t) - F_T(0)) e^{-r(T-t)}, \quad t \in [0, T]. \tag{7.5}$$

*The time-t value of a short forward contract is $-f_T(t)$.*

*Proof.* An idea of proof is to establish two portfolios at time t such that each of them produces a deterministic stream of cash flows as indicated in the table below (where $f_T(t)$ is to be determined).

|                          | At time  t                      | At time  T       |
| ------------------------ | ------------------------------- | ---------------- |
| Cash flows from $\Pi_A$  | $f_T(t)$                        | $F_T(t) - F_T(0)$ |
| Cash flows from $\Pi_B$  | $(F_T(t) - F_T(0)) e^{-r(T-t)}$ | $F_T(t) - F_T(0)$ |

With this idea, the proof is straightforward and left as an exercise to the reader.
□

From (7.5), as what we expected indeed, the initial and terminal values of a forward are

$$f_T(0) = 0,$$
$$f_T(T) = F_T(T) - F_T(0) = S(T) - F_T(0) = \text{ terminal payoff of forward.}$$

Formula (7.5) is also expressed by

$$f_T(t) = (F_T(t) - K_0) e^{-r(T-t)}, \quad t \in [0, T], \tag{7.6}$$

where $K_0 = F_T(0)$.

**Example 7.16.** Given $K_0 = \$252$, $F_T(t) = \$222$, $T - t = 6$ months, and $r = 3\%$, determine the value of the forward contract in an arbitrage-free market.

**Solution.**

$$f_T(t) = (222 - 252) e^{-0.03 \times 0.5} = -29.55,$$

which (a negative value) means that if the forward contract were to be closed out currently, the buyer would compensate the writer $29.55.
□

As a final note of this section, we emphasize that

➤ The forward price (or contract price) and forward value (or contract value) are conceptually different. One should not confuse the two.

## 7.3 Futures

Futures contracts are standardized forward contracts.

### 7.3.1 Evolution from Forwards to Futures

Forward trading has survived for a few hundred years. The forward contract was created to stabilize grain prices by farmers on top of the already centralized grain trade.

In 1848, the Chicago Board of Trade (CBOT) was formed, and trading was originally in forward contracts.

Although the forward contract is customized to meet the user's special needs, it has illiquidity due to the lack of market exposure to potential buyers or sellers,[16] and it has counterparty risk, the risk that their counterparties fail to meet their obligations (e.g., default in the payment or even bankrupt). Thus, counterparties must check each others' creditworthiness before a forward contract is entered into. This is why the end users of forward contracts are mainly big institutions.

Illiquidity and counterparty risk are the inherent limitations of forward contracts. A way to overcome these shortcomings is to standardize forwards. The standardized forward contract is called the futures contract. In 1972, the Chicago Mercantile Exchange (CME) started to offer futures contracts.

To mitigate or even remove credit risks, regulations made in accordance with laws impose mark-to-market and daily settlement on futures traders' margin accounts (to be explained shortly). The daily balance on such an account is calculated based on the settlement price defined by an exchange.

To increase liquidity, regulations made in accordance with laws impose standardizations on terms of futures contracts including what can be delivered, when it can be delivered, how it can be delivered, where it can be delivered,

---

[16] To compare to a familiar environment, simply consider, in a housing market, the difference between the market exposure of "for sale by owner" and that of "for sale by real-estate agency." Generally speaking, the bigger the market exposure, the higher the level of liquidity.

and so on. In otherwords, standardize underlying assets, contract size, expiration days, and method of settlement (cash or physical at what location[17]), and so forth.

In this way, futures contractors deal with an exchange rather than each other. That is, in futures trading, your counterparty is an exchange.

**Example 7.17.** A forward contract allows a farmer to sell 1234 bushels of wheat next February, whereas a futures contract does not because the size of one futures contract is 5000 bushels and the expiration months are March, May, July, September, and December. However, the farmer does not expose himself to any credit risk by entering into a futures contract.                                    □


### 7.3.2 Basic Concepts

**Definition 7.5.** A *futures contract* is an exchange-traded standardized agreement between two parties to buy or sell an asset at a specified future time at a price agreed today. The contract is mark to market daily and guaranteed by the clearinghouse.

The definitions of futures price, long position, short position, expiration date, etc. are parallel to their counterparts for forwards. For example, the *futures price* is the agreed price of an asset in a futures contract.

A futures contract is similar to a forward contract except that contractors deal with a third party, i.e., an exchange, rather than each other. By doing so, the inherent credit risk of a forward contract is mitigated because both parties must meet margin account requirements under the *mark-to-market* accounting rule, which makes a futures contract like a sequence of daily forward contracts until maturity.

Although by definition futures and forward contracts are similar in terms of the final results, the mark-to-market accounting rule requires futures contractors to settle up daily (or make *daily settlement*) through the exercise day, whereas by contrast there is no settlement for forward contracts until the exercise day.

Since margin account requirements and the mark-to-market accounting rule differentiate the futures contract from the forward contract, to understand futures we begin with explaining these two concepts. To do so intuitively, we illustrate margin account requirements and compute the daily margin balance in the example below.

Before trading a futures contract, the prospective trader must deposit funds with a broker. This deposit serves as a performance bond and is referred to as the *initial margin*, the level of which is based on a function of the price volatility of the underlier (e.g., a commodity).

---

[17] Futures contracts allow fewer delivery options than forward contracts.

**Example 7.18. (Computing Daily Margin Balance Under Mark-to-Market Rule)** Suppose an investor takes a long position in two October gold futures contracts with contract size of 100 (i.e., the forward price is for 100 oz of gold). The initial margin requirement for a futures trader by the clearinghouse is $5,000 per contract with a maintenance margin level of $4,000.

Suppose that on day one the futures price moves up from $1200 to $1220. Then the margin account balance will change from $10,000 (2 × $5000) to $14,000 (change = 2 × 100 × $20 = 4,000). More daily margin account balance fluctuations are shown in the table[18] below:

| Day | Futures price | Daily price change | Contract No. | Gain/loss | Margin Bal. |
|-----|---------------|--------------------|--------------|-----------|-------------|
| 0 | 1,200 | | | | 10,000 |
| 1 | 1,220 | +20 | 2 | 4,000 | 14,000 |
| 2 | 1,190 | −30 | 2 | −6,000 | 8,000 |
| 3 | 1,180 | −10 | 2 | −2,000 | 6,000 |

The margin account owner receives a margin call for $6,000 < $8,000 and has lost 40% of the account value in only 3 days. Suppose that the exercise date is still a month away; the buyer of the futures contract still needs to send at least $4,000 to the broker before the beginning of the next business day to bring the margin balance to $10,000, the initial margin level, or close positions. Otherwise, the broker can and will cancel the contract.

Although margin required is used as collateral to cover losses, margin calls can only mitigate, not eliminate, the risk in futures markets, particularly in FX (foreign exchange) markets. For example, the Swiss Franc fiasco in January 2015 brought many traders' margin balance to negative levels over night.

Note that the table shows that the margin balance column fully reflects all the daily losses from the long position of 2 futures contracts, and observe that when the initial margin is cut off 40%, the futures price has lost only 1.7% of its original value. □

### 7.3.3 Impact of Daily Settlement: A Brief Discussion

We let $\hat{F}_T(t)$ be the futures price at time $t$ and $\hat{\mathbf{f}}_T(t)$ be the futures value at time $t$ and close the futures section by the following remark.

---

[18] Precisely speaking, the futures price in the second column should be the *daily settlement price* or simply *settlement price*, which is defined by the exchange. There are different types of settlement procedures. Each derivative exchange has a set of procedures used to calculate the settlement price. Margin requirements are based on the daily settlement price, not the daily closing price. For our purpose, we consider the settlement price to be essentially the closing price on that day.

**Remark 7.5.**

1. Since the daily settlement makes a futures contract like a sequence of daily forward contracts until maturity, and the value of a forward is zero at the initial time, $\hat{\mathscr{F}}_T(t) = 0$ for $t < T$ (although the terminal payoff of a futures contract is the same as that of its forward counterpart), the futures value is different from the value of its forward counterpart (as the latter needs not equal zero except at the time the contract is entered into).

2. Because of the daily settlement and the exchange-treated nature of a futures contract, $\hat{F}_T(t)$, a futures price for delivery of an asset at time $t < T$, is an agreed price between the trader and the exchange (i.e., a price determined by the exchange rather) and, therefore, behaves more like a market price. It has been proved that if futures interest rates are deterministic, then under the no-arbitrage assumption, $\hat{F}_T(t) = F_T(t)$. Otherwise the equality may not hold. More description about the behavior of $\hat{F}_T(t)$ can be found in the literature (e.g., Hull [11]).
   For more impact of daily mark to market on futures contracts such as the correlation between the futures price movements and interest rate movements, we refer the reader to the literature (e.g., Cox, Ingersoll, and Ross [4] and Duffie and Stanton [8]).

3. For the following reasons:

   ➢ Futures have lower transaction cost, since the evolution from OTC-traded forwards to exchange-traded futures encourages liquidity;
   ➢ With futures, it is equally easy to go short as to go long (a convenience inherited from forwards);
   ➢ The value of forward kept at zero (i.e., $\hat{\mathscr{F}}_T(t) = 0$) allows a futures contractor (either buyer or seller) to close out his position at any time;

   it is easier to use futures to hedge, particularly easier to go short with futures market than with the spot market because of the uptick rule[19] imposed on the spot market.

   □

## 7.4 Swaps

Swap contracts are basically a series of cash-settled forward contracts that require action to be taken by investors periodically over time.

The tailor-made structures of swaps create a wide variety of financial instruments that financial institutions trade in order to hedge against risk.

---

[19] The SEC requires that every short sale transaction be entered at a price that is higher than the price of the last trade.

### 7.4.1 A Brief Introduction

**Definition 7.6.** A *swap contract*, or simply a *swap*, is an OTC agreement between two parties to exchange (or swap) two streams of cash flows at specified future times according to certain specified rules.

These cash flows are most commonly the interest payments associated with debt service. Financial institutions and companies dominate the swaps market with almost no individual participation.

By definition, swaps are basically sequential cash-settled forward contracts that require action to be taken by the counterparties on periodic dates. Consequently, the initial value of a swap should be zero.

**Example 7.19.** Consider a forward contract on 100 ounces of gold at expiration in 6 months at the price $1281 per ounce. If the gold is viewed as a currency, this forward can be viewed as a simple example of a swap: a party entering into the forward contract to buy 100 ounces of gold for $1281 in 6 months is equivalent to a party entering into a swap to use $128,100 to exchange for 100 ounces of gold in 6 months.                                                                □

Popular swaps are classified into different categories.

An *interest rate swap* is a transfer of interest rate cash flows without transferring underlying debt and requires both cash flow streams to be in the same currency. The predetermined amounts on which the exchange interest payments are based are called the *notional principal*.

A *currency swap* requires the principal to be specified in each of the two currencies and two parties to exchange cash flows as well as the principals at the beginning and end of the life of the swap.

An interest rate swap is called a *plain vanilla swap*[20] if one stream of cash flow consists of floating interest rate payments and another consists of fixed interest rate payments. The former stream is called the *floating leg* and the latter the *fixed leg*. By tradition, the floating rate payer is called a *buyer* of the swap and the fixed rate payer is a *seller* of the swap. During the life of a plain vanilla swap, every period there is an exchange of payments. The payments under the fixed leg are always known, since the fixed rate is set on the pricing date of the swap, but the payments under the floating leg are known only one period prior to the exchange of payments. For example, if the payments are made at the end of each period, then the floating rate payment will be based on the floating rate that prevailed at the beginning of the period, since the floating rate is reset periodically based on market levels after the rate for the first payment is set at market level on the pricing date (see Example 7.20).

---

[20] The name *plain vanilla swap* reflects that those swaps do not possess any special or unusual features.

Other than the above categories of swaps, there are also *credit default swaps (CDS)*, which can be used as a protection against credit loss, *commodity swaps*, which can be used by commodity producers to manage their exposure to price fluctuations, and other variations of swaps, since tailor-made swap structures create a wide variety of financial instruments that financial institutions trade in order to hedge against risk.

Among all different types and variations of swaps, the most widely used are interest rate swaps, which are used mostly to reduce borrowing costs:

1. Based on changes in its long-term or short-term assets and its credit rating, an institution with an existing debt service obligation faces higher than expected borrowing costs because of a change in its initial interest rate outlook. To avoid these costs, the institution wants to swap to a different exposure (see Example 7.20).

2. To receive lower borrowing costs than those available by directly accessing the fixed-rate or floating-rate markets, two institutions may work together by exploiting their comparative advantages as borrowers in different markets and swapping the proceeds (see Example 7.21).

The interest rate swaps market is the largest and fastest growing financial derivative market in the world.

It is worth noting that in the next two examples, we will display only the mechanics of interest rate swaps and ignore credit risk differences, even though counterparty credit risks are very important for swap traders to evaluate.

**Example 7.20. (Plain Vanilla Swap)** Consider the scenario of two companies $A$ and $B$. Company $A$ has taken a loan at a six-month $LIBOR$[21] plus 1% floating,[22] but now would like to have the loan at a fixed rate of 3% since the company expects interest rates to rise above 3% soon, whereas company $B$ has a loan at a fixed annual rate of 3% and expects interest rates to drop below 3% soon.

Since for a company to pay off an old loan and apply for a new loan or to refinance a loan can be not only costly but also legal document-intensive due to regulations, companies $A$ and $B$ decide to enter into a swap contract, which is a much simpler way of exchange of interest rates between fixed and floating.

Assume that the terms of the swap contract include the following:

➤ The notional principal is two million dollars,
➤ The life of the contract is 4 years,
➤ $A$ pays $B$ six-month $LIBOR + 1\%$,

---

[21] See Section 1.1.2 on page 4.

[22] Such a loan is called a *floating rate loan* or a *variable* or *adjustable rate loan*, which is a debt, such as a bond, mortgage, or credit, that does not have a fixed interest rate over the life of the debt. The interest rate on a floating rate loan is referred to as a *floating interest rate*, or *variable* or *adjustable rate*.

➤ B pays A 3% fixed and

➤ There is an exchange of payments every 6 months from the initialization.

Given the *LIBOR* rates in the table below, since the floating leg payment is always known 6 months before the exchange of interest payments, company B receives $2,000,000 \times (1\% + 1\%)/2 = \$20,000$ from company A as the first period interest payment (out of a total of eight variable payments), and the rest of the floating cash flow and fixed cash flow of the swap is illustrated in the table:

| Time (months) | Six-month LIBOR | Payment from A to B | Payment from B to A |
|---|---|---|---|
| 0 | 1% | | |
| 6 | 2% | $20,000 | $30,000 |
| 12 | 3% | $30,000 | $30,000 |
| 18 | 2% | $40,000 | $30,000 |
| 24 | 3% | $30,000 | $30,000 |
| 30 | 5% | $40,000 | $30,000 |
| 36 | 4% | $60,000 | $30,000 |
| 42 | 2% | $50,000 | $30,000 |
| 48 | | $30,000 | $30,000 |

A straightforward verification of the interest payments in the above table is left as an exercise for the reader. □

**Example 7.21. (Mechanics of Interest Rate Swaps)** Suppose that both companies X and Y need to borrow US dollars and that company X would like to borrow at fixed rate, whereas company Y would like to borrow at floating rate. The cost to each company of accessing either the fixed rate or the floating rate market for a new debt issue is given below[23]:

| Borrower | Fixed rate | Floating rate |
|---|---|---|
| X | 5.50% | LIBOR $+0.40\%$ |
| Y | 4.30% | LIBOR $+0.20\%$ |
| Difference (margin) | 1.2% (120bp) | 0.2% (20bp) |

Given the differences in rates indicated in the table above, companies X and Y realize that they could achieve a combined 100 basis point (i.e., 1%) savings and decide to enter a swap with a *swap bank*[24] B.

We claim that X, Y, and B can all benefit financially if X borrows at the floating rate, Y borrows at the fixed rate, X and Y swap interest payments, and B as the intermediary charges 0.1% of the notional principal. To prove our

---

[23] Company Y enjoys a lower borrowing cost in both markets because we assume that company Y has a better credit rating than company X.

[24] A *swap bank* is a generic term for a financial institution that facilitates swaps between counterparties and serves as a broker or a dealer for the trading.

claim to be true, we demonstrate the simple mechanics of the swap with the calculations below:

1. $X$ pays 4.9% fixed to $B$ and $B$ pays 4.8% fixed to $Y$, and
2. $Y$ pays $LIBOR$ floating to $B$ and $B$ pays the same $LIBOR$ floating to $X$.
3. The borrowing cost for $X$ after swapping proceeds becomes at the fixed rate

$$4.9\% - LIBOR + (LIBOR + 0.4\%) = 5.3\%,$$

which is less than 5.5%. Thus, the savings for $X$ is 0.2%.
4. The borrowing cost for $Y$ after swapping proceeds becomes at the floating rate

$$LIBOR - 4.8\% + 4.3\% = LIBOR - 0.5\%,$$

which is less than $LIBOR + 0.2\%$. Thus, the savings for $Y$ is 0.7%.

□

Although payments paid in a foreign currency may entail currency risk, a currency swap can be used to transform loans or cash flows from one currency to another. The next example briefly explains certain purposes of doing so.

**Example 7.22. (Purpose of Currency Swaps)** Consider the scenario of two companies $A$ and $B$ in two different countries. Company $A$ is an US-based company, whereas company $B$ is a European company. Each company has plan to enter global markets and needs to take loans to fund capital expenditures. Moreover, company $A$ needs to have Eurodollars to carry out its plan and company $B$ needs to have US dollars.

Because of a variety of reasons such as different regulations in different countries and geopolitical concerns, it is often advantageous for a company to borrow from its domestic banks and disadvantageous to borrow from foreign banks. To get around the financial difficulties of borrowing from foreign banks, company A could take a loan from an US bank and company B from a European bank, and then the two companies could enter a swap contract.   □

**Remark 7.6.**

1. A variation of swap is called a *variance swap* which is an *OTC* instrument that allows one to speculate on or hedge risks associated with volatility. Variance swaps on major stock indexes such as S&P 500 are actively traded.

2. Since an interest rate swap contract is basically a series of forward contracts, intuitively, its value is zero at the time the swap is entered into (although some of these forward contracts may have nonnegative values and others may have negative values, the sum of all of them is zero) and may change over time due to the change of interest rates. For the buyer of the swap, the party who receives fixed and pays floating, the value of the swap is positive if the fixed rate is greater than the floating rate; the value of the

swap is zero if the fixed rate equals the floating rate; and the value of the swap is negative if the fixed rate is less than the floating rate.

Since the floating rates cannot be observed beforehand, the valuation of an interest rate swap may involve calculations of forward rates based on a yield curve in addition to present value techniques.

Generally speaking, the pricing and valuation of swaps can be complex and requires vigorous financial analysis to derive fair values. We refer the reader to the literature on this subject.                                                          □

## 7.5 Options

Options are contingent claims which may be viewed as a variation of forwards[25] that provide an investor with an option, but not an obligation, to complete a transaction on or by a future date.

The most practical elements covered in this section are terminal payoff and profit diagrams as they provide the simplest way to analyze option strategies.

A theoretical discussion on the risk-neutral valuation of options will be provided in the next chapter.

### 7.5.1 Basic Concepts

**Definition 7.7.** An *option contract*, or simply an *option*, is either an individually negotiated *OTC* or exchange-traded agreement between two parties, which grants the *buyer* (or *holder* or *owner*) the *right*, but not the obligation, to buy or sell a specified quantity of a specified asset on or by a specified date at a specified price.

The *premium* of an option contract is the amount that the buyer has to pay and the *seller* (or *writer*) receives at the time when both parties enter into the contract.

The specified asset is called the *underlier* or *underlying asset* of the option.

The specified date is called the *expiration* or *maturity* (date) of the option.

The specified price is called the *strike price* or *exercise price* of the option. Naturally, an expiration date is also called an *exercise date*.

The specified quantity is called the *contract size*, which denotes how much of the underlying asset will change hands if the option is exercised, where *exercise an option* means to put into effect the right in an option contract.

---

[25] We mean the European-style option (to be defined shortly), which is the basic option style from which other styles of options derive.

By definition, an option contract is structured for one to pay money to have a choice in the future. This explains the name "option."

Through the standardization of the underliers, contract sizes, and expiration dates, organized option exchanges such as $CBOE^{26}$ and $NYSE$ provide the advantages of liquidity, low transaction costs, and safety for option trading.

## Option Styles

➤ *American-style options*, or simply *American options*, can be exercised at any time prior to or on expiration day, whereas *European-style options*, or simply *European options*, can only be exercised on expiration day.[27] Therefore, American options are more flexible than European options.

➤ Since the ability to exercise an American option at any time prior to or on expiration makes American options more flexible than European options, *American options are more valuable than European options.*[28]

➤ Although the flexibility and ease of exercise makes American options more valuable, it also makes the *valuations or pricing of the American options more complex and difficult than European options.*[29]

## Option Types

The basic bread-and-butter options fall into one of the following two types, designated according to buying and selling rights:

**Definition 7.8.**

1. A *call contract*, or a *call option* or simply a *call*, is an option contract which grants the buyer the *right to buy* a specified quantity of an underlying asset on or by an expiration date $T$ at a strike price $K$. The *payoff* is $\max\{S(T) - K, 0\}$ when exercised at expiration $T$.

2. A *put contract*, or a *put option* or simply a *put*, is an option contract which grants the buyer the *right to sell* a specified quantity of an underlying asset

---

[26] The Chicago Board Options Exchange ($CBOE$), a spin off from the Chicago Board of Trades, first traded standardized options in 1973 and $NYSE$ in 1982.

[27] There are other option styles such as Asian or Bermuda. We only consider American and European options because they are the most actively traded options.

[28] Indeed, American-style stock options tend to cost more than equivalent European-style options for the same stock in practice. Almost all exchange-traded stock options are American-style options, whereas stock index options can be issued as either American or European options (e.g., S&P 100 index options are American options, and Nasdaq 100 index options are European options).

[29] Option pricing done by the Black-Scholes-Merton model applies to European options, not to American options, and reflects the risk associated with having to wait to exercise the option, which is not appropriate for American options because of the possibility of early exercise.

on or by an expiration date $T$ at a strike price $K$. The *payoff* is max$\{K - S(T), 0\}$ when exercised at expiration $T$.

A call option is called a "call" because the buyer (owner) has the right to "call the underlying asset away" from the seller (writer). A put option is called a "put" because the buyer (owner) has the right to "put the underlying asset to" the seller (writer). The *payoff* may be considered as the option value at exercise. Clearly, an option is a financial instrument with nonnegative value at any time (for it involves no obligation prior to expiration).

Calls and puts are also referred to as *vanilla* options.[30]

Although in reality stock options are American options and American options are more useful, all the examples in this section will be confined to European options hypothetically on stocks. They are easier to understand and give the background needed for studying the option pricing models in Chapter 8 and for following related literature beyond. The next example provides an intuition about what a call option actually means.

**Example 7.23. (European Call)** To make a call option easier to understand, in the example we assume the contract size to be 1, that is, a call on 1 unit of the underlier.

Consider a six-month (i.e., $T = 0.5$) European call written on stock XYZ with strike price $K = \$7$.

By entering this contract as a buyer, you have the right (not the obligation) to buy 100 shares of stock XYZ at the price \$7 per share at time $T$ (in 6 months).

On expiration day $T$ (a Friday), whether you exercise the call (i.e., exercise the right to buy the stock) depends on the stock price in the spot market on that Friday:

*If* the stock price is *above* \$7, say \$10, you exercise the call to buy 1 share of the stock at \$7 and sell it immediately for \$10 to make a profit of \$3 (assuming no commissions). We say that the terminal payoff of the call is \$3 if the stock price is \$10 at expiration.

*If* the stock price is *below* \$7, you do not want to exercise the call because you can buy the stock at a lower price in the spot market. Therefore, you let the call expire worthless by doing nothing. We say that the terminal payoff of the call is \$0 if the stock price is below the strike price at expiration.

More generally speaking, if we denote by $C(S, t)$ the value of European call option at time $t$, then at expiration $T$, the *terminal payoff* of an option position of long a call (or simply, terminal payoff of a call), i.e., $C(S, T)$, is found to be

$$C(S, T) = \max\{S(T) - K, 0\} = \begin{cases} 0 & \text{if } S(T) \leq K \\ S(T) - K & \text{if } S(T) > K. \end{cases} \qquad (7.7)$$

---

[30] The name *vanilla* reflects that calls and puts do not possess any special or unusual features. In contrast, *exotic options* have more complex features.

In words,

> ➤ The *terminal payoff* of a call, $C(S(T), T)$, is the value of the call $C(S(t), t)$ at expiration $T$, which is its market value at expiration.

□

In a similar fashion, if we denote by $P(S, t)$ the value of European put option at time $t$, then at expiration $T$,

$$P(S, T) = \max\{K - S(T), 0\} = \begin{cases} K - S(T) & \text{if } S(T) < K \\ 0 & \text{if } S(T) \geq K. \end{cases} \tag{7.8}$$

Option contracts are defined by their terms, which are standardized by the exchange on which the option is listed. In the next two examples, we explain standardized contract size and expiration dates for practical purposes.

**Example 7.24. (Option Contract Size)** For equity options (underliers are stocks), the contract size (also called the option trading unit or multiplier) is 100. In other words, one contract controls 100 shares of the underlying stock.

Suppose that you want to purchase a call on $XYZ$ stock with strike price \$50 and premium \$1.50. Then you will have to pay \$150 for the right to buy 100 shares of $XYZ$ stock in the contract. Note that, in practice, you would also have to pay commissions to your broker.

For standard index options (underliers are stock indexes), the contract size is also 100. In other words, the notional value underlying each contract equals \$100 multiplied by the index value.

However, for mini options, the contract size is 10 (representing 10 shares of an underlier). For example, 10 Mini-SPX options equal 1 SPX full value contract. That is, the notional value underlying each mini SPX contract equals \$10 multiplied by the $S\&P\,500$ index value.[31]                                    □

**Example 7.25. (Option Contract Expiration Date)** Equity and index options expire at 4 pm EDT on the third Friday[32] of the expiration month in the sense that they no longer trade; however, the official expiration day is the Saturday immediately following that Friday.

Knowing the month in which the option you want to purchase will expire is very important, since it is a natural part of your trading strategy design and will have a significant impact on the outcome of your trade.

Traditionally, there is an option expiration cycle for each equity on which options are written. Each cycle contains 4 months. For example, suppose that today is May 11, 2015 and a stock XYZ has a February option expiration

---

[31] For standard $S\&P\,500$ index futures, the multiplier is 250 (index level $\times$ 250 = price); for E-mini SPX futures (smaller contract), the multiplier is 50. Multiplier varies for indices.

[32] If the third Friday is a market holiday, then those options expire on the third Thursday.

cycle. Then tradable option contracts written on stock XYZ have expiration months (at least) in May (the current or front month), August (the near month), November, and February 2016.

General information on exchange-traded option expiration dates is available on the Options Expiration Calendar at the CBOE website.                                  □

## 7.5.2 How Options Work

Equity options can be used for a variety of purposes such as hedging existing positions and speculating or buying or selling stocks. We illustrate some simple applications in the following examples.

**Example 7.26. (How Call Buying Work)** If the price of stock $XYZ$ is $7 per share today (say, May 11) and you speculate that it could rocket above $15 within 30 days, then you could buy the June (expiration) 15 (strike) call option.

Suppose that the premium is $1. Ignoring commissions, in order for you to break even, the stock price would need to rise to $16. In order for you to make a profit, the stock price would need to exceed $16.

As a call holder, your maximum possible loss is the premium you paid ($100 per contract for 100 shares).

If you exercise the call when the stock price is $20 you immediately have $300 profit (per contract) on paper. You may continue to hold the shares if you think the stock price will continue to rise. Otherwise, you could sell your call contract (i.e., close your position) and nail down the profit which is a return of 300% (much higher than $\frac{20-7}{7} = 185.7\%$ return from buying the stock).

*Further discussion.* You may also buy call options in the situation when you wait for cash coming in, say, from selling stocks other than the underlier.     □

**Example 7.27. (How Call Selling Work)** Suppose you have bought 100 shares of stock $XYZ$ at price $7 per share. Today (say, May 11), you think that, although the stock has a good potential to continue to move much higher in the long run, in the short run, the price may move down before breaking a resistance level at about $10 (e.g., historical stock price movements showed that the price moved down from a level at about $10 a number of times). To hedge your position, you could write 1 June 10 call option.

Suppose that the premium is $1.50. The premium that you receive produces income ($150) on the stock that is already in your portfolio. Ignoring commissions, you will not lose money as long as the price is not below $5.50.

*Further discussion.* In fact, the way of using call options as we just explained in this example is referred to as a *covered call* strategy, which means that you write calls when you have enough shares of the underlying stock in your portfolio.

In contrast, a call contract is referred to as a *naked* or *uncovered* call if it is not backed by an offsetting position of underlying stock. In other words, you write a call without any shares of the underlying stock in your portfolio.

Generally speaking, the covered call strategy is applied for the purpose of holding a stock for long term, possibly for tax or dividend advantages. This strategy allows you to earn the premium from the option writing but caps your upside potential gains.

A covered call creates a portfolio consisting of longing 1 share of underlier and shorting a 1 unit call. The initial value of the portfolio is $S(0) - c$ where c is the premium received by the writer at initial time (so $S(0) - c \geq 0$ by Property 7.1 on page 372). Assume the underlier to be nondividend paying and recall (7.7), the portfolio value at expiration is $S(T) - \max\{S(T) - K, 0\}$, and therefore, the *terminal profit* generated by a covered call is

$$S(T) - \max\{S(T) - K, 0\} - (S(0) - c) = \begin{cases} S(T) + c - S(0) & \text{if } S(T) \leq K \\ K + c - S(0) & \text{if } S(T) > K. \end{cases} \quad (7.9)$$

Indeed, the profit is capped by the constant value $K + c - S(0)$ if $S(T) > K$ as we expected.                                                                          □

**Example 7.28. (How Put Buying Work)** Again, suppose that you bought 100 shares of stock $XYZ$ at price $7 per share. Today (say, May 11), you think that, although the stock has a good potential to continue to move much higher in the long run, in the short run, the price may move down before breaking a resistance level at about $10.

Aside from engaging in the covered call strategy in the last example, you can also consider buying puts if you think the stock may go down much lower than $5.50 in the short term. If you are wrong, you lose the premium that you paid when you entered into the contract. In a way, put options are parallel to insurance policies.

*Further discussion.* One can make a real profit in a big downward movement of stock price by either buying puts or shorting the stock directly. Ignoring margin requirements and commissions, in order to make a profit from buying puts, one needs to be right about both direction and timing of the price movement; when shorting stocks, one needs to be right only about the direction. However, the advantage of buying puts over shorting stock is that buying puts allows you to determine and prepare for a worst-case scenario as you know that your loss cannot exceed the premium paid when you entered into the contract.   □

**Example 7.29. (How Put Selling Work)** One strategy of buying stocks is by selling puts.

Suppose you would like to buy 100 shares of stock XYZ at about $6 and the price of the stock is $7 today (May 11). You could sell a June $6 put option on the stock and earn premium income $25 (i.e., premium is $0.25) immediately. If the price of the stock drops below $6, the put buyer will exercise the put, and you will have to honor your commitment to buy the stock at $6 (even if the stock price plunges to $2 unexpectedly). If the price of the stock stays above or at $6, then the put buyer will not exercise the put. Although you will not get a chance to buy the stock, you still keep the premium income.

In fact, one way for speculative traders to earn premium income is selling puts by thinking (being so confident) that the stock will not reach the strike price minus the premium.

*Further discussion.* By definition, a put contract grants the buyer *a right, not an obligation,* to sell the underlying stock. However, by selling this put, the put writer assumes *an obligation, not a right,* to buy 100 shares of the stock at the strike price if the buyer of the put wants to sell (i.e., if the buyer exercises the put), regardless of the price of the stock in the spot market. Because of this obligation, when writing a naked put, the put writer should not rely on a wishful thinking but be prepared to take a loss or to own the stock for a while.

The same, if not higher, level of prudence should apply to consideration of writing a naked call (see rationale in Exercise 7.11).                                   □

### 7.5.3 *Terminal Payoff and Profit Diagrams*

For a portfolio of options on the same underlier and with the same expiration, an option terminal payoff diagram provides an extremely useful visualization that traders rely on to analyze a portfolio strategy. This diagram illustrates how a right combination of option positions can form a portfolio that has a risk exposure to almost any chosen kind of market volatility scenarios. In addition, a terminal profit diagram, a chart obtained by a simple translation of the corresponding terminal payoff diagram, provides a clear visual presentation of the range of profit and loss and the break-even point as outcomes of a chosen trading strategy.

We emphasize that one should not confuse the two.

Just to make these "tools of the trade" easier to underhand, *our discussion in this section will focus on European options only,* since, prior to expiration, the payoff diagrams of American options may be more complicated.

Also, without loss of generality, we assume that *all options in a given portfolio are written on the same stock and have the same expiration T.*

## Terminal Payoff Diagrams

Again, let $C(S,t)$ and $P(S,t)$ respectively denote the value of European call and European put options at time $t$, $t \in [0,T]$, where $T$ is the expiration.

**Definition 7.9.** An option *terminal payoff diagram* is a graph of the value of the option position (e.g., long a call or short a put) at expiration $T$ as a function of the underlier price at $T$.

Recall (7.7) and (7.8) on page 355; the terminal payoffs of an option position of a long call and that of a long put are represented respectively by

$$C(S,T) = \max\{S(T) - K,\, 0\} = \begin{cases} 0 & \text{if } S(T) \leq K \\ S(T) - K & \text{if } S(T) > K, \end{cases}$$

and

$$P(S,T) = \max\{K - S(T),\, 0\} = \begin{cases} K - S(T) & \text{if } S(T) < K \\ 0 & \text{if } S(T) \geq K. \end{cases}$$

**Example 7.30.**

1. If we let $x = S(T)$ and $y = f(S(T)) \equiv C(S(T),T)$, then the graph of $y = f(x)$ on $xy$-plan is the graph of $C(S(T),T)$ against $S(T)$ given in Fig. (3). That is, Fig. (3) is the terminal payoff diagram of an option position of long a call by Definition 7.9.

2. By Definition 7.9, the graph of $P(S(T),T)$ against $S(T)$ is the terminal payoff diagram of long-a-put. Can you sketch this graph? Does your graph coincide with the one in Fig. (6)?

3. Since the value of a short position on a call at expiration $T$ is $-C(S(T),T)$, Fig. (4) gives the terminal payoff diagram of short-a-call.

4. Similarly, we see that the terminal payoff diagram of short a put is given by Fig. (7).

□

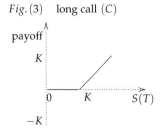

*Fig.* (3)   long call $(C)$

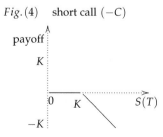

*Fig.* (4)   short call $(-C)$

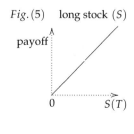

*Fig.* (5)   long stock $(S)$

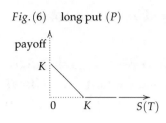

*Fig.* (6)  long put (*P*)

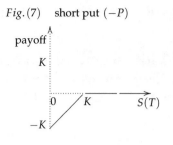

*Fig.* (7)  short put (−*P*)

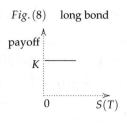

*Fig.* (8)  long bond

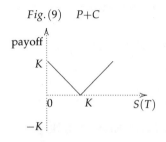

*Fig.* (9)  *P*+*C*

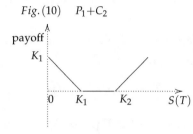

*Fig.* (10)  *P*₁+*C*₂

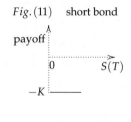

*Fig.* (11)  short bond

➤ Notice that the result of superimposing Fig. (3) and Fig. (7) is the graph of translating Fig. (5) to the right by $K$ units, i.e., $C - P = S - K$ at time $T$.

➤ Also notice that a bond generates only vertical translations of payoff diagrams. Therefore, taking a position in a bond does not play any hedging role.

Recall that a portfolio is a linear combination of securities. If we let $\Pi$ represent a portfolio consisting of longing a call, shorting two puts, and longing three shares of the underlier, then we write $\Pi = C - 2P + 3S$. Also, as in the past, $\Pi(t) = C(S(t), K) - 2P(S(t), K) + 3S(t)$ represents the value of the portfolio at time $t$.

Let $\Pi$ be a portfolio consisting of long or short positions of a stock (S) calls (Cs) and puts (Ps) on the same underlier (S) with the same expiration and zero-coupon bonds (Bs); we introduce the terminology of *portfolio terminal payoff diagram*:

**Definition 7.10.** A portfolio $\Pi$ terminal payoff diagram is a graph of $\Pi(T)$ as a function of $S(T)$.

**Example 7.31.**

1. If $\Pi$ consists of long one share of stock (S), then the terminal payoff diagram of portfolio $\Pi$ is the payoff diagram of long a stock. Can you sketch this graph? Does your graph coincide with Fig. (5) on page 360?

2. If $\Pi$ consists of simultaneous long a call (C) and a put (P), which are on the same stock and with same expiration, then $\Pi = C + P$, and the terminal payoff diagram of portfolio $\Pi$ is given by Fig. (9).

Portfolio $\Pi = C + P$ is called a straddle, which will be revisited in Section 7.5.5.

□

We provide an intuitive way to derive the *put-call parity* in the next example.

**Example 7.32. (Put-Call Parity of European Options)** Verify that superimposing Fig. (4), Fig. (5), and Fig. (6) on the previous page yields Fig. (8). It follows that $S + P - C = K$ at time $T$. To discount $K$ back to time $t \in [0, T)$, we obtain the *put-call parity*:

$$S(t) + P(t) - C(t) = Ke^{-r(T-t)}, \tag{7.10}$$

where we assume that the underlier is nondividend paying and r is the risk-free interest rate.

If the underlier is "an asset with annual dividend yield $q$" (see page 341), the put-call parity takes the form

$$S(t)e^{-q(T-t)} + P(t) - C(t) = Ke^{-r(T-t)}. \tag{7.11}$$

Further discussion about the put-call parity for European options will be given in Section 7.5.6 and Section 7.5.7.                                                          □

**Remark 7.7.** *Four Keys to Reading Option Payoff Diagrams:*

1. Each sharp-corner point corresponds to a strike price (on S(T)-axis).
2. Each vertical parallel shift can be done by using a bond.
3. Each call-put can be converted to a put-call under the put-call parity.
4. Each of the eight most basic diagrams (long call, short call, long put, short put, long stock, short stock, long bond, short bond) must be kept in mind.

                                                                                          □

**Terminal Profit Diagrams**

Since the profit generated by a portfolio is the difference between the terminal payoff and the initial price you pay (e.g., the option premium), ignoring commissions, the *terminal profit diagram* of a portfolio is the graph that is the vertical translation of the corresponding terminal payoff diagram by the initial price you pay in the same coordinate system.

**Example 7.33.** A portfolio consisting of simultaneously buying a call and a put is called a straddle if the call and put are on the same underlier with the same expiration ($T$) and strike ($K$). The terminal payoff of a straddle is provided by Fig. (9) on page 360, and the terminal profit diagram is provided below:

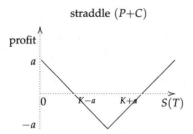
straddle $(P+C)$

where $a$ is the net premium of the straddle. Then the range of profit is given by $S(T) \in [0, K - a) \cup (K + a, \infty)$ and the range of loss by $S(T) \in (K - a,\ K + a)$, and the break-even point is $S(T) = K - a$ or $K + a$. $\qquad\square$

**Remark 7.8.** More precisely speaking, *the terminal profit of a portfolio is defined as the difference between the terminal payoff and the future value at time T of the initial cost.* For example, by this definition, the profit for a call buyer at the expiration $T$ is $C(S(T), K) - c\,e^{rT} = \max\{S(T) - K,\ 0\} - c\,e^{rT}$, where we assume that the buyer purchases the call at time $t = 0$ by paying premium $\$c$.

However, taking into consideration the two facts:

1. $1 - e^{rT}$ is relatively too small and
2. The terminal profit diagram is only used for visualization of option strategy outcomes,

we may ignore the compounding factor $e^{rT}$ in the future value of the initial cost. $\qquad\square$

### 7.5.4 Market Sentiment Terminologies and Option Moneyness

In this section, we introduce some often used jargon in option trading.

*Market sentiment* is a terminology used to describe the overall attitude of investors toward a particular security or a financial market.

The words bullish and bearish are often used to describe a market sentiment.

➤ The *bullish* sentiment on a security (or a market) means that the security (or the market) price is believed to be going up.
➤ The *bearish* sentiment on a security (or a market) means that the security (or the market) price is believed to be going down.

**Example 7.34.** The statement that investors who buy calls are bullish on the underlying stock is equivalent to the statement that investors who buy calls think that the underlying stock price will go up.

In short, we can say that buying a call is a bullish bet on the underlier. $\qquad\square$

**Example 7.35.** The statement that investors who buy puts are bearish on the underlying stock is equivalent to the statement that investors who buy puts think that the underlying stock price will go down.

In short, we can say that buying a put is a bearish bet on the underlier. $\qquad\square$

*Moneyness* is a term describing the condition of an option in terms of its strike price in relation to the underlier's spot price. Again, let $S(t)$ be the spot price of the underlier at time $t$ and $K$ be the strike of the option.

➤ An option is said to be *in-the-money* at time $t$ if the option has a positive payoff if it is exercised at time $t$. More specifically,

    ➤ A call is *in-the-money* at time $t$ if $S(t) > K$,
    ➤ A put is *in-the-money* at time $t$ if $S(t) < K$.

➤ An option is said to be *at-the-money* at time $t$ if the payoff of the option is zero if it is exercised at time $t$. More specifically,

    ➤ A call is *at-the-money* at time $t$ if $S(t) = K$,
    ➤ A put is *at-the-money* at time $t$ if $S(t) = K$.

➤ An option is said to be *out of the money* at time $t$ if the option has a negative payoff if it is exercised at time $t$. More specifically,

    ➤ A call is *out of the money* at time $t$ if $S(t) < K$,
    ➤ A put is *out of the money* at time $t$ if $S(t) > K$.

**Example 7.36.** In option trading, the statement that a call is *deep in the money* at time $t$ means that the underlier spot price $S(t)$ is well above the strike price of the call.

In practice, the deep in-the-money condition is a condition in which a call value changes dollar for dollar with the spot price movement of the underlier; for an at-the-money call, its value changes only about 50% of the spot price change.

A parallel statement can be made for a deep in-the-money put.     □

**Example 7.37.** In option trading, the statement that a call is *deep out of the money* at time $t$ means that the underlier spot price $S(t)$ is well below the strike price of the call.

A parallel statement can be made for a deep out-of-the-money put.

In practice, deep out-of-the-money options are always worth something because there is always a probability that the condition may change. They may be used for two trading strategies—hedging and speculation. In some ways, deep out-of-the-money options are almost like purchasing lottery tickets, i.e., they present an opportunity for profits but with a low probability of success.     □

**Remark 7.9.** From a pure financial theory point of view, one would define the condition for a call being *in the money* at time $t$ to be $S(t) > Ke^{-r(T-t)}$. However, this definition would not make any real difference in terms of the purpose which the option moneyness terminologies serve, since the difference between $K$ and $Ke^{-r(T-t)}$ is very small. An expression like $S(t) > K$ is much simpler, and therefore more convenient to use, than $S(t) > Ke^{-r(T-t)}$.

Similar argument applies to the definitions of *out of the money* and *at the money*.                                                                          □

### 7.5.5 Option Strategies: Straddle, Strangle, and Spread

A *straddle* is a combination of two positions—one is long on a call and the other is long on a put, where the call and put have the same underlier $(S)$, strike $(K)$, and expiration $(T)$.

The *long straddle*, or *buy straddle* or simply *straddle*, is a (underlier spot) market-neutral[33] option strategy that involves simultaneously buying a call and a put. It is a bullish bet on the underlier's volatility and one of the least sophisticated option strategies.

**Example 7.38. (Straddle)** Let us consider a portfolio $\Pi = C + P$ where the underlier is modeled by one period binomial tree on $[0,T]$, with

$$S(0) = \$7, \quad S(T) = \begin{cases} \$13 & \text{with probability } \frac{3}{5} \\ \$1 & \text{with probability } \frac{2}{5}, \end{cases}$$

and

$$c = \$1, \quad p = \$1, \quad K = \$7.$$

What is the expected return of $\Pi$? What is the terminal profit diagram of the straddle?

**Solution.** Note that

$$\mathbb{E}(\Pi(T)) = \mathbb{E}(C(S(T),K)) + \mathbb{E}(P(S(T),K)) = \frac{3}{5} \times 6 + \frac{2}{5} \times 6 = 6.$$

Therefore, the expected return is $\frac{6-2}{2} = 200\%$.

The terminal profit diagram of long a straddle (C+P) can be found in Example 7.33 on page 362 with $K = 7$ and $a = 2$. The break-even spot price of the underlier is either \$5 or \$9. The range of loss is $[-2,0)$ corresponding to $S(T) \in (5,9)$.

We let the reader observe the potential profits when either the call or the put in the straddle is deep in the money.

*Further discussion.* If you are confident that a stock price will change dramatically, but not sure about the direction of the price move, you may consider the long straddle. However, in practice, it is often the situation that both call

---

[33] A *neutral* option strategy is a strategy that is designed to profit from either a rise or fall (non-directional) in the underlier price.

and put prices become much higher than expected when there is a major news pending.                                                                          □

A *strangle* is a combination of two positions, one is long on a put with strike $K_1$ and another is long on a call with strike $K_2$ ($K_2 > K_1$), where the call and put have the same underlier ($S$) and expiration ($T$).

The *long strangle*, or *buy strangle* or simply *strangle*, is a (underlier spot) market-neutral option strategy that involves simultaneously buying an out-of-the-money put and an out-of-the-money call. It is also a bullish bet on the underlier's volatility.

**Example 7.39. (Strangle)** Typically, when you long a strangle, the two strike prices $K_2 > K_1$ are near the money and out of the money (thus, close to in the money). Note that you are betting on volatility rather than on the underlying stock alpha when you long either a straddle or a strangle and the one with lower net premium is preferred. The terminal payoff of a strangle is provided on page 360, Fig. (10), and the terminal profit diagram is provided below:

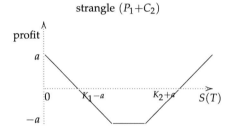

where $a$ is the net premium of the strangle. Identifying the range of profit, loss and break- even point as outcomes of the corresponding portfolio strategy is left as an exercise for the reader.                                                       □

A *spread strategy* corresponds to a portfolio that consists of two or more options of the same type to achieve a certain level of hedging effect. Spreads are the basic building blocks of many option strategies although we will only briefly introduce three of them in the following.

**Example 7.40. (Price Spread or Vertical Spread)** One way to construct *a price spread*, or *vertical spread*, is to use two calls (or two puts) on the same underlying asset ($S$) and the same expiration date ($T$) but with different strike prices[34] ($K_i, i = 1, 2$). One is bought and another is sold in order to achieve a level of hedging effect.

A spread is called a *bull spread* if it is designed to profit from an upward movement of the underlier's price (see payoff diagram below).

A spread is called a *bear spread* if it is designed to profit from a downward movement of the underlier's price (see payoff diagram below).

---

[34] This explains the name "price spread" (more precisely, "strike price spread"). Since the strike prices are listed vertically by the news media, a price spread is also referred to as a vertical spread.

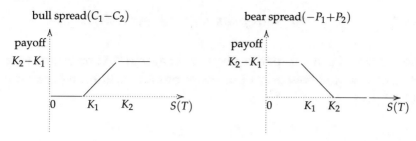

bull spread$(C_1 - C_2)$          bear spread$(-P_1 + P_2)$

Using a formula to express the terminal payoff of each spread shown above is left as an exercise for the reader. □

**Example 7.41. (Calendar Spread or Time Spread or Horizontal Spread)** One way to construct *a calendar spread*, also called *time spread* or *horizontal spread*, is to use two calls (or two puts) on the same underlier ($S$) and with the same strike price ($K$) but on different expiration dates[35] ($T_i, i = 1, 2$).

If you predict (as a pure speculation) that the underlying stock price will rise above \$45 in a few months, you may want to consider the following bullish calendar call spread.

Suppose that XYZ 45 (American) calls are priced below.

| Expiration month | May | June | July | August |
|---|---|---|---|---|
| Premium of XYZ 45 call | \$0.25 | \$2 | \$3 | \$5 |

By buying 1 XYZ Aug 45 call only, you will lose \$500 if the stock XYZ drops to, say, \$35. In contrast, by making the transaction of buying 1 XYZ Aug 45 call and writing 1 XYZ July 45 call, you will loss only \$200 if the stock drops to \$35. The hedging effect created by the spread cuts the loss by more than half. □

**Example 7.42. (Butterfly Spread)** A *butterfly spread* is a portfolio consisting of four options of the same type on the same underlier ($S$) with the same expiration date ($T$) but three different strike prices ($K_1 < K_2 < K_3$). Generally, $K_2 = \frac{1}{2}(K_1 + K_3)$ and is close to the current underlier price. The graph below is the terminal payoff diagram of a butterfly spread consisting of long 1 call with strike $K_1$, write (i.e., short) 2 calls with strike $K_2$, and long 1 call with strike $K_3$.

butterfly spread

The butterfly spread is designed to profit when the underlier price movement stays close to the current price. Using a formula to express the terminal payoff of the butterfly spread indicated above is left as an exercise for the reader. □

---

[35] This explains the name "calendar spread" or "time spread." Since the expiration months are listed across the top of the newspaper page horizontally, a calendar spread is also referred to as a horizontal spread.

### 7.5.6 Put-Call Parity for European Options Revisited

The basic forms of put-call parity given in Example 7.32 on page 362 provide not only a conversion between a European put and a European call but also a perfect hedged portfolio[36] $\Pi = S + P - C$ for

$$S(t)e^{-q(T-t)} + P(t) - C(t) = Ke^{-r(T-t)},$$

from (7.11) on page 362, with dividend yield rate $q \geq 0$, of which a proof was given by using option terminal payoff diagrams.

In this section, we provide a different proof of the put-call parity by using the law of one price.

We denote by $C^E(t)$ and $P^E(t)$ the price of a European call and the price of a European put options at time $t \in [0, T]$, where time 0 is current date and both the call and put are on the same underlying asset and have the same expiration date $T$ and strike price $K$. Let $S(t)$ be the price of the asset at time $t \in [0, T]$ and let r be the risk-free interest rate compounded continuously.

**Theorem 7.3. (Put-Call Parity for European Options)** *The current price relation between European put and call options on the same "asset with annual dividend yield q"(see page 341) and with the same expiration T and strike price K is given by*

$$C^E(0) + Ke^{-rT} = P^E(0) + S(0)e^{-qT}, \tag{7.12}$$

*where r is the risk-free interest rate.*

*Proof.* We consider two portfolios A and B, denoted by $\Pi_A$ and $\Pi_B$, respectively, having positions established at time 0 as indicated below:

> $\Pi_A$ : long 1 European call, and long zero-coupon bond with $Ke^{-rT}$.
> $\Pi_B$ : long 1 European put, and long $e^{-qT}$ units of the asset.

We denote by $\Pi_A(t)$ the time-$t$ value of portfolio $A$ and $\Pi_B(t)$ the time-$t$ value of portfolio $B$ and compute $\Pi_A(T)$ and $\Pi_B(T)$:

$$\Pi_A(T) = C^E(T) + Ke^0 = \max\{S(T) - K, 0\} + K = \max\{S(T), K\}$$
$$\Pi_B(T) = P^E(T) + e^0 S(T) = \max\{K - S(T), 0\} + S(T) = \max\{K, S(T)\}.$$

Since the two portfolios have identical values at a future time, it follows from the definition of arbitrage and the law of one price that they must have identical values today:

$$\Pi_A(0) = \Pi_B(0),$$

---

[36] A perfect hedged portfolio is a portfolio with complete risk elimination.

which is equivalent to

$$C^E(0) + Ke^{-rT} = P^E(0) + S(0)e^{-qT}.$$

$\square$

Replacing 0 by $t < T$ in (7.12), we obtain (7.11) again

$$C^E(t) - P^E(t) = S(t)e^{-q(T-t)} - Ke^{-r(T-t)}.$$

### 7.5.7 Relation Among Put, Call, and Forward

Observe that a superposition of Fig. (3) and Fig. (7) on page 360 yields Fig. (1) on page 340. It follows that at time $T$, $C - P = \mathbf{f}_T(T)$, where we assume that the put, call, and forward are on the same underlier and expiration $T$ and with the same strike $K$ (thus $F_T(0) = K$; also see page 344). Under the law of one price, we obtain an equivalent form of the *put-call parity* in (7.12):

$$C^E(t) - P^E(t) = \mathbf{f}_T(t), \qquad 0 \le t \le T, \tag{7.13}$$

where the forward value $\mathbf{f}_T(t) = (F_T(t) - K)e^{-r(T-t)}$ by (7.6) on page 344. In words, (7.13) says that the difference between the current values of the European call and put is the current value of the forward.

**Example 7.43.** Using the forward price formula $F_T(t) = S(t)e^{(r-q)(T-t)}$, the put-call parity (7.13) can also be derived by straightforward computation as follows:

$$\begin{aligned}(F_T(t) - K)e^{-r(T-t)} &= (S(t)e^{(r-q)(T-t)} - K)e^{-r(T-t)} \\ &= S(t)e^{-q(T-t)} - Ke^{-r(T-t)} = C^E(t) - P^E(t),\end{aligned}$$

where the last equal sign holds due to (7.12). $\square$

The new version of put-call parity expressed by (7.13) says that

➤ The payoff from long a forward equals the payoff from simultaneously long a call and short a put provided the forward, call, and put are on the same underlier and with the same strike and expiration.

Equivalently,

➤ A forward can be replicated by simultaneously long a call and short a put on the same underlier and with the same strike and expiration as the forward.

**Example 7.44.** A call can be replicated by a forward and a put, and a put can be replicated by a forward and a call. $\square$

### 7.5.8 Intrinsic Value and Time Value

An option valuation is a procedure for assigning a market value to an option. Different models generate different procedures, which, in turn, produce different valuations for different purposes. The procedure generated by the Black-Scholes-Merton model in the next chapter is a theory-based approach for calculating the fair value of an option. The procedure to be introduced in the current section is based on a simple decomposition of the market value of an option into two components, one of which is considered to be the value of that option that most professional trading is based upon.

We begin with the following observations.

*Observation 1.*

$$\max\{(F_T(t) - K)e^{-r(T-t)}, 0\} = e^{-r(T-t)}\max\{F_T(t) - K, 0\}$$

$$= e^{-r(T-t)} \begin{cases} F_T(t) - K & \text{if } F_T(t) > K \\ 0 & \text{if } F_T(t) \leq K \end{cases}$$

$$= \begin{cases} e^{-r(T-t)}(F_T(t) - K) & \text{if } F_T(t) > K \\ 0 & \text{if } F_T(t) \leq K \end{cases}$$

$$= \begin{cases} S(t) - e^{-r(T-t)}K & \text{if } S(t) > e^{-r(T-t)}K \\ 0 & \text{if } S(t) \leq e^{-r(T-t)}K \end{cases}$$

$$= \begin{cases} (S(t) - K) + K(1 - e^{-r(T-t)}) & \text{if } S(t) > e^{-r(T-t)}K \\ 0 & \text{if } S(t) \leq e^{-r(T-t)}K. \end{cases}$$

Keeping Remark 7.9 (see page 364) in mind, the expression right after the last equal sign represents the degree to which the call option is in the money at time $t$, as does the expression $e^{-r(T-t)}\max\{F_T(t) - K, 0\}$.

*Observation 2.*

$$\max\{-(F_T(t) - K)e^{-r(T-t)}, 0\} = e^{-r(T-t)}\max\{K - F_T(t), 0\}$$

$$= \begin{cases} -(S(t) - e^{-r(T-t)}K) & \text{if } S(t) < e^{-r(T-t)}K \\ 0 & \text{if } S(t) \geq e^{-r(T-t)}K \end{cases}$$

represents the degree to which the put option is in the money at time $t$.

*Observation 3.*

$$e^{-r(T-t)}\max\{F_T(t) - K, 0\} - e^{-r(T-t)}\max\{K - F_T(t), 0\}$$
$$= (F_T(t) - K)e^{-r(T-t)} = C^E(t) - P^E(t).$$

Now, we define the *intrinsic value* of an option at time $t$ to be the degree to which it is in the money at $t$ and define the *time value* of a call option at time $t$, written $TV_c(t)$, by

$$TV_c(t) = C^E(t) - e^{-r(T-t)} \max\{F_T(t) - K, 0\}. \tag{7.14}$$

Then, a call option price can be expressed by a sum of its intrinsic value and its time value:

$$C^E(t) = e^{-r(T-t)} \max\{F_T(t) - K, 0\} + TV_c(t).$$

Similarly, we define the *time value* of a put option at time $t$, written $TV_p(t)$, by

$$TV_p(t) = P^E(t) - e^{-r(T-t)} \max\{K - F_T(t), 0\}. \tag{7.15}$$

Then, a put option price can be expressed by a sum of its intrinsic value and its time value:

$$P^E(t) = e^{-r(T-t)} \max\{K - F_T(t), 0\} + TV_p(t).$$

Since definitions (7.14) and (7.15) and Observation 3 yield

$$TV_c(t) - TV_p(t) = C^E(t) - P^E(t) - (C^E(t) - P^E(t)) \equiv 0, \quad t \in [0, T],$$

we obtain $TV_c(t) = TV_p(t)$ during the life of the call and put. This result allows us to simply denote by $TV(t)$ the time value of a European option and write

$$C^E(t) = e^{-r(T-t)} \max\{F_T(t) - K, 0\} + TV(t),$$
$$P^E(t) = e^{-r(T-t)} \max\{K - F_T(t), 0\} + TV(t).$$

In words,

$$\text{Option Value} = \text{Intrinsic Value} + \text{Time Value}.$$

Notice that an out-of-the-money option has time value only. As a result, the value of an out-of-the-money option erodes quickly with time as it gets closer to its expiration.

Because out-of-the-money options have time values only, they are significantly cheaper and offer great leverage and, therefore, have better liquidity (are more actively traded). For these reasons, most professional option traders trade the time value only with confidence in turning time value decay into potential profits; for trading purpose, the time value of an option is where the professional traders see the value of an option.

**Remark 7.10.**

1. Time value is subject to several factors, primarily time to expiration and implied volatility. The latter concept will be discussed in Chapter 8.

2. The rate at which time value decays is represented by $\Theta$, one of the Greeks to be introduced along with Black-Scholes-Merton model, again in Chapter 8.

□

**Example 7.45.** SPY is trading at $213 on May 18, 2015. Call options with strike prices below $213 are in-the-money calls. Call options with strike prices above $213 are out-of-the-money calls. Call options with strike prices equal $213 are at-the-money calls. Given the following option information,

| Expiration month | May | June | July |
|---|---|---|---|
| Premium of SPY 212.5 call | $1.28 | $2.92 | $3.67 |
| Time value of SPY 212.5 call | $0.75 | $2.42 | $3.17 |

observe the time value decay in action: the time value drops from June to May much faster than that from July to June.

□

**Example 7.46. (Trading on Time Value)** Option writers attempt to benefit from the time value decay. They collect time value premiums paid by option buyers. Such premiums can become steady cash flows if the underlying security is stationary.

□

### 7.5.9 Some General Relations of Options

We assume that the underlying asset pays a constant, continuous proportional annual dividend yield rate $q$ that is continuously reinvested to buy more units of the asset and that holding the asset neither incurs cost of carry nor provides other convenience yields in the interval $[0, T]$ where time 0 is current date. We also assume that r is the risk-free interest rate compounded continuously.

We denote by $C^A(t)$ and $C^E(t)$ the American call price and the European call price at time $t$, where both the American call and the European call are on the same asset and with the same terms.

Similarly, we denote by $P^A(t)$ and $P^E(t)$ the American put price and European put price at time $t$, where both the American put and the European put are on the same asset and with the same terms.

Assuming that notation $S(t)$, $T$, and $K$ are defined as usual, we have the following property and provide a partial proof.

**Proposition 7.1.** *The following relations of options hold.*

*1.* $S(0) \geq C^A(0) \geq C^E(0) \geq \max\{S(0)e^{-qT} - Ke^{-rT}, 0\}.$

*2.* $K \geq P^A(0) \geq P^E(0) \geq \max\{Ke^{-rT} - S(0)e^{-qT}, 0\}.$

*Proof.* We provide a proof for the first set of inequalities. A proof of the second set is left as an exercise for the reader.

**Step 1.** If $S(0) < C^A(0)$, then construct a portfolio $\Pi$ at time 0, by shorting 1 American call and using the proceeds to immediately long 1 unit of the underlier. We obtain $\Pi(0) = C^A(0) - S(0) > 0$, an immediate profit with the cash amount $a = C^A(0) - S(0)$ at time 0. Thus, $\Pi(T) > 0$ is guaranteed regardless of the fluctuations of the asset price, even if the cash amount \$a is kept under mattress for $\Pi$ taking a short position on a covered call means that the possession of the asset can always cover the exercise made by the buyer in case it happens. Therefore, $\Pi$ is an arbitrage!

**Step 2.** The ability to exercise an American option at any time prior to or at expiration makes American options more flexible than European options; thus, $C^A(0) \geq C^E(0)$.

**Step 3.** Applying the put-call parity (7.12) and $P^E(0) \geq 0$, we obtain

$$C^E(0) = P^E(0) + S(0)e^{-qT} - Ke^{-rT}$$
$$\geq S(0)e^{-qT} - Ke^{-rT}.$$

It follows from $C^E(0) \geq 0$ that $C^E(0) \geq \max\{S(0)e^{-qT} - Ke^{-rT}, 0\}$. $\qquad\square$

**Proposition 7.2.** *One should not prematurely exercise an American call on a nondividend-paying asset if the risk-free interest rate* r > 0. *That is,*

$$C^A(0) = C^E(0).$$

*Proof.* It follows from Proposition 7.1, 1 with $q = 0$ and $r > 0$, that

$$C^A(0) \geq \max\{S(0) - Ke^{-rT}, 0\} > S(0) - K.$$

Note that $S(0) - K$ is the payoff if the call is exercised currently. $\qquad\square$

### 7.5.10 Put-Call Parity for American Options

We denote by $C^A(t)$ and $P^A(t)$ the price of the American call and the price of the American put options at time $t \in [0, T]$, where time 0 is current date and both the call and put are on the same underlying asset and have the same expiration date $T$ and strike price $K$. Let $S(t)$ be the price of the asset at time $t \in [0, T]$ and let r be the risk-free interest rate compounded continuously.

**Theorem 7.4. (Put-Call Parity for American Options)** *The current price relation between American put and call options on the same "asset with annual dividend yield q"(see page 341) and with the same expiration T and strike price K is given by*

$$S(0)e^{-qT} - K \leq C^A(0) - P^A(0) \leq S(0) - Ke^{-rT}, \tag{7.16}$$

*where r is the risk-free interest rate.*

*Proof.*
**Step 1.** To prove the inequality $C^A(0) - P^A(0) \leq S(0) - Ke^{-rT}$, we construct two portfolios A and B, denoted by $\Pi_A$ and $\Pi_B$, respectively:

> $\Pi_{\mathbf{A}}$ : long 1 American call and long zero-coupon bond with $Ke^{-rT}$ at rate r.
> $\Pi_{\mathbf{B}}$ : long 1 American put and long 1 unit of the underlying asset.

We denote by $\Pi_A(t)$ the value of portfolio $A$ and $\Pi_B(t)$ the value of portfolio $B$ at time $t$ and by $D_{PV}$ the time-0 value, or present value, of the dividends.

Notice that the American call cannot be exercised early because there is not enough cash to buy 1 unit of the asset until time $T$ and

$$\Pi_A(T) = \max\{S(T) - K, 0\} + K = \max\{S(T), K\}.$$

With the possibility that the American put could be exercised at $\forall t \in [0, T]$,

$$\Pi_B(t) = \begin{cases} K + D_{PV}e^{rt} & \text{if } S(t) < K \\ S(t) + D_{PV}e^{rt} & \text{if } S(t) \geq K \end{cases} = \max\{S(t), K\} + D_{PV}e^{rt},$$

which implies that

$$\Pi_B(T) = \max\{S(T), K\} + D_{PV}e^{rT} \geq \Pi_A(T).$$

It follows from the definition of arbitrage and the law of one price that

$$\Pi_A(0) \leq \Pi_B(0).$$

Therefore, the desired inequality $C^A(0) + Ke^{-rT} \leq P^A(0) + S(0)$ is obtained.

**Step 2.** To show the other inequality

$$S(0)e^{-qT} - K \leq C^A(0) - P^A(0), \tag{7.17}$$

we write $D_{PV} = S(0)(1 - e^{-qT})$ for $D_{PV}e^{qT} = S(0)(e^{qT} - 1)$ being the total dividends of 1 unit of the underlier over the interval $[0, T]$. Then

$$S(0)e^{-qT} = S(0) - D_{PV}$$

implies that (7.17) is equivalent to

$$C^A(0) + K + D_{PV} \geq P^A(0) + S(0). \tag{7.18}$$

We construct two portfolios X and Y, denoted by $\Pi_X$ and $\Pi_Y$, respectively:

> $\Pi_{\mathbf{X}}$ : long 1 European call and long zero-coupon bond with $K + D_{PV}$ at r.
> $\Pi_{\mathbf{Y}}$ : long 1 American put and 1 unit of the asset simultaneously.

Since the European call cannot be exercised early, we only need to consider whether the American put is exercised early.

If the American put is not exercised early, then

$$\begin{aligned}
\Pi_Y(T) &= \max\{K - S(T), 0\} + S(T) + D_{PV}e^{rT} \\
&= \max\{K, S(T)\} + D_{PV}e^{rT}, \\
\Pi_X(T) &= \max\{S(t) - K, 0\} + D_{PV}e^{rT} + Ke^{rT} \\
&= \max\{S(t) - K, 0\} + K + D_{PV}e^{rT} + Ke^{rT} - K \\
&= \max\{S(T), K\} + D_{PV}e^{rT} + K(e^{rT} - 1) \\
&> \Pi_Y(T).
\end{aligned}$$

If the put is exercised early at time $t \in [0, T)$ (i.e., sell stock to receive $K$), then

$$\begin{aligned}
\Pi_Y(t) &\leq K + D_{PV}e^{rt} \\
&\leq D_{PV}e^{rt} + Ke^{rt} \leq \Pi_X(t).
\end{aligned}$$

Thus, regardless of the early exercise, $\Pi_X(0) \geq \Pi_Y(0)$, which is equivalent to

$$C^E(0) + K + D_{PV} \geq P^A(0) + S(0).$$

Since $C^A(0) \geq C^E(0)$, we obtain

$$C^A(0) + K + D_{PV} \geq C^E(0) + K + D_{PV} \geq P^A(0) + S(0),$$

which implies (7.18).

$\square$

Replacing 0 by $t < T$ in (7.16), we obtain

$$S(t)e^{-q(T-t)} - K \leq C^A(t) - P^A(t) \leq S(t) - Ke^{-r(T-t)}.$$

**Example 7.47. (Intuition of the Put-Call Parity for American Options)** To interpret the put-call parity (bounds) for American options geometrically, we let

$$x = S(0) \quad \text{and} \quad y = C^A(0) - P^A(0)$$

and visualize the region in the xy-plane, which is bounded by two straight lines,

$$L_1: y = e^{-qT}x - K \quad \text{and} \quad L_2: y = x - Ke^{-rT}.$$

Notice that for nondividend-paying underliers, these two lines are parallel.

The relation between $S(0)$ and $C_A(0) - P_A(0)$ has a geometric interpretation in terms of the points in the region bounded by $L_1$ and $L_2$ as shown below:

put-call parity for American options when $q = 0$

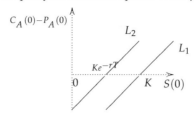

The geometric interpretation of the put-call parity for American option in the case of $q > 0$ is left as an exercise for the reader. (*Hint*: sketch a graph!)  □

**Example 7.48.** Given the current price of a stock and the current price of a put option on the stock with expiration $T$ and strike $K$ along with the stock's dividend yield $q$ and risk-free rate r, find the upper and the lower price bounds for the American call option with the same expiration and strike on the stock.

**Solution.** The lower bound is $P^A(0) + S(0)e^{-qT} - K$.
The upper bound is $P^A(0) + S(0) - Ke^{-rT}$.

□

### 7.5.11 Boundary Conditions for European Options

The next proposition gives the boundary conditions on calls and puts; they are needed later in our study of the Black-Scholes-Merton p.d.e.
  Again, we let $C^E(S,t) = C^E(S(t), t)$ and $P^E(S,t) = P^E(S(t), t)$ for $t \leq T$. Boundary conditions for $C^E(S,t)$ and $P^E(S,t)$ are applied for $S \to 0$ and $S \to \infty$.
  The next proposition can be established by applying Proposition 7.1 (see page 372) and the put-call parity (see (7.12) on page 368) along with following observations:

1. $C^E(S,t) \to 0$ as $S \to 0$ since the call is unlikely to be exercised when the underlier's price is sufficiently small.
2. $C^E(S,t) \to Se^{-q(T-t)} - Ke^{-r(T-t)}$ as $S \to \infty$ since the call is likely to be exercised when the underlier's price is sufficiently large.

**Proposition 7.3. (European Call and Put Boundary Conditions)**

1. $\lim_{S \to 0} C^E(S,t) = 0$ *and* $\lim_{S \to \infty} C^E(S,t) = \lim_{S \to \infty} Se^{-q(T-t)}$.
2. $\lim_{S \to 0} P^E(S,t) = Ke^{-r(T-t)}$ *and* $\lim_{S \to \infty} P^E(S,t) = 0$.

A proof of Proposition 7.3, 2, can also be done directly (without applying the put-call parity). A geometric interpretation of the properties can be done by sketching the graphs of $C^E(S,t)$ and $P^E(S,t)$ against $S$.
  There are many more nice discussions in the literature related to the topics presented in this chapter, e.g., [2, 3, 5, 6, 7, 9, 10, 12, 13, 14, 15, 16, 17, 18, 19].

## 7.6 Exercises

Unless stated otherwise, an option is either American or European.

### 7.6.1 Conceptual Exercises

**7.1.** Is a forward a contingent claim? Is an option a contingent claim?

**7.2.** There are mainly three types of derivative traders: hedgers, speculators, and arbitrageurs. What are their definitions?

**7.3.** Sketch the terminal payoff diagram of a forward with expiration $T$ and forward price $K$. If you short this forward, what is the terminal payoff diagram?

**7.4.** What are the key features of futures that differ from forwards?

**7.5.** If you believe that the market price of a stock will stay at approximately the same price for a period of time, can you still make money from the stock if your hunch is correct? Explain your answer.

**7.6.** What is the possible maximum gain or loss if you sell a call?

**7.7.** What is the possible maximum gain or loss if you buy a call?

**7.8.** Consider the call options given in Example 7.45 (see page 372). Are they all in the money? Identify their intrinsic value(s).

**7.9.** If a call is in the money sufficiently close to the expiration date, then the call price will rise dollar for dollar with the stock price. Agree or disagree? Explain.

**7.10.** An at-the-money American call with a strike price of $80 is being sold for $200. Assume that the stock goes up to $84 per share on the day of expiration.

a) If you bought the option, what is your return from exercising the call and liquidating your stock position? If you did not buy the option, but had bought 100 shares of the stock in the market at $80 per share and then sold them on the option's expiration date at $84 per share, what would be your return? Do the two scenarios have equivalent gain/loss?
b) If you do not exercise the option, what is your approximate return from selling the call right before expiration?
c) Which would you then prefer? Exercise the call or sell the call?

**7.11.** You sell an American call on 1 round lot of a stock at $40 per share. A month later, the market value of that stock is $46 per share. If the buyer exercises the option, you will be obligated to deliver 100 shares at $6 below current market value.

a) If you own those shares, what is your gain/loss from settling the position?
b) If you had naked short sold the American call, what is your gain/loss from settling the position?

**7.12.** You paid $300 for an American call on a stock several months ago. It will expire next month and is now worth only $100. What are the feasible actions that you can take? What are the consequences of your actions?

**7.13.** Argue how American put buying/selling works.

a) How American put buying works.
Buyers of a put expect the underlying stock to fall in value. In each of the following cases, what are the feasible actions of a buyer and their outcomes in terms of monetary gain or loss?
Case 1: The price of the stock increases after the buyer purchased the put.
Case 2: The price of the stock almost does not change.
Case 3: The price of the stock decreases and the exercise price of the put is higher than the price of the stock at the expiration date.
b) How American put selling works.
Under the plan of selling puts, you grant someone else the right to sell 100 shares to you at the exercise price. At the time you sell, you receive a premium. Like the call seller, you do not have much control over the outcome of your investment since the buyer will decide whether to exercise the put you sold him. In each of the following cases, what are the feasible actions of a put seller and their monetary results?
Case 1: The price of the stock increases.
Case 2: The price of the stock remains stable till the expiration date.
Case 3: The price of the stock decreases and the put is in the money at the expiration.

**7.14.** Hedge/hedging is a strategy used to offset investment risk. A perfect hedge is one eliminating the possibility of future gain or loss.

A stockholder worried about declining stock prices, for instance, can hedge his or her holdings by buying a put on the stock or by selling a call.

a) How does each case work?
b) Which type of hedging is preferable?

**7.15. (Call Time Spread Bearish)** Recall Example 1. Given XYZ 40 call price table:

| Expiration | Nov | Dec | Jan |
|---|---|---|---|
| Premium | 2 | 3 | 5 |

If one expected XYZ stock to decline, one might establish a bear spread by taking a position opposite of a bullish one.

Make two transactions to establish a spread in the hope of making a profit if XYZ stock's price declines and of limiting the loss if the expectation turns out to be wrong.

**7.16. (Price Put Spread Bearish)** Open a bear spread by using the following puts:

XYZ Dec 40 Put at 3
XYZ Dec. 45 Put at 7

in the hope of making a profit if XYZ stock declines in price. What is the possible maximum gain or loss? Justify your answers.

## 7.6.2 Application Exercises

**7.17. (Forward Price and Arbitrage)** Suppose that the current spot price of a continually paying dividend asset is $222, the interest rate is $r = 3\%$ and the dividend yield is $q = 2\%$.

a) What are the one-month and eight-month forward prices for the asset in an arbitrage-free market?
b) Let $\Pi$ be a portfolio on time interval $[0, T]$ consisting of three positions starting from time 0: borrow $222 at the rate 3%, long 1 unit of the asset, and short the three-month forward at $F_T(0) = \$222.56$. Is $\Pi$ an arbitrage portfolio? If your answer is no, show a proof. If your answer is yes, explain how you can make a profit by taking the arbitrage opportunity.

**7.18. (Forward Value)** Given $K_0 = \$222$, $F_T(t) = \$252$, $T - t = 6$ months, and $r = 3\%$, determine the value of the forward in an arbitrage-free market and interpret the result.

**7.19. (Swaps)** Assume that the terms of the swap contract include the following:

a) The notional principal is one million dollars,
b) The life of the contract is 2 years,
c) $A$ pays $B$ three-month $LIBOR + 0.2\%$,
d) $B$ pays $A$ 1.5% fixed,
e) There is an exchange of payments every 3 months from the initialization.

Given the $LIBOR$ rates in the table below, calculate both the floating cash flow and fixed cash flow of the swap.

| Period | LIBOR | Payment from A to B | Payment from B to A |
|--------|-------|---------------------|---------------------|
| 0 | 1% | | |
| 1 | 0.8% | | |
| 2 | 1% | | |
| 3 | 1.2% | | |
| 4 | 1.06% | | |
| 5 | 1.1% | | |
| 6 | 1.2% | | |
| 7 | 1.4% | | |
| 8 | | | |

**7.20. (Swaps)** Suppose that both companies $X$ and $Y$ need to borrow US dollars and that company $X$ would like to borrow at a fixed rate, whereas company $Y$ would like to borrow at a floating rate. If $X$ can borrow at 6.00% fixed and $LIBOR + 0.60\%$ floating, and $Y$ can borrow at 5.00% fixed and $LIBOR + 0.20\%$ floating, what is the range of possible cost savings that company $X$ can realize through an interest rate swap with company $Y$? Use an example of swap mechanics to demonstrate how a cost saving to be done for either company (ignoring credit risk differences).

**7.21.** Identify the range of profit, loss, and break-even point as outcomes of the corresponding strangle strategy given in Example 7.39 (page 366).

**7.22.** Use a formula to express the terminal payoff of each spread strategy given in Example 7.40 (page 366).

**7.23.** Use a formula to express the terminal payoff of the butterfly spread given in Example 7.42 (page 367).

### 7.6.3 Theoretical Exercises

**7.24. (Arbitrage)**

a) Suppose that the price of a stock at time $t$, denoted by $S(t)$, is modeled by a one-step binomial tree over the time period $[0, T]$ with

$$S(T) = \begin{cases} S_b & \text{with probability } p \\ S_a & \text{with probability } 1 - p, \end{cases}$$

where $S_b > S_a$.
Show that $S_b > S(0) > S_a$ is a necessary condition for a non-arbitrage opportunity for any investor (assuming $r_f = 0$).

b) Show that there exists a (risk-neutral) probability $p > 0$ holding the equation

$$S(0) = pS_b + (1 - p)S_a.$$

**7.25. (Forward Value)** Prove (7.5) (page 344).

**7.26.** Show that if the price of the underlier of a forward contract follows a geometric Brownian motion, so does the forward price process (see Example 7.15).

**7.27. (General Property of Options)** Let $C_A(0, K, T_1)$ and $C_A(0, K, T_2)$ be two American call options on the same terms except that they have different expirations with $T_1 < T_2$. Show that $C_A(0, K, T_1) \leq C_A(0, K, T_2)$.

**7.28.** Establish the following bounds for American puts on nondividend-paying underliers:

$$K \geq P^A(0) \geq \max\{K - S(0), 0\}.$$

**7.29.** Establish the following relation between American and European puts on the same nondividend-paying underlier and with the same expiration $T$ and strike $K$:

$$P^A(0) \geq P^E(0).$$

**7.30.** Is it never advantageous to exercise early an American put on a nondividend-paying stock? Justify your answer.

**7.31.** Establish the following put-call parity bounds for American options:

$$S(t_0) e^{-q\tau} - K \leq C^A(t_0) - P^A(t_0) \leq S(t_0) - K e^{-r\tau},$$

where $\tau = T - t_0$.

# References

[1] Anderson, G., Kercheval, A.: Lectures on Financial Mathematics: Discrete Asset Pricing. Morgan and Claypool, San Rafael (2010)

[2] Bingham, N.H., Kiesel, R.: Risk-NeutralValuation: Pricing and Hedging of Financial Derivatives. Springer Science and Business Media, New York (2004)

[3] Björk, T.: Arbitrage Theory in Continuous Time. Oxford University Press, Oxford (2009)

[4] Cox, J.C., Ingersoll, J.E., Ross, S.A.: The relation between forward prices and futures prices. J. Financ. Econ. **9**. (1981) [Columbia Business School, New York]

[5] Delbaen, F., Schachermayer, W.: A general version of the fundamental theorem of asset pricing. Math. Ann. **300**, 463(1994)

[6] Delbaen, F., Schachermayer, W.: The fundamental theorem of asset pricing for unbounded stochastic processes. Math. Ann. **312**, 215(1998)

[7] Delbaen, F., Schachermayer, W.: The Mathematics of Arbitrage. Springer, Berlin/Heidelberg (2006)

[8] Duffie, D., Stanton, R.: Pricing continuously resettled contingent claims. J. Econ. Dyn. Control **16**, 561–573 (1992)

[9] Epps, T.W.: Pricing Derivative Securities. World Scientific, River Edge (2007)

[10] Harrison, J., Pliska, S.: Martingales and stochastic integrals in the theory of continuous trading. Stoch. Process. Appl. **11**, 215–260 (1981)

[11] Hull, J.C.: Options, Futures, and Other Derivatives. Pearson Princeton Hall, Upper Saddle River (2015)

[12] Jacod, J., Protter, P.: Probability Essentials. Springer, Berlin/Heidelberg (2004)

[13] Kolb, R.W.: Financial Derivatives. New York Institute of Finance, New York (1993)

[14] Korn, R., Korn, E.: Option Pricing and Portfolio Optimization. American Mathematical Society, Providence (2001)

[15] Kreps, D.M.: Arbitrage and equilibrium in economies with infinitely many commodities. J. Math. Econ. **8**(1), 15(1981)

[16] Musiela, M., Rutkowsk, M.: Martingale Methods in Financial Modelling. Springer, New York (2004)

[17] Reilly, F.K., Brown, K.C.: Investment Analysis and Portfolio Management. South-Western Cengage Learning, Mason (2009)

[18] Whaley, R.: Derivative: Markets, Valuation, and Risk Management. Wiley, Hoboken (2006)

[19] Wilmott, P., Dewynne, N., Howison, S.: Mathematics of Financial Derivatives: a Student Introduction. Cambridge University Press, Cambridge (1995)

# Chapter 8
# The BSM Model and European Option Pricing

The *Black-Scholes-Merton (BSM) model*, also known as the *Black-Scholes model*, is one of the pillars of finance, providing a powerful theoretical framework that is widely applicable in financial engineering and corporate finance.

The BSM model is a partial differential equation (p.d.e.) approach to pricing derivatives, not just call and put options. The model is presented in Section 8.1, which includes the marketplace assumptions, self-financing, replicating portfolios, and a derivation of the BSM p.d.e. using such portfolios. Applications of the BSM model to the special case of European calls and puts are given in Section 8.2. An outline is presented for how to solve the BSM p.d.e. for a European call price. The associated put price is obtained via put-call parity, and the deltas and rate of change relative to strike price are explored for European calls and puts. A corporate finance application of the BSM model is given in Section 8.3, where warrants are priced. Section 8.4 then takes up an alternative approach to pricing derivatives. It prices derivatives using a risk-neutral probability measure, i.e., by discounting the conditional expectation of future prices at the risk-free rate minus the dividend yield rate. The alternative approach provides a deep link between the BSM p.d.e. and existence of risk-neutral measures. Exploration of the latter gives rise to the fundamental theorems of asset pricing. These results show how the no-arbitrage condition and market completeness are related to the existence and uniqueness of risk-neutral probability measures.

The pricing of derivatives can be done not only in a continuous-time setting with a continuum of possible prices of the underlying security but also in discrete time with discrete possible underlier prices. The latter approach is illustrated in Section 8.5 using binomial trees, which provide excellent intuition into the pricing process. Using a European call, we give a discrete-time version of the self-financing, replicating portfolio of the BSM model to price the call as well as a discrete-time version of the risk-neutral probability measure approach to pricing.

© Arlie O. Petters and Xiaoying Dong 2016                                              383
A.O. Petters, X. Dong, *An Introduction to Mathematical Finance with Applications*, Springer
Undergraduate Texts in Mathematics and Technology, DOI 10.1007/978-1-4939-3783-7_8

A core challenge is managing risk in a portfolio when the diversification approach of the Markowitz model does not apply. One of the significant applications of the BSM model is managing portfolio risk through what is known as delta and gamma hedging. Section 8.6 details the theoretical framework for dynamical delta hedging, while Section 8.6.2 illustrates how to use the framework to manage the risk from selling European call options. Section 8.7 extends these ideas to a portfolio of options, explaining how to make the portfolio delta- and gamma-neutral, i.e., stable against movements in the underlying security price.

In Section 8.8, we investigate how the assumptions of the BSM model stand up against data using the S&P 500 and an IBM European call option for illustration. Contrary to the specific BSM assumption of a geometric Brownian motion underlier, the underlier price can exhibit a jump discontinuity, and its log return can show skewness and kurtosis. In addition, the implied volatility of a European call when plotted as a function of strike price (and time to expiration) is not constant as predicted by the BSM model, but is curved and can even reveal shapes that look like a volatility smile.

The issue then is how to amend the BSM model. Section 8.9 gives a detailed introduction to the Merton jump diffusion (MJD) model as an example addressing the above concerns with the BSM model. The MJD model extends the underlier price process to one that is a mixture of geometric Brownian motion and jump discontinuities. We show that adding jump discontinuities to geometric Brownian motion produces log returns with skewness and kurtosis. In this sense, the current price of a European call with a jump-diffusion underlier is viewed as a better approximation to the volatility smiles in market data. We shall also see that the MJD model is in an arbitrage-free, incomplete market. This means that there is not a unique price for a derivative.

The chapter ends in Section 8.10 with a glimpse beyond the theory of derivatives presented in the book. We highlight that though the Merton jump-diffusion model incorporates several features not present in the BSM model, it does assume that the volatility of the underlier's price is deterministic, while data supports a stochastic volatility. This brings us to issues currently of interest in the research on derivatives: pricing derivatives in incomplete markets with underliers having stochastic volatility with jumps.

## 8.1 The BSM Model

The *Black-Scholes-Merton (BSM) model* gives a p.d.e. (i.e., partial differential equation) approach to pricing a derivative. We shall overview the basic assumptions of the model and derive the BSM p.d.e. It is important to emphasize that the BSM p.d.e. applies not only to European call options but also to a rather general class of derivative contracts. Our approach is in parallel with the treatments by Björk [5], Epps [15], and Privault [34].

**Notation (Current time).** The current time is typically set at time 0. In this case, since we are considering risky securities, their prices at a future time $t > 0$ are random. However, in certain settings, we shall need to treat the current time as advancing and shall denote it by the variable $t$. For such situations, assume that time $t + dt$ is in the future relative to $t$. Overall, the context in which the current time is intended should be clear to the reader.

### 8.1.1 Marketplace Assumptions

The marketplace is assumed to be idealized, i.e., the following holds:

➤ *Equilibrium:* supply equals demand.

➤ *No arbitrage:* no arbitrage opportunity exists, i.e., there is no costless, riskless profit. See Section 7.1.4 (page 334) for a precise treatment.

➤ *Access to information:* there is immediate availability of accurate information on all securities.

➤ *Efficiency:* a security's price adjusts instantly to new information, so its current price reflects all known information concerning the security, which includes information about the past and expected future behavior of the security.

➤ *Liquidity:* any number of units—even a fractional amount—of a security can be bought and sold instantly.

➤ *No transaction costs and no bid-ask spreads:* transaction costs, which include fees and margin account requirements, are ignored.

➤ *No taxes:* this includes no taxes on capital gains, interest, or dividends.

➤ *Borrowing/lending:* borrowing and lending are at the risk-free rate r, and there is no limit to how much one can borrow or lend.

➤ *Short selling:* short selling is allowed without restriction; in particular, the funds from a short sale can be used immediately to trade.

**Remark 8.1. (Number of Days in a Year)** For simplicity, a 365-day year is used in this chapter (e.g., McDonald [27]). However, in Chapter 5 (see page 223), when estimating a security's *historical volatility*, the non-trading days were ignored, and a year was treated as a *trading year*, which consists of 252 trading days; see Hull [22, p. 328] for more. On the other hand, a daily-compounded loan will compound on the non-trading days during the term of the loan, so such days are not omitted in this case.  □

### 8.1.2 Money Market Account and the Underlier Model

The BSM model prices a derivative using two securities that act as fundamental drivers: a money market account and the security serving as underlier of the derivative.

Define a *money market account* to be a riskless security that has a value $B_0$ at the current time 0 and grows by continuous compounding at the risk-free rate r. Its value at a general time $t$ is then

$$B_t = B_0 e^{rt},$$

where $B_0$ is an arbitrary constant. By default, we choose $B_0 = 1$ (one dollar). The instantaneous change is

$$dB_t = r\, B_t\, dt \qquad\qquad (B_0 = 1). \qquad\qquad (8.1)$$

**Remark 8.2.** The value $B_t$ is often used as a *numéraire* for an asset, which means that the value of the asset at $t$ can be expressed as a multiple of $B_t$, i.e., in units of $B_t$.

The underlying security of a derivative is assumed to pay a continuous dividend at yield rate $q$ and to be a geometric Brownian motion with *drift parameter* $\mu_{\mathrm{RW}}$ and *volatility parameter* $\sigma$:

$$S_t = S_0 e^{\mu_{\mathrm{RW}} t + \sigma \mathfrak{B}_t} \qquad\qquad (0 \le t \le T),$$

where $S_0$ is the known current price of the underlying security and

$$\mu_{\mathrm{RW}} = m - q - \frac{\sigma^2}{2}.$$

The instantaneous change is given by the following s.d.e.:

$$dS_t = (m - q)\, S_t\, dt + \sigma S_t\, d\mathfrak{B}_t. \qquad\qquad (8.2)$$

The underlier's cash dividend will be continuously reinvested either in the underlying security or the money market account.[1] Our default assumption is to reinvest the cash dividend in the underlier. In this case, we must distinguish between the underlier's *ex-dividend* (without dividend) *unit price* $S_t$, which is the market price of the security at $t$, and its *cum-dividend* (with dividend) *unit price* $S_t^c$; see page 18. Starting with 1 unit of the underlier at 0, the 1 unit will increase to $e^{qt}$ units at time $t$. The cum-dividend unit price of the security will then be (see (2.28) on page 31)

---

[1] Reinvesting the cash dividend in the security is then simply the investor acquiring more units of the security.

$$S_t^c = S_t e^{qt} \qquad (t \geq 0).$$

In other words, as time advances the market price $S_t = e^{-qt} S_t^c$ is a continuous downward adjustment of the cum-dividend price. This is because when a security pays out a cash dividend, value is flowing out of the security.

Finally, we restate *Itô's formula*[2] for easy reference. Suppose that $\{X_t\}_{t \geq 0}$ is an *Itô process*, i.e.,

$$dX_t = u(X_t, t) \, dt + v(X_t, t) \, d\mathcal{B}(t), \quad 0 \leq t \leq T, \tag{8.3}$$

where $u(x,t)$ and $v(x,t)$ are deterministic functions and $\{X_t\}_{t \geq 0}$ is an adapted process (e.g., a stochastic process that is a deterministic function of standard Brownian motion). Assume that $Y_t = g(X_t, t)$, where $g(x,t)$ is a deterministic function that is twice continuously differentiable in $x$ and once continuously differentiable in $t$. Then:

$$dY_t = \left( \frac{1}{2} v^2(X_t, t) \frac{\partial^2 g}{\partial x^2}(X_t, t) + u(X_t, t) \frac{\partial g}{\partial x}(X_t, t) + \frac{\partial g}{\partial t}(X_t, t) \right) dt$$

$$+ v(X_t, t) \frac{\partial g}{\partial x}(X_t, t) \, d\mathcal{B}(t). \tag{8.4}$$

For example, to obtain the instantaneous change in the cum-dividend price, apply (8.4) to $S_t^c = S_t e^{qt}$ with $g(x,t) = x e^{qt}$, which gives the s.d.e. for cum-dividend security prices (Exercise 8.14):

$$dS_t^c = m S_t^c \, dt + \sigma S_t^c \, d\mathcal{B}_t. \tag{8.5}$$

Finally, in keeping with our intuitive approach, we abide by the following:

> *Unless stated to the contrary, we shall freely apply Itô's formula and stochastic integration when necessary and so assume that all processes considered satisfy the properties that allow for such actions.*

For instance, a stochastic process $\{X_t\}_{t \geq 0}$ is always assumed to be adapted to standard Brownian motion, meaning $X_t$ is a deterministic function of $\mathcal{B}_v$ for all $0 \leq v \leq t$. When integrating, say, $\int_0^t X_v \, d\mathcal{B}_v$, we also always assume that the process is square integrable, i.e., $\int_0^t \mathbb{E}(X_v^2) \, dv < \infty$. See Björk [5, Chap 4], Mikosch [32, Chap. 2], and Privault [34, Chaps. 4–6] for an introduction and Elliot and Kopp [14, Chaps. 6, 7] and Korn and Korn [24, Chaps. 2, 3] for an advanced treatment.

---

[2] See (6.57) on page 304.

### 8.1.3 Self-Financing, Replicating Portfolio

In this section, we shall employ a certain trading strategy $(n_t, b_t)$—i.e., take a position with $n_t$ units of the underlying security and $b_t$ units of the money market account—to construct a portfolio whose value replicates the price of a derivative in a self-financing manner.

Before carrying out this strategy, we fix a derivative and assume that its price based on 1 unit of its underlying security is a stochastic process $\{f_t\}_{0 \leq t \leq T}$ that is a deterministic function of the underlier's market price (i.e., ex-dividend price) $S_t$ and time $t$,

$$f_t = f(S_t, t), \tag{8.6}$$

where $f(x, t)$ is assumed to be at least twice continuously differentiable in $x$ and once continuously differentiable in $t$ for $x > 0$ and $0 < t < T$. In particular, the current price of the derivative is $f(S_0, 0)$. *We assume that the derivative does not pay a cash dividend.*

Now, assume that we have an initial capital $V_t$ at $t$. With this money, create a portfolio using a trading strategy $(n_t, b_t)$, i.e., hold $n_t$ units of the cum-dividend underlying security and $b_t$ units of the money market account. Think of $(n_t, b_t)$ as a stochastic process with values evolving in $\mathbb{R}^2$. The value of the portfolio at $t$ is

$$V_t = n_t S_t^c + b_t B_t.$$

As $t$ advances and $(n_t, b_t)$ evolves, a key issue will be how to pay for these changes in the number of units of the underlying security and the money market account.

At time $t + dt$, suppose that $(n_{t+dt}, b_{t+dt})$ replicates the price of the derivative:

$$V_{t+dt} = n_{t+dt} S_{t+dt}^c + b_{t+dt} B_{t+dt} = f(S_{t+dt}, t + dt).$$

After the initial capital at $t$, the portfolio is *self-financing* if the trading strategy $(n_{t+dt}, b_{t+dt})$ at $t + dt$ is funded without withdrawing or adding any external funds to the portfolio. In other words, during the transition from time $t$ to $t + dt$, the value of the portfolio at $t + dt$ arises only from an increase, decrease, or neither in the values of the underlying security and/or the money market account. The original strategy $(n_t, b_t)$, along with the possibly new values of the underlier and money market account at $t + dt$, must then fund the portfolio's replication of the derivative:

$$V_{t+dt} = n_t S_{t+dt}^c + b_t B_{t+dt}.$$

Explicitly, the *self-financing condition* is then

$$n_t S_{t+dt}^c + b_t B_{t+dt} = n_{t+dt} S_{t+dt}^c + b_{t+dt} B_{t+dt},$$

which can be written more compactly as[3]

$$\left(dn_t\right) S^c_{t+dt} + \left(db_t\right) B_{t+dt} = 0. \tag{8.7}$$

Equation (8.7) is equivalent to (Exercise 8.15)

$$dV_t = n_t \, dS^c_t + b_t \, dB_t. \tag{8.8}$$

If the self-financing portfolio replicates the price of the derivative at $t + dt$, then by the law of one price, the portfolio's value at $t$ also replicates the price of the derivative at $t$:

$$V_t = n_t \, S^c_t + b_t B_t = f(S_t, t). \tag{8.9}$$

Let us then determine the trading strategy $(n_t, b_t)$ that will make (8.9) possible. The self-financing condition (8.8) and replicating condition (8.9) will allow us to determine the desired $(n_t, b_t)$. By (8.9),

$$b_t = \frac{f(S_t, t) - n_t \, S^c_t}{B_t}.$$

Substituting into (8.8) and employing (8.1) and (8.5) give

$$dV_t = n_t \, dS^c_t + \left(f(S_t, t) - n_t \, S^c_t\right) \frac{dB_t}{B_t}$$
$$= \left(r f(S_t, t) + n_t \, (m - r) \, S^c_t\right) dt + n_t \sigma S^c_t \, d\mathcal{B}_t. \tag{8.10}$$

On the other hand, Itô's formula (8.4) yields

$$dV_t = df(S_t, t) = \left(\frac{1}{2}\sigma^2 S_t^2 \frac{\partial^2 f}{\partial x^2}(S_t, t) + (m - q) S_t \frac{\partial f}{\partial x}(S_t, t) + \frac{\partial f}{\partial t}(S_t, t)\right) dt$$
$$+ \sigma S_t \frac{\partial f}{\partial x}(S_t, t) \, d\mathcal{B}_t. \tag{8.11}$$

Now, an *Itô process has a unique representation*; see, for example, Korn and Korn [24, p. 77]. This means that if

$$dX_t = a_t \, dt + b_t \, d\mathcal{B}_t = \tilde{a}_t \, dt + \tilde{b}_t \, d\mathcal{B}_t,$$

where $a_t$, $b_t$, $\tilde{a}_t$, and $\tilde{b}_t$ are functions of standard Brownian motion, then

$$a_t = \tilde{a}_t, \qquad b_t = \tilde{b}_t.$$

---

[3] Recall: $dx_t = x_{t+dt} - x_t$.

Equating the $d\mathfrak{B}_t$ coefficients in (8.10) and (8.11), we obtain:

$$n_t = \left(\frac{S_t}{S_t^c}\right)\frac{\partial f}{\partial x}(S_t, t). \tag{8.12}$$

The partial derivative of $f$ in (8.12) is our first encounter with a financial derivative's *Greek*, which we define loosely as a partial derivative of $f$ with respect to $x$, $t$, or any of the parameters used in modeling the financial derivative. Specifically, the partial of $f$ relative to $x$ is called the *delta* of the financial derivative and denoted by

$$\Delta_f(S_t, t) = \frac{\partial f}{\partial x}(S_t, t).$$

Delta is perhaps the most popular of the Greeks and will appear many times in this chapter.

Thus, the price $f(S_t, t)$ of the derivative at any time $t$ can be replicated by a self-financing strategy $(n_t, b_t)$, where

$$f(S_t, t) = n_t S_t^c + b_t B_t \tag{8.13}$$

with

$$n_t = \left(\frac{S_t}{S_t^c}\right)\Delta_f(S_t, t), \qquad b_t = \frac{f(S_t, t) - S_t \Delta_f(S_t, t)}{B_t}. \tag{8.14}$$

### 8.1.4 Derivation of the BSM p.d.e.

Interestingly, the price of a derivative in the BSM model will arise from solving a partial differential equation (p.d.e.). A second order p.d.e. in two independent variables $(x, t)$ is an equation of the form

$$A(x,t)\frac{\partial^2 y}{\partial x^2}(x,t) + B(x,t)\frac{\partial^2 y}{\partial x \partial t}(x,t) + C(x,t)\frac{\partial^2 y}{\partial t^2}(x,t) + D(x,t)\frac{\partial y}{\partial x}(x,t)$$
$$+ E(x,t)\frac{\partial y}{\partial t}(x,t) + F(x,t)y(x,t) = 0, \tag{8.15}$$

where the coefficients and $y(x,t)$ are deterministic functions. The p.d.e. (8.15) is called

| | |
|---|---|
| *hyperbolic* | if $B^2 - 4AC > 0$ |
| *parabolic* | if $B^2 - 4AC = 0$ |
| *elliptic* | if $B^2 - 4AC < 0$. |

**Example 8.1. (Heat Equation)** The following p.d.e. is well known in physics:

$$\frac{\partial y}{\partial t}(x,t) = c\frac{\partial^2 y}{\partial x^2}(x,t), \tag{8.16}$$

where $c$ is a positive constant and $y(x,t)$ designates temperature at position $x$ and time $t$. It describes how heat diffuses (spreads) along a rod over time, where there is no heat source along the rod. The heat Equation (8.16) is also called a *diffusion equation*. Since $B = C = 0$, we have $B^2 - 4AC = 0$, and so the heat equation is a parabolic p.d.e. □

Returning to the self-financing, replicating portfolio of Section 8.1.3, we already did all the work needed to obtain the p.d.e. of the BSM model. By the unique representation of Itô processes (page 389), we can equate the $dt$ coefficients in (8.10) and (8.11), which along with (8.12) gives

$$\frac{1}{2}\sigma^2 S_t^2 \frac{\partial^2 f}{\partial x^2}(S_t,t) + (r-q)S_t \frac{\partial f}{\partial x}(S_t,t) + \frac{\partial f}{\partial t}(S_t,t) - rf(S_t,t) = 0. \quad (8.17)$$

In general, Equation (8.17) is not a deterministic p.d.e. because $S_t$ is random for $t > 0$. However, for each $t$ in $(0,T)$, the possible values of the lognormal random variable $S_t$ range over $(0,\infty)$ independent of $t$. In other words, Equation (8.17) holds at all points $(x,t)$ in $(0,\infty) \times (0,T)$. The associated deterministic p.d.e. is then

$$\frac{1}{2}\sigma^2 x^2 \frac{\partial^2 f}{\partial x^2}(x,t) + (r-q)x \frac{\partial f}{\partial x}(x,t) + \frac{\partial f}{\partial t}(x,t) - rf(x,t) = 0, \quad (8.18)$$

where $0 < x < \infty$ and $0 < t < T$. Equation (8.18) is called the *BSM p.d.e.* Because $B = C = 0$, we get $B^2 - 4AC = 0$, i.e., the BSM p.d.e. is parabolic.

Assuming a solution $f(x,t)$ of the BSM p.d.e. exists, the derivative's price is given by $f(S_t,t)$. The issue is that if $f(x,t)$ is a solution, then we can construct infinitely many other solutions, which creates an infinity of derivative prices. For example, if $f(x,t)$ is a solution, then for every positive real number $c > 0$ the function $f_c(x,t) = f(cx,t)$ is also a solution (Exercise 8.17). The existence and uniqueness of a derivative price require additional constraints on $f(x,t)$:

*Final condition:*                  $f(x,T)$ is given.

*Boundary conditions:*         $f(0,t)$ is given and the growth behavior of $f(x,t)$ as $x \to \infty$ is given.

Note that the explicit nature of these conditions cannot be stated a priori because they depend on the contractual structure of the derivative.

*For sufficiently well-behaved final and boundary conditions, the theory of parabolic p.d.e.'s yields that the BSM p.d.e. will have a unique solution $f(x,t)$ and, hence, the derivative will have a unique price $f(S_t,t)$.* See Korn and Korn [24, Sec. 3.3] and Miersemann [31, Chap. 6] for more.

## 8.2 Applications of BSM Pricing to European Calls and Puts

Below we state the final and boundary conditions for European calls and puts as well as present the associated unique solution of the BSM p.d.e. and the derivative's price.

**Notation.** For European calls and puts, write the solutions of the BSM p.d.e. as $C^E(x,t)$ and $P^E(x,t)$, respectively, rather than $f(x,t)$.

### 8.2.1 Solving the BSM p.d.e. for European Calls

For a European call with strike $K$ and expiration at $T$, the BSM p.d.e. is

$$\frac{1}{2}\sigma^2 x^2 \frac{\partial^2 C^E}{\partial x^2}(x,t) + (r-q)x\frac{\partial C^E}{\partial x}(x,t) + \frac{\partial C^E}{\partial t}(x,t) - rC^E(x,t) = 0, \qquad (8.19)$$

the final condition is

$$C^E(x,T) = \max\{x - K,\, 0\}, \qquad (8.20)$$

and boundary conditions are

$$C^E(0,t) = 0 \quad \text{and} \quad C^E(x,t) \to e^{-q(T-t)}x \quad \text{as} \quad x \to \infty. \qquad (8.21)$$

Note that (8.20) and (8.21) follow from properties of European calls using no arbitrage and put-call parity.

We outline how to solve the p.d.e. along the lines of Wilmott, Dewynne, and Howison [42, Sec. 5.4], leaving the computational details as exercises:

➢ Transform the p.d.e. (8.19) to a form without dividend. It can be shown (Exercise 8.20) that with the new underlier price process

$$\bar{S}_t = S_t e^{-q(T-t)},$$

Equation (8.19) transforms to a form without dividend:

$$\frac{1}{2}\sigma^2 \bar{x}^2 \frac{\partial^2 C^E}{\partial \bar{x}^2}(\bar{x},t) + r\bar{x}\frac{\partial C^E}{\partial \bar{x}}(\bar{x},t) + \frac{\partial C^E}{\partial t}(\bar{x},t) - rC^E(\bar{x},t) = 0, \qquad (8.22)$$

where $\bar{x} = xe^{-q(T-t)}$ and the call price is now viewed as a function of $(\bar{x},t)$. The associated terminal and boundary conditions are

$$C^E(\bar{x},T) = \max\{x - K,\, 0\}, \quad C^E(0,t) = 0, \quad C^E(\bar{x},t) \to \bar{x} \text{ as } \bar{x} \to \infty. \qquad (8.23)$$

Note that $\bar{x} = x$ at $t = T$.

➤ Convert to the following convenient variables:

$$\tilde{x} = \ln\left(\frac{x}{K}\right), \quad \tilde{\tau} = \frac{\sigma^2}{2}(T - t), \quad v(\tilde{x}, \tilde{\tau}) = \frac{C^E(\tilde{x}, t)}{K}, \quad \tilde{k} = \frac{r}{(\sigma^2/2)}.$$

Note that $-\infty < \tilde{x} < \infty$ and $\tilde{\tau} > 0$, where the initial time $\tilde{\tau} = 0$ corresponds to the expiration date $T$ (i.e., the final value problem becomes an initial value one). Equations (8.22) and (8.23) then become (Exercise 8.21):

$$\frac{\partial v}{\partial \tilde{\tau}}(\tilde{x}, \tilde{\tau}) = \frac{\partial^2 v}{\partial \tilde{x}^2}(\tilde{x}, \tilde{\tau}) + (\tilde{k} - 1)\frac{\partial v}{\partial \tilde{x}}(\tilde{x}, \tilde{\tau}) - \tilde{k}v(\tilde{x}, \tilde{\tau}) \tag{8.24}$$

and

$$v(\tilde{x}, 0) = \max\{e^{\tilde{x}} - 1, 0\}, \quad \lim_{\tilde{x} \to -\infty} v(\tilde{x}, \tilde{\tau}) = 0, \quad v(\tilde{x}, \tilde{\tau}) \to e^{\tilde{x}} \text{ as } \tilde{x} \to \infty. \tag{8.25}$$

➤ Using a trial solution

$$v(\tilde{x}, \tilde{\tau}) = \tilde{u}(\tilde{x}, \tilde{\tau})e^{a\tilde{x} + b\tilde{\tau}}, \tag{8.26}$$

it can be shown (Exercise 8.22) that the choices

$$a = -\frac{1}{2}(\tilde{k} - 1), \quad b = -\frac{1}{4}(\tilde{k} + 1)^2, \tag{8.27}$$

transform (8.24) into the *heat equation*

$$\frac{\partial \tilde{u}}{\partial \tilde{\tau}}(\tilde{x}, \tilde{\tau}) = \frac{\partial^2 \tilde{u}}{\partial \tilde{x}^2}(\tilde{x}, \tilde{\tau}) \tag{8.28}$$

and (8.25) to

$$\tilde{u}(\tilde{x}, 0) = \max\left\{e^{\frac{1}{2}(\tilde{k}+1)\tilde{x}} - e^{\frac{1}{2}(\tilde{k}-1)\tilde{x}}, 0\right\}, \quad \lim_{|\tilde{x}| \to \infty} \tilde{u}(\tilde{x}, \tilde{\tau})e^{-c\tilde{x}^2} = 0, \tag{8.29}$$

where $c > 0$.

The heat equation has been extensively studied in physics and mathematics. The key result is that Equations (8.28) and (8.29) have a unique solution given by

$$\tilde{u}(\tilde{x}, \tilde{\tau}) = \frac{1}{2\sqrt{\pi\tilde{\tau}}} \int_{-\infty}^{\infty} \tilde{u}(s, 0)\, e^{-\frac{(\tilde{x}-s)^2}{4\tilde{\tau}}}\, ds, \tag{8.30}$$

where $\tilde{u}(s, 0)$ is given in (8.29). After some change of variables and completing the square, the solution (8.30) can be transformed (see Wilmott, Dewynne, and Howison [42, Sec. 5.4]) to

$$\tilde{u}(\tilde{x}, \tilde{\tau}) = e^{\frac{1}{2}(\tilde{k}+1)\tilde{x} + \frac{1}{4}(\tilde{k}+1)^2\tilde{\tau}}\, N(d_+) - e^{\frac{1}{2}(\tilde{k}-1)\tilde{x} + \frac{1}{4}(\tilde{k}-1)^2\tilde{\tau}}\, N(d_-), \tag{8.31}$$

where $N(\cdot)$ is the standard normal c.d.f. and

$$d_+ = \frac{\tilde{x} + (\tilde{k}+1)\,\tilde{\tau}}{\sqrt{2\tilde{\tau}}}, \qquad d_- = d_+ - \sqrt{2\tilde{\tau}}.$$

➤ Inserting (8.31) and (8.27) in (8.26), the call price becomes (Exercise 8.23)

$$C^E(\tilde{x},t) = K v(\tilde{x},\tilde{\tau}) = \tilde{x}\, N(d_+) - K\,e^{-r(T-t)}\,N(d_-). \tag{8.32}$$

Transforming back to the original variables $(x,t)$, Equation (8.32) shows that the unique solution of the BSM p.d.e. (8.19) subject to (8.20) and (8.21) is

$$C^E(x,t) = x\,e^{-q(T-t)}\,N\big(d_+(x,T-t)\big) - K\,e^{-r(T-t)}\,N\big(d_-(x,T-t)\big), \tag{8.33}$$

where[4]

$$d_\pm(x,T-t) = \frac{1}{\sigma\sqrt{T-t}}\left(\ln\left(\frac{x}{K\,e^{-(r-q)\,(T-t)}}\right) \pm \frac{1}{2}\sigma^2(T-t)\right). \tag{8.34}$$

Note that Equation (8.34) can be written as follows:

$$d_+(x,\tau) = \frac{\ln\left(\frac{x}{K}\right) + \left(r - q + \frac{1}{2}\sigma^2\right)\tau}{\sigma\sqrt{\tau}} \tag{8.35}$$

$$d_-(x,\tau) = \frac{\ln\left(\frac{x}{K}\right) + \left(r - q - \frac{1}{2}\sigma^2\right)\tau}{\sigma\sqrt{\tau}} = \frac{\ln\left(\frac{x}{K}\right) + \mu_*\,\tau}{\sigma\sqrt{\tau}} \tag{8.36}$$

$$d_+(x,\tau) = d_-(x,\tau) + \sigma\sqrt{\tau},$$

where $\tau = T - t$ and $\mu_* = r - q - \frac{1}{2}\sigma^2$ (which appeared in our study of the CRR tree (see Equation (5.52) on page 234)).

Equation (8.33) gives the *BSM formula* for pricing European call options. Explicitly, the price of a European call on 1 unit of the underlying security is then a stochastic process given by the value of the function $C^E(x,t)$ at $(x,t) = (S_t,t)$ for all $t \geq 0$:

$$C^E(S_t,t) = S_t\,e^{-q(T-t)}\,N(d_+(S_t,T-t)) - K\,e^{-r(T-t)}\,N\big(d_-(S_t,T-t)\big), \tag{8.37}$$

or, more compactly,

$$C_t^E = S_t\,e^{-q\tau}\,N(d_+) - K\,e^{-r\tau}\,N(d_-) \qquad\qquad (\tau = T - t). \tag{8.38}$$

The current price of the European call is then

$$C_0^E = C^E(S_0,0).$$

---

[4] Some authors use the alternative notation $d_1 = d_+$ and $d_2 = d_-$.

*It is important to emphasize that European call prices in the real world are determined by market forces, not by the BSM formula.*

The BSM pricing formula (8.37) is also applied when the underlying security is replaced by a risky portfolio of securities. In this situation, the total value of the portfolio at a general time $t$ is $S_t$, and the portfolio is assumed to follow geometric Brownian motion for $t \geq 0$. Like the underlying security of a derivative, the portfolio is also assumed to be tradable.

**Remark 8.3.** The original 1973 formula by Black and Scholes for a European call's price assumed $q = 0$, while Merton's 1973 paper extended the result to $q > 0$. □

**Notation.** The price of a European call will be written in several different ways, depending on the degree of notational simplicity needed and the dependence we wish to highlight:

$$C_t^E = C^E(S_t, t) = C^E(S_t, K, \sigma, r, \tau, q) \qquad (\tau = T - t).$$

The rightmost expression emphasizes the call's dependence on all the inputs $S_t$, $K$, $\sigma$, $r$, $\tau$, and $q$.

**Example 8.2.** What is the fair current price of 500 European calls on an index with current dollar value \$1,100, strike price \$1,100, volatility 15%, two months to expiration, and dividend yield of 2.5%? Assume a risk-free rate of 2%.

**Solution.** The fair price of a derivative is its no-arbitrage price, which in the case of a call is the BSM price. At the current time $t = 0$, the inputs are:

$$S_0 = \$1,100, \quad K = \$1,100, \quad \sigma = 0.15, \quad r = 0.02, \quad \tau = 0.166667, \quad q = 0.025.$$

Then the European call formula (8.37) yields $C_0^E = \$26.31437$.[5] Each call is based on 100 indexes, so the total cost of the 500 calls is
$$500 \times 100 \times \$26.31437 = \$1,315,718.50.$$
□

## 8.2.2 BSM Pricing Formula for European Puts

Inserting the call price (8.37) into the put-call parity formula, namely,

$$P^E(S_t, t) = K e^{-r\tau} - S_t e^{-q\tau} + C^E(S_t, t),$$

immediately yields the price process of a European put:

$$P^E(S_t, t) = K e^{-r\tau} N\big(-d_-(S_t, \tau)\big) - S_t e^{-q\tau} N(-d_+(S_t, \tau)). \qquad (8.39)$$

---

[5] Many free online calculators are available for computing the prices of calls and puts.

The current price is then $P_0^E = P^E(S_0, 0)$. *As noted earlier for calls, the actual prices of European puts are dictated by the marketplace and not by (8.39).*

The pricing formula (8.39) can also be obtained by solving the BSM p.d.e. with final condition

$$P^E(x, T) = \max\{K - x, 0\}$$

and boundary conditions $P^E(0, t) = Ke^{-r(T-t)}$ and $P^E(x, t) \to 0$ as $x \to \infty$. The final and boundary conditions also follow from no arbitrage and put-call parity. The corresponding unique solution is

$$P^E(x, t) = Ke^{-r(T-t)} N(-d_-) - xe^{-q(T-t)} N(-d_+),$$

which yields the put price (8.39) at $t$ via $(x, t) = (S_t, t)$.

**Example 8.3.** What is the fair current price of 1,000 European puts on a stock with current price \$82, strike price \$82, volatility 10%, and six months to expiration? Assume a risk-free rate of 3% and that the stock pays no dividend.

**Solution.** The inputs at the current time $t = 0$ are:

$$S_0 = \$82, \quad K = \$82, \quad \sigma = 0.10, \quad r = 0.03, \quad \tau = T = 0.5, \quad q = 0.$$

The BSM pricing formula (8.39) gives $P_0^E = \$1.7365$. Hence, the fair cost of the 1,000 puts is: $1,000 \times 100 \times \$1.7365 = \$173,650$.

$\square$

### 8.2.3 Delta and the Partial Derivative Relative to Strike Price

The European call price naturally involves partial derivatives relative to the underlier price (delta) and the strike price. In fact, Equation (8.37) can be expressed more compactly in terms of partial derivatives:

$$C^E(S_t, t) = S_t \Delta_C(t) + K \frac{\partial C^E}{\partial K}(t), \tag{8.40}$$

where the *delta of the call* is

$$\Delta_C(t) = \left. \frac{\partial C^E}{\partial x} \right|_{(x,t)=(S_t,t)} = e^{-q\tau} N(d_+(S_t, \tau)) > 0 \tag{8.41}$$

and

$$\frac{\partial C^E}{\partial K}(t) = -e^{-r\tau} N(d_-(S_t, \tau)) < 0, \tag{8.42}$$

where the strike price $K$ is treated as a variable.

**Notation.** When confusion is less likely, we shall also write

$$\Delta_C(t) = \frac{\partial C^E}{\partial S}(t).$$

The partial derivative $\frac{\partial C^E}{\partial S}(t)$ should be understood as the partial of $C^E(x,t)$ with respect to $x$ with evaluation at $(x,t) = (S_t, t)$:

$$\frac{\partial C^E}{\partial S}(t) = \left.\frac{\partial C^E}{\partial x}\right|_{(x,t)=(S_t,t)}.$$

A similar notation will be used for European puts.

An important consequence of (8.40), (8.42), and Itô's formula is that *the price of a European call is more volatile than that of its underlying security* (Exercise 8.25). In addition, Equation (8.41) yields that the delta of a European call is always positive, so *the price of a European call increases as the price of the underlying security increases.* Furthermore, Equation (8.42) shows that $\frac{\partial C^E}{\partial K}(t) < 0$, i.e., *the price of a European call decreases as the strike price increases.* In other words, *an out-of-the-money European call is cheaper than an at-the-money or in-the-money call with the same inputs:*

$$\underbrace{C^E(S_t, K_\circ, \sigma, r, \tau, q)}_{\text{out-of-the-money}} < \underbrace{C^E(S_t, K_\bullet, \sigma, r, \tau, q)}_{\text{at-the-money or in-the-money}} \qquad \text{for} \quad K_\bullet \leq S_t < K_\circ.$$

$$(8.43)$$

For European puts, Equation (8.39) becomes

$$P^E(S_t, t) = S_t \Delta_P(t) + K\frac{\partial P^E}{\partial K}(t),$$

where

$$\Delta_P(t) = \frac{\partial P^E}{\partial S}(t) = -e^{-q\tau}N(-d_+(S_t, \tau)) < 0 \qquad (8.44)$$

and

$$\frac{\partial P^E}{\partial K}(t) = e^{-r\tau}N(-d_-(S_t, \tau)) > 0. \qquad (8.45)$$

The put's delta is negative by (8.44), i.e., *the price of a European put decreases as the price of the underlying security increases.* Moreover, since (8.45) gives $\frac{\partial P^E}{\partial K}(t) > 0$, *the price of a European put increases as the strike price increases.* It follows that *an out-of-the-money European put is cheaper than an at-the-money or in-the-money put with the same inputs:*

$$\underbrace{P^E(S_t, K_\circ, \sigma, r, \tau, q)}_{\text{out-of-the-money}} < \underbrace{P^E(S_t, K_\bullet, \sigma, r, \tau, q)}_{\text{at-the-money or in-the-money}} \qquad \text{for} \quad K_\circ < S_t \leq K_\bullet.$$

$$(8.46)$$

### *8.2.4 European Call and Put Deltas at Expiration*

We conclude with the behavior of the deltas of European calls and puts as time approaches expiration, i.e., as $\tau = T - t \to 0$. First, recall that (see page 394)

$$d_{\pm}(S_t, \tau) = \frac{\ln\left(\frac{S_t e^{-q\tau}}{K e^{-r\tau}}\right) \pm \frac{1}{2}\sigma^2 \tau}{\sigma\sqrt{\tau}} = \frac{\ln\left(\frac{S_t}{K}\right)}{\sigma\sqrt{\tau}} + \frac{(r - q \pm \frac{1}{2}\sigma^2)}{\sigma}\sqrt{\tau}$$

$$d_{-}(S_t, \tau) = d_{+}(S_t, \tau) - \sigma\sqrt{\tau}.$$

Then the following holds as $t$ approaches expiration $T$:

- If $S(T) = K$, then $d_{\pm}(S_t, \tau) \to 0$ as $\tau \to 0$.
- If $S(T) > K$, then $d_{\pm}(S_t, \tau) \to +\infty$ as $\tau \to 0$.
- If $S(T) < K$, then $d_{\pm}(S_t, \tau) \to -\infty$ as $\tau \to 0$.

Using (8.41), (8.44), and the above, we find

$$\text{If } S(T) = K, \text{ then }\quad \Delta_C(t) \to 1/2 \quad \text{and} \quad \Delta_P(t) \to -1/2 \quad \text{as} \quad \tau \to 0. \quad (8.47)$$

$$\text{If } S(T) > K, \text{ then }\quad \Delta_C(t) \to 1 \quad \text{and} \quad \Delta_P(t) \to 0 \quad \text{as} \quad \tau \to 0. \quad (8.48)$$

$$\text{If } S(T) < K, \text{ then }\quad \Delta_C(t) \to 0 \quad \text{and} \quad \Delta_P(t) \to -1 \quad \text{as} \quad \tau \to 0. \quad (8.49)$$

$$\text{If } t < T, \text{ then }\quad \Delta_C(t) < 1 \quad \text{and} \quad \Delta_P(t) > -1. \quad (8.50)$$

## 8.3 Application to Pricing Warrants

*Warrants* are call options issued (i.e., sold) by a company on its own stock. They provide a way for companies to raise money. When the warrants are exercised, the company issues new shares of its stock and sells them to the holders at the strike price. Note that issuing these new shares dilutes the share price of the stock.

*Suppose that at the current time, $t = 0$, a company has $N_{out}$ outstanding shares with each of price $S(0)$ and issues $N_w$ warrants, where each warrant is a European call on 1 share of the company's stock with strike price $K$ and expiration $T$.* The number of outstanding shares will not change until the warrants are exercised. The *equity value* of the company, i.e., the value of the company's asset minus the value of its debt, at 0 is denoted by $V(0)$. It consists of the current value $N_{out} S(0)$ of the $N_{out}$ outstanding shares and the proceeds $N_w \mathcal{W}(0)$ from selling the $N_w$ warrants, where $\mathcal{W}(0)$ is the current value of each warrant at 0:

$$V(0) = N_{out}\, S(0) \, + \, N_w\, \mathcal{W}(0). \quad (8.51)$$

*Assume that all $N_W$ warrants are exercised at expiration $T$.* Denote by $V_a(T)$ the equity value of the company immediately after the warrants are exercised. It is the sum of the company's equity value $V(T)$ just before the warrants are exercised and the proceeds from selling $N_W$ new shares at unit price $K$ to settle the warrants being exercised:

$$V_a(T) = V(T) + N_W K.$$

The price of a share of the company's equity instantly after the warrants are exercised is

$$\frac{V_a(T)}{N_{out} + N_W},$$

where $N_{out} + N_W$ is the number of outstanding shares after exercise.

To value the warrant today, construct two portfolios as follows:

➤ *Portfolio A:* long 1 warrant on 1 share of the company stock. The current payoff of portfolio A is then the current value $\mathcal{W}(0)$ of the warrant.

➤ *Portfolio B:* long $\frac{N_{out}}{N_{out}+N_W}$ European calls with current underlier price $\frac{V(0)}{N_{out}}$, strike price $K$, expiration $T$, and no dividend. The price of the underlier at a general time $t$ is the value $\frac{V(t)}{N_{out}}$ of a share of company equity before exercise. *We assume that $\frac{V(t)}{N_{out}}$ follows a geometric Brownian motion with volatility parameter $\sigma_V$.* The current payoff of Portfolio B is then determined using the BSM European call pricing formula:

$$\frac{N_{out}}{N_{out} + N_W} C^E\left(\frac{V(0)}{N_{out}}, K, \sigma_V, r, T, q\right).$$

The payoffs of portfolios A and B at $T$ are:

$$\text{(Payoff of A at } T) = \max\left\{\frac{V_a(T)}{N_{out} + N_W} - K, 0\right\} = \frac{N_{out}}{N_{out} + N_W}\max\left\{\frac{V(T)}{N_{out}} - K, 0\right\},$$

$$\text{(Payoff of B at } T) = \frac{N_{out}}{N_{out} + N_W}\max\left\{\frac{V(T)}{N_{out}} - K, 0\right\}.$$

Because both portfolios have the same payoff at expiration, the law of one price yields that they have the same payoff today, which gives the current value of each warrant to be

$$\mathcal{W}(0) = \frac{N_{out}}{N_{out} + N_W} C^E\left(\frac{V(0)}{N_{out}}, K, \sigma_V, r, T, q\right). \tag{8.52}$$

Equation (8.52) assumes that $\frac{V(0)}{N_{out}}$ and $\sigma_V$ are known. If $\frac{V(0)}{N_{out}}$ is unknown, then employing (8.51) we find:

$$W(0) = \frac{N_{\text{out}}}{N_{\text{out}} + N_{\text{w}}} C^E \left( S(0) + \frac{N_{\text{w}}}{N_{\text{out}}} W(0), K, \sigma_V, r, T, q \right). \qquad (8.53)$$

This is a BSM-type pricing formula for the current price of the warrant in terms of the current price of the warrant. In other words, we can numerically solve (8.53) for $W(0)$ implicitly.

**Example 8.4.** Suppose that a company has 3 million outstanding shares, the current value of a share of the company's equity is \$110, and the company's equity per share has a volatility parameter of 20% per annum. The company plans to issue 500,000 European warrants. Each warrant is based on 1 share of the company's stock, the strike price is \$125, and the expiration date is 3 years away. Determine the fair price of the issuance, i.e., the BSM price. Assume the company pays no dividend and the risk-free rate is 2.5%.

**Solution.** Applying Equation (8.52) with $N_{\text{out}} = 3 \times 10^6$, $N_{\text{w}} = 0.5 \times 10^6$, $\frac{V(0)}{N_{\text{out}}} = \$110$, $\sigma_V = 0.2$, $K = \$125$, $r = 0.025$, $T = 3$, and $q = 0$, the price per warrant is:

$$W(0) = \frac{3}{3.5} C^E (\$110, \$125, 0.025, 3, 0) = 0.85714286 \times \$12.73093 = \$10.91.$$

The fair total price of the issuance is:

$$500,000 \times \$10.91 = \$5,455,000.$$

This is how much money the company would raise if it sold all the warrants at the BSM price. $\qquad \square$

## 8.4 Risk-Neutral Pricing

In this section, we shall restrict our presentation to European style derivatives and give an intuitive, informal presentation. The case of American-style derivatives requires a more extended, complex discussion and is beyond the scope of this text; see, for example, Björk [5, Chaps. 7, 21] and Epps [15, Chap. 7].

### 8.4.1 Review of Conditional Expectation

Conditional expectations $\mathbb{E}(X_t|\mathcal{F}_s)$ will play an important role in our discussion of risk-neutral pricing, so we shall review the concept intuitively starting with the $\sigma$-algebra $\mathcal{F}_s$. See Section 6.2 (page 268) for more.

First, the "$\sigma$" in the name "$\sigma$-algebra" is not the volatility parameter of a security price modeled as geometric Brownian motion. The term "$\sigma$-algebra" was developed in mathematics independent of finance, and its usage is ingrained in the literature, so we abide by its customary usage.

Second, *unless otherwise stated, the $\sigma$-algebra $\mathcal{F}_s$ refers to the collection of all events generated by standard Brownian motion $\mathfrak{B} = \{\mathfrak{B}_t\}_{t\geq 0}$ on the interval $[0, s]$:*

$$\mathcal{F}_s = \mathcal{F}_s^{\mathfrak{B}} = \sigma(\mathfrak{B}_v, 0 \leq v \leq s).$$

In other words, the events in $\mathcal{F}_s$ are typically in terms of ranges of values of $\mathfrak{B}$. Some examples are:

$$A = \{\omega: \mathfrak{B}_v(\omega) \leq 10, \ 0 \leq v \leq s\}, \qquad B = \{\omega: \mathfrak{B}_s(\omega) > 15\},$$
$$C = \{\omega: -1 \leq \mathfrak{B}_v(\omega) \leq 1, \ 0 \leq v \leq s\}.$$

The collection $\mathcal{F}_s$ also includes the sample space $\Omega$, which consists of all possible paths of standard Brownian motion, the empty set $\varnothing$, complements of events in $\mathcal{F}_s$, and countable unions of events in $\mathcal{F}_s$. *Each event in $\mathcal{F}_s$ then carries a piece of information about standard Brownian motion $\mathfrak{B}$ on $[0, s]$.* For example, the sample space $\Omega$ is the event that the actual path standard Brownian motion will follow will be one of the possible paths in $\Omega$, i.e., one of the possible paths of $\Omega$ will occur (superficial information), while $C$ is the event that the values of $\mathfrak{B}$ are between $-1$ and $1$ from time 0 to time $s$. The $\sigma$-algebra $\mathcal{F}_s$ is then an information set about the "history" of $\mathfrak{B}$ up to and including time $s$. Bear in mind that the word "history" should not be interpreted literally to imply that $s$ is the current time or a past time; the time $s$ can be in the future.

If standard Brownian motion is observed on $[0, s]$, then we know the actual path it took on $[0, s]$, i.e., we can confirm which events in $\mathcal{F}_s$ occurred or not. Since the current time is at 0 and $s > 0$ is in the future, we may not yet know which events in $\mathcal{F}_s$ have occurred. However, all events in $\mathcal{F}_s$ are still confirmable in the sense that when the current time reaches $s$, we shall know for each event in $\mathcal{F}_s$ whether or not it occurred. On the other hand, the event

$$\tilde{A} = \{\omega: \mathfrak{B}_v(\omega) \leq 10, \ 0 \leq v \leq s+1\}$$

is not in $\mathcal{F}_s$ because it involves a portion of the possible Brownian motion paths beyond time $s$, namely, on $(s, s+1]$, and so its occurrence cannot be tested based on observations of $\mathfrak{B}$ on $[0, s]$. Event $\tilde{A}$ is not confirmable relative to the information set $\mathcal{F}_s$, but confirmable relative to $\mathcal{F}_{s+1}$. Note that some events in $\mathcal{F}_{s+1}$ may still be in $\mathcal{F}_s$, though not all.

In summary, *we think of the information set $\mathcal{F}_s$ as the set of all confirmable events about standard Brownian motion on $[0, s]$.* As time increases, we assume that no information is lost, i.e., $\mathcal{F}_s \subseteq \mathcal{F}_t$ for $s \leq t$. The collection $\{\mathcal{F}_t\}_{t\geq 0}$, where $\mathcal{F}_s \subseteq$

$\mathcal{F}_t$ for every $s \leq t$, is called a *filtration* and is viewed as capturing the "entire history" of standard Brownian motion.

Third, a stochastic process $X = \{X_t\}_{t \geq 0}$ is said to be $\mathcal{F}_s$-*measurable* if all events generated by X on $[0,s]$ are also events in $\mathcal{F}_s$:

$$\mathcal{F}_s^X = \sigma(X_v, \ 0 \leq v \leq s) \subseteq \mathcal{F}_s.$$

Analogous to $\mathcal{F}_s$, the elements of $\mathcal{F}_s^X$ include $\Omega, \varnothing$, the events

$$X_v^{-1}\big((-\infty, b]\big) = \{\omega : \ X_v(\omega) \leq b\}$$

for all $b$ in $\mathbb{R}$ and $0 \leq v \leq s$, and complements and countable unions of events in $\mathcal{F}_s^X$. If X is $\mathcal{F}_s$-measurable, then all the information in $\mathcal{F}_s^X$ is contained in $\mathcal{F}_s$, i.e., each event in $\mathcal{F}_s^X$ is confirmable. This implies that when the path of standard Brownian motion is observed on $[0,s]$, we shall know the actual path the stochastic process X took on $[0,s]$ and can confirm the occurrence of each event in $\mathcal{F}_s^X$. In a nutshell, X is $\mathcal{F}_s$-measurable if all confirmable events about X on $[0,s]$ are determined by the confirmable events about $\mathfrak{B}$ on $[0,s]$. The stochastic process X is called *adapted to standard Brownian motion* or, simply, an *adapted process*, if X is $\mathcal{F}_s$-measurable for all $s \geq 0$. Intuitively, to say that a stochastic process $X = \{X_t\}_{t \geq 0}$ is an *adapted process* means that for any $t$, the values of the random variable $X_t$ are determined (almost surely) by the confirmable events about $\mathfrak{B}$ up to and including time $t$. A deterministic function of standard Brownian motion is then an adapted process. On the other hand, the stochastic process $\{Y_t\}_{t \geq 0}$, where $Y_t = \mathfrak{B}_{t+1}$, is not adapted since at a given $t$ the random variable $Y_t$ requires information about $\mathfrak{B}$ on $(t, t+1]$ and so is not completely determined by the information about $\mathfrak{B}$ on $[0,t]$. The process $\{Y_t\}$ looks into the future of $\mathfrak{B}$ at each $t$. We shall be interested primarily in adapted stochastic processes.

Our geometric Brownian motion model $\{S_t\}_{t \geq 0}$ of an underlying security is an adapted process since

$$S_t = g(\mathfrak{B}_t, t), \qquad\qquad g(x, t) = S_0 \, \mathrm{e}^{(m-q-\frac{1}{2}\sigma^2)t + \sigma x}.$$

Moreover, given that the constants $S_0$, $m$, $q$, and $\sigma$ are assumed known, we see that for every fixed $t$ the quantity $\mathfrak{B}_t$ uniquely determines $S_t$ through $g(\mathfrak{B}_t, t)$ and $S_t$ uniquely determines $\mathfrak{B}_t$ via

$$\mathfrak{B}_t = \tilde{g}(S_t, t), \qquad\qquad \tilde{g}(x,t) = \frac{1}{\sigma} \ln\left(\frac{x}{S_0}\right) - \frac{1}{\sigma}\left(m - q - \frac{1}{2}\sigma^2\right)t.$$

For this reason, the $\sigma$-algebra $\mathcal{F}_s$ generated by $\{\mathfrak{B}_t\}_{t \geq 0}$ is the same as the one generated by the security price process $\{S_t\}_{t \geq 0}$:[6]

---

[6] Note: $\sigma(X) \subseteq \sigma(Y)$ if and only if $X = f(Y)$; see Mikosch [32, p. 66] for more.

$$\mathcal{F}_s = \sigma(\mathcal{B}_v, \ 0 \le v \le s) = \sigma(S_v, \ 0 \le v \le s). \tag{8.54}$$

Finally, we now turn to the conditional expectation $\mathbb{E}\left(X_t\middle|\ \mathcal{F}_s\right)$. Intuitively, it computes the expectation of $X_t$ relative to some probability measure from the vantage point of time $s$, taking into account the information set $\mathcal{F}_s$ and looking toward time $t$. In other words, the conditional expectation $\mathbb{E}\left(X_t\middle|\ \mathcal{F}_s\right)$ takes the expectation of $X_t$ given the history of $\mathcal{B}$ up to and including time $s$. Note (as pointed out earlier) that if $t > s$, then knowing the history of $\mathcal{B}$ up to time $s$, i.e., given the information set $\mathcal{F}_s$, is not enough to determine $X_t$.

The conditional expectation $\mathbb{E}\left(X_t\middle|\ \mathcal{F}_0\right)$ computes the expectation of $X_t$ given the information set $\mathcal{F}_0 = \{\Omega, \varnothing\}$, which merely tells you that one of the paths in the sample space $\Omega$ will occur, while a path not in $\Omega$ will not occur. With such insignificant information, the expectation of $X_t$ has to be computed by taking into account all the possible sample paths of X. This yields the unconditional expectation:

$$\mathbb{E}\left(X_t\middle|\ \mathcal{F}_0\right) = \mathbb{E}\left(X_t\right).$$

When $s > 0$, the quantity $\mathbb{E}\left(X_t\middle|\ \mathcal{F}_s\right)$ is typically random since for each event in $\mathcal{F}_s$, we use the event's information when computing the expectation of $X_t$. This usually yields different possible values of $\mathbb{E}\left(X_t\middle|\ \mathcal{F}_s\right)$ as we range through the events in $\mathcal{F}_s$. Overall, *the conditional expectation $\mathbb{E}\left(X_t\middle|\ \mathcal{F}_s\right)$ is a random variable that can be thought of, in a certain sense, as the best approximation of $X_t$ given the history of $\mathcal{B}$ up to time $s$.* With our best-approximation interpretation, if $\{X_t\}_{t\ge 0}$ is an adapted process, then $\mathbb{E}\left(X_v\middle|\ \mathcal{F}_s\right) \overset{\text{a.s.}}{=} X_v$ for all $0 \le v \le s$. Moreover, for adapted processes $\{X_t\}_{t\ge 0}$ and $\{Y_t\}_{t\ge 0}$, we have:

$$\mathbb{E}\left(X_v\, Y_s \ \middle|\ \mathcal{F}_v\right) \overset{\text{a.s.}}{=} X_v\, \mathbb{E}\left(Y_s\middle|\ \mathcal{F}_v\right) \qquad (0 \le v \le s).$$

We also see that if $X_t$ is independent of the information contained in $\mathcal{F}_s$, then we are back to the unconditional expectation:

$$\mathbb{E}\left(X_t\middle|\ \mathcal{F}_s\right) = \mathbb{E}\left(X_t\right) \qquad\qquad (X_t \text{ is independent of } \mathcal{F}_s). \tag{8.55}$$

Another useful result is the *tower property*:

$$\mathbb{E}\left(\mathbb{E}\left(X_t\middle|\mathcal{F}_s\right)\middle|\mathcal{F}_v\right) \overset{\text{a.s.}}{=} \mathbb{E}\left(X_t\middle|\mathcal{F}_v\right) \qquad\qquad (0 \le v \le s \le t). \tag{8.56}$$

For $v = 0$, the tower property (8.56) implies that

$$\mathbb{E}\left(\mathbb{E}\left(X_t\middle|\mathcal{F}_s\right)\right) = \mathbb{E}\left(X_t\right). \tag{8.57}$$

## 8.4.2 From BSM Pricing to Risk-Neutral Pricing

First, a basic assumption is that there is a market probability measure $\mathbb{P}$ that determines the true probabilities of all traded securities. Of course, no one knows the exact nature of $\mathbb{P}$, and all market participants try to approximate $\mathbb{P}$ as closely as possible. Each investor then has his or her own subjective probability measure to forecast the expected security prices. To illustrate this explicitly, recall from Section 3.6.2 that investors always maximize their expected utility and can be divided into three categories according to risk preference: risk averse, risk seeking, and risk neutral. Let $\mathbb{P}_{RA}$, $\mathbb{P}_{RS}$, and $\mathbb{P}_{RN}$ be their respective subjective probabilities. The lognormal nature of the underlying security yields that investors maximizing their expected utility will hold the underlier if their subjective probabilities determine the same expected utility for the underlier and a risk-free security. Equations (3.88),[7] (3.89), and (3.90) imply:

Risk-averse investor:    $\mathbb{E}_{\mathbb{P}_{RA}}(S_t) = S_0 e^{(m-q)t} > S_0 e^{(r-q)t} \iff m > r$

Risk-seeking investor:    $\mathbb{E}_{\mathbb{P}_{RS}}(S_t) = S_0 e^{(m-q)t} < S_0 e^{(r-q)t} \iff m < r$

Risk-neutral investor:    $\mathbb{E}_{\mathbb{P}_{RN}}(S_t) = S_0 e^{(m-q)t} = S_0 e^{(r-q)t} \iff m = r,$

where $m$ is the instantaneous mean return of the underlier and $r$ is the risk-free rate.

Turning to the BSM model, it determines the price of a derivative as the solution of the BSM p.d.e. under appropriate final and boundary conditions. A simple but profound property that is easy to gloss over is that the coefficients of the BSM p.d.e. are independent of the instantaneous expected return rate $m$. In other words, given a derivative with well-behaved final and boundary conditions, the solution of the BSM p.d.e. gives a unique derivative price process that is independent of $m$. *This means that the investor's risk preference is irrelevant!*

Denote the value of the instantaneous expected return rate $m$ more explicitly by $m_{\widetilde{\mathbb{P}}}$, where $\widetilde{\mathbb{P}}$ is the probability measure used to compute $m$. The above observation suggests that a derivative can be priced in a world of risk-neutral investors, where each investor has a probability measure $\mathbb{P}_{RN}$ such that $m_{\mathbb{P}_{RN}} = r$.

**Remark 8.4.** The insight for pricing a derivative in a risk-neutral world is due to Cox and Ross [9].                                                                    □

*Two critical points need to be emphasized:*

➢ We have assumed the mathematical existence of a risk-neutral world, i.e., the existence of a probability measure $\mathbb{P}_{RN}$ for which $m_{\mathbb{P}_{RN}} = r$.

➢ We have assumed that there is a single probability measure $\mathbb{P}_{RN}$ for the risk-neutral world.

---

[7] See page 135.

How do we know that a risk-neutral probability measure exists and there is only one? It may be possible for different such probability measures to produce different prices of the same derivative. We shall address below the existence of a risk-neutral measure in the context of the BSM model, which assumes that all underlying securities follow a geometric Brownian motion. In general, the existence and uniqueness of these probability measures is a deep issue that culminates with the fundamental asset pricing theorems (Sections 8.4.3). These theorems will bring into sharper focus some of the core market assumptions we are making in the BSM model.

Given the real market probability measure $\mathbb{P}$, we shall now discuss the issue of whether there exists a risk-neutral probability measure $\mathbb{Q}$ relative to which derivatives can be priced.

## Risk-Neutral Pricing of Underliers

A key assumption of the BSM model is that any underlying security of a derivative follows geometric Brownian motion:

$$dS_t = (m - q) S_t dt + \sigma S_t d\mathfrak{B}_t. \tag{8.58}$$

Assume that $\mathfrak{B}_t$ is standard Brownian motion with respect to $\mathbb{P}$. In particular, the instantaneous expected return is $m_{\mathbb{P}} = m$ and the expected capital-gain return $(m - q) dt$ over the instant $dt$ is not due to the subjective probability of some risk-averse investor, but is relative to the true market probability measure $\mathbb{P}$.

Now, recall *Girsanov theorem:*[8] *there is a probability measure $\mathbb{Q}$ that is equivalent to $\mathbb{P}$ and given by*

$$d\mathbb{Q} = e^{\left(-\left(\frac{m-r}{\sigma}\right)\mathfrak{B}_T - \frac{1}{2}\left(\frac{m-r}{\sigma}\right)^2 T\right)} d\mathbb{P} \qquad (0 \le t \le T) \tag{8.59}$$

*such that*

$$\mathfrak{B}_t^{\mathbb{Q}} = \mathfrak{B}_t + \left(\frac{m-r}{\sigma}\right) t$$

*is a standard Brownian motion on $(\Omega, \mathcal{F}_T, \mathbb{Q})$.*[9] Note that $\frac{m-r}{\sigma}$ is a Sharpe ratio. Under the probability measure $\mathbb{Q}$, underlying security price, which by assumption is a geometric Brownian motion, has instantaneous expected return given by the risk-free rate $r$. To see this, rewrite (8.58) as

$$dS_t = (r - q) S_t dt + \sigma S_t d\mathfrak{B}_t^{\mathbb{Q}}, \tag{8.60}$$

where

---

[8] See Privault [34, Thm. 6.1].

[9] $\mathbb{Q}(A) = \int_A D_T(\omega) \, d\mathbb{P}(\omega)$, where $\omega \in A$, $A \in \mathcal{F}_T$, and $D_T(\omega) = e^{\left(-\left(\frac{m-r}{\sigma}\right)\mathfrak{B}_T(\omega) - \frac{1}{2}\left(\frac{m-r}{\sigma}\right)^2 T\right)}$.

$$d\mathcal{B}_t^Q = d\mathcal{B}_t + \left(\frac{m-r}{\sigma}\right) dt.$$

Since $\mathcal{B}_t^Q$ is a standard Brownian motion with respect to $Q$, the underlier price $S_t$ in (8.60) is also a geometric Brownian motion relative to $Q$, and its instantaneous expected return is the risk-free rate, i.e., $m_Q = r$. The latter reveals the risk neutrality of $Q$.

Let us express the risk neutrality of $Q$ more explicitly by showing the link to martingales. Equation (8.60) is solved by

$$S_t = S_0 e^{(r-q-\frac{1}{2}\sigma^2)t + \sigma \mathcal{B}_t^Q} \qquad\qquad (0 \le t \le T),$$

where $S_0$ is the known initial price of the underlying security. Consider the *discounted price process*

$$\widetilde{S}_t = e^{-rt} S_t^c = e^{-(r-q)t} S_t,$$

where $S_t^c$ is the cum-dividend price of the underlier. Taking the conditional expectation under $Q$ yields (Exercise 8.28)

$$\mathbb{E}_Q\left(\widetilde{S}_t | \mathcal{F}_v\right) = \widetilde{S}_v \qquad\qquad (0 \le v \le t \le T),$$

where the conditioning is relative to $\mathcal{F}_v$ since $\mathcal{B}^Q$ is a function of $\mathcal{B}$. It follows that *the discounted price process* $\{\widetilde{S}_t\}_{t\ge0}$ *is a martingale*. In other words, the value of $\widetilde{S}_v$ is obtained by continuously discounting $\mathbb{E}_Q(S_t|\mathcal{F}_v)$ at the net risk-free rate $r - q$:

$$S_v = e^{-(r-q)(t-v)} \mathbb{E}_Q(S_t|\mathcal{F}_v) \qquad\qquad (0 \le v \le t \le T). \qquad (8.61)$$

For this reason, the probability measure $Q$ is called *risk-neutral probability measure*, i.e., for every underlying security price process $\{S_t\}_{t\ge0}$, the discounted process $\{e^{-(r-q)t}S_t\}_{t\ge0}$ is a martingale relative to $Q$. Equation (8.61) is the risk-neutral price of the underlying security at time $v$. Note that the market probability measure $\mathbb{P}$ is risk neutral only when $m = r$, in which case $\mathbb{P} = Q$.

### Risk-Neutral Pricing of Derivatives

We now show how to price a derivative relative to $Q$. In the BSM model, we obtained the price of a derivative using a self-financing, replicating portfolio:[10]

$$f(S_t, t) = V_t = n_t S_t^c + b_t B_t, \qquad (8.62)$$

where $V_t$ is the value of the portfolio at time $t$ and

---

[10] See (8.13) on page 390.

$$n_t = \left(\frac{S_t}{S_t^c}\right)\Delta_f(S_t,t), \qquad b_t = \frac{f(S_t,t) - S_t\Delta_f(S_t,t)}{B_t}. \qquad (8.63)$$

In fact, we saw that the self-financing, replicating nature of the portfolio determines $f$ as a solution of the BSM p.d.e.

Consider the discounted portfolio value process

$$\tilde{V}_t = e^{-rt}\,V_t,$$

where there is no downward adjustment by $q$ since the derivative is assumed to pay no dividend. As noted earlier, we always assume that $n_t$ and $b_t$ satisfy the properties needed to apply stochastic integration and Itô's formula.

Since (Exercise 8.24)

$$d\tilde{S}_t = \sigma\tilde{S}_t\,d\mathfrak{B}_t^Q, \qquad dV_t = r\,V_t\,dt + n_t(m-r)\,S_t^c\,dt + n_t\sigma S_t^c\,d\mathfrak{B}_t,$$

we find

$$d\tilde{V}_t = \sigma n_t\tilde{S}_t\,d\mathfrak{B}_t^Q.$$

Integrating yields

$$\tilde{V}_t = \tilde{V}_0 + \sigma\int_0^t n_v\tilde{S}_v\,d\mathfrak{B}_v^Q. \qquad (8.64)$$

An important result from stochastic calculus is that stochastic processes like those in (8.64) are martingales under the given measure and, conversely, a (square integrable) martingale can be expressed in such a form (*martingale representation theorem*).[11] Since $\tilde{V}_t$ is a martingale under $Q$, we have

$$\mathbb{E}_Q\left(\tilde{V}_t|\mathcal{F}_v\right) = \tilde{V}_v \qquad (0 \le v \le t \le T),$$

which is equivalent to

$$V_v = e^{-r(t-v)}\,\mathbb{E}_Q\left(V_t|\mathcal{F}_v\right) \qquad (0 \le v \le t \le T).$$

Because the portfolio value $V_s$ replicates the price of the derivative at $s$ for all $0 \le s \le T$, we obtain

$$f(S_v,v) = e^{-r(t-v)}\,\mathbb{E}_Q\left(f(S_t,t)|\mathcal{F}_v\right) \qquad (0 \le v \le t \le T).$$

In particular, the derivative's price at time $t$ can be expressed relative to the expiration date $T$:

$$f(S_t,t) = e^{-r(T-t)}\,\mathbb{E}_Q\left(f(S_T,T)|\mathcal{F}_t\right) = e^{-r(T-t)}\,\mathbb{E}_Q\left(f(S_T,T)\right) \qquad (0 \le t \le T), \qquad (8.65)$$

where

---

[11] See, for example, Björk [5, Sec. 4.4] and Elliot and Kopp [14, p. 176] for details.

$$S_T = S_t\, e^{\mu_*\,(T-t)} + \sigma\,(\mathcal{B}^Q_T - \mathcal{B}^Q_t) \stackrel{d}{=} S_t\, e^{\mu_*\,(T-t)} + \sigma \mathcal{B}^Q_{T-t}\,(0 \le t \le T)$$

and

$$\mu_* = r - q - \frac{1}{2}\sigma^2.$$

To obtain the rightmost equality in (8.65), recall that the independence of Brownian motion increments implies that $\mathcal{B}^Q_T - \mathcal{B}^Q_t$ is independent of $\mathcal{B}^Q_v$ for all $0 \le v \le t$. Consequently, the increment $\mathcal{B}^Q_T - \mathcal{B}^Q_t$ is independent of $\mathcal{F}_t$. Since the $f$ is a deterministic function of $S_t$ with $S_t$ an invertible deterministic function of standard Brownian motion, Equations (8.54) (page 403) and (8.55) yield the rightmost equality in (8.65).

*Equation (8.65) is a stochastic process giving the risk-neutral price process of a European-style derivative under the probability measure* Q *during the time interval* $[0,T]$. Since at the current time, $t = 0$, we know the entire history of the underlying security's prices up to and including time $t_0$, Equation (8.65) implies that the derivative's current risk-neutral price is the following constant:

$$f(S_0,0) = e^{-rT}\, \mathbb{E}_Q\left(f(S_T,T)\right), \tag{8.66}$$

where $T$ is the expiration date. Because of the conditional expectation in (8.66), risk-neutral pricing lends itself naturally to numerical simulations.

In summary, the function $f$ on the left-hand side of (8.66) is determined by solving the BSM p.d.e. under appropriate final and boundary conditions, and the value $f(S_{t_0},t_0)$ of $f$ at the current security price $S_{t_0}$ and current time $t_0$ is exactly the value obtained from discounting the risk-neutral conditional expectation in accordance with the right-hand side of (8.66). The important lesson is that *we can price a derivative either by solving the BSM p.d.e. under appropriate final and boundary conditions or by computing a discounted risk-neutral conditional expectation.* This connection between partial differential equations and probability theory is not coincidental and is far reaching.

**Remark 8.5.** Readers interested in more about the link between p.d.e.'s and stochastic differential equations should explore the Feynman-Kac formula.

$\square$

### 8.4.3 The Fundamental Theorems of Asset Pricing

Two issues were raised in Section 8.4.2 (see page 404): does there exist a risk-neutral probability measure relative to which a European-style derivative can be priced and, if it exists, is there a unique such probability measure? In Section 8.4.2, we employed Girsanov theorem to invoke the existence of such a measure $Q$ and derived a formula (8.66) for a derivative's risk-neutral price under $Q$. In this section, we shall address which assumption of the BSM model gives rise to the existence of a risk-neutral measure.

Consider a more general market setting where we do not enforce a priori the no-arbitrage condition. The theorem below gives a link between the no-arbitrage condition and the existence of a risk-neutral probability measure.

**Theorem 8.1 (First Fundamental Theorem of Asset Pricing).** [12] *A market has no arbitrage if and only if there is a risk-neutral probability measure $Q$ that is equivalent[13] to the market probability measure $\mathbb{P}$.*

The theorem reveals that fundamentally it is actually the no-arbitrage assumption of the BSM model which grants us the existence of a risk-neutral probability. The remaining outstanding item is the uniqueness of the risk-neutral probability measure.

A derivative is called *attainable* if there is a self-financing portfolio that replicates its price. A securities market is called *complete* if all its derivatives are attainable.[14] The next theorem takes up the issue of market completeness and the uniqueness of risk-neutral measures.

**Theorem 8.2 (Second Fundamental Theorem of Asset Pricing).** *Suppose that a market has no arbitrage. Then the market is complete if and only if there is a unique risk-neutral probability $Q$ that is equivalent to the market probability measure $\mathbb{P}$.*

**Notation.** For simplicity, we shall often use an asterisk to indicate when the risk-neutral probability is being employed. For example, if $f_Y^Q$ is the p.d.f. of $Y$ relative to $Q$, we shall also write $f_Y^*$ instead of $f_Y^Q$. In some cases, we shall also write $\mathbb{E}_*(X)$ instead of $\mathbb{E}_Q(X)$ for efficiency. This carries over from the study of binomial trees. It should be clear from the context whether $\mathbb{E}_*(X)$ is an expectation relative to the risk-neutral uptick probability $p_n^*$ of a binomial tree or the risk-neutral measure $Q$ of a continuous-time setting.

We saw in Section 8.1.3 (page 388) that all the European-style derivatives are attainable in the BSM model. Since the marketplace for the BSM model has no arbitrage and is complete, the BSM model has a unique risk-neutral probability measure and so only one price for a derivative. Indeed, the unique risk-neutral probability measure is the one given via Girsanov theorem by (8.59). This is consistent with the BSM p.d.e. having a unique solution under appropriate final and boundary conditions.

---

[12] Some authors call Theorem 8.1 the *Fundamental Theorem of Asset Pricing*.

[13] Probability measures $\mathbb{P}$ and $Q$ are called *equivalent* when $\mathbb{P}(A) = 0$ if and only if $Q(A) = 0$.

[14] See Chapter 12 (e.g., Appendices A and B) by Staum in Birge and Linetsky's handbook [4] for a further discussion of the financial meaning and mathematical modeling of a complete/incomplete market.

Later, we shall study the Merton jump diffusion model (Section 8.9) and encounter a no-arbitrage market that is incomplete. Moreover, since the Merton model reduces to the BSM one, it also yields a unique risk-neutral probability measure for the BSM model. See Section 8.9.4 on page 458.

**Remark 8.6.** The papers by Ross [37], Harrison and Kreps [19], Harrison and Pliska [20], Kreps [25], and Delbaen and Schachermayer [11] laid the mathematical foundation for the two fundamental theorems of asset pricing. See Schachermayer [39] for a history and discussion as well as Epps [15] for an insightful summary. The lecture notes by Privault [34, Chaps. 2, 5, 15] give an excellent accessible introduction to martingale pricing and the fundamental asset pricing theorems in discrete and continuous time.

### 8.4.4 Risk-Neutral Pricing of European Calls and Puts

In Section 8.4, we saw that it suffices to price European-style derivatives in the BSM model using its unique risk-neutral probability measure. This will provide a relatively fast way of obtaining the price of a European call and, via put-call parity, the price of a European put. In general, however, risk-neutral pricing does not lead to a simple analytical derivation of the price of a derivative. Oftentimes it has to be computed by numerical simulation.

By (8.65), the continuous-time risk-neutral current price of a European call is:

$$C^E(0) = e^{-rT} \, \mathbb{E}_*(C^E(T)).$$

We compute the risk-neutral expectation as follows:

$$\mathbb{E}_*(C^E(T)) = \mathbb{E}_*(\max\{S_T - K, 0\}) = \int_K^\infty (v - K) f_{S(T)}^*(v) \, dv$$
$$= \int_K^\infty v f_{S_T}^*(v) \, dv - K \int_K^\infty f_{S_T}^*(v) \, dv,$$

where $f_{S_T}^*$ is the risk-neutral lognormal density of $S_T$.

The integrals are in terms of the risk-neutral probability density of the underlying security's price at expiration $T$, where in a risk-neutral world, we have

$$S_T = S_0 \, e^{\mu_* T + \sigma \mathcal{B}_T^*}, \qquad\qquad \mu_* = r - q - \frac{\sigma^2}{2}. \qquad (8.67)$$

The following theorem yields the desired evaluation of the integrals:

**Theorem 8.3.** *Suppose* $Y_T = Y_0 e^X$, *where* $Y_0$ *is a known constant and* $X \sim \mathcal{N}_*(\mu_0, \sigma_0^2)$ *is the price of a security. Let* $f_{Y_T}^*$ *be the risk-neutral lognormal density of* $Y_T$. *Then:*

$$\int_K^\infty f_{Y_T}^*(v) \, dv = N(d_2) \qquad\qquad (8.68)$$

*and*

$$\int_K^\infty v f^*_{Y_T}(v)\,dv = Y_0 e^{\mu_0 + \sigma_0^2/2}\,N(d_1),\tag{8.69}$$

*where* $N(\cdot)$ *is the standard normal cumulative distribution function and*

$$d_+ = \frac{\ln(Y_0/K) + \mu_0 + \sigma_0^2}{\sigma_0}, \qquad\qquad d_- = \frac{\ln(Y_0/K) + \mu_0}{\sigma_0}.$$

Apply Theorem 8.3 with $Y_T = S_T$, $Y_0 = S_0$, and

$$X = \mu_* T + \sigma \mathcal{B}^*_T \sim \mathcal{N}(\mu_0, \sigma_0^2).$$

Here $\mu_0 = \mu_* T$, $\sigma_0^2 = \sigma^2 T$, and

$$d_+ = \frac{\ln\left(\frac{S_0}{K}\right) + (r - q + \frac{1}{2}\sigma^2)T}{\sigma\sqrt{T}}, \qquad d_- = \frac{\ln\left(\frac{S_0}{K}\right) + (r - q - \frac{1}{2}\sigma^2)T}{\sigma\sqrt{T}}.$$

That is, $d_\pm = d_\pm(S_0, T)$; see Equations (8.35) and (8.37) (page 394). It follows:

$$\mathbb{E}_*(C^E(T)) = \int_K^\infty v f^*_{S_T}(v)\,dv - K \int_K^\infty f^*_{S_T}(v)\,dv$$

$$= Y_0 e^{\mu_0 + \sigma_0^2/2} N(d_+(S_0, T)) - KN(d_-(S_0, T))$$

$$= S_t e^{(r-q)T} N(d_+(S_0, T)) - KN(d_-(S_0, T)).$$

Hence, we obtain the BSM formula for *the current price of a European call*:

$$C(0) = S_0 e^{-qT} N(d_+(S_0, T)) - Ke^{-rT} N(d_-(S_0, T)).\tag{8.70}$$

## 8.5 Binomial Approach to Pricing European Options

The binomial trees of Chapter 5 provide a useful and pedagogically insightful way to explore the pricing of options. Basically, this approach approximates the standard BSM formula in discrete time using binomial trees with a sufficiently large number $n$ of time steps and recovers the exact BSM formula in the continuous-time limit $n \to \infty$.

We shall first summarize the binomial-tree strategy for pricing European call options and then determine explicitly the option pricing formula for *one-* and *two*-period trees and present the *n*-period case in Section 8.5.4. Note that, in applications with $n$ sufficiently large, a specific binomial tree is chosen—e.g., a CRR tree—so we can determine the values of the uptick factor $u_n$, downtick factor $d_n$, and uptick probability $p_n$ from the inputs $r - q$, $\sigma$, and $h_n$ (time-period size).

> For simplicity, suppress notationally the dependence of $u_n$, $d_n$, and $p_n$ on the number $n$ of time steps:
>
> $$u_n = u, \qquad d_n = d, \qquad p_n = p.$$
>
> Additionally, write the n-period binomial tree price of a European call at time $t_i$ as
> $$C^E(t_i; n),$$
> where $t_i = t_{i-1} + h$ for $i = 1, \ldots, n$. The current time is $t_0 = 0$.

### 8.5.1 One-Period Binomial Pricing by Self-Financing Replication

For a one-period binomial tree from time $t_0$ to $t_1$, we shall determine the current price $C^E(t_0; 1)$ of a European call option with strike $K$ and expiration $T$ in terms of a replicating portfolio that finances itself after an initial capital.

Assume an initial capital $V_0$ at the current time $t_0$, and create a portfolio as follows:

➤ Take a position consisting of $n_0$ units of a risky security, where the unit price is $S(t_0)$ and the security pays a continuous cash dividend at annual yield rate $q$. Recall that, unless otherwise stated, we assume that the cash dividend is continuously invested back in the security by purchasing more units.

➤ Take a position consisting of $b_0$ units of the money market fund, which is $B(t_0)$ per unit at time $t_0$.

We do not yet know the explicit nature of the trading strategy $(n_0, b_0)$—e.g., whether we have a long position in the security ($n_0 > 0$) or short position in the money market account ($b_0 < 0$). However, since only the initial capital is used to form these positions, we must have:

$$V(t_0) = n_0 S(t_0) + b_0 B(t_0),$$

where the choice of $B(t_0)$ is an arbitrary positive number that we have simply set to be \$1. The value of the portfolio at $t_1$ arises from changes in the values of the security and the money market fund:

$$V(t_1) = n_0 S^c(t_1) + b_0 B(t_1), \tag{8.71}$$

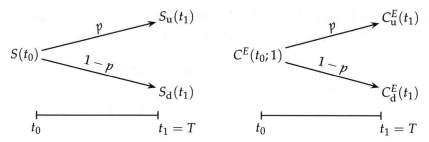

**Fig. 8.1** One-period binomial tree models of the prices of the security (left tree) and European call option (right tree). Here $p$ is the probability of an upward movement in the tree

where $S^c(t_1)$ is the value of security cum dividend (with dividend). In other words, due to dividend reinvesting, the original one unit of the security has grown to $e^{qh}$. Consequently, the cum-dividend unit price of the security at $t_1$ is

$$S^c(t_1) = S(t_1)e^{qh}, \tag{8.72}$$

where $S(t_1)$ is the security's ex-dividend (without dividend) unit price. Note that the prices shown in Figure 8.1 are ex-dividend prices. The value of the money market account at $t_1$ is

$$B(t_1) = B(t_0)e^{rh}. \tag{8.73}$$

Now, suppose that at time $t_1$ we have a trading strategy $(n_1, b_1)$. Then the value of the trading strategy is

$$n_1 S^c(t_1) + b_1 B(t_1).$$

However, for the portfolio to be self-financing, any trading strategy $(n_1, b_1)$ at $t_1$ must be financed from the portfolio value (8.71) at $t_1$ resulting from the original trading strategy $(n_0, b_0)$:

$$V(t_1) = n_0 S^c(t_1) + b_0 B(t_1) = n_1 S^c(t_1) + b_1 B(t_1). \tag{8.74}$$

That is, after using the initial capital, no outside funds can be added, and no funds can be taken out the portfolio.

We want to find a trading strategy $(n_1, b_1)$ at expiration $t_1$ such that the portfolio's value replicates the call's price

$$V(t_1) = n_1 S^c(t_1) + b_1 B(t_1) = C^E(t_1; 1). \tag{8.75}$$

If this is the case, then by the law of one price, the current price of the European call will equal the current value of the portfolio. Basically, the goal is to show that the European call is attainable, i.e., there is a self-financing portfolio that replicates its price.

Let us first solve for the trading strategy $(n_1, b_1)$ using the self-financing condition (8.74). The two possible values of the $S(t_1)$ are (Figure 8.1):

$$S(t_1) = \begin{cases} S_u(t_1) = S(t_0)\,u & \text{with probability } p \\ \\ S_d(t_1) = S(t_0)\,d & \text{with probability } 1 - p, \end{cases} \tag{8.76}$$

where

$$u > 1, \qquad 0 < d < 1, \qquad 0 < p < 1.$$

By (8.72) and (8.76), Equation (8.74) becomes

$$n_0\,e^{qh}\,S_u(t_1) + b_0\,B(t_1) = n_1\,e^{qh}\,S_u(t_1) + b_1\,B(t_1)$$

$$\tag{8.77}$$

$$n_0\,e^{qh}\,S_d(t_1) + b_0\,B(t_1) = n_1\,e^{qh}\,S_d(t_1) + b_1\,B(t_1).$$

This is solved by

$$n_1 = n_0, \qquad b_1 = b_0. \tag{8.78}$$

Turning to the replicating condition (8.75), the possible prices of the call at $t_1$ are (Figure 8.1):

$$C^E(t_1; 1) = \begin{cases} C_u^E(t_1) = \max\{S_u(t_1) - K, 0\} & \text{with probability } p \\ \\ C_d^E(t_1) = \max\{S_d(t_1) - K, 0\} & \text{with probability } 1 - p. \end{cases} \tag{8.79}$$

Note that in (8.79), we made use of the fact that the terminal value of a European call is its payoff:

$$C^E(T) = \max\{S(T) - K, 0\}, \tag{8.80}$$

where $K$ is the strike price. Employing (8.78), we see that (8.75) becomes

$$n_0\,e^{qh}\,S_u(t_1) + b_0\,e^{rh}\,B(t_0) = C_u^E(t_1)$$

$$\tag{8.81}$$

$$n_0\,e^{qh}\,S_d(t_1) + b_0\,e^{rh}\,B(t_0) = C_d^E(t_1).$$

The coefficient matrix of this linear system of two equations in two unknowns is invertible:

$$\det \begin{bmatrix} e^{qh}\,S_u(t_1) & e^{rh}\,B(t_0) \\ \\ e^{qh}\,S_d(t_1) & e^{rh}\,B(t_0) \end{bmatrix} = e^{(r+q)h}\,S(t_0)\,B(t_0)\,(u - d) > 0.$$

Hence, the unique solution of (8.81) is

$$n_0 = e^{-qh} \left( \frac{C_u^E(t_1) - C_d^E(t_1)}{S_u(t_1) - S_d(t_1)} \right), \qquad b_0 = -\frac{e^{-rh}}{B(t_0)} \left( \frac{C_u^E(t_1)\, d - C_d^E(t_1)\, u}{u - d} \right).$$

$$(8.82)$$

**Remark 8.7.** The expression for $n_0$ in (8.82) includes a discrete version of a European call's delta, namely, the partial difference at $t_1$ of the call price with respect to the underlier price:

$$\frac{\delta C^E}{\delta S}(t_1) = \frac{C_u^E(t_1) - C_d^E(t_1)}{S_u(t_1) - S_d(t_1)}.$$

□

In summary, since the dividends are reinvested in the underlying security, there is a unique trading strategy $(n_0, b_0)$ given by (8.82) that can replicate the call value at $t_1$ in a self-financing way: $(n_1, b_1) = (n_0, b_0)$. In other words, the call is attainable. By the law of one price, the portfolio and the call have the same price today:

$$C^E(t_0, 1) = V(t_0) = n_0\, S(t_0) + b_0\, B(t_0),$$

where $n_0$ and $b_0$ are given (8.82). Explicitly, the current price of the European call can be expressed as

$$C^E(t_0; 1) = e^{-rh} \left[ \left( \frac{e^{(r-q)h} - d}{u - d} \right) C_u^E(t_1) + \left( \frac{u - e^{(r-q)h}}{u - d} \right) C_d^E(t_1) \right], \quad (8.83)$$

where (8.76) and (8.80) with $T = t_1$ give

$$C_u^E(t_1) = \max\{S(t_0)\, u - K, 0\}, \qquad C_d^E(t_1) = \max\{S(t_0)\, d - K, 0\}.$$

Equation (8.83) yields that *the option's current price (8.83) is independent of the underlier's instantaneous expected return rate m!* Hence, the one-period call price is independent of investors' view on $m$. This is consistent with the earlier observation (page 404) from the BSM p.d.e. that the price of a derivative is independent of $m$.

### 8.5.2 One-Period Binomial Pricing by Risk Neutrality

As noted above, the current price of the European call is independent of the instantaneous expected return $m$ of the underlier. In other words, even if the marketplace were a risk-neutral world, i.e., one where the value of $m$ is $r$, the current price of the call would be unchanged. Since $r$ is known, while estimating $m$ from market data is problematic, it is simpler to price the call in a risk-neutral world, i.e., price the call using a risk-neutral uptick probability.

How do we know that there is a risk-neutral uptick probability? First, let us assume that there is one, i.e., there is a probability $p_*$,

$$0 < p_* < 1, \tag{8.84}$$

such that, relative to $p_*$, the expected price at $t_1$ of the underlying security is given by

$$\mathbb{E}_* \left( S(t_1) \right) = S(t_0) e^{rh}, \tag{8.85}$$

where (as always) the current price $S(t_0)$ is assumed known. Equation (8.85) is independent of $S(t_0)$ and equivalent to

$$p_* u + (1 - p_*) d = e^{(r-q)h}. \tag{8.86}$$

The risk-neutral uptick probability $p_*$ is then characterized by (8.84) and (8.86). Recall that for a risk-neutral binomial tree, we always assume[15]

$$e^{(r-q)h} \neq d, \qquad e^{(r-q)h} \neq u.$$

Otherwise, we get $p_* = 0$ or $p_* = 1$, which are excluded in our binomial-tree analysis. Now, since $u - d > 0$, we then obtain the following unique solution of (8.86):

$$p_* = \frac{e^{(r-q)h} - d}{u - d}. \tag{8.87}$$

Furthermore, recall that if there is no arbitrage, then (see page 232):

$$d < e^{(r-q)h} < u. \tag{8.88}$$

Since $u - d > 0$, Equations (8.87) and (8.88) yield

$$0 < p_* < 1.$$

Hence, $p_*$ given by (8.87) is a unique risk-neutral uptick probability.

The probability $p_*$ allows us to compute a unique current price for the call. In fact, the expected future value of the European call at $t_1$ is:

$$\mathbb{E}_* \left( C^E(t_1; 1) \right) = C^E(t_0; 1) e^{rh}, \tag{8.89}$$

where the call does not pay a dividend (by assumption). By (8.89), the current call price is:

$$C^E(t_0; 1) = e^{-rh} \left( p_* C_u^E(t_1) + (1 - p_*) C_d^E(t_1) \right), \tag{8.90}$$

---

[15] See the constraints (5.49) on page 232.

where

$$C_u^E(t_1) = \max\{S(t_0)u - K, 0\}, \qquad C_d^E(t_1) = \max\{S(t_0)d - K, 0\}.$$

Comparing the risk-neutral price (8.90) and the replicating-portfolio price (8.83) immediately shows that they are identical. In other words, *it suffices to price a European call using the risk-neutral approach*, which we shall do going forward for binomial trees.

**Example 8.5.** Suppose that the risk-free rate is 2% per annum and a stock with current price of $50. Assume that the stock pays no dividend and its price 3 months from now is either $53.8900 or $46.3850, where four decimal places are used to minimize rounding-off errors. Using a binomial tree, compute the one-period current price of a 3-month European call on this stock given a strike price of $50.

**Solution.** The formula for the one-period European call price at the current time $t_0$ is

$$C^E(t_0; 1) = e^{-rh}\left(p_* \max\{S(t_0)u - K, 0\} + (1 - p_*) \max\{S(t_0)d - K, 0\}\right),$$

where $t_0 = 0$ and

$$p_* = \frac{e^{(r-q)h} - d}{u - d}.$$

The needed inputs are:

$$h = 0.25, \quad r = 0.02, \quad q = 0, \quad S(t_0) = \$50, \quad K = \$50$$

$$u = \frac{\$53.8900}{\$50} = 1.0778, \quad d = \frac{\$46.3850}{\$50} = 0.9277, \quad p_* = 0.5151.$$

Direct calculation then yields the current call price: $C^E(t_0; 1) = \$1.99$. $\qquad\square$

### 8.5.3 Two-Period Binomial Pricing

Consider the two-period risk-neutral binomial tree in Figure 8.2, where the time remaining on the European call is $2h$ and runs from $t_0$ to $t_2$. We shall employ risk-neutral pricing to determine the current price of the call:

$$C^E(t_0; 2) = e^{-r(2h)}\, \mathbb{E}_*\left(C^E(t_2; 2)\right).$$

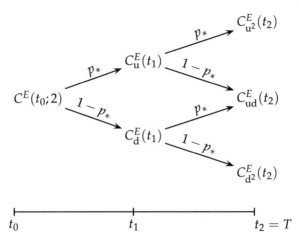

**Fig. 8.2** Possible European call prices for a two-period binomial tree

**Direct approach.** To compute the expectation $\mathbb{E}_*\left(C^E(t_2;2)\right)$ directly, Figure 8.2 shows that there are four paths leading to the three possible prices of the call. The probability of a given price is the sum of the probabilities of each path leading to the price, where the probability along a path is the product of the probabilities along each section. It follows:

$$C^E(t_0;2) = e^{-r(2h)}\left(p_*^2\,C_{u^2}^E(t_2) \;+\; 2p_*(1-p_*)\,C_{ud}^E(t_2) \;+\; (1-p_*)^2\,C_{d^2}^E(t_2)\right). \tag{8.91}$$

Using (8.20) and (8.76), we get the explicit forms:

$$\begin{aligned}
C_{u^2}^E(t_2) &= \max\{S_u(t_1)\,u - K, 0\} = \max\{S(t_0)\,u^2 - K, 0\}\\
C_{ud}^E(t_2) &= \max\{S_u(t_1)\,d - K, 0\} = \max\{S(t_0)\,ud - K, 0\}\\
C_{d^2}^E(t_2) &= \max\{S_d(t_1)\,d - K, 0\} = \max\{S(t_0)\,d^2 - K, 0\}.
\end{aligned}$$

$$\tag{8.92}$$

Note that when a binomial tree has many paths, our direct method to compute $\mathbb{E}_*\left(C^E(t_2;2)\right)$ becomes nontrivial quickly—e.g., a 20-step binomial tree already has over 1 million paths.

**Algorithmic approach.** We can also obtain (8.91) using a pedagogically simple algorithm. The strategy is to divide $[t_0, t_2]$ into two equal-length subintervals, namely, $[t_1, t_2]$ and $[t_0, t_1]$, and then apply the one-period risk-neutral result (8.90) to $[t_1, t_2]$ and $[t_0, t_1]$. That is, we work backward through the tree. Let us now begin with the one-period subinterval $[t_1, t_2]$. To apply the one-period analysis to $[t_1, t_2]$, think of $t_1$ and $t_2$ as now playing the respective roles of the present time $t_0$ and the expiration time $t_1$ in (8.90). Since the call price at time $t_1$ has two possible values $C_u^E(t_1)$ and $C_d^E(t_1)$, apply (8.90) to each of them with

$t_0$ and $t_1$ in (8.90) replaced by $t_1$ and $t_2$, respectively:

$$C_u^E(t_1) = e^{-rh}\, \mathbb{E}_*\left(C_u^E(t_2) \mid S(t_1) = S_u(t_1)\right)$$

$$C_d^E(t_1) = e^{-rh}\, \mathbb{E}_*\left(C_d^E(t_2) \mid S(t_1) = S_d(t_1)\right),$$

where

$$C_u^E(t_2) = \begin{cases} C_{u^2}^E(t_2) & \text{with probability } p_* \\[2mm] C_{ud}^E(t_2) & \text{with probability } 1 - p_* \end{cases}$$

and

$$C_d^E(t_2) = \begin{cases} C_{ud}^E(t_2) & \text{with probability } p_* \\[2mm] C_{d^2}^E(t_2) & \text{with probability } 1 - p_*. \end{cases}$$

The possible call prices at time $t_1$ are then:

$$C^E(t_1;2) = \begin{cases} C_u^E(t_1) = e^{-rh}\left(p_* C_{u^2}^E(t_2) + (1 - p_*) C_{ud}^E(t_2)\right) \\[3mm] C_d^E(t_1) = e^{-rh}\left(p_* C_{ud}^E(t_2) + (1 - p_*) C_{d^2}^E(t_2)\right). \end{cases} \tag{8.93}$$

Moving backward in the tree, consider the interval $[t_0, t_1]$. Application of (8.90) yields the current call price as:

$$C^E(t_0;2) = e^{-rh}\, \mathbb{E}_*\left(C^E(t_1;2)\right) = e^{-rh}\left(p_* C_u^E(t_1) + (1 - p_*) C_d^E(t_1)\right),$$

where $C_u^E(t_1)$ and $C_d^E(t_1)$ are now given by (8.93). The European call option price over the two-period interval $[t_0, t_2]$ is then given as follows:

$$C^E(t_0;2) = e^{-r(2h)}\left(p_*^2 C_{u^2}^E(t_2) + 2p_*(1 - p_*) C_{ud}^E(t_2) + (1 - p_*)^2 C_{d^2}^E(t_2)\right), \tag{8.94}$$

where $C_{u^2}^E(t_2)$, $C_{ud}^E(t_2)$, and $C_{d^2}^E(t_2)$ are given by (8.92). Equation (8.94) agrees with (8.91).

**Filtration approach.** We can also obtain Equation (8.94) via risk-neutral pricing using the $\sigma$-algebra $\mathcal{F}_t$ for discrete time $t = t_j$; see Example 6.17 (page 269) for more on the filtration. In our binomial tree setting, the $\sigma$-algebra $\mathcal{F}_{t_j} \equiv \mathcal{F}_j$ is generated by the underlying security's prices up to time $t_j$. For example, $\mathcal{F}_1$ is generated by $S_0, S_1$ and given by

$$\mathcal{F}_1 = \{\varnothing,\, A_U,\, A_D,\, \Omega\},$$

where

$$A_U = \{UU, UD\}, \qquad A_D = \{DU, DD\}, \qquad \Omega = \{UU, UD, DU, DD\}.$$

Note that each element in the events $A_U$, $A_D$, and $\Omega$ is an entire price path from $t_0$ from $t_2$ (not just to $t_1$). In particular, the path $\omega = UU$ has an uptick at $t_1$ followed by another uptick at $t_2$ and $A_U$ consists of all price paths with an uptick at $t_1$, while $A_D$ are those with a downtick at $t_1$. The risk-neutral price of the European call is

$$C^E(t_0;2) = e^{-r(2h)} \mathbb{E}_* \left( C^E(t_2;2) \right), \tag{8.95}$$

where the tower property yields (see (8.57) on page 403):

$$\mathbb{E}_* \left( C^E(t_2;2) \right) = \mathbb{E}_* \left( \mathbb{E}_* \left( C^E(t_2;2) \mid \mathcal{F}_1 \right) \right). \tag{8.96}$$

The conditional expectation $\mathbb{E}_* \left( C^E(t_2;2) \mid \mathcal{F}_1 \right)$ is relative to each event in $\mathcal{F}_1$. Since the sample space $\Omega$ merely tells us that one of the four price paths will occur, the possible values of $\mathbb{E}_* \left( C^E(t_2;2) \mid \mathcal{F}_1 \right)$ will come from the information carried by $A_U$ and $A_D$. Figure 8.2 shows that

$$\mathbb{E}_* \left( C^E(t_2;2) \mid \mathcal{F}_1 \right) = \begin{cases} \mathbb{E}_* \left( C^E_u(t_2) \mid S(t_1) = S_u(t_1) \right) & \text{with probability } p_* \\ \\ \mathbb{E}_* \left( C^E_u(t_2;2) \mid S(t_1) = S_d(t_1) \right) & \text{with probability } 1 - p_* \end{cases}$$

$$= \begin{cases} p_* C^E_{u^2}(t_2) \; + \; (1 - p_*) C^E_{ud}(t_2) & \text{with probability } p_* \\ \\ p_* C^E_{ud}(t_2) \; + \; (1 - p_*) C^E_{d^2}(t_2) & \text{with probability } 1 - p_*. \end{cases}$$

Consequently,

$$\mathbb{E}_* \left( C^E(t_2;2) \right) = p_*^2 C^E_{u^2}(t_2) \; + \; 2p_*(1 - p_*) C^E_{ud}(t_2) \; + \; (1 - p_*)^2 C^E_{d^2}(t_2).$$

Inserting the above into (8.95) yields the same call price as in (8.94).

**Example 8.6.** Suppose that a stock with current price of $50 pays not dividend. Assume that at 1.5 months from now, the price of the stock is either $52.7250 or $47.4150, where four decimal places are used for pedagogical reasons to minimize rounding errors. Employing a binomial tree and risk-free rate of 2% per annum, compute the two-period current price of a 3-month European call option on this stock given a strike price of $50.

**Solution.** The necessary inputs are as follows: $h = \frac{0.25}{2} = 0.125$, $r = 0.02$, $q = 0$, $S(t_0) = \$50$, $K = \$50$, and

$$u = \frac{\$52.7250}{\$50} = 1.0545, \quad d = \frac{\$47.4150}{\$50} = 0.9483, \quad p_* = \frac{e^{(r-q)h} - d}{u - d} = 0.5104.$$

Hence, the two-period pricing formula (8.94) yields $C^E(t_0;2) = \$1.45$.  □

### 8.5.4 $n$-Period Binomial Pricing

For the $n$-period case $0 = t_0 < t_1 < \cdots < t_n$, an induction argument yields the current price of a European call as given by (Exercise 8.32):

$$C^E(t_0;n) = e^{-r(nh)} \mathbb{E}_*(C^E(t_n,n)) = e^{-r(nh)} \sum_{i=0}^{n} \binom{n}{i} p_*^i (1-p_*)^{n-i} C_{u^i d^{n-i}}^E(t_n),$$

(8.97)

where

$$C_{u^i d^{n-i}}^E(t_n) = \max\{S(t_0)u^i d^{n-i} - K, 0\}$$

for $i = 0,1,\ldots,n$. The formula (8.97) can be expressed more simply as (Exercise 8.33):

$$C^E(t_0;n) = S(t_0)e^{-q\tau} \mathrm{N}(n,k_*,\hat{p}_*) - Ke^{-r\tau} \mathrm{N}(n,k_*,p_*),$$

(8.98)

where $k_*$ is the smallest value of $i$ for which $S(t_0)u^i d^{n-i} - K > 0$ and

$$\tau = t_n - t_0, \quad \mathrm{N}(n,k_*,p_*) = \sum_{i=k_*}^{n} \binom{n}{i} p_*^i (1-p_*)^{n-i}, \quad \hat{p}_* = \frac{p_* u}{e^{(r-q)h}}.$$

The reader may have noticed that the pricing formula (8.98) looks similar to the BSM pricing formula (8.70) on page 411. In the continuous-time limit $n \to \infty$, it can be shown (see Hsia [21]) that (8.98) converges to (8.70):

$$\lim_{n \to \infty} C^E(t_0;n) = C_0^E = S_0 e^{-qT} \mathrm{N}(d_+(S_0,T)) - Ke^{-rT} \mathrm{N}(d_-(S_0,T)).$$

**Example 8.7.** Suppose that a stock paying no dividend has a current price of \$50 and annual volatility of 15%. For a risk-free rate of 2% per annum, compute the 100-period current price of a 3-month European call on this stock with a strike price of \$50.

**Solution.** Given the large number of periods, we employ the CRR formulas for $u$, $d$, and $p_*$, namely, $u \approx e^{\sigma\sqrt{h}}$, $d \approx e^{-\sigma\sqrt{h}}$, and $p_* \approx \frac{e^{(r-q)h}-d}{u-d}$. Applying (8.97) with the values $n = 100$, $h = \frac{0.25}{100} = 0.0025$, $\sqrt{h} = 0.05$, $r = 0.02$, $q = 0$, $\sigma = 0.15$, $S(t_0) = \$50$, $K = \$50$, and

$$u = 1.007530, \quad d = 0.992528, \quad p_* = 0.501458,$$

we obtain a current call price of $C^E(t_0;100) = \$1.62$. This coincides with the result of the continuous-time BSM pricing formula (8.70). In fact, for an 80-period tree, the two prices already agree to the given decimal places; see Figure 8.3.

□

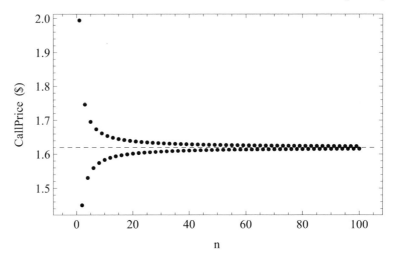

**Fig. 8.3** The per share prices of a 3-month European call option, where the strike price is $K = \$50$ per share, risk-free rate is 0.02 per annum, and underlying stock has current share price of $50 and volatility of 15%. The bullets show the call prices given by the $n$-period binomial option-pricing formula using a CRR tree for $n = 1, \ldots, 100$. The horizontal dashed line is the per share call price of $1.62 obtained using the BSM formula. Even values of $n$ give call prices below the BSM line, while odd values are above. The binomial per share call price is $1.99 for $n = 1$ and $1.45 for $n = 2$. The price to two decimal places is already $1.62 for $n = 80$

**Remark 8.8.** The original binomial-tree approach to deriving the continuous-time BSM price for European calls used the CRR formulas for u, d, and $p_*$. However, it was shown by Hsia [21] that such specificity is unnecessary to take the continuous-time limit. He gave a simpler proof for general u, d, and $p_*$ using the de Moivre-Laplace Theorem. The lecture notes by Chance [8] also give an accessible treatment of the proof in [21].                                    □

## 8.6 Delta Hedging

In this section, we develop the theoretical infrastructure for delta hedging of European calls (Section 8.6.1) and apply the framework to an example (Section 8.6.2).

### 8.6.1 Theoretical Delta Hedging for European Calls

When selling European calls, the risk to the seller is to be able to meet the obligation should the calls be exercised. For instance, if you sell 500 European calls on a stock and they are exercised at expiration, then you need to have

50,000 shares of the stock available at expiration. The risk lies, for example, in not having them available at expiration. Or, if you buy the 50,000 shares right after selling the calls to be assured of covering your obligations in the event of exercise, and the calls expire out of the money, you are left with a lot of shares that may have even lost value.

Our core strategy for managing the risk from selling a European call is to create simultaneously an offsetting position in a synthetic European call. The synthetic call will be constructed using the funds from the short sale and borrowing a certain amount. The core of the strategy is the process of *delta hedging* and will result in the loan being paid off at expiration. Note that the cash dividend from the underlier can be used either to buy more units of the underlier, which is assumed by default, or to pay toward the loan. To make the presentation different, instead of employing our default assumption, *assume that the cash dividend is used continuously as payment toward the loan.*

We now detail the theoretical framework for hedging.

## Time $t$

At time $t < T$, construct a costless portfolio as follows:

a) **(Short Call)** Short sell 1 European call on 1 unit of an underlying security. This brings in proceeds of $C_t^E$.

b) **(Long Synthetic Call)** Using the proceeds $C_t^E$ as start-up capital and borrowing the amount

$$L_t = K e^{-r\tau} N\big(d_-(S_t, \tau)\big),$$

we can buy

$$\Delta_C(t) = e^{-q\tau} N(d_+(S_t, \tau))$$

units of the underlying security. This is because (8.38) (page 394) yields

$$C_t^E + L_t = \Delta_C(t) S_t.$$

The portfolio with $\Delta_C(t)$ units of the underlier and loan $L_t$ is called a *synthetic call* because its value $C_{syn}^E(t)$ replicates the call price:

$$C_{syn}^E(t) = \Delta_C(t) S_t - L_t = C_t^E.$$

## Remark 8.9.

1. Since $C_{syn}^E(t) = n_t S_t^c + b_t B_t$, where $n_t = \left(\frac{S_t}{S_t^c}\right) \Delta_C(t)$ and $b_t = -\frac{L_t}{B_t}$, the synthetic call can be viewed as a portfolio with a position in $n_t$ units of the

cum-dividend underlier and $b_t$ units of the money market account. Equation (8.14) then yields that the strategy $(n_t, b_t)$ is self-financing and replicates the call price for all times $t \geq 0$.

2. By (8.50), the number $\Delta_C(t)$ of units we purchased of the underlying security is less than 1 since $t < T$. We did not purchase 1 unit of the underlier to cover the call and neither did we take a naked position on the call.  □

The portfolio with positions (a) and (b) has a value:

$$0 = \mathcal{V}_C(t) = - C_t^E + C_{\text{syn}}^E(t) = 0. \tag{8.99}$$

Since

$$C_{\text{syn}}^E(t) = \Delta_C(t) S_t - L_t \qquad \text{with} \qquad L_t = \Delta_C(t) S_t - C_t^E, \tag{8.100}$$

the value of the portfolio at $t$ can then be expressed as

$$0 = \mathcal{V}_C(t) = \underbrace{- C_t^E}_{\text{short call}} + \underbrace{I_t}_{\text{risk-free investment}} - \underbrace{\mathbb{L}_t}_{\text{loan}} + \underbrace{\Delta_C(t) S_t}_{\text{long } \Delta_C(t) \text{ units}} ,$$

$$\tag{8.101}$$

where

$$I_t = C_t^E, \qquad \mathbb{L}_t = \Delta_C(t) S_t.$$

Equation (8.101) has a natural financial interpretation: *the portfolio with positions (a) and (b) at t is equivalent in value to a costless portfolio where one short sells 1 European call for $C_t^E$, invests the proceeds $C_t^E$ in a risk-free investment, borrows the amount $\mathbb{L}_t = \Delta_C(t) S_t$, and uses the loan to buy $\Delta_C(t)$ units of the underlying security.*

**Time $t + dt$**

As time $t$ advances, the positions on the right-hand side of (8.101) will change. We shall have to update continuously our earlier positions and will do so in a costless, self-financing way to maintain the equation at zero, i.e., maintain the replication by the synthetic call. Recall that self-financing means the change in the value of the portfolio comes strictly from the change in the value of the securities in the portfolio, which includes a loan. For instance, any purchases of additional units of the security will be funded by increasing the loan, i.e., more borrowing, and any proceeds from selling units of the security will not be withdrawn, but paid toward the loan. Note that borrowing is at the risk-free rate.

Let us look closely at this process. At time $t + dt$, the portfolio's value has the form:

$$0 = \mathcal{V}_C(t+dt) = \underbrace{-C^E_{t+dt}}_{\text{short call}} + \underbrace{\mathbb{I}_{t+dt}}_{\text{risk-free investment}} - \underbrace{\mathbb{L}_{t+dt}}_{\text{loan}} + \underbrace{\Delta_C(t+dt)\,S_{t+dt}}_{\text{long } \Delta_C(t) \text{ units}}.$$

$$(8.102)$$

We consider each term:

➤ The short position $-C^E_t$ has changed to $-C^E_{t+dt}$.

➤ The original investment of $\mathbb{I}_t = C^E_t$ has grown at the risk-free rate r to $\mathbb{I}_{t+dt} = C^E_t\,e^{r\,dt}$.

➤ **Delta hedging.** The previous two positions change on their own. The actual adjusting occurs in maintaining $\Delta_C(t)$ units of the underlier as time $t$ advances. This adjusting is called *delta hedging* and will be reflected in the balance of the loan. In fact, at time $t + dt$, the new loan balance $\mathbb{L}_{t+dt}$ is the net result of using the security's cash dividend to pay toward the loan and adjusting, if necessary, the number of units of the underlier from $\Delta_C(t)$ to $\Delta_C(t + dt)$. We then consider the difference

$$d\Delta_C(t) = \Delta_C(t+dt) - \Delta_C(t).$$

If $d\Delta_C(t) = 0$, then $\Delta_C(t)$ suffices and the loan's balance is

$$\mathbb{L}_{t+dt} = \mathbb{L}_t\,e^{(r-q)\,dt}.$$

If $d\Delta_C(t) > 0$, then $\Delta_C(t)$ units are not enough, so we purchase $d\Delta_C(t)$ units by borrowing $\left(d\Delta_C(t)\right) S_{t+dt}$. The loan's balance becomes

$$\mathbb{L}_{t+dt} = \mathbb{L}_t\,e^{(r-q)\,dt} + \left(d\Delta_C(t)\right) S_{t+dt}.$$

If $d\Delta_C(t) < 0$, then $\Delta_C(t)$ units are too much. We then sell $-d\Delta_C(t)$ units and receive $\left(-d\Delta_C(t)\right) S_{t+dt}$. We pay this amount toward the loan, which decreases the change in value of the loan to

$$\mathbb{L}_{t+dt} = \mathbb{L}_t\,e^{(r-q)\,dt} - \left(-d\Delta_C(t)\right) S_{t+dt}.$$

All three cases are captured by the single equation,

$$\mathbb{L}_{t+dt} = \mathbb{L}_t\,e^{(r-q)\,dt} + \left(d\Delta_C(t)\right) S_{t+dt}, \qquad (8.103)$$

where $d\Delta_C(t)$ can be zero, positive, or negative. The delta hedging process is repeated during every instant until expiration.

**Expiration Time $T$**

At expiration $T$, the short position in the call will be $-C_T^E$, the initial investment $C_t^E$ would have grown at the risk-free rate r to $C_t^E e^{r\tau}$ over the period $\tau = T - t$, the loan's balance will be $\mathbb{L}_T$, and the long position in the underlying asset will consist of $\Delta_C(T)$ units and have value $\Delta_C(T) S_T$. The value of the portfolio at expiration is then:

$$0 = \mathcal{V}_C(T) = -C_T^E + \mathbb{I}_T - \mathbb{L}_T + \Delta_C(T) S_T,$$

where $C_T^E = \max\{S_T - K, 0\}$ by the final boundary condition, $\mathbb{I}_T = C_t^E e^{r\tau}$, and the loan's balance is

$$\mathbb{L}_T = \mathbb{L}_{T-dt} e^{(r-q)dt} + \left( d\Delta_C(T - dt) \right) S_T \tag{8.104}$$

by (8.103). There are two possibilities to consider at expiration: either $S_T > K$ or $S_T \le K$.

**In-the-money European call at expiration.** Suppose $S_T > K$, i.e., the call is exercised.[16] Then the obligation of the short sale of the call can be fulfilled. In fact, Equation (8.48) shows that $\Delta_C(T) = 1$ for $S_T > K$, i.e., the delta hedging process results in 1 unit of the underlying asset. It is this 1 unit that is sold to the call holder at price $K$ to fulfill the obligation of the exercised call. The outstanding item now is whether the loan can be paid off. For the case $S_T > K$, we have $C_T^E = S_T - K$ and $\Delta_C(T) = 1$, which yield

$$0 = -S_T + K + C_t^E e^{r\tau} - \mathbb{L}_T + S_T.$$

Consequently,

$$K + C_t^E e^{r\tau} = \mathbb{L}_T.$$

Hence, *if the call is exercised, then the loan's balance $\mathbb{L}_T$ can also be paid off* by using the proceeds $K$ from selling the 1 unit of the underlying security to the call holder and the cash $C_t^E e^{r\tau}$ from liquidating the risk-free investment.

**At-the-money or out-of-the-money European call at expiration.** Assume $S_T \le K$, i.e., the call is not exercised. Then even though the issuer has no obligation to the call holder, the balance on the loan still has to be settled. For $S_T \le K$, we have $C_T^E = 0$ and so

$$0 = C_t^E e^{r\tau} - \mathbb{L}_T + \Delta_C(T) S_T,$$

i.e.,

$$C_t^E e^{r\tau} + \Delta_C(T) S_T = \mathbb{L}_T.$$

---

[16] We assume that a European call is exercised if and only if $S_T > K$ and not exercised if and only if $S_T \le K$.

In other words, *if the call is not exercised, then the loan's balance* $\mathbb{L}_T$ *can still be paid off* by the cash inflow $C_t^E e^{r\tau} + \Delta_C(T)S_T$ from liquidating the risk-free investment and the long position of $\Delta_C(T)$ units of the underlying security. Note that (8.47) yields $\Delta_C(T) = \frac{1}{2}$ for $S_T = K$ and (8.49) implies $\Delta_C(T) = 0$ for $S_T < K$. The latter, along with (8.104) and

$$d\Delta_C(T - dt) = \Delta_C(T) - \Delta_C(T - dt),$$

implies that

$$\mathbb{L}_T = \begin{cases} \mathbb{L}_{T-dt}\, e^{(r-q)\,dt} + \left(\frac{1}{2} - \Delta_C(T - dt)\right) S_T & \text{at-the-money at } T \\[2ex] \mathbb{L}_{T-dt}\, e^{(r-q)\,dt} - \Delta_C(T - dt)\, S_T & \text{out-of-the-money at } T. \end{cases}$$

$$(8.105)$$

In summary, we see that employing delta hedging enables one to meet the obligations of selling European calls using a costless, self-financing, replicating process. Even though the process requires no initial capital, it involves borrowing, but the loan is paid off at expiration.

## 8.6.2 Application of Delta Hedging to Selling European Calls

A major concern to a European call seller is fulfilling the obligations of the call if it is exercised. To get a sense of this risk, we give a simple example:

**Example 8.8.** Suppose that a firm sells 500 European calls for $2.264248 per share of a stock currently trading at $75 per share. Since each call involves 100 shares, the firm receives

$$500 \times 100 \times \$2.264248 = \$113{,}212.40.$$

Assume that there are 80 days to expiration, the strike price is $75, and the firm invests the $113,212.40 proceeds in a risk-free investment growing at 2% annually. Suppose that there are 365 days in a year; see Remark 8.1 (page 385).

Should the calls be exercised at expiration, the firm must sell $500 \times 100 = 50{,}000$ shares of the stock to the call holder for $75 per share. But if the firm does not own any share of the stock and plans to buy the 50,000 shares only when the calls are exercised, then the firm is taking a *naked position* and exposing itself to potential loss. For instance, if the share price at expiration is $87.98, then the calls will be exercised and it will cost the firm $4,399,000 to buy the shares to satisfy the obligations of the calls. Though firm would receive $3,750,000 from selling the shares at the strike price and the firm's proceeds from selling the calls would have grown to $e^{0.02 \times (80/365)} \times \$113,212.40 =$ $\$113,709.76$ at expiration, the firm would still have a loss that more than quadruples the gain from selling the calls:

$$\$113,709.76 - \$4,399,000 + \$3,750,000 = -\$535,290.24.$$

On the other hand, the firm can take a *covered position* by buying the 50,000 shares at the time it sells the calls. If the stock price falls to $65 at expiration, i.e., drops in value to $500,000, then the call will not be exercised and, despite the growth in the call-sale proceeds to $113,709.76, the firm will experience a loss in value that more than triples this gain:

$$\$113,709.76 - \$500,000 = -\$386,290.24.$$

□

How many shares should the firm then hold to hedge against the risk from selling a European call? The answer is actually not a fixed number of shares. The number will have to change as time advances. It is called *delta hedging*. This tool is actually contained in the construction of the costless, self-financing, riskless portfolio employed in Section 8.1.4. We apply the ideas and results of that section to illustrate, in principle, how delta hedging works. For simplicity, the discussion will focus on the issuance of 1 call on 1 unit of an underlier. The application in Section 8.6.2 and, in particular, Example 8.9 (page 428) will illustrate delta hedging using the example above.

We now illustrate the theoretical framework for delta hedging European calls by applying it to Example 8.8.

**Example 8.9.** Suppose that a firm sells 500 European calls for $113,212.40 on a stock with share price of $75 and annual volatility of 15%. All of the calls then involve 50,000 shares of the stock. Assume that there are 80 days to expiration, the strike price is $75, and the firm invests the $113,212.40 in a risk-free investment growing at 2% per annum. How can the firm manage the risk from the call sale without taking a naked or covered position? We illustrate how to manage the risk using daily delta hedging. Since the risk-free investment will earn interest even on non-trading days, assume 365 days in a year and, for simplicity, suppose that trading occurs on each of the 80 days remaining till expiration.

Let us now delta hedge day by day for 80 days using Equations (8.101) and (8.102) as a guide, which we rewrite for convenience:

$$0 = \mathcal{V}_C(t) = \underbrace{-C_t^E}_{\text{short call}} + \underbrace{I_t}_{\text{risk-free investment}} - \underbrace{\mathbb{L}_t}_{\text{loan}} + \underbrace{\Delta_C(t)S_t}_{\text{long }\Delta_C(t)\text{ units}},$$

$$(8.106)$$

and

$$0 = \mathcal{V}_C(t + dt) = \underbrace{-C^E_{t+dt}}_{\text{short call}} + \underbrace{\mathbb{I}_{t+dt}}_{\text{risk-free investment}} - \underbrace{\mathbb{L}_{t+dt}}_{\text{loan}} + \underbrace{\Delta_C(t + dt) S_{t+dt}}_{\text{long } \Delta_C(t+dt) \text{ units}},$$

(8.107)

where

$$\mathbb{I}_{t+dt} = C^E_t e^{r\,dt}, \qquad \mathbb{L}_{t+dt} = \mathbb{L}_t e^{(r-q)\,dt} + (d\Delta_C(t)) S_{t+dt}$$

(8.108)

with

$$d\Delta_C(t) = \Delta_C(t + dt) - \Delta_C(t).$$

Note that Equations (8.106) and (8.107) are based on one share of the stock.

The current time is Day 0, each instantaneous time change is approximated by a day, and expiration is at $T$:

$$t = 0, \qquad dt \approx h = \frac{1}{365}, \qquad T = 80h.$$

Because we cannot perfectly delta hedge, i.e., we cannot hedge every moment of time (e.g., $dt$ is replaced by a day) and cannot work to infinitely many decimal places, our results will not yield $\mathcal{V}_C(t) = 0$ and $\mathcal{V}_C(t + dt) = 0$ as the current time $t$ advances day by day. However, though a perfect hedge would yield zero loss at expiration, our approximate hedge will significantly reduce any losses compared to the naked and covered positions in Example 8.8 (page 427). The results are summarized in Table 8.1.

**Remark 8.10. (Rounding Errors)** Rounding errors are unavoidable whenever results are expressed to a fixed number of decimal places. The delta hedging in Table 8.1 is based on a MATLAB code that outputs results to much more than six decimal places. We summarized the results in the table by rounding off the MATLAB output at six decimal places to avoid making the table look too dense. Naturally, the rounding off will produce errors. For instance, in the first row of Table 8.1, we have $\Delta_C(0) S_0 = \$40.413671$. However, using the initial stock price $S_0 = \$75$ and the rounded value $\Delta_C(0) = 0.538849$, the product gives a different answer to six decimal places: $\Delta_C(0) S_0 = \$40.413675$.

As noted, the actual MATLAB value of $\Delta_C(0)$ has far more than six decimal places. To nine decimal places, it yields $\Delta_C(0) = 0.538848947$ and so produces $\Delta_C(0) S_0 = \$40.413671025$, which to six decimal places yields the entry $\$40.413671$ in the table. The latter entry is more accurate. Note that at the end of the delta hedging, the total net value of the portfolio, i.e., the per-share value $\mathcal{V}(T)$ times 50,000 shares, will be rounded off to cents. □

We shall carry out the delta hedging on a per-share basis. Each per-share position can then be multiplied by 50,000 to obtain the total size of the position.

**Table 8.1** Delta hedging to mitigate against risk from selling 500 calls. The time to expiration is 80 days. Multiply the per-share values by 50,000 to get the total values. Though the European call expires in the money, delta hedging enables the seller to have the required 50,000 shares and minimizes the losses to only $2,326.10. Simulation is based on a MATLAB code, and we truncated the output at six decimal places. The portfolio's value at time $t$ is $\mathcal{V}_C(t) = -C_t^E + \mathbb{I}_t - \mathbb{L}_t + \Delta_C(t) S_t$

| Delta hedging values per share | | | | | | | |
|---|---|---|---|---|---|---|---|
| $t$ | $S_t$ | $\Delta_C(t)$ | $-C_t^E$ short call | $\mathbb{I}_t$ investment | $-\mathbb{L}_t$ loan balance | $\Delta_C(t) S_t$ long $\Delta_C(t)$ shares | $\mathcal{V}_C(t)$ port. value |
| Day 0 | $75.000000 | 0.538849 | -$2.264248 | $2.264248 | -$40.413671 | $40.413671 | $0 |
| Day 1 | $75.806609 | 0.598789 | -$2.707914 | $2.264373 | -$44.959756 | $45.392182 | -$0.011116 |
| $\vdots$ | $\vdots$ | $\vdots$ | $\vdots$ | $\vdots$ | $\vdots$ | $\vdots$ | $\vdots$ |
| Day 79 | $92.155473 | 1 | -$17.159583 | $2.274071 | -$77.316481 | $92.155473 | -$0.046519 |
| Day 80 | $93.561474 | 1 | -$18.561474 | $2.274196 | -$77.320717 | $93.561474 | -$0.046522 |

## Day 0

The stock price is $75 and there are 80 days until expiration. We employ the per-share Equation (8.106). The firm short sells each European call on one share at the BSM price. With the starting time $t = 0$, the BSM price per share is

$$C_0^E = C^E(S_0, K, \sigma, r, T, q) = C^E\left(\$75, \$75, 0.15, 0.02, 80h, 0\right) = \$2.264248445.$$

We shall employ at least six decimal places to minimize rounding errors. The short sale creates a negative position with per-share value

$$-C_0^E = -\$2.264248445.$$

The firm invests the proceeds from the sale in a risk-free account paying 2% per annum:

$$\mathbb{I}_0 = C_0^E = \$2.264248445 \text{ (per share)}.$$

Now, the current delta of the call to nine decimal places is

$$\Delta_C(0) = 0.538848947,$$

which is rounded in Table 8.1 to 0.538849 merely for ease of presentation. In other words, on a per-share basis, the firm needs to long 0.538848947 shares of the stock, which results in longing $50{,}000 \times \Delta_C(0)$ shares of the stock when considering the total. The firm takes out a loan at the risk-free rate to buy the 0.538848947 shares. This creates a negative position, i.e., a per-share loan amount of

$$\mathbb{L}_0 = 0.538848947 \times \$75 = \$40.413671.$$

Buying the shares also creates a positive position:

$$\Delta_C(0)\, S_0 = \$40.413671 \text{ (per share)}.$$

The per-share value of the portfolio at the start is then zero:

$$\mathcal{V}_C(0) = -\$2.264248445 + \$2.264248445 - \$40.413671$$
$$+ \$40.413671 = 0.000000.$$

## Day 1

We shall use (8.107). Suppose that the stock price is \$75.80660915. There are 79 days until expiration and the per-share value of the short sale position is now

$$-C_h^E = -C^E\left(\$75.80660915, \$75, 0.15, 0.02, 79\,h, 0\right) = -\$2.707914261.$$

and the investment has grown to a per-share value of

$$I_h = I_0\, e^{rh} = \$2.264372517,$$

which follows from the more accurate value $I_0 = 2.264248445$.

Since the stock price increased, the delta also increased:

$$\Delta_C(h) = 0.598789243.$$

The extra number of shares needed relative to a per-share basis is

$$d\Delta_C(0) = \Delta_C(h) - \Delta_C(0) = 0.598789243 - 0.538848947 = 0.059940296.$$

To buy these shares, borrow

$$\left(d\Delta_C(0)\right) S_h = \$4.543870582.$$

The original loan with interest has grown to

$$\mathbb{L}_0\, e^{rh} = \$40.413671 \times e^{\frac{0.02}{365}} = \$40.415885508.$$

The per-share loan balance on Day 1 is then

$$\mathbb{L}_h = \mathbb{L}_0\, e^{rh} + \left(d\Delta_C(0)\right) S_h = \$40.415885508 + \$4.543870582 = \$44.959756090.$$

The new value of the long position in the stock is

$$\Delta_C(h)\, S_h = 0.598789243 \times \$75.80660915 = \$45.392182107 \quad \text{(per share)}.$$

The net value of all the positions is then

$$\mathcal{V}_C(h) = -\$2.707914261 + \$2.264372516 - \$44.9597561 + \$45.392182107$$
$$= -\$0.011116 \quad \text{(per share)},$$

which rounds off at six decimal places to the value in the table.

The process from Day 0 to Day 1 is repeated across consecutive days. We then skip ahead to the day before expiration.

## Day 79

Suppose that on Day 79 we have the rounded values given in Table 8.1.

## Day 80

The stock price is \$93.56147415, which means that the call expires in the money and will be exercised. The short sale position now has value

$$-C_T^E = -(S_T - K) = -(\$93.56147415 - \$75) = -\$18.56147415 \quad \text{(per share)}.$$

The value of the investment is

$$I_T = I_{79h}e^{rh} = \$2.274195703 \quad \text{(per share)}.$$

Note that the above value, which is exposed to more rounding errors due to the day-to-day delta hedging, still gives the same result to six decimal places as the more accurate value

$$I_0 e^{rT} = \$2.264248445 \times e^{0.02 \times \frac{80}{365}} = \$2.274195704.$$

Because the call is in the money, delta is exactly unity:

$$\Delta_C(T) = 1.$$

In other words, *at expiration the firm has the required 50,000 shares to meet the obligations of the call being exercised.* Since

$$d\Delta_C(T - h) = d\Delta_C(79h) = \Delta_C(80h) - \Delta_C(79h) = 1 - 1 = 0,$$

the balance on the loan to nine decimal places is

$$\mathbb{L}_T = \mathbb{L}_{79h}e^{rh} + (d\Delta_C(79h))S_T = \mathbb{L}_{79h}e^{rh} = \$77.32071734 \quad \text{(per share)}.$$

The long position in the stock is \$87.98 per share since $\Delta_C(T) = 1$.

The net value to six decimal places of all the positions is

$$-\$0.046522 = \mathcal{V}_C(T) = -C_T^E + I_T - \mathbb{L}_T + \Delta_C(T)\,S_T$$
$$= -S_T + K + I_T - \mathbb{L}_T + S_T$$
$$= (K + I_T) - \mathbb{L}_T \quad \text{(per share)}.$$

In other words, the proceeds $K$ received from selling the call at strike plus the amount $I_T$ from the risk-free investment are not enough to cover the balance $\mathbb{L}_T$ of the loan. The negative value corresponds to a total loss of

$$50{,}000 \times \mathcal{V}_C(T) = -\$2{,}326.10.$$

This loss is insignificant compared to the total monies involved in the sale.

In the idealized case of continuously delta hedging, the firm would net zero when the calls are sold at the BSM price. Note that if the firm had sold the calls sufficiently higher than the BSM price, then it can even make a profit.

Exercise 8.12 explores a case where the European call in this example expires out of the money. In this case, we would make use of (8.105). □

In practice, each transaction has a variety of costs, including bid-ask spreads, brokerage commissions, and taxes. Delta hedging can be expensive and difficult to execute since it requires frequent rebalancing.

## 8.7 Option Greeks and Managing Portfolio Risk

The price of a European call is more volatile than that of its underlying security (Exercise 8.25). A 1% price movement of the underlier can lead to a price movement in the call that is significantly larger (Exercise 8.8). This section explores how to make the value of a portfolio of options with the same underlying security more stable against small price movements in the underlier. To accomplish this, we first discuss option Greeks and then apply these ideas to the construction of delta- and gamma-neutral portfolios.

> *Unless stated to the contrary, in this section assume for simplicity*
> *that the underlying security pays no dividend ($q = 0$).*

### 8.7.1 Option Greeks for Portfolios and the BSM p.d.e.

Before generalizing delta neutrality to portfolios, we introduce option Greeks for a portfolio consisting of options and a security and show that the portfolio's value satisfies the BSM p.d.e.

Consider a portfolio consisting of $k$ options (or derivatives) with the same underlying security, where the per-unit value of the $i$th option at time $t$ is

$f_i(S_t,t)$. Let $N_i$ be the number, based on each unit of the underlier, of the $i$th option in the portfolio. Here $N_i > 0$ indicates a long position, $N_i = 0$ no position, and $N_i < 0$ a short position in the $i$th option. For example, suppose that the first position in the portfolio is long 300 calls on a stock. Since a call involves 100 shares of the stock, the number of calls on a per-share basis is $N_1 = 300 \times 100 = 30{,}000$. The value of the portfolio at time $t$ is

$$\mathbb{V}(S_t,t) = N_1 f_1(S_t,t) + \cdots + N_k f_k(S_t,t).$$

The portfolio also has what are called *option Greeks*, which define the rate of change of the portfolio relative to various parameters.

We shall first introduce these rates of change for the options in the portfolio. Recall that the value of an option is a function of $(x,t)$ with $x$ representing the possible prices of the underlier at time $t$ (see page 391). Using rates of change in $x$ and $t$, we then define the following three *option Greeks* of the $i$th option in the portfolio :

$$Delta: \quad \Delta_i(S_t,t) = \frac{\partial f_i}{\partial S}(S_t,t) = \left.\frac{\partial f_i}{\partial x}\right|_{(x,t)=(S_t,t)}$$

$$Gamma: \quad \Gamma_i(S_t,t) = \frac{\partial^2 f_i}{\partial S^2}(S_t,t) = \left.\frac{\partial^2 f_i}{\partial x^2}\right|_{(x,t)=(S_t,t)}$$

$$Theta: \quad \Theta_i(S_t,t) = \frac{\partial f_i}{\partial t}(S_t,t) = \left.\frac{\partial f_i}{\partial t}\right|_{(x,t)=(S_t,t)}.$$

The formulas for these option Greeks in the case of calls and puts are given in Exercise 8.35 on page 473.

Now, viewing the portfolio value $\mathbb{V}$ as a function of $(x,t)$, the option Greeks extend naturally to the portfolio:

$$\Delta(S_t,t) = \frac{\partial \mathbb{V}}{\partial S}(S_t,t) = \left.\frac{\partial \mathbb{V}}{\partial x}\right|_{(x,t)=(S_t,t)} = \sum_{i=1}^{k} N_i \Delta_i(S_t,t)$$

$$\Gamma(S_t,t) = \frac{\partial^2 \mathbb{V}}{\partial S^2}(S_t,t) = \left.\frac{\partial^2 \mathbb{V}}{\partial x^2}\right|_{(x,t)=(S_t,t)} = \sum_{i=1}^{k} N_i \Gamma_i(S_t,t)$$

$$\Theta(S_t,t) = \frac{\partial \mathbb{V}}{\partial t}(S_t,t) = \left.\frac{\partial \mathbb{V}}{\partial t}\right|_{(x,t)=(S_t,t)} = \sum_{i=1}^{k} N_i \Theta_i(S_t,t).$$

Note that if the gamma of a portfolio stays sufficiently small, then delta changes a little as the underlier price changes, which reduces the need for frequent rebalancing in delta hedging.

Turning to the BSM p.d.e., we know that each option in the portfolio satisfies this p.d.e. ($q = 0$):

$$\frac{1}{2}\sigma^2 S_t^2 \frac{\partial^2 f_i}{\partial x^2}(S_t,t) + r S_t \frac{\partial f_i}{\partial x}(S_t,t) + \frac{\partial f_i}{\partial t}(S_t,t) - r f_i(S_t,t) = 0.$$

This can then be expressed in terms of option Greeks as follows:

$$\frac{1}{2}\sigma^2 S_t^2 \Gamma_i(S_t,t) + r S_t \Delta_i(S_t,t) + \Theta_i(S_t,t) - r f_i(S_t,t) = 0.$$

Summing over the $k$ options in the portfolio, it follows that $\mathbb{V}$ satisfies the BSM p.d.e.:

$$\frac{1}{2}\sigma^2 S_t^2 \Gamma(S_t,t) + r S_t \Delta(S_t,t) + \Theta(S_t,t) - r\mathbb{V}(S_t,t) = 0. \tag{8.109}$$

As mentioned earlier (page 391), this equation holds for all $0 \le t \le T$ and every $0 < S_t < \infty$.

Now, expand the portfolio to include $N_S$ units of the underlying security:

$$\widetilde{\mathbb{V}}(S_t,t) = \mathbb{V}(S_t,t) + N_S S_t.$$

The expanded portfolio has option Greeks

$$\widetilde{\Delta}(S_t,t) = \frac{\partial \widetilde{\mathbb{V}}}{\partial S}(S_t,t) = \Delta(S_t,t) + N_S, \qquad \widetilde{\Gamma}(S_t,t) = \frac{\partial^2 \widetilde{\mathbb{V}}}{\partial S^2}(S_t,t) = \Gamma(S_t,t),$$

and

$$\widetilde{\Theta}(S_t,t) = \frac{\partial \widetilde{\mathbb{V}}}{\partial t}(S_t,t) = \Theta(S_t,t).$$

Since the portfolio value $\mathbb{V}$ and the security price $S_t$ satisfy the BSM p.d.e., the value of the expanded portfolio does as well:

$$\frac{1}{2}\sigma^2 S_t^2 \widetilde{\Gamma}(S_t,t) + r S_t \widetilde{\Delta}(S_t,t) + \widetilde{\Theta}(S_t,t) - r\widetilde{\mathbb{V}}(S_t,t) = 0. \tag{8.110}$$

### 8.7.2 Delta-Neutral Portfolios

We saw that at the core of managing risk from short selling, an option (or derivative) is *delta hedging*, i.e., maintaining a long position in a delta number of units of the underlying security. This process makes the portfolio with a short position in the option and delta long position in the security *delta neutral*, meaning the portfolio value is robust against sufficiently small price movements in the underlier. For instance, suppose that at the current time $t$, a portfolio is short 1 European call, and denote its value by

$$\mathbb{V}(S_t,t) = -C^E(S_t,t).$$

To make this portfolio insensitive to a small movement in the underlying security price over the next instant, long $N_S$ shares of the underlying stock. We now determine the value of $N_S$ that will accomplish the desired insensitivity. Denote the value of the expanded portfolio by

$$\widetilde{\mathbb{V}}(S_t,t) = \mathbb{V}_P(S_t,t) + N_S\, S_t = -C^E(S_t,t) + N_S\, S_t.$$

Using Equation (8.33) on page 394 to express the call price as a function of two variables $(x,t)$ with the security price as $x$, we have

$$\widetilde{\mathbb{V}}_P(x,t) = -C^E(x,t) + N_S\, x.$$

Then

$$\frac{\partial \widetilde{\mathbb{V}}}{\partial S}(S_t,t) = \frac{\partial \widetilde{\mathbb{V}}}{\partial x}\bigg|_{(x,t)=(S_t,t)} = -\frac{\partial C^E}{\partial x}\bigg|_{(x,t)=(S_t,t)} + N_S(t) = -\Delta_C(t) + N_S(t).$$

By choosing a delta hedged position, i.e., selecting $N_S = \Delta_C(t)$, we obtain

$$\frac{\partial \widetilde{\mathbb{V}}}{\partial S}(S_t,t) = 0.$$

In other words, a sufficiently small movement in the security price during the instant from $t$ to $t + dt$ causes little change in the portfolio value. The portfolio is said to be *delta neutral* at time $t$ in the sense that it is neutral on whether the stock has a small price movement up or down. Of course, the number of shares $N_S = \Delta_C(t)$ that was bought at time $t$ to make the expanded portfolio delta neutral at $t$ will change as the current time $t$ advances. This means that the expanded portfolio will have to be continuously rebalanced.

Let us now turn to delta neutrality of the general portfolio of options considered in Section 8.7.1. We are interested in how an instantaneous change $dS$ in the price of a security impacts the instantaneous change $d\widetilde{\mathbb{V}}$ in the value of the expanded portfolio. Itô's formula yields:

$$d\widetilde{\mathbb{V}}(S_t,t) = \frac{1}{2}\sigma^2 S_t^2\, \frac{\partial^2 \widetilde{\mathbb{V}}}{\partial x^2}(S_t,t)\, dt + \frac{\partial \widetilde{\mathbb{V}}}{\partial x}(S_t,t)\, dS_t + \frac{\partial \widetilde{\mathbb{V}}}{\partial t}(S_t,t)\, dt.$$

In terms of option Greeks, we obtain:

$$d\widetilde{\mathbb{V}}(S_t,t) = \widetilde{\Delta}(S_t,t)\, dS_t + \frac{1}{2}\widetilde{\Gamma}(S_t,t)\,(dS_t)^2 + \widetilde{\Theta}(S_t,t)\, dt, \tag{8.111}$$

where

$$\widetilde{\Gamma}(S_t,t) = \Gamma(S_t,t), \qquad \widetilde{\Theta}(S_t,t) = \Theta(S_t,t), \qquad (dS_t)^2 = \sigma^2 S_t^2\, dt.$$

For sufficiently small incremental changes $\delta\widetilde{\mathbb{V}}$, $\delta S$, and $\delta t$, Equation (8.111) yields

$$\delta \widetilde{V}(S_t,t) \approx \widetilde{\Delta}(S_t,t)\, \delta S_t + \frac{1}{2}\widetilde{\Gamma}(S_t,t)\, (\delta S_t)^2 + \widetilde{\Theta}(S_t,t)\, \delta t. \tag{8.112}$$

Equation (8.112) shows that the impact of a price movement $\delta S_t$ on the change $\delta \widetilde{V}(S_t,t)$ in the value of the expanded portfolio can be reduced by making the portfolio *delta neutral*, i.e., make delta vanish:

$$\widetilde{\Delta}(S_t,t) = \Delta(S_t,t) + N_S = 0.$$

This is accomplished by having a position of

$$N_S = -\Delta(S_t,t) \tag{8.113}$$

units in the underlying security. In the special case of a short position in 1 option on 1 unit of the underlier, we have

$$\mathbb{V}(S_t,t) = -f_1(S_t,t), \qquad \Delta(S_t,t) = -\Delta_1(S_t,t),$$

which gives

$$N_S = -\Delta(S_t,t) = \Delta_1(S_t,t).$$

In other words, the portfolio is made delta neutral by expanding it to include a long position with $\Delta_1(S_t,t)$ units of the underlier. In general, *a portfolio of options is made delta neutral by expanding it to include $N_S = -\Delta(S_t,t)$ units of the underlying security.*

**Example 8.10. (Delta Neutrality)** A portfolio has a short position in 500 European calls and long position in 300 puts, all on the same underlying stock, which is assumed to pay no dividend. The per-share deltas of each call and put are 0.5389 and 0.7584, respectively. How to make the portfolio delta neutral?

**Solution.** On a per-share basis, the number of calls is

$$N_1 = -500 \times 100 = -50{,}000$$

and puts is

$$N_2 = 300 \times 100 = 30{,}000.$$

By (8.113), the portfolio can be made delta neutral by expanding it to include 4,193 shares of the stock:

$$\begin{aligned} N_S &= -\Delta(S_t,t) = -\left(N_1\,\Delta_1(S_t,t) + N_2\,\Delta_2(S_t,t)\right) \\ &= -\left(-50{,}000 \times 0.5389 + 30{,}000 \times 0.7584\right) \\ &= 4{,}193. \end{aligned}$$

$\square$

A delta-neutral portfolio also has an interesting link between its gamma and theta. By the BSM p.d.e. (8.110), when the expanded portfolio is delta neutral, we have

$$\frac{1}{2}\sigma^2 S_t^2 \widetilde{\Gamma}(S_t, t) + \widetilde{\Theta}(S_t, t) = r\widetilde{V}(S_t, t).$$

Since the right-hand side $r\widetilde{V}(S_t, t)$ is a fixed value at $t$, if theta $\widetilde{\Theta}(S_t, t)$ has a sufficiently large positive (resp., negative) value, then it forces gamma $\widetilde{\Gamma}(S_t, t)$ to have a sufficiently large negative (resp., positive) value to maintain the fixed value on the right. For this reason, *theta $\widetilde{\Theta}(S_t, t)$ is also interpreted intuitively as a proxy for gamma $\widetilde{\Gamma}(S_t, t)$ in a delta-neutral portfolio*; see Hull [22, Sec. 19.7]. Note that theta measures the impact of the change in time on the change in value of the portfolio, while our discussion is focused on the impact of the underlying security's price change on the portfolio's change in value. Readers are referred to Hull [22, Sec. 19.5] and Kwok [26, Sec. 2.1.3] for more on theta.

### 8.7.3 Delta-Gamma-Neutral Portfolios

A delta-neutral portfolio is robust against a sufficiently small price change $\delta S$ of the underlier over a sufficiently small time $\delta t$. Nonetheless, Equation (8.112) shows that even for a delta-neutral portfolio, a sufficiently nontrivial price change $\delta S$ can still impact the change $\delta \widetilde{V}$ in value of the portfolio through gamma:

$$\delta\widetilde{V}(S_t, t) \approx \frac{1}{2}\widetilde{\Gamma}(S_t, t)\, (\delta S_t)^2 + \widetilde{\Theta}(S_t, t)\, \delta t. \qquad (8.114)$$

However, a delta-neutral portfolio can also be made *gamma neutral* (i.e., gamma vanishes) by adding an appropriate number of options.

Explicitly, consider a portfolio of options with the same underlying security. Assume that the portfolio has value $\widetilde{V}$, delta $\widetilde{\Delta} = 0$, and gamma $\widetilde{\Gamma}$. Also, identify an option with gamma $\Gamma_o \neq 0$ on the same underlier. Assume that the option is tradable to allow us to take a long or short position in the option. Create a new portfolio by expanding the delta-neutral portfolio to include $N_o$ units of the option. As before, we count the number of options on a per unit of the underlier; e.g., 10 calls on a stock refer to $N_o = 1{,}000$ calls counting on a per-share basis. This gives a new portfolio value at $t$ of

$$\widetilde{\widetilde{V}}(S_t, t) = \widetilde{V}(S_t, t) + N_o\, f_o(S_t, t),$$

where $f_o(S_t, t)$ is the value of the option at $t$. The gamma of the new portfolio is

$$\widetilde{\widetilde{\Gamma}}(S_t, t) = \frac{\partial^2 \widetilde{\widetilde{V}}}{\partial S^2}(S_t, t) = \widetilde{\Gamma}(S_t, t) + N_o\, \Gamma_o(S_t, t) = 0.$$

The gamma $\widetilde{\widetilde{\Gamma}}(S_t, t)$ vanishes exactly when

$$N_o = -\frac{\widetilde{\Gamma}(S_t,t)}{\Gamma_o(S_t,t)}. \tag{8.115}$$

Note that we cannot accomplish this gamma-neutrality by adding a certain number of units of the underlying security since the gamma of the underlier is zero.

The reader may have noticed from (8.114) that though $N_o$ in (8.115) makes the new portfolio gamma neutral, it causes the delta of the new portfolio to be nonzero:

$$\widetilde{\widetilde{\Delta}}(S_t,t) = \widetilde{\Delta}(S_t,t) + N_o\,\Delta_o(S_t,t) = N_o\,\Delta_o(S_t,t),$$

where $\widetilde{\Delta}(S_t,t) = 0$ and $\Delta_o(S_t,t)$ is the delta of the option. To make the new portfolio delta neutral, we take a position

$$N_{\mathrm{mod}} = -N_o\,\Delta_o(S_t,t) \tag{8.116}$$

in the underlying security. The new portfolio then has the following value at $t$:

$$\widetilde{\widetilde{V}}_{\mathrm{mod}}(S_t,t) = \widetilde{V}(S_t,t) + N_o\,f_o(S_t,t) + N_{\mathrm{mod}}\,S_t, \tag{8.117}$$

where $N_o$ is given by (8.115) and $N_{\mathrm{mod}}$ is fixed. View the value of the modified portfolio as a function of $(x,t)$, namely,

$$\widetilde{\widetilde{V}}_{\mathrm{mod}}(x,t) = \widetilde{V}(x,t) + N_o\,f_o(x,t) + N_{\mathrm{mod}}\,x.$$

Then by (8.115), (8.116), and (8.117), the modified portfolio is both delta neutral and gamma neutral at $(x,t) = (S_t,t)$:

$$\widetilde{\widetilde{\Delta}}_{\mathrm{mod}}(S_t,t) = \widetilde{\Delta}(S_t,t) + N_o\,\Delta_o(S_t,t) + N_{\mathrm{mod}} = 0$$

and

$$\widetilde{\widetilde{\Gamma}}_{\mathrm{mod}}(S_t,t) = \widetilde{\Gamma} + N_o\Gamma_o = 0.$$

**Example 8.11. (Gamma Neutrality)** A delta neutral portfolio of options on the same stock has a gamma $\widetilde{\Gamma} = -2{,}390$ (per \$), and a tradable call on the exact stock has a delta $\Delta_C = 0.4681$ and gamma $\Gamma_C = 0.3467$ (per \$). How can we construct a delta- and gamma-neutral portfolio using the original portfolio and the call?

**Solution.** By (8.117), we expand the original delta-neutral portfolio to include $N_o$ units of the call per share and $N_{\mathrm{mod}}$ shares of the underlying stock. By (8.115) and (8.116), we have:

$$N_o = -\frac{\widetilde{\Gamma}(S_t,t)}{\Gamma_C(S_t,t)} = -\frac{-2{,}390}{0.3467} = 6{,}893.57$$

and

$$N_{\mathrm{mod}} = -N_{\mathrm{o}} \Delta_{\mathrm{o}}(S_t, t) = -6{,}893.57 \times 0.4681 = -3{,}226.88.$$

In other words, buy 6,893.57 calls on a per-share basis (or 68.9357 calls in round lots) and short sell 3,226.88 shares of the stock and include these positions in the original portfolio. The resulting portfolio will be delta- and gamma-neutral.                                                                      □

Readers are referred to Hull [22, Chap. 19] for more on option Greeks and their applications.

## 8.8 The BSM Model Versus Market Data

We shall take a closer look at some of the assumptions of the BSM model to see how they hold up against market data. A fundamental assumption of the BSM model is that security prices follow geometric Brownian motion. Two consequences are that security prices are continuous with probability 1, i.e., there are almost surely no jump discontinuities, and the log returns of security prices are normal. We shall use the S&P 500 index as an example to illustrate that real-world security prices can jump and exhibit log return behavior that deviates from normality.

### 8.8.1 Jumps in Security Prices

Our first observation from the market data is that security prices can have jumps. In particular, we consider the daily closing prices of the S&P 500 index from January 3, 1950, to January 2, 2015.[17] These prices include the stock market crash of October 19, 1987, which was marked by a negative jump (a drop) in the price of the index that day. The opening price was 282.70, which was also the closing price on the previous trading date of October 16, 1987. The S&P 500 closed at 224.84 on October 19, 1987, creating a drastic fractional percentage drop of -20.4669% in the price or a log return drop of -22.8997%. The negative jump is shown in Figure 8.4.

The daily log returns for the data in the top left panel of Figure 8.4 are shown in the bottom panel of the figure. The longest negative spike is due to the crash of October 19, 1987. Note that the second longest spike and the accompanying volatility reflect the 2008 financial crisis, which actually began in 2007 and peaked in the latter part of 2008. Significant damaging effects spilled into early 2009, and an economic slowdown continued into 2012.

---

[17] See Yahoo's historical prices for ^GSPC.

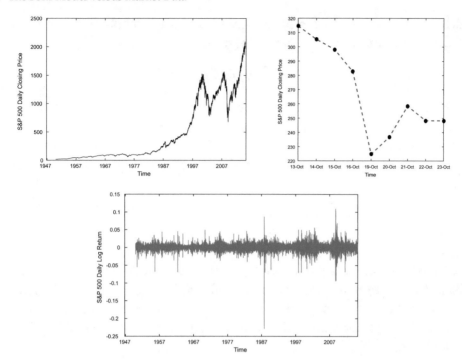

**Fig. 8.4** Top panels: S&P 500 index closing prices from January 3, 1950, to January 2, 2015 (top left) and from October 13, 1987, to October 23, 1987 (top right). The stock market crash of October 19, 1987, is clearly seen in the middle panel. The S&P 500 index closed with a negative jump of about -20.5% relative to the opening price that day, which is the same as the closing price on the previous trading day of October 16. Note that there was no trading on October 17 and 18 (a weekend). Bottom panel: daily log returns of the S&P 500 based on prices in the top left panel. The longest negative spike is due to the stock market crash on October 19. The next longest negative spike is connected with the 2008 financial crisis that became highly pronounced in the latter part of 2008

The evidence for jumps is not restricted to the S&P 500 index, but shows up in other securities and even in intraday trading. The study of jumps has become a significant area of research; e.g., see the discussion by Taylor [41, Sec. 13.6] and references therein.

## 8.8.2 Skewness and Kurtosis in Security Log Returns

Our second observation is that the log return of security prices is not necessarily normal. Before showing this with S&P 500 data, let us review some of the moments of a random variable. Let $X$ be a random variable with a p.d.f. $f(x)$. Denote the first moment (i.e., mean) of $X$ by

$$\mu_X = \mathbb{E}(X),$$

which measures the probability-weighted center of the possible values of $X$, and the second moment (i.e., variance) of $X$ by

$$\sigma_X^2 = \text{Var}(X),$$

which measures the dispersion or spread of the possible values of $X$ to the left and right of the mean $\mathbb{E}(X)$. In the BSM model, the log return of the price of a security is normal and so is completely characterized by its first and second moments. However, we shall see from data that all the higher moments cannot be ignored.

## Skewness

Given that the log return of a security in the BSM model is normal, the shape of its p.d.f. is symmetric about the vertical line through its center (i.e., mean). Deviations of the graph of $f(x)$ from a symmetric shape about its center can be measured through the standardized third moment of $X$. It is called the *skewness* of $X$ and defined by

$$\text{skew}(X) = \mathbb{E}\left(\left(\frac{X - \mu_X}{\sigma_X}\right)^3\right) = \frac{1}{\sigma_X^3} \int_{-\infty}^{\infty} (x - \mu_X)^3 f(x)\,dx.$$

When the p.d.f. is symmetric about the center, we have $\text{skew}(X) = 0$.[18] Equivalently, if $\text{skew}(X) \neq 0$, then there is a break in the symmetry about the center. Intuitively, if $\text{skew}(X) > 0$, then a unimodal (single peak) p.d.f. will have a more elongated right tail area like the dashed p.d.f. in Figure 8.5. In this case, we say that $f(x)$ is *positively skewed*. The opposite happens for $\text{skew}(X) < 0$, in which case we call $f(x)$ *negatively skewed*; note the stretched left tail area of the dotted p.d.f. in Figure 8.5.

## Kurtosis

The standardized fourth moment of $X$, called the *kurtosis* of $X$, measures the flatness or sharpness of the central peak of $f(x)$ relative to that of a normal p.d.f. determined by the mean $\mu_X$ and standard deviation $\sigma_X$ of $X$. The *kurtosis* of $X$ is defined by

$$\text{kurt}(X) = \mathbb{E}\left(\left(\frac{X - \mu_X}{\sigma_X}\right)^4\right) = \int_{-\infty}^{\infty} \left(\frac{x - \mu_X}{\sigma_X}\right)^4 f(x)\,dx.$$

---

[18] This is because the portion of $\text{skew}(X)$ from $-\infty$ to $0$ cancels the part from $0$ to $\infty$. In fact, all odd power moments $\mathbb{E}\left((X - \mu_X)^{2n+1}\right)$ vanish if the p.d.f. is symmetric about the line $x = \mu_X$.

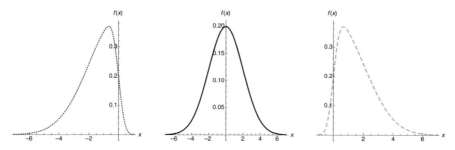

**Fig. 8.5** The three p.d.f.s are due to a skew-normal distribution with location parameter 0, scale parameter 2, and shape parameters -6 (dotted graph), 0 (solid graph), and 6 (dashed graph). The dotted graph has a negative skewness of -0.891159, the solid graph has skewness 0 (the skew-normal distribution reduces to a normal in this case), and the dashed graph has a skewness of 0.891159

Since the kurtosis of all normal random variables is 3, it is convenient to use that kurtosis as a benchmark and define an *excess kurtosis* by

$$\text{ekurt}(X) = \text{kurt}(X) - 3.$$

For the cases $\text{ekurt}(X) < 0$, $\text{ekurt}(X) = 0$, and $\text{ekurt}(X) > 0$, the p.d.f. $f(x)$ is called *platykurtic* ("platy" means flat; think platypus), *mesokurtic* ("meso" means middle), and *leptokurtic* ("lepto" means thin), respectively. In interpreting kurtosis of a random variable $X$, we shall always compare the p.d.f. $f(x)$ of $X$ with the p.d.f. of the normal random variable determined by the mean $\mu_X$ and standard deviation $\sigma_X$ of $X$. Various kurtosises are illustrated in the left graphs of Figure 8.6 using three symmetric, unimodal p.d.f.'s with identical mean 0 and standard deviation 1. The middle p.d.f. (solid curve) is mesokurtic since it is a standard normal. The leptokurtic p.d.f. (dashed curve) has a thinner peak and thicker (heavier) tails than the standard normal, while the platykurtic p.d.f. (dotted curve) has a flatter peak and thinner tail than the standard normal. In addition, observe that the leptokurtic and platykurtic p.d.f.'s intersect the corresponding normal p.d.f. twice on the right of the mean and twice on the left of the mean. The lower crossing on each side results in a tail thicker than that of the associated normal; see the right graphs in Figure 8.6. These properties are not uncommon for unimodal p.d.f.'s. See DeCarlo [10] and Balanda and MacGillivray [2] for more.

**Remark 8.11.** It is important not to confuse effects due to variance with that due to kurtosis. For example, Figure 8.7 shows the two normal p.d.f.'s in comparison with the standard normal. Though the dashed p.d.f. has a narrower peak and the dotted one a flatter peak, all three p.d.f.'s have the same excess kurtosis of 0. Consult DeCarlo [10] for some of the pitfalls in the interpretation of kurtosis. □

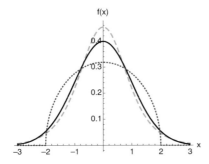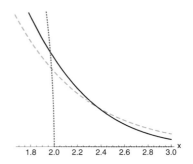

**Fig. 8.6** Left plots: the p.d.f.'s all have the same mean 0 and standard deviation 1 and are symmetric about the center (zero skewness). The solid p.d.f. is the standard normal, which has excess kurtosis 0 (mesokurtic). The dotted p.d.f. has excess kurtosis -1 (platykurtic) with a flatter peak and thinner tail than the standard normal. It is given by a Wigner semicircle density with radius 2 centered at the origin. The dashed p.d.f. has excess kurtosis 1.2 (leptokurtic) with a thinner peak and thicker tails than the standard normal. It is given by the logistic density with mean 0 and scale parameter $\sqrt{3}/\pi$. Right plots: zoom-in of the right tails of the p.d.f.s in the left plots. Observe that the leptokurtic (dashed) and platykurtic (dotted) p.d.f.'s have tails that are thicker and thinner, respectively, than that of the associated normal

### Skewness and Kurtosis in the S&P 500 Index Daily Log Returns

Turning now to the daily log returns of the S&P 500 based on prices from January 3, 1950, to January 2, 2015, the frequency histogram of the log returns reveals both asymmetry and leptokurtosis. Figure 8.8 shows the frequency histogram. The skewness is -1.0281, which is reflected in the left tail having a more elongated area than the right tail.

The excess kurtosis is 27.6769, which is significantly above that of the associated normal. Indeed, the tails of the S&P 500 daily log returns are thicker (heavier) than those of the corresponding normal. In other words, there is a higher probability (than in the case of a normal) of having extreme values in the daily log return. Figure 8.9 shows the thicker tails by zooming in on his-

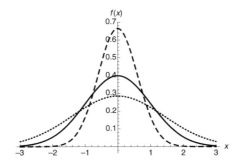

**Fig. 8.7** The peakedness and flatness of the dashed and dotted p.d.f.'s relative to the standard normal (solid p.d.f.) should not be confused with leptokurtosis and platykurtosis, respectively. All three graphs have the same excess kurtosis 0, but different standard deviations: 0.6 (dotted graph), 1 (solid graph), and 1.4 (dashed graph)

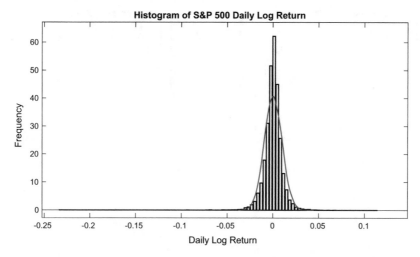

**Fig. 8.8** The histogram shows the frequency of the log returns of the S&P 500 index based on prices from January 3, 1950, to January 2, 2015. The mean and standard deviation are 0.0002945 and 0.0097, respectively. The solid curve is the p.d.f. of a normal distribution with mean 0.0002945 and standard deviation 0.0097. The skewness is -1.0281 and the excess kurtosis is 27.6769

togram and using a QQ-plot. A point $(x, y)$ on the QQ-plot is a pair of quantiles, where $x$ is a quantile of the standard normal and $y$ is the corresponding quantile in the standardized S&P 500 daily return data. For example, if $x$ is the 30% quantile of the standard normal distribution, which means that there is at most a 30% probability that $X < x$ and 70% probability that $X > x$, then the corresponding $y$-coordinate will be the 30% quantile of the standardized S&P daily log returns.[19] Indeed, if the standardized S&P daily log returns are standard normal, then its quantiles will be the same as for the standard normal, i.e., the QQ-plot will be the 45° line $x = y$. The deviations of the S&P daily log returns from normality are evident in Figure 8.9.

Overall, the data in Figures 8.4, 8.8, and 8.9 support that the *the daily log returns of the S&P 500 are not normal*, i.e., the prices do not follow geometric Brownian motion. Deviations of the log returns from normality show up not only in the S&P 500 but also in common stocks, currencies, etc.; see the introduction by Epps [15, Chaps. 8, 9] and references within.

### 8.8.3 Volatility Skews

A central assumption of the BSM model is that the inputs $S_t, K, \sigma, r, \tau, q$ are known. However, the volatility $\sigma$ is not market observable. Nonetheless, it can be inferred from the BSM pricing formula. For example, if we assume that the BSM call price (8.37), i.e.,

---

[19] See Section 4.2.4 (page 174) for more on quantiles.

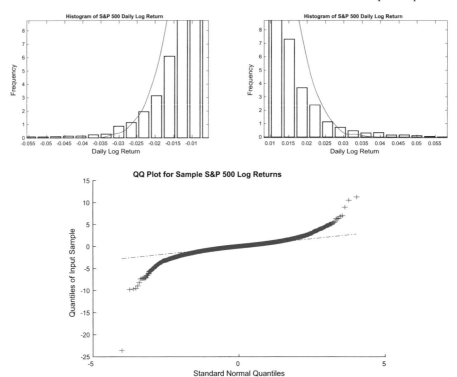

**Fig. 8.9** Top: The histograms' zoom-in on portion of the left and right tails of the histogram in Figure 8.8. The solid curve is the p.d.f. of a normal distribution with mean and standard deviation given by the sample data of the S&P 500 daily log returns. Notice that the tails are thicker than the tails of of corresponding normal. Bottom: QQ-plot of the standardization of the log returns of the S&P 500 log returns in Figure 8.8. The left and right tails are heavier than those of the standard normal. The excess kurtosis is 27.6208

$$C^E(S_t, K, \sigma, r, \tau, q) = S_t e^{-q\tau} \mathsf{N}(d_+(S_t, \tau)) - K e^{-r\tau} \mathsf{N}(d_-(S_t, \tau)),$$

truly models the prices of European calls in the marketplace, then given the current market price $C^E_{\text{market}}(t)$ of a European call, we can solve the equation

$$C^E_{\text{market}}(t) = C^E(S_t, K, \sigma, r, \tau, q)$$

implicitly for $\sigma$. The resulting value of $\sigma$ is called the *implied volatility* and denoted by $\sigma_{\text{im}}$. In other words, the implied volatility is the volatility that makes the theoretical BSM call price equal to the market price.

How do we know that one and only one implied volatility corresponds to the market price of a European call? There is actually a 1-1 correspondence between the possible prices of a European call and the possible volatilities of its underlying security. To see this, observe that the *vega* of the call, which is defined by

$$\wedge_C = \frac{\partial C^E}{\partial \sigma}(S_t, t),$$

is positive:

$$\wedge_C = S_t e^{-q\tau} \sqrt{\tau} N'(d_+(S_t,t))) > 0.$$

The European call price is a strictly increasing function of $\sigma$. This implies that for each possible European call price, there is a unique volatility and vice versa. For this reason, one can freely switch between volatilities and European call prices. A similar result holds for European puts, which have the same vega as a European call. Hence, for each market price of a European call (or put), there is a unique implied volatility, and for each implied volatility, only one market price can correspond to it.

Unfortunately, implied volatility opens up several concerns with the BSM model. The BSM model assumes that the volatility $\sigma$ of the underlying security is an inherent property of the underlier that is constant during the life of a call or put. In other words, according to the BSM model, the value of $\sigma$ is not only unchanging; it is also independent of contractual elements like the option's strike price $K$, expiration date $T$, and type (call or put). Explicitly, if the BSM formulas (8.37) and (8.39) truly model the respective prices of European calls and puts in the marketplace, then:

➤ The implied volatility $\sigma_{im}$ is the same for a European call (or put) with inputs $S_t, r, \tau, q$, but different strike prices $K_i$. In the BSM model, the graph of implied volatility as a function of the $K_i$'s is a horizontal line through the $y$-value $\sigma_{im}$.

➤ The implied volatility $\sigma_{im}$ is the same for a European call (or put) with inputs $S_t, K, r, q$, but different times $\tau_i$ to expiration. In the BSM model, the graph of implied volatility as a function of the $\tau_i$'s is a horizontal line through the $y$-value $\sigma_{im}$.

➤ The implied volatility $\sigma_{im}$ is the same for a European call and put with the same underlying security and identical inputs $S_t, r, q$, but different strike prices $K_i$ and times $\tau_i$ to maturity.

Unfortunately, all three of the properties above are known to be inconsistent with data. For example, data on equity options show that the implied volatility $\sigma_{im}$ varies as the strike price and time to expiration change. The resulting plot of $\sigma_{im}$ as a function of $(K_i, \tau_i)$ does not lie on a plane as predicted by the BSM model, but on a curved discrete surface, called a *volatility surface*. Moreover, for a fixed time to expiration, some of the plots of $\sigma_{im}$ as a function of $K_i$ even show shapes resembling smiles—hence, the name *volatility smiles*; see Figure 8.10. In addition, the graph of the implied volatility as a function of strike price changes as one switches from a European call to a European put with the same underlier and inputs.

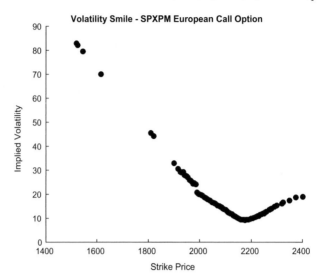

**Fig. 8.10** Implied volatility in percent versus strike price for the S&P 500 index European call option with PM settlement (SPXPM). The data is from Yahoo! Finance and based on values at 4:49 p.m. EDT on June 26, 2015, when the index was at 2,101.59. The expiration date is July 17, 2015. The data shows that the implied volatility as a function of strike price is not constant as predicted by the BSM model. The graph has a volatility smile

There is a vast literature on implied volatility. See, for example, the texts by Hull [22, Chap. 20] and McDonald [27, Chap. 23] and the lecture by Rachev[35]. An extensive introduction to the volatility surface is given by Gatheral [16].

## 8.9 A Step Beyond the BSM Model: Merton Jump Diffusion

Several proposed modifications of the BSM model allow for more heavy tails, peakedness, and volatility skews in security prices. By mixing geometric Brownian motion (which is a diffusion) with jump discontinuities, Merton [29] introduced in 1976 a model that naturally extends the BSM theoretical framework and addresses some of the issues facing the BSM model. We give an introduction to this model.

For easy reference in the discussion to follow, we recap the BSM theoretical prices for European calls and puts given in Equations (8.37) and (8.39):

$$C^E(S_t, K, \sigma, r, \tau, q) = S_t e^{-q\tau} N(d_+(S_t, \tau)) - K e^{-r\tau} N(d_-(S_t, \tau)),$$

and

$$P^E(S_t, K, \sigma, r, \tau, q) = K e^{-r\tau} N(-d_-(S_t, \tau)) - S_t e^{-q\tau} N(-d_+(S_t, \tau)),$$

where $\tau = T - t$.

## 8.9.1 Poisson Processes

The number of jumps in the future price of a security is random and assumed to follow a Poisson process, which we now introduce—see, for example, Privault [34, Chap. 14]. First, we can view the possible outcomes of a Poisson process at a fixed moment of time in terms of successes and failures. In our context, define a *success* at time $t$ as the arrival of information that causes the price of the security to jump in value. Such information can be important news pertaining to earnings, sector outlook, serious macroeconomic concerns, etc. We assume a 1-1 correspondence between successes and the security's price jumps and so freely identify a success with a price jump.

Let the current time be 0 and let $\mathfrak{N}_t$ be the number of price jumps during $[0,t]$, i.e., over the next $t$ years. The increment $\mathfrak{N}_t - \mathfrak{N}_x$, where $0 \leq x < t$, is then the number of price jumps during the time interval $(x,t]$. Note that the times when future price jumps occur are not known a priori. We assume that the stochastic process $\{\mathfrak{N}_t\}_{t\geq 0}$ is a *Poisson process*, which means that the following properties hold:

➤ The number of jumps at the starting time 0 is zero, i.e., $\mathfrak{N}_0 = 0$, and the mean number of price jumps per year[20] is known and denoted by $\lambda$. The parameter $\lambda$ is called the *intensity* of the Poisson process.

➤ The increments are stationary: for all $0 \leq x < t$ and all u such that $x + u \geq 0$ and $t + u \geq 0$, we have

$$\mathfrak{N}_t - \mathfrak{N}_x \overset{\mathrm{d}}{=} \mathfrak{N}_{t+u} - \mathfrak{N}_{x+u}. \tag{8.118}$$

In other words, the probability of $n$ price jumps during a time period[21] $\mathfrak{T}$ starting now is the same as during a time period of the same length $\mathfrak{T}$ starting at any other time. By choosing $u = -x$ in (8.118), we can always shift to an interval starting out at 0:

$$\mathfrak{N}_t - \mathfrak{N}_x \overset{\mathrm{d}}{=} \mathfrak{N}_{t-x}. \tag{8.119}$$

➤ The increments are independent: for every sequence of times $0 = t_0 < t_1 < \cdots < t_k$, the increments

$$\mathfrak{N}_{t_1} - \mathfrak{N}_{t_0}, \quad \mathfrak{N}_{t_2} - \mathfrak{N}_{t_1}, \quad \ldots, \quad \mathfrak{N}_{t_k} - \mathfrak{N}_{t_{k-1}},$$

---

[20] Recall that our assumed unit of time is a year.

[21] Recall that in our usage a *time period* is the length of a time interval.

are independent. In other words, the number of price jumps during a given time interval I is independent of the number of price jumps during a time interval that does not overlap[22] with I.

➣ At most one jump can occur during $[t, t + dt]$:

$$d\mathfrak{N}_t = \mathfrak{N}_{t+dt} - \mathfrak{N}_t = \begin{cases} 1 & \text{if a jump occurs during } [t, t + dt] \\ \\ 0 & \text{if no jump occurs during } [t, t + dt]. \end{cases} \tag{8.120}$$

In addition, during any instant $dt$, the probability of one price jump is $\lambda\, dt$ and the probability of more than one jump is zero:

$$d\mathfrak{N}_t = \begin{cases} 1 & \text{with probability } \lambda\, dt \\ \\ 0 & \text{with probability } 1 - \lambda\, dt \\ \\ k > 1 & \text{with probability } 0. \end{cases}$$

Using the Itô multiplication rule $(dt)^a = 0$ for $a > 0$, we see:

$$\mathbb{E}_{\mathbb{P}_\lambda}(d\mathfrak{N}_t) = \lambda\, dt = \operatorname{Var}(d\mathfrak{N}_t), \tag{8.121}$$

where $\mathbb{P}_\lambda$ is the probability measure of the Poisson process.

In general, for a fixed but arbitrary interval $(u, t]$, where $u \geq 0$, the probability that the number $\mathfrak{N}_t - \mathfrak{N}_u$ of price jumps during $(u, t]$ is $n$ is given by a Poisson distribution with parameter $\lambda(t - u)$ (e.g., Durrett [13, Sec. 3.6.3]). In other words, the probability measure is

$$\mathbb{P}_\lambda\left(\mathfrak{N}_t - \mathfrak{N}_u = n\right) = e^{-\lambda(t-u)}\frac{\left(\lambda(t - u)\right)^n}{n!} \qquad (n = 0, 1, 2, \dots). \tag{8.122}$$

Note that though a Poisson process has discrete values, it is a continuous-time parameter stochastic process. Additionally, the inter-arrival times of the jumps of a Poisson process must be independent exponential random variables with parameter $\lambda$.

### 8.9.2 The MJD Stochastic Process

Merton models the instantaneous change in the price at a general time $t$ as having contributions from a no-jump component determined by geometric Brownian motion and a jump component.

---

[22] Two intervals are *nonoverlapping* if their interiors are disjoint.

## No-Jump Case

When a security's price has no possibility of jumping, denote its price at $t$ by $\widetilde{S}_t$. This case corresponds to $\lambda = 0$ (i.e., the mean number of jumps per year is zero) and has an instantaneous capital-gain return determined by geometric Brownian motion:

$$\frac{d\widetilde{S}_t}{\widetilde{S}_t} = (m - q)\,dt + \sigma\,d\mathfrak{B}_t. \tag{8.123}$$

This is solved by

$$\widetilde{S}_t = S_0\, e^{\mu_{\text{RW}}\, t + \sigma \mathfrak{B}_t} \qquad\qquad (t \geq 0), \tag{8.124}$$

where the current time is at $0$, $S_0$ is the known current price, and

$$\mu_{\text{RW}} = m - q - \frac{\sigma^2}{2}.$$

Given that the history of the security price is known up to time $t$, the expected instantaneous capital-gain return at $t$ is then:

$$\mathbb{E}\left( \frac{d\widetilde{S}_t}{\widetilde{S}_t} \,\Big|\, \mathcal{F}_t \right) = (m - q)\,dt. \tag{8.125}$$

The $\sigma$-algebra $\mathcal{F}_t$ in (8.125) is generated by standard Brownian motion $\mathfrak{B}$ up to time $t$, so in the conditional expectation we know $\widetilde{S}_t$, but not $\widetilde{S}_{t+dt}$, which appears in $d\widetilde{S}_t = \widetilde{S}_{t+dt} - \widetilde{S}_t$.

## The Merton Jump-Diffusion s.d.e.

Assume it is possible for the security price to have jumps and, for this situation, denote its price at $t$ by $S_t$. Let us explore the instantaneous capital-gain return when jumps are possible, but not guaranteed. First, if there is no jump during $[t, t + dt]$, then the Merton model assumes that the capital-gain return is determined by a geometric Brownian motion:

$$\frac{dS_t}{S_t} = (\widehat{m} - q)\,dt + \sigma\,d\mathfrak{B}_t \qquad\qquad \text{(no jump during } [t, t + dt]). \tag{8.126}$$

The instantaneous total mean return $\widehat{m}$ of the security should not be confused with $m$, which is the instantaneous total mean return of a security with no possibility of jumps. Second, suppose there is a jump (i.e., discontinuity) in the price at $t$. By our assumptions, it is the only jump during $[t, t + dt]$. Then the price has two values at $t$ determined by the left and right limits at $t$. Let $S_{t-}$ be the price of the security just before the jump, i.e., the left-hand limit price at $t$:

$$S_{t^-} = \lim_{t \to t^-} S_t.$$

The price of the security at $t$, given a jump at $t$, is defined to be the right-hand limit of the price at $t$ and denoted by $S_t$:

$$S_t = S_{t^+} = \lim_{t \to t^+} S_t.$$

Express the price jump as a multiple $J_t > 0$ of $S_{t^-}$:

$$S_t = J_t\, S_{t^-} \qquad \text{(a jump occurs at } t, \text{ where } J_t > 0). \qquad (8.127)$$

We shall call $J_t$ the *jump factor*. The percentage change in the price at $t$ due only to the jump is then

$$\frac{S_t - S_{t^-}}{S_{t^-}} = J_t - 1 \qquad \text{(due only to the jump at } t).$$

We can also think of $J_t - 1$ as the fractional size of the jump at $t$, where a negative value is a downward jump. On the other hand, if we do not know that a jump occurs at $t$, but only that it is possible, then the percentage change contribution coming only from the jump is modeled by

$$\frac{S_t - S_{t^-}}{S_{t^-}} = (J_t - 1)\, d\mathfrak{N}_t \qquad \text{(contribution only from a possible jump at } t).$$

$$(8.128)$$

*We assume that $J_t$ and $d\mathfrak{N}_t$ are independent.* By this assumption and (8.121), the expected capital gain only from the jump is then

$$\mathbb{E}\Big( (J_t - 1)\, d\mathfrak{N}_t \Big) = \mathbb{E}(J_t - 1)\, \mathbb{E}(d\mathfrak{N}_t) = \lambda \kappa\, dt, \qquad \kappa = \mathbb{E}(J_t - 1), \quad (8.129)$$

where $\kappa$ is the mean gain factor in the price due to the jump.

The Merton model assumes that the instantaneous capital-gain return at $t$ is given by the sum of a no-jump component (8.126) just before $t$ and a possible jump component (8.128) at $t$. The two equations can then be combined at a general time $t$ as follows:

$$\frac{dS_t}{S_{t^-}} = (\widehat{m} - q)\, dt + \sigma\, d\mathfrak{B}_t + (J_t - 1)\, d\mathfrak{N}_t \qquad (t > 0). \qquad (8.130)$$

*Another key assumption of the Merton model is that the risk due to jumps is idiosyncratic* (see page 151). For example, it can be due to company-specific events like the sudden finding of corruption among the senior management that threatens to bring down the company. The model assumes there is no reward or risk premium for jumps since, as we learned in the Markowitz theory, such risk can be diversified away. Consequently, *the expected instantaneous capital-gain return at any time $t$, conditioned on the security price being known up to $t^-$, is given by the* *case (8.125) when there is no jump possibility:*

$$\mathbb{E}\left(\frac{dS_t}{S_{t-}}\,\middle|\,\mathcal{F}_{t-}^{\mathrm{MJD}}\right) = (m - q)\,dt \qquad (t > 0). \qquad (8.131)$$

In other words, whether the mean jump factor $\mathbb{E}(J_t)$ is positive or negative will not impact the expected capital gain. The $\sigma$-algebra $\mathcal{F}_{t-}^{\mathrm{MJD}}$ is generated by standard Brownian motion, the Poisson process, and the jump-factor process.

Computing the left-hand side of (8.131) using (8.130), it follows:

$$\widehat{m} = m - \lambda\kappa.$$

Hence,

$$\frac{dS_t}{S_{t-}} = (m - q - \lambda\kappa)\,dt + \sigma\,d\mathfrak{B}_t + (J_t - 1)\,d\mathfrak{N}_t \qquad (t > 0). \quad (8.132)$$

We call (8.132) a *Merton jump-diffusion s.d.e.* Note that if there is no jump at $t$, then $S_{t-} = S_t$ and $d\mathfrak{N}_t = 0$, while for a jump at $t$, we have $d\mathfrak{N}_t = 1$. In other words,

$$dS_t = \begin{cases} (m - q - \lambda\kappa)\,S_t\,dt + \sigma\,S_t\,d\mathfrak{B}_t & \text{given no jump at } t \\[2mm] (m - q - \lambda\kappa)\,S_{t-}\,dt + \sigma\,S_{t-}\,d\mathfrak{B}_t + (J_t - 1)\,S_{t-} & \text{given a jump at } t. \end{cases}$$
$$(8.133)$$

Note that if there is no jump at $t$, then (8.133) reduces to a geometric Brownian motion with drift parameter $\mu_{\mathrm{RW}} - \lambda\kappa$ and volatility parameter $\sigma$. If there is no possibility for the security price to jump, then $\lambda = 0$ and $d\mathfrak{N}_t = 0$, and so (8.133) reduces to geometric Brownian motion with drift $\mu_{\mathrm{RW}}$ and volatility $\sigma$.

### Solving the Merton Jump-Diffusion s.d.e.

We shall solve the Merton jump-diffusion s.d.e. by drawing on the fact that the price between jumps is a geometric Brownian motion, while the price at a jump time $t$ is $J_t$ times the price just before $t$.

Suppose that there are price jumps at times $\mathbb{T}_\ell$ with

$$0 < \mathbb{T}_1 < \cdots < \mathbb{T}_\ell < \mathbb{T}_{\ell+1} < \cdots < \mathbb{T}_{\mathfrak{N}_t} \leq t.$$

Recall that $\mathfrak{N}_t$ is the random number of price jumps in $[0, t]$ and so $\mathbb{T}_{\mathfrak{N}_t}$ is the time of the last jump in $[0, t]$. Note that for times $0 \leq v < \mathbb{T}_1$ there are no jumps in $[0, v]$, while for $\mathbb{T}_\ell \leq v < \mathbb{T}_{\ell+1}$ the intervals $[0, \mathbb{T}_\ell]$ and $[0, v]$ have the same number of jumps, namely, $\mathfrak{N}_v$. Denote the jump factor at $\mathbb{T}_\ell$ by $J_\ell = J_{\mathbb{T}_\ell}$ for $\ell = 1, \ldots, \mathfrak{N}_t$.

*Assume that the jump factors* $J_1, \ldots, J_{\mathfrak{N}_t}$ *are i.i.d. and*

$$\kappa = \mathbb{E}(J_\ell) - 1 \qquad\qquad (\ell = 1, \ldots, \mathfrak{N}_t).$$

Let us determine the security's price as $v$ varies across $[0, t]$:

➤ $0 \le v < \mathbb{T}_1$

Since no jump occurs during this interval, the security price follows a geometric Brownian motion

$$S_v = S_0 e^{(\mu_{\text{RW}} - \lambda \kappa) v + \sigma \mathfrak{B}_v} \qquad\qquad (0 \le v < \mathbb{T}_1). \qquad (8.134)$$

➤ $v = \mathbb{T}_1$

The price is $S_{\mathbb{T}_1} = J_1 S_{\mathbb{T}_1^-}$. But the price $S_{\mathbb{T}_1^-}$ just before $\mathbb{T}_1$ has no jumps and so is given by taking the left-hand limit of (8.134) at $\mathbb{T}_1$:

$$S_{\mathbb{T}_1^-} = S_0 e^{(\mu_{\text{RW}} - \lambda \kappa) \mathbb{T}_1 + \sigma \mathfrak{B}_{\mathbb{T}_1}}.$$

Consequently:

$$S_{\mathbb{T}_1} = S_0 e^{(\mu_{\text{RW}} - \lambda \kappa) \mathbb{T}_1 + \sigma \mathfrak{B}_{\mathbb{T}_1}} J_1. \qquad (8.135)$$

➤ $\mathbb{T}_1 < v < \mathbb{T}_2$

Since there is no jump during this interval, the price at $v$ is a geometric Brownian motion with initial price $S_{\mathbb{T}_1}$:[23]

$$S_v = S_{\mathbb{T}_1} e^{(\mu_{\text{RW}} - \lambda \kappa)(v - \mathbb{T}_1) + \sigma(\mathfrak{B}_v - \mathfrak{B}_{\mathbb{T}_1})} \qquad\qquad (\mathbb{T}_1 < v < \mathbb{T}_2).$$

Equation (8.135) yields:

$$S_v = S_0 e^{(\mu_{\text{RW}} - \lambda \kappa) v + \sigma \mathfrak{B}_v} J_1 \qquad\qquad (\mathbb{T}_1 < v < \mathbb{T}_2). \qquad (8.136)$$

➤ $v = \mathbb{T}_2$

We have $S_{\mathbb{T}_2} = J_2 S_{\mathbb{T}_2^-}$, where $S_{\mathbb{T}_2^-}$ is obtained by taking the left-hand limit of (8.136) at $\mathbb{T}_2$:

$$S_{\mathbb{T}_2^-} = S_0 e^{(\mu_{\text{RW}} - \lambda \kappa) \mathbb{T}_2 + \sigma \mathfrak{B}_{\mathbb{T}_2}} J_1.$$

Hence:

$$S_{\mathbb{T}_2} = S_0 e^{(\mu_{\text{RW}} - \lambda \kappa) \mathbb{T}_2 + \sigma \mathfrak{B}_{\mathbb{T}_2}} J_1 J_2. \qquad (8.137)$$

➤ $\mathbb{T}_{\mathfrak{N}_t - 1} < v < \mathbb{T}_{\mathfrak{N}_t}$

Continuing across the remaining jump times, we have geometric Brownian motion during the given interval:

$$S_v = S_{\mathbb{T}_{\mathfrak{N}_t - 1}} e^{(\mu_{\text{RW}} - \lambda \kappa)(v - \mathbb{T}_{\mathfrak{N}_t - 1}) + \sigma(\mathfrak{B}_v - \mathfrak{B}_{\mathbb{T}_{\mathfrak{N}_t - 1}})} \qquad\qquad (\mathbb{T}_{\mathfrak{N}_t - 1} < v < \mathbb{T}_{\mathfrak{N}_t}),$$

where

---

[23] Note: if $S(t) = S(0) e^{\mu t + \sigma \mathfrak{B}_t}$, then $S(t) = S(v) e^{\mu(t - v) + \sigma(\mathfrak{B}_t - \mathfrak{B}_v)}$, where $t \ge 0$ and $v \ge 0$.

$$S_{\mathbb{T}_{\mathfrak{N}_t-1}} = S_0\, e^{(\mu_{\text{RW}}-\lambda\kappa)\,\mathbb{T}_{\mathfrak{N}_t-1}+\sigma\mathfrak{B}_{\mathbb{T}_{\mathfrak{N}_t-1}}}\, J_1 J_2 \cdots J_{\mathfrak{N}_t-1}.$$

Equation (8.136) then extends to:

$$S_\nu = S_0\, e^{(\mu_{\text{RW}}-\lambda\kappa)\nu+\sigma\mathfrak{B}_\nu}\, J_1 J_2 \cdots J_{\mathfrak{N}_t-1} \qquad (\mathbb{T}_{\mathfrak{N}_t-1} < \nu < \mathbb{T}_{\mathfrak{N}_t}). \quad (8.138)$$

➤ $\nu = \mathbb{T}_{\mathfrak{N}_t}$

This is the time of the last price jump in $[0,t]$. Taking the left limit of (8.138) at $\mathbb{T}_{\mathfrak{N}_t}$, we get

$$S_{\mathbb{T}_{\mathfrak{N}_t}^-} = S_0\, e^{(\mu_{\text{RW}}-\lambda\kappa)\,\mathbb{T}_{\mathfrak{N}_t}+\sigma\mathfrak{B}_{\mathbb{T}_{\mathfrak{N}_t}}}\, J_1 J_2 \cdots J_{\mathfrak{N}_t-1}.$$

Equation (8.137) generalizes to:

$$S_{\mathbb{T}_{\mathfrak{N}_t}} = J_{\mathbb{T}_{\mathfrak{N}_t}} S_{\mathbb{T}_{\mathfrak{N}_t}^-} = S_0\, e^{(\mu_{\text{RW}}-\lambda\kappa)\,\mathbb{T}_{\mathfrak{N}_t}+\sigma\mathfrak{B}_{\mathbb{T}_{\mathfrak{N}_t}}}\, J_1 J_2 \cdots J_{\mathfrak{N}_t-1} J_{\mathfrak{N}_t}. \qquad (8.139)$$

➤ $\mathbb{T}_{\mathfrak{N}_t} < \nu \leq t$

For this situation, a jump does not occur at $\nu$ since the last jump in $[0,t]$ is at $\mathbb{T}_{\mathfrak{N}_t}$. The price at $\nu$ is then given by a geometric Brownian motion with initial price $S_{\mathbb{T}_{\mathfrak{N}_t}}$:

$$S_\nu = S_{\mathbb{T}_{\mathfrak{N}_t}}\, e^{(\mu_{\text{RW}}-\lambda\kappa)(\nu-\mathbb{T}_{\mathfrak{N}_t})+\sigma(\mathfrak{B}_\nu-\mathfrak{B}_{\mathbb{T}_{\mathfrak{N}_t}})} \qquad (\mathbb{T}_{\mathfrak{N}_t} < \nu \leq t).$$

By (8.139), we obtain:

$$S_\nu = S_0\, e^{(\mu_{\text{RW}}-\lambda\kappa)\nu+\sigma\mathfrak{B}_\nu}\, J_1 J_2 \cdots J_{\mathfrak{N}_\nu-1} J_{\mathfrak{N}_\nu} \qquad (\mathbb{T}_{\mathfrak{N}_t} < \nu \leq t). \quad (8.140)$$

Equation (8.140) shows that for an interval $[0,t]$ containing a random number $\mathfrak{N}_t$ of jumps at times $0 < \mathbb{T}_1 < \cdots < \mathbb{T}_{\mathfrak{N}_t} \leq t$, the security price at $t$ is given by

$$S_t = S_0\, e^{(\mu_{\text{RW}}-\lambda\kappa)t+\sigma\mathfrak{B}_t} \prod_{\ell=1}^{\mathfrak{N}_t} J_\ell \qquad (t \geq 0), \qquad (8.141)$$

where $\{\mathfrak{N}_t\}_{t\geq 0}$ is a Poisson process with intensity $\lambda$ and $\kappa = \mathbb{E}(J_\ell) - 1$. As expected, the underlier's price process is a mix of jumps and geometric Brownian motion with drift $\mu_{\text{RW}} - \lambda\kappa$ and volatility $\sigma$. The stochastic process (8.141) solves the s.d.e. (8.132) and is called a *Merton jump diffusion* (MJD).

For $0 \leq u \leq t$, Equation (8.141) and $\mathfrak{B}_{t-u} \stackrel{d}{=} \mathfrak{B}_t - \mathfrak{B}_u$ imply:

$$\frac{S_t}{S_u} \stackrel{d}{=} \exp\left( (\mu_{\text{RW}} - \lambda\kappa)(t-u) + \sigma\mathfrak{B}_{t-u} + \sum_{\ell=\mathfrak{N}_u+1}^{\mathfrak{N}_t} X_\ell \right),$$

where $X_\ell = \ln(J_\ell)$. Shifting the summation index to $\ell \to \ell' = \ell - \mathfrak{N}_u$ and employing (8.119) (page 449) yield:

$$S_t \overset{\mathrm{d}}{=} S_u \exp\left((\mu_{\mathrm{RW}} - \lambda\kappa)(t - u) + \sigma\mathfrak{B}_{t-u} + \sum_{\ell'=1}^{\mathfrak{N}_t - \mathfrak{N}_u} X_{\ell'}\right)$$

$$\overset{\mathrm{d}}{=} S_u \exp\left((\mu_{\mathrm{RW}} - \lambda\kappa)(t - u) + \sigma\mathfrak{B}_{t-u} + \sum_{\ell'=1}^{\mathfrak{N}_{t-u}} X_{\ell'}\right), \qquad (8.142)$$

where $0 \le u \le t$. Note that $S_u$ in (8.142) is random for $u > 0$.

### Assumptions About the MJD Process

When studying options with underlier process (8.141), we add several assumptions about the MJD process to make the analysis more tractable:

➤ The jump factors $J_1, \ldots, J_{\mathfrak{N}_t}$ are i.i.d. lognormal random variables with

$$J_\ell = e^{X_\ell}, \qquad X_\ell \sim \mathcal{N}(\mu_J, \sigma_J^2), \qquad \mathbb{E}(J_\ell) = e^{\mu_J + \frac{1}{2}\sigma_J^2} = \kappa + 1, \qquad (8.143)$$

where $X_\ell$ is normal with mean $\mu_J$ and variance $\sigma_J^2$ for $\ell = 1, \ldots, \mathfrak{N}_t$. We also refer to $\mu_J$ as the *jump-factor drift parameter* and $\sigma_J$ as the *jump-factor volatility parameter*. The mean percentage change $\kappa$ of the jump factor is the same for each jump.

➤ The jump factors are independent of the jump times.

➤ The random variables $\mathfrak{B}_t$, $\mathfrak{N}_t$, and $J_\ell$ are independent, where $t > 0$ and $\ell = 1, \ldots, \mathfrak{N}_t$. In particular, $\mathfrak{B}_t$ is independent of $X_\ell$ for $\ell = 1, \ldots, \mathfrak{N}_t$.

**Notation.**  Denote the probability measure for the MJD price process (8.141) by $\mathbb{P}$ and, of course, assume that the above properties are enforced.

**Remark 8.12.** The natural log, $\ln\left(\prod_{\ell=1}^{\mathfrak{N}_t} J_\ell\right) = \sum_{\ell=1}^{\mathfrak{N}_t} X_\ell$, is an example of a compound Poisson process. More generally, a *compound Poisson process* is a sum $\sum_{\ell=1}^{\mathfrak{N}_t} X_\ell$, where $\mathfrak{N}_t$ is a Poisson process and $X_1, \ldots, X_{\mathfrak{N}_t}$ are i.i.d. random variables that are independent of $\mathfrak{N}_t$. Furthermore, the log return of the MJD price process in Equation (8.141), i.e., $\ln\left(\frac{S_t}{S_0}\right) = (\mu_{\mathrm{RW}} - \lambda\kappa)t + \sigma\mathfrak{B}_t + \sum_{\ell=1}^{\mathfrak{N}_t} X_\ell \equiv L_t$, is an example of a *Lévy process*, while the MJD price process, i.e., $S_t = S_0 e^{L_t}$, is an example of an *exponential Lévy process*.  □

### 8.9.3 Illustration of MJD Jump, Skewness, and Kurtosis

Consider the MJD security-price process (8.141) over 1 trading year or 252 trading days. The current time is 0, time step is $\frac{1}{252}$ (trading day), final time is $t = 1$, and inputs are:

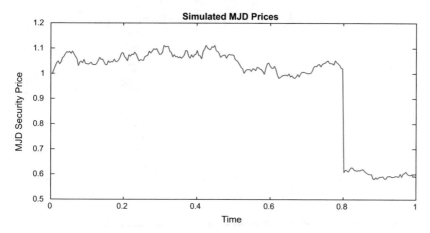

**Fig. 8.11** Simulated daily MJD security prices over 1 trading year. A negative price jump occurs in the transition from trading day 202 (0.8016 years) to trading day 203 (0.8056 years). The price fell from $1.0207 to $0.6077, which is a 40.46% drop. The inputs of the simulation are given in Equation (8.144)

$$S_0 = \$1, \quad \mu_{\mathrm{RW}} = 0.1, \quad \sigma = 0.15, \quad \lambda = 0.5, \quad \mu_J = -0.05, \quad \sigma_J = 0.08, \quad (8.144)$$

where $\kappa = e^{\mu_J + \frac{1}{2}\sigma_J^2} - 1$. Figure 8.11 depicts a sample path of $\{S_t\}_{0 \le t \le 1}$ over 1 trading year, i.e., 252 trading days. A very pronounced negative jump occurs in the transition from trading day 202 (0.8016 years), when the price was $1.0207, to trading day 203 (or 0.8056 years), when the price dropped to $0.6077. It is a percentage drop of:

$$J_t - 1 = \frac{S_t - S_{t^-}}{S_{t^-}} = \frac{0.6077 - 1.0207}{1.0207} = -40.46\%.$$

The MJD security price model can also produce skewness and kurtosis. Figure 8.12 shows a histogram of the log returns of a simulated MJD price process running over 65 years and with the same inputs as in (8.144). The skewness and excess kurtosis of the simulated MJD prices are -1.3325 and 23.9169, respectively. In other words, the MJD security price model has daily log return behavior that deviates from normality. Notice the qualitative similarities between Figure 8.12 and Figure 8.8 (page 445), which shows log returns of the S&P 500 for prices over a 65-year period.

A QQ-plot of the standardized MJD log returns is shown in Figure 8.13. The figure gives a clear depiction of the MJD log returns deviating from normality. It is interesting to observe the qualitative features that Figure 8.13 shares with the QQ-plot of the standardized S&P 500 log returns shown in Figure 8.9 (page 446). Notice the outlier in the bottom left in both figures. These qualitative comparisons show that the MJD security model can address the difficulties faced by the geometric Brownian motion model of underliers. Naturally, a

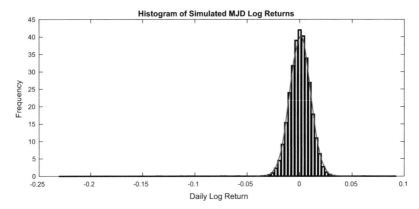

**Fig. 8.12** Simulated daily log returns for an MJD price process over 65 years. The mean is 0.000424 (practically zero) and standard deviation is 0.0099. The skewness is -1.3325. In particular, the left tail has a more elongated area (due to more histograms on that side) than the right tail; see the dark lines along the $x$-axis. The excess kurtosis is 23.9169, which is much higher than that of the corresponding normal with mean 0.000424 and standard deviation 0.0099. Compare with Figure 8.8 (page 445)

proper fitting of the MJD model to security prices will not depend on qualitative comparisons, but will involve a detailed statistical investigation.

### 8.9.4 No-Arbitrage Condition and Market Incompleteness

We shall see in this section that a market with an MJD underlier has no arbitrage, but is incomplete, i.e., not all its derivatives are attainable. In other words, there is at least one derivative whose payoff cannot be replicated using a self-financing trading strategy in other securities.

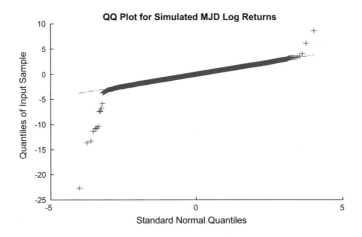

**Fig. 8.13** QQ-plot of the standardization of the simulated MJD log returns in Figure 8.12. The left and right tails are thicker than that of the standard normal. Compare with Figure 8.9 (page 446)

First, write the MJD security price process in (8.142) as

$$S_t \stackrel{d}{=} S_u \, e^{(m-q-\lambda\kappa-\frac{1}{2}\sigma^2)(t-u)+\sigma\mathfrak{B}_{t-u}} \prod_{\ell=1}^{\mathfrak{N}_{t-u}} J_\ell, \qquad (8.145)$$

where the index $\ell'$ was relabeled to $\ell$ for simplicity. This price is relative to a "real-world" probability measure $\mathbb{P}_{\lambda,\gamma}$, which incorporates our assumptions on page 456 and takes into account not only geometric Brownian motion but also the intensity $\lambda$ of the jumps and the probability measure $\gamma$ of the jump factors. In particular, $\mathfrak{B}_s$ is a standard Brownian motion relative to $\mathbb{P}_{\lambda,\gamma}$. The conditional expectation of $S_t$ with respect to $\mathbb{P}_{\lambda,\gamma}$ is (Exercise 8.37):

$$\mathbb{E}_{\mathbb{P}_{\lambda,\gamma}}(S_t \mid \mathcal{F}_u^{\mathrm{MJD}}) = S_u \, e^{(m-q)(t-u)} \qquad (0 \le u \le t). \qquad (8.146)$$

The $\sigma$-algebra $\mathcal{F}_u^{\mathrm{MJD}}$ is generated by the standard Brownian motion $\{\mathfrak{B}_s\}_{s\geq 0}$, the Poisson process $\{\mathfrak{N}_s\}_{s\geq 0}$, and the jump-factor process $\{J_\ell\}$, where $\ell = 1,\ldots,\mathfrak{N}_s$.

If $\mathbb{P}_{\lambda,\gamma}$ is a risk-neutral measure, then

$$\mathbb{E}_{\mathbb{P}_{\lambda,\gamma}}(S_t \mid \mathcal{F}_u^{\mathrm{MJD}}) = S_u \, e^{(r-q)(t-u)} \qquad (0 \le u \le t), \qquad (8.147)$$

and the price $C^E(S_t,t)$ of a European call on the security would satisfy

$$\mathbb{E}_{\mathbb{P}_{\lambda,\gamma}}\left(C^E(S_t,t) \mid \mathcal{F}_u^{\mathrm{MJD}}\right) = C^E(S_u,u) \, e^{r(t-u)} \qquad (0 \le u \le t). \quad (8.148)$$

This allows us to obtain the current price of the call by setting $u = t_0$ and $t = T$. However, comparing (8.146) and (8.147), we see that this is only possible if

$$m = r.$$

Since the risk-free rate $r$ is fixed independent of the security, the above constraint will not hold for all $\mathbb{P}_{\lambda,\gamma}$.

There is a transformation from $\mathbb{P}_{\lambda,\gamma}$ to a risk-neutral probability measure that allows us to price the European call. Though the details are beyond the scope of this text, we shall sketch the basic idea and refer readers to Privault [34, Chaps. 14, 15] for more. Choose $c_*$ and $\lambda_* > 0$ as well as a probability measure $\gamma_*$ for the jump factors such that

$$c_*\sigma - \lambda_*\kappa_* = m - q - \lambda\kappa - (r - q), \qquad (8.149)$$

where

$$\kappa_* = \mathbb{E}_{\gamma_*}(J_\ell^* - 1).$$

We write the jump factor as $J_\ell^*$ instead of $J_\ell$ to indicate that the probability measure being used is $\gamma_*$. Then the following can be shown using *Girsanov theorem for jump processes* (see Privault [34, Theorem 14.3, Chap. 15]): *there exists a* *risk-neutral probability measure, denoted* $\mathbb{Q}_{c_*, \lambda_*, \gamma_*}$, *associated with the choice* $(c_*, \lambda_*, \gamma_*)$ *in (8.149) that is equivalent to* $\mathbb{P}_{\lambda, \gamma}$ *and such that* $\lambda_*$ *is now the mean annual number of jumps for the Poisson process,* $\gamma_*$ *is the probability measure for the jump factors, and*

$$\mathcal{B}_t^* = \mathcal{B}_t + c_* t$$

*is a standard Brownian motion.*

The existence of $\mathbb{Q}_{c_*, \lambda_*, \gamma_*}$ immediately implies via the First Fundamental Theorem of Asset Pricing that *a market with MJD underliers has no arbitrage.* In addition, using (8.149) and $\mathcal{B}_t^*$, the MJD security price (8.145) becomes

$$S_t \overset{\mathrm{d}}{=} S_u \, e^{\left(r - q - \lambda_* \gamma_* - \frac{1}{2}\sigma^2\right)(t-u) \, + \, \sigma \mathcal{B}_{t-u}^*} \prod_{\ell=1}^{\mathfrak{N}_{t-u}^*} J_\ell^*. \tag{8.150}$$

An $*$ on $\mathfrak{N}_t^*$ and $J_\ell^*$ is a reminder that their probabilities are relative to $\mathbb{Q}_{c_*, \lambda_*, \gamma_*}$, i.e., the Poisson process now has intensity $\lambda_*$ and the jump factor distribution is now $\gamma_*$.

As a consistency check, taking the expectation of (8.150) relative to $\mathbb{Q}_{c_*, \lambda_*, \gamma_*}$ gives the risk-neutral condition:[24]

$$\mathbb{E}_{\mathbb{Q}_{c_*, \lambda_*, \kappa_*}} \left( S_t \mid \mathcal{F}_u^{\mathrm{MJD}} \right) = S_u \, e^{(r-q)(t-u)} \qquad (0 \le u \le t). \tag{8.151}$$

Given the risk-neutral measure $\mathbb{Q}_{c_*, \lambda_*, \kappa_*}$, the price at $t$ of a European call with respect to it is then:

$$C^E_{\mathbb{Q}_{c_*, \lambda_*, \gamma_*}}(t) = e^{-r(T-t)} \, \mathbb{E}_{\mathbb{Q}_{c_*, \lambda_*, \gamma_*}} \left( C^E(S_T, T) \mid \mathcal{F}_t \right) \qquad (0 \le t \le T). \tag{8.152}$$

This price depends on $\mathbb{Q}_{c_*, \lambda_*, \kappa_*}$, which in turn is based on a choice $(c_*, \lambda_*, \gamma_*)$ satisfying (8.149). *Unfortunately, there are infinitely many such choices, and so there is no unique call price in the MJD model, though there is no arbitrage.* The Second Fundamental Theorem of Asset Pricing then implies that *a market with an MJD-underlier model is incomplete.*

**Remark 8.13.** We saw that for each solution $(c_*, \lambda_*, \gamma_*)$ of (8.149), namely, $c_* \sigma - \lambda_* \kappa_* = m - q - \lambda \kappa - (r - q)$, there exists a risk-neutral probability measure $\mathbb{Q}_{c_*, \lambda_*, \gamma_*}$. If there is no possibility of a price jump, which is the setting of the BSM model, then $\lambda = \lambda_* = \gamma_* = 0$ and the constraint equation reduces to

$$c_* \sigma = m - r.$$

Consequently, there is a unique solution given by the Sharpe ratio, i.e.,

$$c_* = \frac{m - r}{\sigma}.$$

---

[24] For $S_t \overset{\mathrm{d}}{=} S_u \, e^{a_0 (t-u) \, + \, \sigma \mathcal{B}_{t-u}^*} \prod_{\ell=1}^{\mathfrak{N}_{t-u}^*} J_\ell^*$, the desired expectation is $e^{\left(a_0 + \frac{1}{2}\sigma^2 + \lambda_* \kappa_*\right)(t-u)}$; see Exercise 8.37.

Therefore, there is only one risk-neutral measure, namely, $Q_{\frac{m-r}{\sigma}, 0, 0}$, and so only one call (or derivative) price. In other words, the BSM model has no arbitrage and its market is complete. $\square$

### Further Reading on Pricing Derivatives in an Incomplete Market

For the MJD model, we saw above that though the market has no arbitrage, there is not a unique risk-neutral probability to price a European call option with MJD underlier. In fact, there are infinitely many such probability measures, and so the market is incomplete. How then can one price a derivative in an arbitrage-free, incomplete market? A practical approach is to have the market "choose" the risk-neutral probability measure, i.e., fit the model to market data (market calibration), and employ the resulting measure to price the derivative. Another approach is to take into account an investor's utility function and then use an associated utility maximization to determine the risk-neutral probability measure and, hence, price the derivative.

Properly addressing the above important and deep issue is surely beyond the scope of our text. Readers are referred to the highly informative survey article by Staum in Birge and Linetsky's handbook [4]. In Chapter 12 of [4], Staum discusses the meaning[25] of market incompleteness and its causes, different approaches to derivative pricing in incomplete markets (including market calibration and expected utility maximization), hedging in incomplete markets, and many other pertinent topics.

### 8.9.5 Pricing European Calls with an MJD Underlier

Analogous to Section 8.4.4, risk-neutral pricing gives a relatively quick method of deriving the price of a European call with an MJD underlier.

Consider pricing a European call with expiration $T$, strike price $K$, and MJD underlier. Assume that a risk-neutral probability measure $Q_{c_*, \lambda_*, \kappa_*}$ is chosen (e.g., after market calibration). The price at time $t$ of the call is computed using Equation (8.152). For notational simplicity, we shall write the formula as

$$C^E_{\text{MJD}}(t) = e^{-r\tau} \, \mathbb{E}_*(C^E(T) \mid \mathcal{F}_t) = e^{-r\tau} \, \mathbb{E}_*(\max\{S_T - K, 0\} \mid \mathcal{F}_t), \qquad (8.153)$$

where $\tau = T - t$. By (8.150), the underlier price at expiration is

$$S_T \stackrel{\text{d}}{=} S_t \exp\left( (\mu_* - \lambda_* \kappa_*)\tau + \sigma \mathcal{B}^*_\tau + \sum_{\ell=1}^{\mathfrak{N}^*_\tau} X^*_\ell \right),$$

---

[25] Appendices A and B of Chapter 12 in reference [4] treat the meaning of market incompleteness.

where

$$\mu_* = r - q - \frac{\sigma^2}{2}$$

and, analogous to (8.143),

$$J_\ell^* = e^{X_\ell^*}, \qquad X_\ell^* \sim \mathcal{N}(\mu_J^*, (\sigma_J^*)^2), \qquad \mathbb{E}_*(J_\ell^*) = e^{\mu_J^* + \frac{1}{2}(\sigma_J^*)^2} = \kappa_* + 1. \quad (8.154)$$

We now evaluate (8.153) with the understanding that it is relative to $Q_{c_*, \lambda_*, \kappa_*}$. Assume $\mathfrak{N}_\tau^* = n$ and let $J_0^* = 1$, so $X_0^* = \ln J_0^* = 0$. Since $X_1^*, \ldots, X_n^*$ are i.i.d. with $X_\ell^* \sim \mathcal{N}(\mu_J^*, (\sigma_J^*)^2)$ and because $\mathfrak{B}_\tau^* \sim \mathcal{N}(0, \tau)$ is independent of each $X_\ell^*$, we see that

$$(\mu_* - \lambda_* \kappa_*)\tau + \sigma \mathfrak{B}_\tau^* + \sum_{\ell=0}^{n} X_\ell^*$$

is a normal random variable. Its risk-neutral expectation and variance are

$$\mathbb{E}_* \left( (\mu_* - \lambda_* \kappa_*)\tau + \sigma \mathfrak{B}_\tau^* + \sum_{\ell=0}^{\mathfrak{N}_\tau^*} X_\ell^* \,\Big|\, \mathfrak{N}_\tau^* = n \right) = (\mu_* - \lambda_* \kappa_*)\tau + n\mu_J^*$$

and

$$\mathrm{Var}_* \left( (\mu_* - \lambda_* \kappa_*)\tau + \sigma \mathfrak{B}_\tau^* + \sum_{\ell=0}^{\mathfrak{N}_\tau^*} X_\ell^* \,\Big|\, \mathfrak{N}_\tau^* = n \right) = \sigma^2 \tau + n (\sigma_J^*)^2.$$

Consequently, the security price at expiration $T$ conditioned on $\mathfrak{N}_\tau^* = n$ jumps occurring during $[t, T]$ is[26]

$$\left( S_T \mid \mathfrak{N}_\tau^* = n \right) \stackrel{\mathrm{d}}{=} S_t e^{m_n^* \tau + \mathfrak{s}_n^* \mathfrak{B}_\tau^*}, \quad (8.155)$$

where $m_n^*$ and $\mathfrak{s}_n^*$ are defined by

$$m_n^* = \mu_* - \lambda_* \kappa_* + \frac{n}{\tau} \mu_J^*, \qquad (\mathfrak{s}_n^*)^2 = \sigma^2 + \frac{n}{\tau} (\sigma_J^*)^2. \quad (8.156)$$

Hence, the conditional security price (8.155) is a lognormal random variable with mean parameter $m_n^*$ and variance parameter $(\mathfrak{s}_n^*)^2$.

To cast the quantity $m_n^* \tau$ in (8.155) in a form analogous to $r - q - \frac{1}{2}\sigma^2$ in the risk-neutral lognormal security price (8.67) on page 410, define $\imath_n^*$ such that

$$m_n^* = \imath_n^* - q - \frac{(\mathfrak{s}_n^*)^2}{2}.$$

By (8.143) and (8.156), we get

$$\imath_n^* = r - \lambda_* \kappa_* + \frac{n}{\tau} \ln(1 + \kappa_*). \quad (8.157)$$

---

[26] If $X \sim \mathcal{N}(a_*, b_*^2)$, where * indicates a risk-neutral setting, then $e^X \stackrel{\mathrm{d}}{=} e^{a_* + b_* Z^*}$. Compare with (5.69) on page 243.

Equation (8.155) then becomes

$$\left(S_T \,|\, \mathfrak{N}^*_\tau = n\right) \overset{\text{d}}{=} S_t \exp\left[\left(\mathbf{z}^*_n - q - \frac{1}{2}(\mathbf{s}^*_n)^2\right)\tau + \mathbf{s}^*_n \mathfrak{B}^*_\tau\right],$$

where $\mathbf{z}^*_n$ and $\mathbf{s}^*_n$ play the roles of the risk-free rate and volatility, respectively.

Proceeding as in Section 8.4.4, we apply Theorem 8.3 to evaluate

$$\mathrm{e}^{-\mathbf{z}^*_n \tau}\, \mathbb{E}_* \left(\max\{S_T - K, 0\} \,|\, \mathfrak{N}^*_\tau = n\right).$$

The result is the price at $t$ of a European call, conditioned on $n$ jumps occurring during the time remaining until expiration, with strike price $K$, volatility $\mathbf{s}^*_n$, and risk-free rate $\mathbf{z}^*_n$:

$$\mathrm{e}^{-\mathbf{z}^*_n \tau}\, \mathbb{E}_* \left(\max\{S_T - K, 0\} \,|\, \mathfrak{N}^*_\tau = n\right) = C^E_{\text{BSM}}(S_t, K, \mathbf{s}^*_n, \mathbf{z}^*_n, \tau, q). \qquad (8.158)$$

Since the possible values of $\mathfrak{N}^*_\tau$ are $0, 1, 2, \ldots$, the unconditional expectation (8.153) can be expressed as

$$C^E_{\text{MJD}}(t) = \mathrm{e}^{-r\tau} \sum_{n=0}^{\infty} \mathbb{E}_* \left(\max\{S_T - K, 0\} \,|\, \mathfrak{N}^*_\tau = n\right) \mathbb{Q}\left(\mathfrak{N}^*_\tau = n\right).$$

Employing (8.158) and the Poisson distribution with intensity $\lambda_*$, it follows:

$$C^E_{\text{MJD}}(t) = \sum_{n=0}^{\infty} \frac{(\lambda_* \tau)^n}{n!}\, \mathrm{e}^{(-\lambda_* + \mathbf{z}^*_n - r)\tau}\, C^E_{\text{BSM}}(S_t, K, \mathbf{s}^*_n, \mathbf{z}^*_n, \tau, q).$$

By (8.157), we then get the following theorem:

**Theorem 8.4 (MJD Price for a European Call).** *The price at time $t$ of a European call with an MJD underlier is given as follows relative to the risk-neutral measure* $\mathbb{Q}_{c_*, \lambda_*, \kappa_*}$:

$$C^E_{\text{MJD}}(t) = \sum_{n=0}^{\infty} \frac{\mathrm{e}^{-\lambda'_* \tau}\,(\lambda'_* \tau)^n}{n!}\, C^E_{\text{BSM}}(S_t, K, \mathbf{s}^*_n, \mathbf{z}^*_n, \tau, q), \qquad (8.159)$$

*where $0 \leq t \leq T$ and $\tau = T - t$. The current price relative to the choice* $\mathbb{Q}_{c_*, \lambda_*, \kappa_*}$ *is obtained at $t = 0$.*

In Theorem 8.4, the quantities with an * are computed relative to $\mathbb{Q}_{c_*, \lambda_*, \kappa_*}$. To summarize, the mean number of jumps per year is $\lambda_*$ with

$$\lambda'_* = \lambda_*(1 + \kappa_*).$$

The jump factors $J^*_\ell$ are i.i.d., and each is a lognormal random variable $J^*_\ell = \mathrm{e}^{X^*_\ell}$, where $X_\ell \sim \mathcal{N}\left(\mu^*_J, (\sigma^*_J)^2\right)$, and

$$\kappa_* = \mathbb{E}(J^*_\ell) - 1 = \mathrm{e}^{\mu^*_J + \frac{1}{2}(\sigma^*_J)^2} - 1,$$

with

$$(\mathcal{J}_n^*)^2 = \sigma^2 + \frac{n}{\tau}(\sigma_j^*)^2, \qquad \mathcal{I}_n^* = r - \lambda_* \kappa_* + \frac{n}{\tau}\ln(1 + \kappa_*).$$

We shall call (8.159) simply the *MJD European call price*. Put-call parity immediately yields the associated *MJD European put price*:

$$P_{\text{MJD}}^E(t) = K e^{-r\tau} - S_t e^{-q\tau} + C_{\text{MJD}}^E(t).$$

In addition, *the BSM European call price can be recovered from the MJD European call price*. Separate out the first-order term in (8.159):

$$C_{\text{MJD}}^E(t) = e^{-\lambda_*'\tau} C_{\text{BSM}}^E(S_t, K, \mathcal{J}_0^*, \mathcal{I}_0^*, \tau, q)$$
$$+ \sum_{n=1}^{\infty} \frac{e^{-\lambda_*'\tau}\left(\lambda_*'\tau\right)^n}{n!} C_{\text{BSM}}^E(S_t, K, \mathcal{J}_n^*, \mathcal{I}_n^*, \tau, q).$$

Since $\mathcal{I}_0^* = r - \lambda_* \kappa_*$ and $\mathcal{J}_0^* = \sigma$, we see that when there is no possibility of a price jump, i.e., $\lambda_* = 0$, the BSM formula follows:

$$C_{\text{MJD}}^E(t) = C_{\text{BSM}}^E(S_t, K, \sigma, r, \tau, q).$$

Let us illustrate the MJD European call price (8.159) in comparison with the CRR and BSM prices in Example 8.7 on page 421.

**Example 8.12.** Consider a 3-month European call with strike price of $50 on a stock whose price follows a Merton jump diffusion. Suppose that the stock is a nondividend-paying stock with the current price $50 and annual volatility 15%. Let the risk-free rate be 2% per annum. We saw in Example 8.7 that the BSM price and the 100-period CRR tree price of the European call on one share is $1.62. What is the MJD European call price if the stock has a mean number of jumps per year of 0.25, a jump-factor drift parameter value of -1.5%, and a jump-factor volatility parameter value of 4%?

**Solution.** The inputs are

$$S_0 = \$50, \quad K = \$50, \quad \sigma = 0.15, \quad r = 0.02, \quad \tau = T = 0.25, \quad q = 0,$$

with jump parameters

$$\lambda_* = 0.25, \quad \mu_J^* = -1.5\%, \quad \sigma_J^* = 4\%.$$

Using a software, Equation (8.159) can be computed to give

$$C_{\text{MJD}}^E(0) = \$1.63 > C_{\text{BSM}}^E(0) = \$1.62.$$

The MJD model gives a higher current price than that of the BSM model.  □

### 8.9.6 MJD Volatility Smile

As noted earlier, the BSM model predicts that the implied volatility $\sigma_{\text{im}}(K)$ is constant as $K$ changes, which does not agree with market data. We also saw (see page 446) that the BSM call price formula produces a one-to-one correspondence between call prices and implied volatilities. In other words, given any current European call price $C(0)$, whether it is the market price or even a theoretical price arising in a model different from the BSM model, we can still assign a unique implied volatility to the call price, i.e., we can assign the value of $\sigma$ obtained by solving the BSM call price formula for $\sigma$ using $C(0)$ as the input price. In this sense, *the unique BSM implied volatility assigned to a call price can be used as a marker or proxy for the call price.*

We now determine the BSM implied volatility associated with the European call price due to the MJD model, i.e., given an MJD call price $C^E_{\text{MJD}}(0)$, we solve the following equation for the BSM implied volatility $\sigma$:

$$C^E_{\text{MJD}}(0) = C^E_{\text{BSM}}(S_0, K, \sigma, r, T, q).$$

We illustrate this below using the call option in Example 8.12.

**Example 8.13. (MJD Volatility Smile)** A European call on a nondividend-paying stock has strike price of \$50 and three months until expiration. The stock price is an MJD process with the current price \$50, mean number of jumps per year of 0.25, jump-factor drift parameter value of -1.5%, and jump-factor volatility parameter value of 4%. Assume a risk-free rate of 2% per annum. In Figure 8.14, the implied volatility found using the MJD European call price (8.159) is plotted as a function of strike price, showing a volatility smile.

□

Naturally, the accuracy of MJD volatility smiles has to be tested against the associated market volatility smiles. See, for example, Epps [15, Sec. 9.3] and references therein, for an introduction.

## 8.10  A Glimpse Ahead

A rigorous assessment of how well the MJD model for option pricing fits market data will require an empirical analysis that is outside the scope of this text. Nonetheless, the MJD model took an important first step beyond the BSM model because it can incorporate jumps, skewness, and kurtosis. We also saw that though there is no arbitrage in the MJD model, its market environment is incomplete, unlike the BSM model.

As with any model, there are always aspects that need to be modified to improve the MJD's fit with data. For instance, the BSM and MJD models

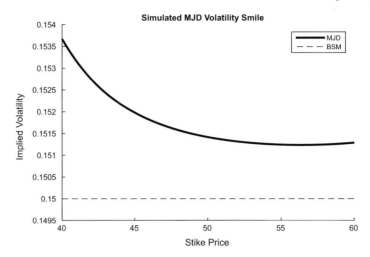

**Fig. 8.14** Implied volatility of a European call as a function of strike price, which the underlying security has price jumps. Each implied volatility was computed using the MJD European call price (8.159). The shape shows a volatility smile. The inputs for the model are current underlier price $S_t =$ $50, risk-free rate $r = 2\%$, time to expiration of $\tau = 0.25$ years, dividend yield rate of $q = 0$, jump-factor drift parameter value of $\mu_J^* = -1.5\%$, and jump-factor volatility parameter value of $\sigma_J^* = 4\%$. The horizontal line with constant implied volatility of 15% is due to the BSM model with a strike price of $K = \$50$

assume that the volatility of the underlying security is deterministic. However, data shows that volatility is stochastic. There are many sources that give introductions to extensions of the BSM and MJD models that include the following (e.g., see Epps [15], Gathereal [16], and Taylor [41], and the references therein):

➤ Models with stochastic volatility, but no price jumps (e.g., Heston model)
➤ Models with stochastic volatility and price jumps
➤ Models with stochastic volatility and with jumps in the price and volatility

On the other hand, jumps and stochastic volatility in security prices may cause market incompleteness.[27] In fact, quite an extensive research literature has been developed around derivative pricing, hedging, expected utility optimization, etc., in incomplete markets. Readers are referred to Staum's survey chapter in the handbook by Birge and Linetsky [4, Chap. 12].

Without a doubt, a whole universe of adventures lies ahead.

---

[27] See Staum's article in [4, Chap. 12, Sec. 3]

## 8.11 Exercises

### 8.11.1 Conceptual Exercises

**8.1.** A modeler who knows nothing about the BSM model is trying to find a formula for the present value $C(0)$ of a European call option, where the underlying security has current price $S(0)$ and the strike price is $K$. She proposes the following formula after considerable experimentation:

$$C(0) = w_1 S^n(0) + w_2 K^m,$$

where the weights $w_1$ and $w_2$ are to be determined. Without using any information about the BSM model, give a two-sentence argument that determines the possible values of $n$ and $m$.

**8.2.** Give a brief intuitive reason why a European call option is more risky than its underlying security.

**8.3.** Express the return rate of a European call during an instant $dt$ as a s.d.e.

**8.4.** If a stock satisfies the CAPM, then does a European call on the stock also satisfy the CAPM? Justify your answer.

**8.5.** Traders often abide by simple intuitive rules concerning volatility. Here are some examples you may have heard:

  "Sell a stock when its volatility is high."
  "Favor puts when volatility is high."
  "Buy a stock when its volatility is low."

Are these rules of thumb captured by the BSM model for underliers? Justify your answer.

**8.6.** Explain why a European call and put have the same implied volatility.

**8.7.** Briefly critique the MJD model's assumption that the risk of price jumps is diversifiable.

### 8.11.2 Application Exercises

**8.8. (Price Change in Options Versus Stocks)** Traders use options for speculation. To get an intuitive feel for why this is the case, we consider an example of how the price of a European call option changes with variations in the underlying security. A financial company's stock currently has a price of $40. The risk-free interest rate is 7% per annum and the stock has volatility parameter of 28%. Consider a European call option on the stock for a strike price of $41 with expiration in 6 months. Let $t$ be the current time and $t + h$ an hour later.

a) From time $t$ to $t + h$, the price of the stock increases by 1%. What is the percentage change in the value of the call? Would the price of the put move by the same percentage?
b) From time $t$ to $t + h$, the price of the stock decreases by 1%. What is the percentage change in the value of the call? Would the price of the put move by the same percentage?

**8.9. (European Calls as Insurance)** After careful research, a fund manager would like to purchase 100,000 shares of a nondividend-paying stock currently trading at $60. The fund manager estimates the stock's volatility at 15% and believes the stock will rise over the coming months. The $6 million needed to buy the shares now will not be available until a month away. He is concerned that if the stock rises over the next month, it will become too expensive to buy. How can the fund manager insure against the risk of the stock price increasing? How much would this insurance cost? Suppose that the risk-free rate is 2%.

**8.10. (European Puts as Insurance)** An investor owns 10,000 shares of a stock paying no dividend and currently trading at $160 per share. The stock has a volatility of 20% and the current risk-free rate is 2.5%. Three months from now she would like to liquidate the shares to purchase an investment property for $1,500,000. She is concerned that if the stock price falls over the next three months, she would not be able to buy the property. On the other hand, she does not want to sell her shares now since there is also the possibility that the stock price will increase over the next three months and so she would miss out on such gains. How can she mitigate against this risk? Note that since the portfolio has a single stock, Markowitz portfolio theory does not directly apply.

**8.11. (Warrants)** Assume that the equity per share of a company satisfies the BSM model and has volatility of 25%. Suppose that the current equity value of the company is $50 million. Assume that its stock pays no dividend and equity is presently $50 per share. The risk-free rate is 6%. The company plans to issue 300,000 warrants with strike price of $70 and maturity in 3 years. Each warrant is based on 1 share of the company's stock. Determine how much money the company will raise if it sells all the warrants at a fair price.

**8.12. (Delta Hedging European Calls That End Out-of-the-Money at Expiration)** Assume that a firm sells 1,000 European calls (in round lots) on a nondividend-paying stock with current price $75, strike $75, annual volatility of 15%, and 80 days to expiration. Suppose that the risk-free rate is 2% and assume 365 days in a year. In Table 8.2, some entries are shown for delta

**Table 8.2** Delta-hedging table for Exercise 8.12.

| | Delta Hedging Values Per Share | | | | | | |
|---|---|---|---|---|---|---|---|
| $t$ | $S_t$ | $\Delta_C(t)$ | $-C_t^E$ short call | $I_t$ investment | $-\mathbb{L}_t$ loan balance | $\Delta_C(t)S_t$ long $\Delta_C(t)$ shares | $V_C(t)$ portf. value |
| Day 0 | $75.000000 | | -$2.264248 | | | | $0.000000 |
| Day 1 | $75.472305 | 0.574123 | | $2.264373 | | | $0.004812 |
| Day 2 | $75.646263 | | | | | | |
| Day 3 | $75.523208 | | | | | | $0.029374 |
| Day 4 | $76.652920 | | | | | | |
| Day 5 | $77.036379 | | | | | | |
| ⋮ | ⋮ | ⋮ | ⋮ | ⋮ | ⋮ | ⋮ | ⋮ |
| Day 79 | $69.932887 | 0.000000 | $0.000000 | $2.274071 | -$2.146712 | $0.000000 | $0.127359 |
| Day 80 | $69.904710 | | | | | | |

hedging based on a MATLAB code that outputs values to at least nine decimal places, but were rounded at the sixth decimal place so the entries appear less congested. If you work to six decimal places only, then there naturally will be rounding errors in the day-to-day delta hedging, and not all your numerical values will exactly match those in the table. See Example 8.9 on page 428 and Remark 8.10 in that example.

a) Complete the values for Days 1–5 and Day 80 in Table 8.2. Assume that the firm sold the European calls at the BSM price. Did the firm experience a profit or a loss? Determine how much.
b) If the firm sold the calls at $3.50 per share of the stock, did the firm have a profit or loss? Determine the amount.

**8.13. (Delta Hedging European Calls That End in-the-Money at Expiration)**
Suppose that a firm sells 800 European calls (in round lots) on a nondividend-paying stock with current price $110, strike price $110, annual volatility of 20%, and 90 days to expiration. Suppose that the risk-free rate is 3% and assume 365 days in a year (see Remark 8.1). Table 8.3 shows a portion of delta hedging using a MATLAB code; see the comment in Exercise 8.12 about rounding errors.

a) Compute the values for Days 1–5 and Day 90 in Table 8.3 under the assumption that the firm sold the European calls at the BSM price. Did the firm experience a profit or a loss? Determine how much.
b) If the firm sold the calls at $5.50 per share of the stock, did the firm have a profit or loss? Determine the amount.

**Table 8.3** Delta-hedging table for Exercise 8.13.

| | Delta Hedging Values Per Share | | | | | | |
|---|---|---|---|---|---|---|---|
| $t$ | $S_t$ | $\Delta_C(t)$ | $-C_t^E$ | $I_t$ | $-\mathbb{L}_t$ | $\Delta_C(t)\,S_t$ | $\mathcal{V}_C(t)$ |
| | | | short call | investment | loan balance | long $\Delta_C(t)$ shares | port. value |
| Day 0 | \$ 110.000000 | | | \$ 4.757733 | | | |
| Day 1 | \$111.276638 | | | | | | -\$0.004965 |
| Day 2 | \$108.680788 | 0.499957 | | | | | - \$0.103196 |
| Day 3 | \$110.757606 | | | | | | |
| Day 4 | \$110.049030 | | | | | | |
| Day 5 | \$111.589771 | | | | | | -\$0.161611 |
| $\vdots$ | $\vdots$ | $\vdots$ | $\vdots$ | $\vdots$ | $\vdots$ | $\vdots$ | $\vdots$ |
| Day 89 | \$125.092736 | 1.000000 | -\$15.101777 | \$4.792664 | - \$ 115.392349 | \$125.092736 | -\$ 0.608726 |
| Day 90 | \$124.937041 | | | | | | |

### 8.11.3 Theoretical Exercises

**8.14.** For a cum-dividend security price $S_t^c = e^{qt} S_t$, where $S_t$ follows geometric Brownian motion, show that $dS_t^c = m\, S_t^c\, dt + \sigma\, S_t^c\, d\mathfrak{B}_t$.

**8.15.** Show that the self-financing condition $(dn_t)\, S_{t+dt}^c + (db_t)\, B_{t+dt} = 0$ is equivalent to $dV_t = n_t\, dS_t^c + b_t\, dB_t$.

**8.16.** Show that if the BSM p.d.e. does not hold, then there is an arbitrage.

**8.17.** If $f(x,t)$ is a solution of the BSM p.d.e., then show that for every positive constant $c > 0$, the function $f_c(x,t) = f(cx,t)$ is also a solution.

**8.18.** Given solutions $f_1(x,t), \ldots, f_n(x,t)$ of the BSM p.d.e., show that all linear combinations $c_1 f_1(x,t) + \cdots + c_n f_n(x,t)$ are also solutions.

**8.19.** If a solution $f(x,t)$ of the BSM p.d.e. has an $n$th partial derivative with respect to $x$, then show that $x^n \dfrac{\partial^n f}{\partial x^n}(x,t)$ is also a solution.

**8.20.** Show that, for the price process $\widehat{S}_t = S_t e^{-q(T-t)}$, the BSM p.d.e. (8.19) on page 392 transforms to a form without dividend:

$$\frac{1}{2}\sigma^2 \bar{x}^2 \frac{\partial^2 C^E}{\partial \bar{x}^2}(\bar{x},t) + r\bar{x}\frac{\partial C^E}{\partial \bar{x}}(\bar{x},t) + \frac{\partial C^E}{\partial t}(\bar{x},t) - rC^E(\bar{x},t) = 0,$$

where $\bar{x} = x e^{-q(T-t)}$.

**8.21.** Using the variables,

$$\tilde{x} = \ln\left(\frac{x}{K}\right), \quad \tilde{\tau} = \frac{\sigma^2}{2}(T - t), \quad v(\tilde{x}, \tilde{\tau}) = \frac{C^E(\tilde{x}, t)}{K}, \quad \tilde{k} = \frac{r}{(\sigma^2/2)},$$

show that (8.22) and (8.23) transform, respectively, to

$$\frac{\partial v}{\partial \tilde{\tau}}(\tilde{x}, \tilde{\tau}) = \frac{\partial^2 v}{\partial \tilde{x}^2}(\tilde{x}, \tilde{\tau}) + (\tilde{k} - 1)\frac{\partial v}{\partial \tilde{x}}(\tilde{x}, \tilde{\tau}) - \tilde{k}v(\tilde{x}, \tilde{\tau})$$

and

$$v(\tilde{x}, 0) = \max\{e^{\tilde{x}} - 1, 0\}, \quad \lim_{\tilde{x} \to -\infty} v(\tilde{x}, \tilde{\tau}) = 0, \quad v(\tilde{x}, \tilde{\tau}) \to e^{\tilde{x}} \text{ as } \tilde{x} \to \infty.$$

**8.22.** Using a trial solution $v(\tilde{x}, \tilde{\tau}) = \tilde{u}(\tilde{x}, \tilde{\tau})e^{a\tilde{x} + b\tilde{\tau}}$, show that for the choices $a = -\frac{1}{2}(\tilde{k} - 1)$ and $b = -\frac{1}{4}(\tilde{k} + 1)^2$, Equation (8.24) transforms into the heat equation

$$\frac{\partial \tilde{u}}{\partial \tilde{\tau}}(\tilde{x}, \tilde{\tau}) = \frac{\partial^2 \tilde{u}}{\partial \tilde{x}^2}(\tilde{x}, \tilde{\tau})$$

and (8.25) into

$$\tilde{u}(\tilde{x}, 0) = \max\left\{e^{\frac{1}{2}(\tilde{k}+1)\tilde{x}} - e^{\frac{1}{2}(\tilde{k}-1)\tilde{x}}, 0\right\}, \quad \lim_{|\tilde{x}| \to \infty} \tilde{u}(\tilde{x}, \tilde{\tau})e^{-c\tilde{x}^2} = 0,$$

where $c > 0$.

**8.23.** Derive Equations (8.31) and (8.32) on page 394 and show that (8.32) equals

$$C^E(x, t) = x e^{-q(T-t)} N\big(d_+(x, T - t)\big) - K e^{-r(T-t)} N\big(d_-(x, T - t)\big).$$

**8.24.** Consider the discounted underlier price process $\{\tilde{S}_t\}_{t \geq 0}$, where $\tilde{S}_t = e^{-rt} S_t^c$ with $S_t^c = e^{qt} S_t$ the cum-dividend price process, and a discounted self-financing, replication portfolio value process $\{\tilde{V}_t\}_{t \geq 0}$, where $\tilde{V} = e^{-rt} V_t$. Show that
a) $d\tilde{S}_t = \sigma \tilde{S}_t d\mathcal{B}_t^Q$, where $d\mathcal{B}_t^Q = d\mathcal{B}_t + \left(\frac{m-r}{\sigma}\right) dt$.
b) $dV_t = r V_t dt + n_t(m - r) S_t^c dt + n_t \sigma S_t^c d\mathcal{B}_t$.
c) $d\tilde{V}_t = n_t d\tilde{S}_t$.

**8.25.** Consider a nonvanishing stochastic process $\{X_t\}_{t \geq 0}$ such that

$$\frac{dX_t}{X_t} = a(X_t, t) dt + b(X_t, t) d\mathcal{B}_t,$$

where $a(x, t)$ and $b(x, t)$ are deterministic functions. The coefficients $a(X_t, t)$ and $b(X_t, t)$ are called the *drift* and *volatility*, respectively, of $\{X_t\}_{t \geq 0}$. For example, a security price following geometric Brownian motion has constant volatility $b(X_t, t) = \sigma$. Show that the volatility of a European call is strictly greater than the volatility of its underlying security.

**8.26.** Assume that a security satisfies the CAPM. Show that the beta of a European call on the security is strictly greater than the beta of the security.

**8.27.** Establish the following:

If $S(T) = K$, then $\Delta_C(t) \to 1/2$ and $\Delta_P(t) \to -1/2$ as $\tau \to 0$.

If $S(T) > K$, then $\Delta_C(t) \to 1$ and $\Delta_P(t) \to 0$ as $\tau \to 0$.

If $S(T) < K$, then $\Delta_C(t) \to 0$ and $\Delta_P(t) \to -1$ as $\tau \to 0$.

If $t < T$, then $\Delta_C(t) < 1$ and $\Delta_P(t) > -1$.

**8.28.** Show that the discounted underlier process $X_t = e^{-(r-q)t} S_t$ and discounted derivative price process $Y_t = e^{-rt} f(S_t, t)$ are martingales relative to the risk-neutral measure $Q$ of Girsanov theorem.

**8.29.** In a continuous-time approach, we saw that the BSM European option pricing formula can be derived as the solution of the BSM p.d.e. On the other hand, the BSM pricing formula can be determined as the continuum limit of the discrete-time binomial tree model. Is there a discrete-time analog of the BSM p.d.e. in the binomial tree framework? If so, then using appropriate discrete-time interpretations, determine the partial difference equation analog of the BSM p.d.e. directly from the binomial tree.

**8.30.** Consider the binomial tree model for option pricing.

a) Give a one-sentence mathematical reason why the constraint $d < e^{(r-q)h} < u$ holds. Do not use a specific binomial tree model such as a CRR tree, JR tree, etc.

b) Give a financial reason why the condition $d < e^{(r-q)h} < u$ holds. If this result does not hold, then is any assumption of the BSM model violated? If so, indicate which one.

**8.31.** Using a three-period binomial tree, show that a European call price is given by

$$C(t_0, 3) = e^{-(3h)r}[p_*^3 C_{u^3} + 3p_*^2(1 - p_*)C_{u^2d} + 3p_*(1 - p_*)^2 C_{ud^2} + (1 - p_*)^3 C_{d^3}].$$

**8.32.** For an $n$-period binomial tree, show that the price of a European call is given by

$$C(t_0) = e^{-r(nh)} \left[ \sum_{i=0}^{n} \binom{n}{i} p_*^i (1 - p_*)^{n-i} C_{u^i d^{n-i}}(t_n) \right],$$

where $C_{u^i d^{n-i}}(t_n) = \max\{S(t_0)u^i d^{n-i} - K, 0\}$ for $i = 0, 1, \ldots, n$.

**8.33.** Show that the $n$-period binomial formula for a European call can be expressed as

$$C(t_0) = S(t_0)e^{-q\tau} N(n, k_*, \hat{p}_*) - Ke^{-r\tau} N(n, k_*, p_*),$$

where $k_*$ is the smallest value of $i$ for which $S(t_0)u^i d^{n-i} - K > 0$ and

$$\tau = t_n - t_0, \quad N(n, k_*, p_*) = \sum_{i=k_*}^{n} \binom{n}{i} p_*^i (1 - p_*)^{n-i}, \quad \hat{p}_* = \frac{p_* u}{e^{(r-q)h}}.$$

**8.34.** Find the deltas of a forward and futures on a nondividend-paying stock.

**8.35.** Show that delta, gamma, and theta of European calls and puts are:

$$\Delta_C(S_{t_0}, t_0) = e^{q\tau} N(d_+(S_{t_0}, \tau)), \qquad \Delta_P(S_{t_0}, t_0) = -e^{-q\tau} N(-d_+(S_{t_0}, \tau))$$

$$\Gamma_C(S_{t_0}, t_0) = \frac{e^{-q\tau} N'(d_+(S_{t_0}, \tau))}{S_{t_0} \sigma \sqrt{\tau}}, \qquad \Gamma_P(S_{t_0}, t_0) = \Gamma_C(S_{t_0}, t_0)$$

$$\Theta_C(S_{t_0}, t_0) = -\frac{S_{t_0} e^{-q\tau} \sigma N'(d_+(S_{t_0}, \tau))}{2\sqrt{\tau}} + q S_{t_0} e^{-q\tau} N(d_+(S_{t_0}, \tau))$$
$$- rKe^{-r\tau} N(d_-(S_{t_0}, \tau))$$

$$\Theta_P(S_{t_0}, t_0) = -\frac{S_{t_0} e^{-q\tau} \sigma N'(d_+(S_{t_0}, \tau))}{2\sqrt{\tau}} - q S_{t_0} e^{-q\tau} N(-d_+(S_{t_0}, \tau))$$
$$+ rKe^{-r\tau} N(-d_-(S_{t_0}, \tau)).$$

**8.36.** Consider a portfolio of derivatives with the same underlying security that pays no dividend. Prove that if the portfolio has zero gamma, then it is theta-market neutral, meaning $\frac{\partial \Theta_P}{\partial S} = \frac{\partial \Theta_P}{\partial x}(S_{t_0}, t_0) = 0$.

**8.37.** Consider an MJD security price process, i.e.,

$$S_t \overset{d}{=} S_u e^{\mu_0(t-u) + \sigma \mathfrak{B}_{t-u}} \prod_{\ell=1}^{\mathfrak{N}_{t-u}} J_\ell,$$

where $0 \leq u \leq t$ and $\mu_0 = m - q - \lambda \kappa - \frac{1}{2}\sigma^2$. Show:

a) $\mathbb{E}\left( \prod_{\ell=1}^{\mathfrak{N}_{t-u}} J_\ell \right) = e^{\lambda \kappa (t-u)}$

b) $\mathbb{E}_{\mathbb{P}_{\lambda, \gamma}}(S_t \mid \mathcal{F}_u^{MJD}) = S_u e^{(\mu_0 + \frac{1}{2}\sigma^2 + \lambda \kappa)(t-u)} = S_u e^{(m-q)(t-u)}$.

## References

[1] Ai, H.: Lecture Notes on Derivatives. Fuqua School of Business, Duke University, Durham (2008)
[2] Balanda, K., MacGillivray, H.: Kurtosis: a critical review. Am. Stat. **42**(2), 111 (1988)

[3] Bemis, C.: The Black-Scholes PDE from Scracth (Lecture Notes). Financial Mathematics Seminar, University of Minnesota (27 November 2006)

[4] Birge, J., Linetsky, V. (ed.): Handbooks in Operations Research and Management Science: Financial Engineering, vol. 15, 1st edn. North-Holland, Amsterdam (2008)

[5] Björk, T.: Arbitrage Theory in Continuous Time. Oxford University Press, Oxford (2009)

[6] Black, F.: How we came up with the option formula. J. Portf. Manag. **15**(2), 4 (1989)

[7] Black, F., Scholes, M.: The pricing of options and corporate liabilities. J. Polit. Econ. **81**, 637 (May/June 1973)

[8] Chance, D.: Lecture Notes on the Convergence of the Binomial to the Black-Scholes Model. Louisiana State University, Baton Rouge (2008)

[9] Cox, J., Ross, S.: The valuation of options for alternative stochastic processes. J. Financ. Econ. **3**, 145 (1976)

[10] DeCarlo, L.: On the meaning and use of kurtosis. Psychol. Methods **2**(3), 292 (1997)

[11] Delbaen, G., Schachermayer, W.: A general version of the fundamental theorem of asset pricing. Math. Ann. **300**, 463 (1994)

[12] Demeterfi, K., Derman, E., Kamal, M., Zou J.: More than you ever wanted to know about volatility swaps. Goldman Sachs, Quantitative Strategies Research Notes (March 1999)

[13] Durrett, R.: Probability: Theory and Examples, 4th edn. Cambridge University Press, Cambridge (2010)

[14] Elliot, R., Kopp, P.: Mathematics of Financial Markets, 2nd edn. Springer, New York (2005)

[15] Epps, T.W.: Pricing Derivative Securities. World Scientific, Hackensack (2007)

[16] Gatheral, J.: The Volatility Surface: A Practitioner's Guide. Wiley, Hoboken (2006)

[17] Gray, D., Malone, S.: Macrofinancial Risk Analysis. Wiley, West Sussex (2008)

[18] Groebner, D., Shannon, P., Fry, P.: Business Statistics: A Decision-Making Approach. Pearson, Boston (2014)

[19] Harrison, J., Kreps, D.: Martingales and arbitrage in multiperiod securities markets. J. Econ. Theory **20**, 381 (1979)

[20] Harrison, J., Pliska, S.: Martingales and stochastic integrals in the theory of continuous trading. Stoch. Process. Appl. **11**(3), 215 (1981)

[21] Hsia, C.-C.: On binomial option pricing. J. Financ. Res. **6**, 41 (1983)

[22] Hull, J.C.: Options, Futures, and Other Derivatives. Pearson Princeton Hall, Upper Saddle River (2015)

[23] Kolb, R.W.: Financial Derivatives. New York Institute of Finance, New York (1993)

[24] Korn, R., Korn E.: Option Pricing and Portfolio Optimization. American Mathematical Society, Providence (2001)

[25] Kreps, D.: Arbitrage and equilibrium economics with infinitely many commodities. J. Math. Econ. **8**, 15 (1981)

[26] Kwok, Y.: Mathematical Models of Financial Derivatives. Springer, New York (1998)

[27] McDonald, R.: Derivative Markets. Addison-Wesley, Boston (2006)

[28] Merton, R.C.: Theory of rational option pricing. Bell J. Econ. Manag. Sci. **4**, 141 (Spring 1973)

[29] Merton, R.C.: Option pricing when underlying stock returns are discontinuous. J. Financ. Econ. **3**, 125 (1976)

[30] Merton, R.C.: Applications of option-pricing theory: twenty-five years later. Am. Econ. Rev. **88**(4), 323 (1998) [Nobel Lecture]

[31] Miersemann, E.: Lecture Notes on Partial Differential Equations. Department of Mathematics, Leipzig University (2012)

[32] Mikosch, R.: Elementary Stochastic Calculus with Finance in View. World-Scientific, Singapore (1998)

[33] Neftci, S.: An Introduction to the Mathematics of Financial Derivatives. Academic, San Diego (2000)

[34] Privault, N.: Notes on Stochastic Finance. Nanyang Technological University, Singapore (2013)

[35] Rachev, S.: Lecture Notes on Option Pricing. Applied Mathematics and Statistics, SUNY-Stony Brook, Stony Brook (2011)

[36] Reilly, F.K., Brown, K.C.: Investment Analysis and Portfolio Management. South-Western Cengage Learning, Mason (2009)

[37] Ross, S.: A simple approach to the valuation of risky streams. J. Bus. **51**, 453 (1978)

[38] Ross, S.: An Elementary Introduction to Mathematical Finance. Cambridge University Press, Cambridge (2011)

[39] Schachermayer, W.: The Fundamental Theorem of Asset Pricing. Faculty of Mathematics, University of Vienna (2009, Preprint). http://www.mat.univie.ac.at/~schachermayer/preprnts/prpr0141a.pdf

[40] Scholes, M.: Derivatives in a dynamic environment. Am. Econ. Rev. **88**(3), 350 (1998) [Nobel Lecture]

[41] Taylor, S.: Asset Price Dynamics, Volatility, and Prediction. Princeton University Press, Princeton (2005)

[42] Wilmott, P., Dewynne, N., Howison, S.: Mathematics of Financial Derivatives: a Student Introduction. Cambridge University Press, Cambridge (1995)

# Index

$\mathcal{F}$-measurable function, 257
$\sigma$-algebra, 254
    generated by random variable, 258
    generated by stochastic process, 270

adapted processes, 273, 402
almost surely continuous stochastic process, 264
alpha, 192
Amortization, 53
Annuity, 46
    application to bond valuation, 66
    application to equity in a house, 61
    application to sinking funds, 62
    application to stock valuation, 63
    applications to saving, borrowing, and
        spending, 59
    future value, 49
    present value, 50
    with varying payments and interest rates, 56
APR, 33
APY, 33
arbitrage, 84, 334
asset, 329
    financial, 329
at-the-money, 364
average value-at-risk, 182

Banker's Rule, 16
basis points, 5
bearish, 363
beta, 158, 160
    for portfolios, 162, 192
    linear factor beta, 195
bid-ask spread, 9

bid/ask price, 8
binomial pricing
    comparison with BSM pricing, 422
    of European calls, 411–422
    of underliers, 209–246
binomial trees
    Cox-Ross-Rubinstein tree, 218
    general, 209–218
    Jarrow-Rudd tree, 251
    probability measure, 214
    recombining property, 211
Black-Scholes-Merton model, *see* BSM model
bonds, 66
    bond prices vs. interest rates, 70
    bond prices vs. YTM, 72
    bond valuation formula, 69
    callable, 66
    convertible, 66
    coupon payment, 68
    coupon rate, 68
    current yield, 68
    issue date, 67
    maturity date, 67
    maturity, par, or face value, 67
    par bonds, 67
    premium bonds, 67
    yield to maturity (YTM), 68
    zero-coupon bond, 68
book value, 196
book-to-market ratio, 197
boundary conditions, 391
    for European calls, 392
    for European puts, 396

Brownian motion, 282–289
  with drift, 289
  with drift and scaling, 284, 314
  with starting point, 284
Brownian path, 284
BSM model, 384–391
  unique risk-neutral probability measure, 461
  versus market data, 440–448
BSM p.d.e., 391
  equivalent to heat equation, 393
  existence and uniqueness of solutions, 391
  solving for European calls, 392
bullish, 363

Cameron-Martin-Girsanov theorem, *see* Girsanov theorem
Capital Allocation Line (CAL), 154
Capital Asset Pricing Model, *see* CAPM
Capital Market Line (CML), 153–157
  tangent point—market portfolio, 156
CAPM, 152, 158–165
  beta versus linear factor beta, 195
  for portfolios, 162
  formula, 159
  risk premium of a security, 158
  risk premium of the market portfolio, 158
  security price, 160
  security risk decomposition, 164
cash market, 337
coherent risk measure, 184
coincident indicators, 10
commercial banking, 3
commodity, 331
commodity swap, *see* swaps
compound interest, 21
  continuous compounding, 31
  formula, 27
  fractional compounding, 28
  fractional vs. simple compounding, 30
  future value, 27
  generalized compounding, 31
  nonnegative integer number of periods, 22
  nonnegative real number of periods, 24
  present value, 27
conditional expectation, 270–273
conditional value-at-risk, 182
contingent claim, 330
continuous stochastic process, 264
continuous-state processes, 262
continuous-time processes, 262

converge
  almost surely, 265
  in distribution, 265
  in mean square, 265
  in probability, 265
correlation of Brownian motion, 295
covered call, *see* options
Cox-Ross-Rubinstein tree, *see* CRR tree
credit default swaps, *see* swaps
CRR tree, 218–246
  continuous-time limit, 237–246
  CRR equations, 226, 234
  real world, 219–230
  real-world uptick probability, 226, 234
  risk-neutral uptick probability, 233, 234
  risk-neutral world, 230–236
  security price formula, 229, 236

dealers, 8
deep in-the-money, 364
deep out-the-money, 364
delivery market, 338
delta, 390
  discrete version, 415
  European call and put deltas at expiration, 398
  of a European call, 396
  of a European put, 397
delta hedging, 422–433
  application, 427–433
  theoretical framework, 422–427
derivatives
  characteristics of its valuation, 332
  commodity, 331
  defined, 330
  financial, 331
  purposes of, 331
Descartes's Rule of Signs, 43
diffusion coefficient, 315
diffusion equation, *see* heat equation
diffusion process, 289
Dirichlet density, 139
discounted price process, 406
discrete-state processes, 262
discrete-time processes, 262
diversifiable risk, 143, 151
diversification, *see* Markowitz portfolio theory
diversified portfolio, *see* Markowtiz portfolio theory
dividend
  continuously reinvested, 340

cum-dividend, 18, 31, 387
ex-dividend, 18, 31
ex-dividend date, 18
yield, 19, 215
dividend discount model, 64
dividend yield, 281
drift parameter, 386
drift process, 302

economic cycle, 10
economic indicator, 10
efficient frontier, *see* Markowitz portfolio theory
equity, 330
equity in a house, 61
equivalent martingale measure, 314
equivalent measures, 313
European call price, 394
    behavior relative to security price, 397
    behavior relative to strike price, 397
European put price, 395
    behavior relative to security price, 397
    behavior relative to strike price, 397
exact time, 16
exchanges, 7
expected short fall, 183
expected tail loss, 183

Fama-French three-factor model, 196–199
feasible portfolios, 125
federal discount rate, 2
federal funds rate, 2
Federal Reserve, 2
filtered probability space, 268
filtrations, 268–270
final condition, 391
    for European calls, 392
    for European puts, 396
financial markets, 1
First Fundamental Theorem of Asset Pricing, 408
forward commitment, 330
    delivery date, 330
forwards, 337–345
    contract size, 337
    delivery price, 338
    delivery, expiry, expiration, exercise, or
        maturity date, 337
    forward or exercise price, 337
    forward price formula, 341
    forward value formula, 344
    long forward, 337
    relation to put-call parity, 369

short forward, 337
spot-forward parity formula, 342
underlier, 337
writer, 337
fundamental factor model, 191
futures, 345–348
    evolution from forwards to futures, 345
    futures contract, 346
    futures price, 346, 347
    futures value, 347
    impact of daily settlement, 347
    maintenance margin, 347
    margin account, 346
    margin requirement, 347
    mark-to-market, 346

geometric Brownian motion, 314–319
Girsanov theorem, 235, 311, 405, 459
global minimum-variance portfolio, *see*
        Markowitz portfolio theory
gradient, 120
Greeks
    delta, 390, 396–398, 434
    for a portfolio, 434
    gamma, 434
    theta, 434
    vega, 446
gross return, 19

heat equation, 390, 393
hedgers, 331
Hessian, 120
hurdle rate, 161

i.i.d., 211
idiosyncratic risk, 151
in-the-money, 364
incomplete market, 458
index rates, 4
innovation process, 279
intensity, 449
interest, 15
    exact, 16
    interest rate per period, 15
    negative interest rate, 15
    ordinary, 16
    quoted rate, 15
    total interest on a loan, 56
intrinsic value, 370
investment banking, 3

IRR, 41, 42
   multiple, 45
   relation to NPV, 44
Itô diffusion, 302
Itô integral, 299–302, 314
Itô process, 302, 387
   unique representation, 389
Itô product rule, 292
Itô's formula, 302–387
Itô's lemma, 304

jumps in security prices, 440

kurtosis, 267, 442
   excess, 443
   in S&P 500 log returns, 446

Lévy process, 456
   exponential, 456
lagging indicators, 10
Lagrange Multiplier Theorem, 122
latent factor, 191
Law of One Price, 334, 389, 399, 413
   consequence of no-arbitrage condition, 335
leading indicators, 10
leptokurtic, 443
leverage ratio, 3
LIBOR, 5, 350
limit buy/sell order, 8
Lindeberg Central Limit Theorem, 237
Lindeberg condition, 239
linear factor models, 185
long-term rates, 5

macroeconomic factor model, 191
maintenance margin, *see* futures
margin requirement, *see* futures
mark-to-market accounting rule, 346
market capitalization, 196
market liquidity, 9
market portfolio, 156, 157
market risk, 151
market sentiment, 363
Markov process, 288
Markov property, 244, 288
Markowitz bullet, 126
Markowitz portfolio theory
   diversification, 138–143
   diversified portfolio, 130
   efficient frontier for $N$ securities, 117–128
   efficient frontier for two securities, 107–117
   expected portfolio return rate, 94

   global minimum-variance portfolio, 124, 128
   model, 83
   multivariate normality, 87
   Mutual Fund Theorem, 130
   one-period assumption, 88
   optimal portfolios, 132
   portfolio log return, 100–103
   portfolio log return versus portfolio return
      rate, 103
   portfolio risk, 96, 119
   return rates, 85
   securities' variances and covariances, 96–100
   two-security portfolio analysis, 104–117
   utility function, 131
   weight vector for minimum-variance
      portfolio, 123
   weights, 89
martingale representation theorem, 407
martingales, 275–278, 406
   necessary condition of efficient market, 277
maximum drawdown, 172
Merton jump-diffusion model, *see* MJD model
mesokurtic, 443
method of least squares, 192
MJD model, 448–465
   assumptions, 456
   European call pricing, 461–464
   recovering BSM price, 464
   solving the MJD s.d.e., 453–455
   volatility smile, 465
money market account, 386
moneyness, 364
multivariate normality, 87
mutual fund theorem, *see* Markowiz portfolio
   theory

naked call, *see* options
NPV, 38, 42
   relation to IRR, 44
numéraire, 386

observable factor, 191
opportunity cost, 16
optimization problem, 120
options, 353–376, 383–466
   American, 354, 372–376
   buyer, holder, or owner, 353
   call option, 354
   contract size, 353, 356
   covered call, 357
   European, 354, 359–372, 383–466

exercise an option, 353

expiration, exercise, or maturity date, 353, 356

final or terminal payoff, 355

how options work, 357–359

moneyness, 364

naked or uncovered call, 358

premium, 353

put option, 354

seller or writer, 353

strike or exercise price, 353

styles, 354

trading strategies, 365

types, 354

underlier, 353

vanilla, 355

order statistic, 176

out-the-money, 364

over-the-counter market (OTC), 7


p-quantile, 174

p.d.e., *see* partial differential equations

partial differential equations, 390–395

connection with probability, 408

parabolic p.d.e., 391

payoff diagram

forward, 339

terminal, 359–363

perpetuity, 51

physical market, 337

platykurtic, 443

Poisson process, 449

compound, 456

portfolio

alpha, 192

beta, 162, 192

delta-gamma-neutral, 438–440

delta-neutral, 435–438

log return, 101

replicating, 388

replicating condition, 389

risk, 96, 119

risk measures, 151, 165–180

self-financing, 388, 412

self-financing condition, 389

trading strategy, 89, 388

weights, 89

positive definite matrix, 86, 105, 119

power set, 254

price discovery, 331

price-to-book ratio, 197

primary market, 6, 67

prime rate, 5

principal, 15

probability measure, 253, 255

probability space, 253

profit diagram

terminal, 359–363

put-call parity

American options, 373

European options, 362, 368, 395

relation to forward, 369


QQ-plot, 445

quadratic covariation, 291

quadratic variation, 289, 290

quantile function, 175

quantiles, 445

quoted interest rate, *see* interest


Radon-Nikodym derivative, 313

random variables, 257

convergence of, 265

independent, 259

independent of $\sigma$-algebra, 260

random walk, 276, 280, 320

simple, 320

symmetric, 320

reserve ratio, 2

return

arithmetic mean return, 36

capital-gain return, 212

geometric mean return, 36

gross return, 211

log return, 212

required return rate, 16

risk-averse investor, 87, 404

risk-free rate, 16

proxy, 17

real, 16

risk-neutral investor, 137, 404

risk-neutral pricing

of European calls and puts, 410

of European-style derivatives, 404, 408

with binomial trees, 415–422

risk-neutral probability measure, 309, 406

and no-arbitrage, 409

for Merton jump-diffusion model, 458

uniqueness for BSM model, 461

risk-seeking investor, 136, 404

rounding errors, 429

s.d.e., *see* stochastic differential equation
sample-continuous stochastic process, 264
scale parameter, 315
Second Fundamental Theorem of Asset Pricing,
    409
secondary market, 6, 67
securities
    basic behavior, 278–282
    cum-dividend price, 386
    debt securities, 330
    definition, 329
    derivative securities, 330
    equity securities, 330
    ex-dividend price, 386
securities markets, 6
    professional participants, 8
Security Market Line (SML), 163
semivariance, 171
Sharpe ratio, 166–170, 244
    as slope of CML, 167
    in BSM model, 460
short selling, 90
short-term rates, 4
simple interest, 20
    formula, 21
    future value, 21
    present value, 21
    return rate, 21
    versus fractional compounding, 30
sinking funds, 62
size premium, 197
skewness, 266, 442
    in S&P 500 log returns, 445
Sortino ratio, 170, 174
speculators, 331
spot market, 337
spot price, 338
spread, 366
    bear, 366
    bull, 366
    butterfly, 367
    calendar, 367
    horizontal, 367
    price, 366
    time, 367
    vertical, 366
statistical factor model, 191
stochastic differential equation
    for cum-dividend security price, 387
    for geometric Brownian motion, 386, 387,
        451

for Merton jump diffusion, 453
stochastic processes
    basics, 260–265
    Merton jump diffusion, 450–458
stock valuation, 63
straddle, 365
strangle, 366
sub-sigma algebra, 255
swap contract, 349
swaps, 348–353
    commodity swaps, 350
    credit default swap, 350
    currency swap, 349
    fixed leg, 349
    floating leg, 349
    interest rate swap buyer, 349
    interest rate swap seller, 349
    interest rate swaps, 349
    mechanics of interest rate swaps, 351
    notional principal, 349
    plain vanilla swap, 349, 350
    swap bank, 351
    variance swap, 352
systematic risk, 143, 151, 165

tail VaR, 183
time value, 370
total variation, 289
tower property, 403, 420
trading costs, 9
trading strategies with options, *see* options

uncovered call, *see* options
unobservable factor, 191
unsystematic risk, 143, 165
utility function, 131–137
    concave, 134
    convex, 136
    marginal utility, 132

value premium, 198
value-at-risk, 178
VaR, 177–180
variance swap, *see* swaps
volatility
    implied, 446
    MJD volatility smile, 465
    parameter, 386
    skews, 445
    smiles, 447
    surface, 447

volatility parameter, 315
volatility process, 302

warrants, 398–400
Weak Efficient Market Hypothesis, 244
weights, 89, 118

white noise, 280
    Gaussian, 280
    independent, 280
    strict, 280

yield curve, 6

Printed in the United States
By Bookmasters